ENCYCLOPEDIA OF
ENVIRONMENTAL ISSUES

ENCYCLOPEDIA OF ENVIRONMENTAL ISSUES

REVISED EDITION

Volume 1

Abbey, Edward—Dichloro-diphenyl-trichloroethane

Editor

Craig W. Allin

Cornell College

SALEM PRESS

Pasadena, California   Hackensack, New Jersey

Editor in Chief: Dawn P. Dawson
Editorial Director: Christina J. Moose   *Photo Editor:* Cynthia Breslin Beres
Development Editor: R. Kent Rasmussen   *Production Editor:* Joyce I. Buchea
Project Editor: Judy Selhorst   *Graphics and Design:* James Hutson
Acquisitions Manager: Mark Rehn   *Layout:* William Zimmerman
Research Supervisor: Jeffry Jensen

Cover photo: ©cultura/CORBIS

Library of Congress Cataloging-in-Publication Data

Encyclopedia of environmental issues / editor, Craig W. Allin. — Rev. ed.
    Planned in 4 v.
  Includes bibliographical references and index.
  ISBN 978-1-58765-735-1 (set : alk. paper) — ISBN 978-1-58765-736-8 (vol. 1 : alk. paper) —
ISBN 978-1-58765-737-5 (vol. 2 : alk. paper) — ISBN 978-1-58765-738-2 (vol. 3 : alk. paper) —
ISBN 978-1-58765-739-9 (vol. 4 : alk. paper)
  1. Environmental sciences—Encyclopedias. 2. Pollution—Encyclopedias.
I. Allin, Craig W. (Craig Willard)
  GE10.E523 2011
  363.7003—dc22

                                      2011004176

3506

Contents

Publisher's Note

For centuries, human attitudes toward the environment were based on the assumption that the planet's resources were infinite and that the earth had an inexhaustible capacity to sustain life. The Industrial Revolution of the nineteenth century, however, inaugurated numerous technologies and practices that greatly increased the pace of environmental degradation and resource exploitation. With these technological advances came a growing number of voices that began expressing uneasiness about the rates at which wilderness was disappearing and the human population was growing, and these voices proliferated during the twentieth century. By the late 1960's, they had coalesced into the environmental movement, a collection of individuals and organizations willing to take action to maintain the health of the planet and the well-being of those species, human and otherwise, that inhabit it.

Environmentalists are concerned with issues ranging from water pollution and storage of nuclear waste at the local level to problems associated with worldwide population growth, the endangerment of plant and animal species, and the ramifications of global warming. Because such environmental hazards have the potential to affect everyone on the planet, it is important that people of all nations have access to information about environmental issues at all points of the learning process, from kindergarten to the highest levels of postsecondary education. Such knowledge helps to create informed citizens who are capable of making sound choices about how to deal with the growing list of environmental hazards.

This four-volume *Revised Edition* of Salem Press's *Encyclopedia of Environmental Issues* is a wide-ranging guide designed to meet the ongoing need for up-to-date information on the environment that is accessible to nonspecialists. The encyclopedia assembles information from numerous fields of knowledge relevant to the study of environmental issues, including biology, geology, anthropology, demographics, genetics, and engineering, and it explains their interrelationships in easily understood terms.

SCOPE AND COVERAGE

The *Encyclopedia of Environmental Issues, Revised Edition*, adds more than 300 new articles to the 470 presented in the first edition, expanding the set to four volumes. All of the original essays have been evaluated for their currency, and 122 of them have been either completely replaced or substantially revised to reflect the latest information available. The Further Reading sections accompanying the essays have all been updated.

The set contains 772 alphabetically arranged articles that range in length from 300 to 3,000 words. All articles are signed by the academics and other experts who wrote them. They cover a wide variety of topics, including endangered animal species, air pollution, national parks, environmental legislation, oil spills, alternative energy sources, and global climate change. In recognition of the fact that people spend an overwhelming proportion of their time in human-made environments, essays also cover such topics as sick building syndrome, noise pollution, smog, and urban planning.

The bulk of the articles consist of overviews of issues, concepts, and terms relevant to the study of environmental matters. The encyclopedia also includes numerous specialized articles covering biographies, events, books, legislation and international treaties, court cases, and organizations.

- *Overviews:* These articles range from broad concepts such as ecology to specific issues such as the conflict between loggers and defenders of the endangered northern spotted owl in the Pacific Northwest of the United States.
- *Biographies:* Among the biographical essays are discussions of John Muir, the American preservationist who founded the Sierra Club; James Watt, U.S. president Ronald Reagan's secretary of the interior, who attempted to undo many of the environmental protections introduced in the decades before the Reagan administration; and former U.S. vice president Al Gore, whose work to raise awareness of the dangers of global warming led to his being awarded the Nobel Peace Prize.
- *Events:* Articles detailing significant events range from disasters such as the toxic gas release from a pesticide plant in Bhopal, India, in 1984 and the BP *Deepwater Horizon* oil spill in the Gulf of Mexico in 2010 to international meetings—such as the 1992

Earth Summit and the 2002 World Summit on Sustainable Development—that have explored global solutions to environmental problems.

- *Treaties, legislation, and court cases:* These essays explore how various international agreements, laws, and court decisions have affected the manner in which humans interact with the environment. Topics include the the U.S. Endangered Species Act of 1973; *Sierra Club v. Morton*, the 1972 U.S. Supreme Court case that first posed the question of whether trees have legal standing; and the 1992 United Nations Framework Convention on Climate Change.
- *Organizations and movements:* Important organizations and movements covered include the antienvironmental Sagebrush Rebellion of the late 1970's; the Chipko Andolan movement, which stopped the clear-cutting of forests in India; the Sea Shepherd Conservation Society, known for its confrontational tactics in defense of marine animals; and the U.S. Fish and Wildlife Service.

ORGANIZATION AND FORMAT

The opening of each article in this set includes the following features:

- Category subheads
- A brief definition or identification of the topic, along with any relevant dates
- A summary of the topic's environmental significance

Following each article of more than 300 words is a list of suggested further reading to aid students who are seeking resources for more in-depth, current information on the topic at hand. All articles are accompanied by cross-references to related articles in the set, and headnote cross-references of alternative terms (for example, "Particulate matter. *See* Airborne particulates") are dispersed throughout. The encyclopedia also includes 300 photographs and more than 200 sidebars, charts, graphs, tables, maps, and other illustrations that illuminate the events and concepts detailed in the essays.

The *Encyclopedia of Environmental Issues, Revised Edition*, ends with several useful appendixes, including a time line of important environment-related events, a directory of environmental organizations, a directory of U.S. national parks, a glossary, a biographical directory of notable figures in the history of environmentalism, and an extensive bibliography listing up-to-date publications that provide new insights into ongoing is-

sues. Also included are a comprehensive subject index and an index of entries arranged according to the following categories: activism and advocacy; agriculture and food; animals and endangered species; atmosphere and air pollution; biotechnology and genetic engineering; disasters; ecology and ecosystems; energy and energy use; forests and plants; human health and the environment; land and land use; nuclear power and radiation; organizations and agencies; philosophy and ethics; places; pollutants and toxins; population issues; preservation and wilderness issues; resources and resource management; treaties, laws, and court cases; urban environments; waste and waste management; water and water pollution; weather and climate.

ONLINE ACCESS

Salem provides access to its award-winning content both in traditional, printed form and online. Any school or library that purchases this four-volume set is entitled to free, complimentary access to Salem's online version of the content through our Salem Science Database. For more information about our online database, please contact our online customer service representatives at (800) 221-1592.

The advantages are clear:
- Complimentary with print purchase
- Fully supported
- Unlimited users at your library
- Full access from home or dorm rooms
- Immediate access via online registration
- A simple, intuitive interface
- User profile areas for students and patrons
- Sophisticated search functionality
- Complete content, including appendixes
- Integrated searches with any other Salem Press product you already have on the Salem Science platform.
- E-books are also available.

ACKNOWLEDGMENTS

Salem Press thanks editor Craig W. Allin of Cornell College for his many contributions to the creation of these volumes. We also thank Karen N. Kähler for her careful and thorough work in revising and updating many of the essays that appeared in the earlier edition of this set. Finally, we thank all the experts in the various fields of environmental studies who wrote the articles; their full names and affiliations appear in the list of contributors that follows.

Contributors

Richard Adler
University of Michigan-Dearborn

Craig W. Allin
Cornell College

Emily Alward
College of Southern Nevada

Robin Attfield
Cardiff University

Anita Baker-Blocker
Ann Arbor, Michigan

Ruth Bamberger
Drury College

Niharika Banerjea
University of Southern Indiana

Grace A. Banks
Chestnut Hill College

David Landis Barnhill
Guilford College

David Barratt
Montreat College

Elizabeth A. Barthelmes
Boston College

Melissa A. Barton
Westminster, Colorado

Robert B. Bechtel
University of Arizona

Raymond D. Benge, Jr.
Tarrant County College-Northeast Campus

Alvin K. Benson
Utah Valley University

Lisa M. Benton
Colgate University

Massimo D. Bezoari
Huntingdon College

Cynthia A. Bily
Macomb Community College

Margaret F. Boorstein
C. W. Post College of Long Island University

Richard G. Botzler
Humboldt State University

Lakhdar Boukerrou
Florida Atlantic University

Pat Brereton
Dublin City University

Victoria M. Breting-García
Houston, Texas

Josephus J. Brimah
Fourah Bay College

Howard Bromberg
University of Michigan

JoEllen Broome
Georgia Southern University

Kenneth H. Brown
Northwestern Oklahoma State University

Jeffrey C. Brunskill
Bloomsburg University of Pennsylvania

Bruce G. Brunton
James Madison University

Michael A. Buratovich
Spring Arbor University

Aubyn C. Burnside
Hickory, North Carolina

Dale F. Burnside
Lenoir-Rhyne College

Welland D. Burnside
Hickory, North Carolina

Joseph P. Byrne
Belmont University

Byron Cannon
University of Utah

Glenn Canyon
Pasadena, California

Roger V. Carlson
Jet Propulsion Laboratory

Robert S. Carmichael
University of Iowa

Robert E. Carver
University of Georgia

Jeff Cervantez
University of Tennessee

Frederick B. Chary
Indiana University Northwest

Dennis W. Cheek
Ewing Marion Kauffman Foundation

Nader N. Chokr
Oslo, Norway

Thomas Clarkin
University of Texas at San Antonio

Kathryn A. Cochran
Longview Community College

Daniel J. Connell
Boston College

Mark Coyne
University of Kentucky

Greg Cronin
University of Colorado at Denver

Ralph D. Cross
University of Southern Mississippi

Robert L. Cullers
Kansas State University

George Cvetkovich
Western Washington University

Roy Darville
East Texas Baptist University

Matt Deaton
University of Tennessee

C. R. de Freitas
University of Auckland

René A. De Hon
Northeast Louisiana University

Jonelle DePetro
Eastern Illinois University

Dennis R. DeVries
Auburn University

Joseph Dewey
University of Pittsburgh

Gordon Neal Diem
*ADVANCE Education and Development
Institute*

John P. DiVincenzo
Middle Tennessee State University

Stephen B. Dobrow
Fairleigh Dickinson University

Gary E. Dolph
Indiana University Kokomo

Colleen M. Driscoll
Villanova University

Andrew P. Duncan
New England College

John M. Dunn
Ocala, Florida

Sohini Dutt
Kansas State University

Timothy C. Earle
Western Washington University

Frank N. Egerton
University of Wisconsin-Parkside

Howard C. Ellis
Millersville University of Pennsylvania

Robert D. Engelken
Arkansas State University

Jess W. Everett
Rowan University

Jack B. Evett
*University of North Carolina
at Charlotte*

Thomas R. Feller
Nashville, Tennessee

George J. Flynn
*State University of New York at
Plattsburgh*

David R. Foran
Michigan State University

Brian J. Gareau
Boston College

Roberto Garza
San Antonio College

Soraya Ghayourmanesh
Bayside, New York

Craig S. Gilman
Coastal Carolina University

James S. Godde
Monmouth College

Lissy Goralnik
Michigan State University

D. R. Gossett
Louisiana State University Shreveport

Daniel G. Graetzer
*University of Washington Medical
Center*

Hans G. Graetzer
South Dakota State University

Jerry E. Green
Miami University

William Crawford Green
Morehead State University

Phillip A. Greenberg
San Francisco, California

Kevin Guilfoy
Carroll University

Wendy Halpin Hallows
Chestnut Hill College

Wendy C. Hamblet
North Carolina A&T State University

Michael S. Hamilton
University of Southern Maine

Randall Hannum
*New York City College of Technology,
CUNY*

Clayton D. Harris
Middle Tennessee State University

Gerald K. Harrison
Massey University

C. Alton Hassell
Baylor University

Jennifer F. Helgeson
Grantham Research Institute on Climate Change and the Environment

Thomas E. Hemmerly
Middle Tennessee State University

Howard V. Hendrix
California State University, Fresno

Mark Henkels
Western Oregon University

Charles E. Herdendorf
Ohio State University

Jane F. Hill
Bethesda, Maryland

Joseph W. Hinton
Portland, Oregon

Laurent Hodges
Iowa State University

John R. Holmes
Franciscan University of Steubenville

Robert M. Hordon
Rutgers University

Louise D. Hose
Westminster College

Ronald K. Huch
University of Papua New Guinea

Allyson Leigh Hughes
Michigan State University

Patrick Norman Hunt
Stanford University

Diane White Husic
East Stroudsburg University

H. David Husic
Lafayette College

Raymond Pierre Hylton
Virginia Union University

Solomon A. Isiorho
Indiana University-Purdue University Fort Wayne

Bernard Jacobson
SciMed Writers

Allan Jenkins
University of Nebraska at Kearney

Albert C. Jensen
Central Florida Community College

Jeffrey A. Joens
Florida International University

Bruce E. Johansen
University of Nebraska at Omaha

Suzanne Jones
Huntingdon College

Karen N. Kähler
Pasadena, California

Karen E. Kalumuck
The Exploratorium

Kara Kaminski
Boston College

Joseph Kantenbacher
University of California, Berkeley

Michael D. Kaplowitz
Michigan State University

Susan J. Karcher
Purdue University

Jamie Michael Kass
University of California, Berkeley

Kyle L. Kayler
Kayler Geoscience, Ltd.

Christopher Kent
Pasadena, California

Carolynn A. Kimberly
University of Dayton

Robert W. Kingsolver
Kentucky Wesleyan College

Christopher C. Kirby
Eastern Washington University

Samuel V. A. Kisseadoo
Hampton, Virginia

Grove Koger
Boise State University

Narayanan M. Komerath
Georgia Institute of Technology

Padma P. Komerath
SCV, Inc.

Andrew Lambert
University of Hawaii at Manoa

Nicholas Lancaster
Desert Research Institute

Timothy Lane
Louisville, Kentucky

Eugene Larson
Los Angeles Pierce College

Thomas T. Lewis
Mount Senario College

Josué Njock Libii
Indiana University-Purdue University Fort Wayne

Victor Lindsey
East Central University

Donald W. Lovejoy
Palm Beach Atlantic University

David C. Lukowitz
Hamline University

Michael Mooradian Lupro
North Carolina A&T State University

Larry S. Luton
Eastern Washington University

R. C. Lutz
CII Group

Fai Ma
University of California, Berkeley

Steven B. McBride
West Virginia University

Joel P. MacClellan
University of Tennessee

Robert McClenaghan
Pasadena, California

David F. MacInnes, Jr.
Guilford College

Francis P. Mac Kay
Providence College

Marianne M. Madsen
University of Utah

Louise Magoon
Fort Wayne, Indiana

Nancy Farm Männikkö
Centers for Disease Control and Prevention

Sergei A. Markov
Austin Peay State University

Chogollah Maroufi
California State University, Los Angeles

Kathleen Rath Marr
Lakeland College

Laurence W. Mazzeno
Alvernia College

Roman Meinhold
Assumption University

Randall L. Milstein
Oregon State University

Richard F. Modlin
University of Alabama in Huntsville

Charles Mortensen
Ball State University

M. Marian Mustoe
Eastern Oregon University

Alice Myers
Bard College at Simon's Rock

Mysore Narayanan
Miami University

Peter Neushul
California Institute of Technology

Anthony J. Nicastro
West Chester University

Martin A. Nie
University of Pittsburgh at Bradford

David L. O'Hara
Augustana College

Dónal P. O'Mathúna
Dublin City University

Oghenekome U. Onokpise
Florida A&M University

G. Padmanabhan
North Dakota State University

Beth Ann Parker
Huntingdon College

Gordon A. Parker
University of Michigan-Dearborn

Barbara Bennett Peterson
University of Hawaii

John Pichtel
Ball State University

Julio César Pino
Kent State University

George R. Plitnik
Frostburg State University

Aaron S. Pollak
Omaha, Nebraska

Oliver B. Pollak
University of Nebraska at Omaha

Noreen D. Poor
University of South Florida

Allison M. Popwell
Huntingdon College

Robert Powell
Avila University

Victoria Price
Lamar University

Syed R. Qasim
University of Texas at Arlington

P. S. Ramsey
Brighton, Michigan

C. Mervyn Rasmussen
Renton, Washington

R. Kent Rasmussen
Thousand Oaks, California

Ronald J. Raven
State University of New York at Buffalo

Donald F. Reaser
University of Texas at Arlington

Claudia Reitinger
Aachen University

John Rickett
University of Arkansas at Little Rock

Raymond U. Roberts
Oklahoma Department of Environmental Quality

Gene D. Robinson
James Madison University

Jacqueline J. Robinson
Huntingdon College

James L. Robinson
University of Illinois at Urbana-Champaign

Charles W. Rogers
Southwestern Oklahoma State University

Donna L. Rogers
Arkansas Tech University

Kenneth A. Rogers
Arkansas Tech University

Carol A. Rolf
Rivier College

Keith E. Rolfe
University of Dayton

Joseph R. Rudolph, Jr.
Towson University

Neil E. Salisbury
University of Oklahoma

Robert M. Sanford
University of Southern Maine

Elizabeth D. Schafer
Loachapoka, Alabama

John Richard Schrock
Emporia State University

Miriam E. Schwartz
Los Angeles, California

Alexander Scott
Pasadena, California

Rose Secrest
Chattanooga, Tennessee

Elizabeth F. Shattuck
Michigan State University

Martha A. Sherwood
Kent Anderson Law Office

R. Baird Shuman
University of Illinois at Urbana-Champaign

Carlos Nunes Silva
University of Lisbon

Paul P. Sipiera
Harper College

Amy Sisson
Houston Community College

Adam B. Smith
University of California, Berkeley

Courtney A. Smith
Parks and People Foundation

Jane Marie Smith
Slippery Rock University

Roger Smith
Portland, Oregon

Daniel L. Smith-Christopher
Loyola Marymount University

Diane Stanitski-Martin
Shippensburg University

Anne Statham
University of Wisconsin-Parkside

Joan C. Stevenson
Western Washington University

Dion Stewart
Adams State College

Robert J. Stewart
California Maritime Academy

Toby Stewart
Duluth, Georgia

Alexander R. Stine
University of California, Berkeley

Mary W. Stoertz
Ohio University

Theresa L. Stowell
Adrian College

Hubert B. Stroud
Arkansas State University

Sandra S. Szegedi
Palm Beach Atlantic University

Rena Christina Tabata
University of British Columbia

Julia Tanner
Palmerston North, New Zealand

John R. Tate
Montclair State College

William R. Teska
Furman University

John M. Theilmann
Converse College

Nicholas C. Thomas
Auburn University at Montgomery

Donald J. Thompson
California University of Pennsylvania

Anh Tran
Wichita State University

Oluseyi A. Vanderpuye
Albany State University

Charles L. Vigue
University of New Haven

Joseph M. Wahome
Mississippi Valley State University

C. J. Walsh
Mote Marine Laboratory

John P. Watkins
Westminster College

Megan E. Watson
Duke University

Shawncey Webb
Taylor University

Lynn L. Weldon
Adams State College

Robert J. Wells
Society for Technical Communication

Winifred O. Whelan
St. Bonaventure University

Edwin G. Wiggins
Webb Institute

Thomas A. Wikle
Oklahoma State University

Kay R. S. Williams
Shippensburg University

Marcie L. Wingfield
Huntingdon College

Brian G. Wolff
Minnesota State Colleges and Universities

Sam Wong
University of Liverpool

William C. Wood
James Madison University

Scott Wright
University of St. Thomas

George Wrisley
George Washington University

Lisa A. Wroble
Redford Township District Library

Robin L. Wulffson
Faculty, American College of Obstetrics and Gynecology

Jay R. Yett
Orange Coast College

Michele Zebich-Knos
Kennesaw State University

Editor's Introduction

Progress is a cultural concept—and an anthropocentric one. When we speak of progress we almost always mean human progress: the management of global resources so as to provide greater benefits to the human species. This should come as no surprise—we humans are unavoidably anthropocentric. All species strive to thrive, and some may do serious damage to their habitats in the process. We differ from other species in the extent of our influence on the natural environment and in our ability to think about what we are doing.

Earth has been home to anatomically modern human beings for about 200,000 years. For most of that time the human population was small and its environmental footprint localized. As recently as 1800 the world's people numbered fewer than one billion, and most were engaged in subsistence agriculture. The Industrial Revolution changed all that. Global population reached two billion in 1927, three billion in 1960, four billion in 1974, five billion in 1987, and six billion in 1999, when the first edition of the *Encyclopedia of Environmental Issues* was published. By midyear 2012 world population is expected to exceed seven billion.

The increase in our numbers and the power of our technology have radically increased the impact of people on planetary resources and shifted the natural balance dramatically in our favor—at least in the short run. The question of how many people Earth can sustain has become a matter of serious debate. For the first time in history, the human race is forced to confront the physical limits of Earth's resources and to wrestle with the increasing probability that our impact on those resources may produce catastrophic results for ourselves and for the other organisms with which we share this planet. In short, we have "environmental issues."

Planet Earth is the only locale in the entire universe where we know humans can live. Although mathematics suggests there are other Earth-like planets somewhere in the cosmos, science holds out little chance of our ever reaching one. We are stuck right here on the only known planet with a natural and—up to a point—self-repairing life-support system. "Environmental issues" are of compelling importance precisely because they threaten that life-support system.

UNDERSTANDING ENVIRONMENTAL ISSUES

Creating an *Encyclopedia of Environmental Issues* is an enormous challenge, and the dimensions of that challenge grow greater every year. Our understanding of environmental issues is uneven, inadequate, and ever changing. Our ability to meet the human challenges presented by environmental issues is very much in doubt. New issues emerge even as older ones remain unresolved, perhaps even unaddressed. An explosion in the breadth of environmental issues that began in the twentieth century continues unabated, from a relatively narrow focus on conservation to a far broader concept that embraces almost everything that provides context for human existence, from acid deposition and aquaculture to zoning and zoos. One result is that this *Revised Edition* has grown in size and scope.

The traditional view of environmental issues focused on negative consequences for the biophysical environment resulting from human behavior, but it is increasingly obvious that this definition is too narrow. Humans are adapted to a very narrow range of environmental circumstances, and we are very likely dependent in ways yet unknown on other species whose ability to tolerate environmental change may be even less than our own. If environmental parameters change more quickly than we can adapt, our survival is threatened whether those changes are caused by human beings or by forces completely outside human decision making.

Climate stability is a case in point. The Intergovernmental Panel on Climate Change is increasingly certain that the current global warming is a result of human activities, primarily the combustion of fossil fuels. It makes little difference, however, whether climate change is anthropogenic. The cascading crises brought about by increasingly severe famines and floods, massive extinction of species, and rising sea levels would be no less catastrophic if we were to discover that humans were not to blame. Whether caused by the combustion of fossil fuels or by an asteroid impact, rapid environmental change is destined to reduce biological diversity and further threaten the critical ecological services on which human beings depend. Humans may very well be responsible for most of the planet's environmental issues, but we have environmental issues whether we are responsible or not.

In 2010, as scores of authors were working on articles for this *Encyclopedia of Environmental Issues*, Americans were riveted by the environmental calamity du jour: the explosion and fire on the British Petroleum (BP) *Deepwater Horizon* drilling platform and the subsequent discharge of some five million barrels of oil into the Gulf of Mexico. It was the largest accidental oil spill in maritime history, easily surpassing the eerily similar explosion and fire in 1979 on the PEMEX drilling rig Ixtoc I, also in the Gulf of Mexico.

The BP oil spills suggests at least two important lessons about how we tend to deal with environmental issues. The first is our failure to learn from our failures. More than three decades after the Ixtoc I disaster, the mechanism of failure in the *Deepwater Horizon* drilling operation was precisely the same: a blowout preventer that failed to work as intended, followed by an explosion, a fire, and the release of millions of gallons of oil into the Gulf. Once the discharge began, BP initiated most of the same strategies for containment that had been used with limited success by PEMEX thirty years earlier. As this is written, the long-term environmental damage to the Gulf of Mexico ecosystem cannot be measured with any degree of precision, but recent reports suggest that the damage might be less severe than originally feared.

That possibility calls attention to a second, and equally obvious, lesson. As a society, we pay far too much attention to the novel or emergent environmental story and far too little attention to the gradual but ultimately more damaging erosion of environmental quality that is going on every day. The BP oil spill was spectacular, but it was just one of a multitude of environmental insults to which the Gulf of Mexico ecosystem has been subjected. Indeed, the Gulf is home to an oxygen-depleted "dead zone" the size of New Jersey, the result not of headline-stealing oil spills but of environmentally damaging agricultural practices that result in fertilizer runoff throughout the Mississippi River basin. Ironically, if the consequences of the BP oil spill should prove to be more damaging than expected, those results will likely get little attention. The well has been capped. The emergency has ended, and the media circus has moved on.

We should not be surprised that modern environmental issues are poorly understood and poorly covered by the mass media. Concern regarding the environment is still relatively new. Modern environmentalism is the most recent and the most far-reaching manifestation of a conservation movement that was born in the nineteenth century and has been growing ever since. Nineteenth century conservation was a response to population growth, urbanization, and industrialization. Late in the century unprecedented human impact on the natural world stimulated public concern about exhausting resources. Forests were being logged faster than they could regrow, and important wildlife species that had once numbered in the millions—such as the passenger pigeon and the American bison—were extinct or in danger of extinction. The resulting efforts at conservation emphasized the areas of greatest national concern: preservation of forests and wildlife. By 1920 the U.S. government had responded with policies creating national forests, national parks, national wildlife refuges, and national monuments.

Like the conservation movement that preceded it, modern environmentalism was a response to shortages brought about by the increasing pace of economic development. Many more resources now seemed in short supply, including clean air and water and unpolluted land. Equally important, the emerging science of ecology emphasized the interconnectedness of human life with all other life. For laypersons, the environmental era may have begun with the publication of Rachael Carson's *Silent Spring* in 1962. Carson's exploration of the consequences of pesticide use drove home the lesson that people are part of the global ecosystem and that in destroying parts of nature, we threaten it all.

The modern concern for environmental issues still centers on the use and abuse of natural resources, but to the historic concerns of conservation—such as forests, wilderness, and wildlife—have been added clean water, clean air, energy supply, hazardous and toxic waste, environmental illness, nuclear safety, and a host of other quality-of-life issues. Beginning in the 1960's and 1970's the U.S. Congress responded to these new concerns with an unprecedented flood of environmental policy laws, which are covered extensively in these volumes.

SCIENCE AND ENVIRONMENTAL ISSUES

In recent decades an explosion in scientific understanding has produced new environmental concerns. The discovery of previously unanticipated problems may spur a search for solutions, but there is no guarantee that science can achieve the requisite level of understanding, at least in the short run.

Global climate change serves as a kind of poster child for the complex relationship between modern science and the human environment. First, global climate change could never have become an environmental issue without sophisticated science. That emissions of "greenhouse" gases from the combustion of fossil fuels could result in global warming could not have been hypothesized without a degree of scientific understanding that is relatively recent. To test this greenhouse hypothesis requires even greater scientific sophistication. We all experience weather in our daily lives, but without high-tech instrumentation globally deployed, even the most astute scientist could not begin to measure global warming.

Second, global climate change became a political issue not because voters or citizens generally were concerned but because scientists were concerned and they voiced those concerns publicly. In the United States, global warming became a political issue on June 23, 1988, when James E. Hansen, director of the National Aeronautics and Space Administration's (NASA) Goddard Institute for Space Studies, testified before the Senate Committee on Energy and Natural Resources. Hansen reported that he was 99 percent certain that global warming was under way. To many it seemed as if NASA had spoken with all the authority of science, and suddenly a scientific concern was a political issue as well.

Third, although science created the climate change issue, contemporary science is far from certain about its existence, its causes, or its consequences. Despite increasing empirical evidence, every aspect of the case is subject to scientific challenge. Conclusions necessarily rest on incomplete modern temperature data, historic temperature data that are even less reliable, and computer models that may be inaccurate in their portrayal of the relationships among variables.

If the facts of global warming are difficult to establish, the causes are even more problematic. We know that Earth's climate changes over time. Evidence of ice ages lasting thousands of years is well established, but recent studies of ice cores recovered from continental sheets in Greenland and Antarctica suggest that during human prehistory global climate may have undergone violent short-term changes as well. Various scientists have speculated that current climate change—if it is occurring—may result from solar activity, from the atmospheric debris of volcanic

eruptions, or from some unidentified natural cycle, as well as from the increased emissions of greenhouse gases associated with large human populations burning fossil fuels, clearing forests, and raising cattle.

The consequences of global warming seem less uncertain. Temperature zones will shift toward the poles. Ocean levels will rise as polar ice caps melt, changing the circulation of ocean currents and displacing billions of people in low-lying areas around the world. Changing patterns of precipitation will require a reorganization of world agriculture. Changes of this magnitude will result in huge social, economic, and political changes for people and in massive extinctions for species unable to adapt to their new environments.

GLOBALIZATION OF ENVIRONMENTAL ISSUES

It has become increasingly obvious that both the causes and the consequences of environmental issues are global in nature, but the international community has been slow to respond. First, where the environmental consequences of human behavior are localized, there is no longer any guarantee that the localized effects will occur nearby. The ecologist Raymond F. Dasmann reminds us that modern technology and a global marketplace have given us the capacity to do environmental damage worldwide. Until recently Earth was populated by "ecosystem people" whose survival depended directly on their care of the ecosystems of which they were a part. If you were a Native American whose people depended on the buffalo, you dared not overhunt because the end of the buffalo would be the end of you. Ecosystem people had no choice but to live with the ecological results of their actions. Today, the world's population is interdependent in countless ways. We are "biosphere people." Resources are mobilized on a global scale—from the entire biosphere—to provide for our food, shelter, comfort, and amusement. The effects of our behavior are often distant, so the link between cause and effect is obscured. Not knowing the effects of our behavior, we have no strong incentive to behave in a way that is protective of Earth resources.

Second, many environmental consequences of human behavior are no longer localized at all—they are global. Nuclear fallout, acid precipitation, and ocean pollution move unimpeded across international boundaries. Stratospheric ozone depletion and climate change threaten the ability of the biosphere to sustain life as we have known it. Declines in global biological diversity may presage an Earth environment that is eventually less hospitable to people as well.

The international community is only slowly becoming aware of its global environmental responsibilities. The history of treaties seeking to address environmental issues parallels the history of environmental issues themselves. Early efforts were aimed almost exclusively at the conservation of threatened fauna and flora. By the 1970's the focus had shifted to the broader concerns associated with the term "environmentalism." Most of the world's industrialized democracies enacted significant environmental protection legislation during the 1970's. Given the limits of national legislation in the face of global environmental problems, it is not surprising that they should have also sought to achieve environmental objectives through international agreements.

A number of international programs have been instituted to reward individual nations for doing the right thing for the environment. In 1965 the Council of Europe established the European Diploma of Protected Areas to recognize effective protection of internationally significant areas. Six years later the United Nations Educational, Scientific, and Cultural Organization (UNESCO) launched its Man and the Biosphere Programme, designed to improve the relationship between human beings and their environment through science and education. A key feature was the Biosphere Programme, which established an international system of biosphere reserves, representative ecological areas established to preserve genetic diversity. Another UNESCO convention, effective in 1975, designates World Heritage Sites. These endeavors resemble the older conservation tradition, and they have had little real effect. The designations generally go to national parks and historical sites that are already well protected. The 1973 Convention on International Trade in Endangered Species of Wild Fauna and Flora (CITES) has been more effective, but poaching and illegal trafficking remain serious problems.

Since the 1970's a series of world conferences have been held on modern environmental issues. The first Conference on the Human Environment was held in Stockholm, Sweden, in 1972 and attracted representatives from 113 nations. It dramatized international environmental concerns and resulted in establishment of the United Nations Environment Programme. Two decades later the Earth Summit in Rio de Janeiro, Brazil, drew delegates from 179 nations and emphasized the linkage between economic development in the developing world and environmental protection. The limitations of the treaty system for achieving international environmental goals have been demonstrated by the refusal of the United States to accept international agreements on reductions of greenhouse gas emissions and preservation of biological diversity.

In 1997 a third international gathering at Kyoto, Japan, produced a draft treaty designed to reduce global emission of greenhouse gases to below 1990 levels by 2012. Industrial nations committed to specific reductions, and developing nations committed to doing what they could. Virtually all of the world's nations (191) have ratified or indicated an intention to ratify the Kyoto Protocol; the United States alone has rejected ratification. Many advanced industrialized nations have made good progress toward meeting their emissions reduction targets, but that progress has been overwhelmed by increasing emissions in rapidly developing countries and the United States. Despite the Kyoto Protocol, numerous subsequent international climate conferences, and ever-increasing clarity about the consequences of failure, global carbon dioxide emissions are growing faster than ever.

INSTITUTIONAL CAPACITY AND ENVIRONMENTAL ISSUES

Dealing effectively with environmental issues will require institutions capable of global coordination over long periods of time. In the first decades of a new millennium this institutional capacity is noticeably absent. As the recent history of international cooperation on environmental issues attests, the international arena lacks the institutions required to make and enforce any global environmental policy. Planet Earth is subdivided into nation-states, each recognizing no power superior to itself. Nation-states are reluctant to give up any portion of their sovereignty or to recognize that environmental survival may require it. Treaties and other agreements among nation-states are regarded as binding only so long as they serve each nation's interest. Weaker nations may occasionally be forced to comply by stronger ones, but this system of international "might makes right" falls far short of government in any meaningful sense of the term.

Domestic institutions are also seriously limited in dealing with environmental issues. With the end of the Cold War, the world became more homogeneous, both politically and economically. Politically, all nations claim to be democratic, though the degree to which they live up to that claim varies widely. Economically, capitalism is ascendent. Governments continue to regulate markets, but increasingly the norm is private ownership of property. In many respects these changes are liberating, but they have not resolved our environmental issues. In some respects, they have made things worse. Around the world, the global economic crisis that began in the fall of 2008 pushed environmental issues into the background. Willingness to make difficult political decisions wilted in the heat of public disapproval, and the prospect of meaningful international cooperation to solve global environmental issues came to seem more remote than it was at the dawn of the twenty-first century, when the first edition of the *Encyclopedia of Environmental Issues* was published.

Both capitalism and democracy are notoriously shortsighted. Economic theory systematically dis-counts the future. Corporate managers are required to maximize short-term profits to please investors. In an ideal world this capitalistic myopia would be balanced by political leadership committed to the long-term interests of society and even to generations yet unborn. In the practical world of democracy, however, political leaders behave just like corporate managers, seeking short-term benefits likely to influence the next election.

Our institutions have often been ineffective in addressing environmental issues, but they are not pre-destined to fail. In the final analysis, both democratic elections and free markets are driven from the bottom up by popular sentiment. Citizens in all nations have important political and economic choices to make. To choose wisely they will require both information and insight. It is my hope that this *Revised Edition* of the *Encyclopedia of Environmental Issues* contributes to fulfilling those requirements.

Craig W. Allin
Cornell College

Complete List of Contents

Volume 1

Volume 2

Contents xxxix
Complete List of Contents xliii

Volume 3

Contents lix
Complete List of Contents lxiii

Volume 4

Contents lxxix
Complete List of Contents lxxxi

ENCYCLOPEDIA OF
ENVIRONMENTAL ISSUES

A

Abbey, Edward

CATEGORIES: Activism and advocacy; preservation and wilderness issues

IDENTIFICATION: American environmental activist and author

BORN: January 29, 1927; Indiana, Pennsylvania

DIED: March 14, 1989; Tucson, Arizona

SIGNIFICANCE: The originality of Abbey's ideas regarding the preservation of nature, expressed with great eloquence in his writings, helped to increase awareness of environmental issues and inspired a radical environmental movement.

Over the course of his lifetime, Edward Abbey produced twenty-one volumes of fiction, essays, speeches, and letters expressing his love for the earth, his hatred of modern technological society, and his fervent belief that development was destroying the American West. When he was twenty-one years old, having spent some time in the military and in college, he left his home in Pennsylvania to see the American West. He hitch-hiked, rode trains, and walked over the mountains and through the desert. He claimed the desert as his spiritual home and lived in or near it for most of the rest of his life. He completed a master's degree at the University of New Mexico and wrote his Ph.D. thesis on anarchism and the morality of violence. During his ten-year college career, which included two years as a Fulbright Fellow at the University of Edinburgh in Scotland, he began a number of writing projects and published his first novel, *Jonathan Troy* (1954).

For fifteen years, during his thirties and forties, Abbey worked as a part-time ranger at various national parks in the American Southwest. The two years in the late 1950's that he spent at Arches National Monument (now a national park) in Utah led to his first important book, *Desert Solitaire: A Season in the Wilderness* (1968). This book combines beautiful descriptive passages, an unflinching look at the violence in nature, and a strong call for the preservation of desert habitats. Reminiscent of Henry David Thoreau's *Walden* (1854) in its ideas and its use of the natural year for its structure, *Desert Solitaire* brought Abbey national attention as an environmental writer.

In 1975 Abbey published *The Monkey Wrench Gang*, a novel about four rebels who set out to destroy the roads, bridges, and power lines that they believe are defacing the southwestern desert. This work is loosely based on the exploits of a friend of Abbey who had committed some of the acts depicted in the novel. Despite the fact that Abbey consistently maintained that he intended the book primarily as humor, it helped inspire the radical environmental group Earth First!, a group that Abbey did come to support, praising its operations although never actually joining it. In fact, Abbey never joined any political or environmental or-

Edward Abbey at work in his Arizona home. (©Buddy Mays/CORBIS)

ganizations, although he participated in political actions, especially those that expressed disapproval of the military or land development.

Though most of his books are set in the wilderness of the Southwest and express his deep love for such spaces, Abbey disliked being called a "nature writer." In fact, students of his work have had difficulty attaching any label to Abbey and making it stick, so idiosyncratic are his ideas and connections. When Abbey died in 1989, he left instructions that he should be buried in the desert, unembalmed, in his sleeping bag. Although this kind of burial was illegal, his friends followed his wishes.

Cynthia A. Bily

FURTHER READING

Bishop, James, Jr. *Epitaph for a Desert Anarchist: The Life and Legacy of Edward Abbey*. New York: Atheneum, 1994.
Cahalan, James M. *Edward Abbey: A Life*. Tucson: University of Arizona Press, 2001.
Pozza, David M. *Bedrock and Paradox: The Literary Landscape of Edward Abbey*. New York: Peter Lang, 2006.

SEE ALSO: Earth First!; Monkeywrenching; Preservation.

Accounting for nature

CATEGORY: Resources and resource management

DEFINITION: The expansion of economic principles and adaptation of financial decisions to take into consideration natural resources, ecosystem services, and values derived from human contact with the natural world

SIGNIFICANCE: In comparisons of the costs and benefits of various actions in relation to the environment, the coordination of ecological and economic expertise can result in a balanced problem-solving approach.

Environmental decisions frequently set ecologists and economists on a collision course: Does society's demand for energy justify the environmental impacts of mining and burning coal? Is the increased economic efficiency of large corporate farms worth the loss of a rural lifestyle? How much should the public sacrifice to protect endangered birds or plants?

These decisions are difficult because they force comparisons of "apples and oranges," pitting one set of values against another. Quantitative models (or accounting systems) that compare the costs and benefits of a course of action are frequently used to guide business and government decisions, but these have generally omitted environmental values. Accounting for nature in these models cannot produce completely objective solutions but can help coordinate ecological and economic expertise in a more balanced problem-solving approach.

Ecologist Edward O. Wilson has identified three kinds of national wealth: economic, cultural, and biological. He has observed that nations frequently create the illusion of a growing economy by consuming their biological or cultural "capital" to create short-term economic prosperity. For example, burning rain forests and replacing them with row crops may temporarily increase farm production in a developing nation, but tropical soils are often nutrient-poor and easily degraded by exposure to the sun and rain. If the biological basis of production is ignored, the population simply transfers wealth from one category to another, ensuring ecological disaster for its children in the process.

Developed nations have also neglected biological wealth in past cost-benefit analyses. Nations have justified the damming of wild rivers to make recreational lakes, for example, by counting the benefits of lumbering trees from the watershed but not the losses of aquatic and forest habitats. Recreational activities have been evaluated according to the money that people pay to participate in them; thus hikers and canoeists in a wilderness area are given less consideration than water-skiers or drivers of recreational vehicles in a developed area because hikers and canoeists spend less money on equipment, fuel, and supplies.

Ecologists Eugene P. Odum and Howard T. Odum addressed this issue by calculating the value of ecosystem services provided by intact biological communities. Their approach was to measure the beneficial work performed by living systems and place a value on that service based on the time and energy required to replicate the service. A living tree, they reasoned, may provide a few hundred dollars in lumber if cut; if left alive, however, the oxygen it produces, carbon dioxide it absorbs, wildlife it feeds and shelters, soil it builds, evaporative cooling it yields, and flood protection it provides are worth far more on an annual basis.

In 1972, economists William Nordhaus and James Tobin refined the concept of national wealth by developing an index of net economic welfare (NEW) to replace the more familiar measurements of economic health such as gross domestic product (GDP). Their criticism of the GDP was that it counts any expenditure as a positive contribution to national wealth, whether or not the spending improves people's lives. A toxic waste dump, for example, contributes to the GDP when the pollutants are produced, again when millions of dollars are spent to clean it, and yet again if medical costs rise because of pollution-related illness. Nordhaus and Tobin's NEW index subtracts pollution abatement and other environmental costs from the value of goods and services that actually improve living standards.

Economist E. F. Schumacher subsequently argued that environmental costs should be "internalized," or charged to the industries that create them. This idea, also called the "polluter pays principle," not only generates funds for environmental cleanup but also encourages businesses to make environmentally sound decisions. The price of recycled paper, for example, would be more competitive if the public costs of deforestation and pollution from pulp mills were added to the price of virgin wood fiber. Proposals for internalizing environmental costs have ranged from centrally planned models, such as a carbon tax on fossil fuels, to free market trading of pollution credits. Debt-for-nature swaps, through which developing nations receive financial benefits for preserving natural ecosystems, represent environmental cost accounting on the asset side of the ledger.

A fundamental difference between economic and ecological worldviews is the time scale under consideration. Business strategies may look five years ahead, but ecological processes can take centuries. Thus economic models that fail to take long-term issues into account are a frequent source of criticism by environmentalists. The U.S. decision to build nuclear fission reactors during the 1960's and 1970's is a case in point. Nuclear power appeared economically attractive over the thirty-five-year life span of a fission reactor, but the twenty-four-thousand-year half-life of radioactive plutonium 239 in spent fuel rods made skeptics wonder who would pay the costs of nuclear waste disposal for generations after the plants were closed.

Debates about growth are especially contentious. Traditional economists view the growth of populations, goods, and services as positive and necessary for economic progress and social stability. As early as 1798, however, economist Thomas Robert Malthus pointed out that on a finite Earth, an exponentially expanding human population would eventually run out of vital resources. In the closing decades of the twentieth century, Paul R. Ehrlich and Anne Ehrlich warned that unless population growth slowed soon, each person would have to consume less space, food, fuel, and other materials to avoid a global population crash. Whether one considers them economic pessimists or environmental realists, Malthus and the Ehrlichs demonstrate that taking a longer view is central to the task of accounting for nature. Sustainable development is the watchword for ecologists, economists, and political leaders attempting to create prosperity today while accounting for the welfare of future generations.

Robert W. Kingsolver

FURTHER READING

Anderson, Terry L., Laura E. Huggins, and Thomas Michael Power, eds. *Accounting for Mother Nature: Changing Demands for Her Bounty.* Stanford, Calif.: Stanford University Press, 2008.

Costanza, Robert, ed. *An Introduction to Ecological Economics.* Boca Raton, Fla.: St. Lucie Press, 1997.

Pearce, Joseph. *Small Is Still Beautiful: Economics As If Families Mattered.* Wilmington, Del.: ISI Books, 2006.

Schumacher, E. F. *Small Is Beautiful: Economics As If People Mattered.* 1973. Reprint. Point Roberts, Wash.: Hartley & Marks, 1999.

Worldwatch Institute. *State of the World.* New York: W. W. Norton, 2010.

SEE ALSO: Debt-for-nature swaps; Ecological economics; Environmental economics; Schumacher, E. F.; Sustainable development; Wilson, Edward O.

Acid deposition and acid rain

Category: Atmosphere and air pollution

Definition: Deposition of acidic gases, particles, and precipitation (rain, fog, dew, snow, or sleet) on the surface of the earth

Significance: Electric utilities, industries, and automobiles emit sulfur dioxide and nitrogen oxides that are readily oxidized into sulfuric and nitric acids in the atmosphere. Long-range transport and dispersion of these air pollutants produce regional acid deposition, which alters aquatic and forest ecosystems and accelerates corrosion of buildings, monuments, and statuary.

In 1872 Robert Angus Smith used the term "acid rain" in his book *Air and Rain: The Beginnings of a Chemical Climatology* to describe precipitation affected by coal-burning industries. Acidity is created when sulfur dioxide (SO_2) and nitrogen oxides (NO_x) react with water and oxidants in the atmosphere to form water-soluble sulfuric and nitric acids. The normal acidity of rain is pH 5.6, which is caused by the formation of carbonic acid from water-dissolved carbon dioxide. The acidity of precipitation collected at monitoring stations around the world varies from pH 3.8 to 6.3 (pH 3.8 is three hundred times as acidic as pH 6.3). Ammonia, as well as soil constituents such as calcium and magnesium that are often present in suspended dust, neutralizes atmospheric acids, which helps explain the geographical variation of precipitation acidity.

Increasing Acidity

Between the mid-nineteenth century and World War II, the Industrial Revolution led to a tremendous increase in coal burning and metal ore processing in both Europe and North America. The combustion of coal, which contains an average of 1.5 percent sulfur by weight, and the smelting of metal sulfides released opaque plumes of smoke and SO_2 from short chimneys into the atmosphere.

Copper, nickel, and zinc smelters inundated nearby landscapes with SO_2 and heavy metals. One of the world's largest nickel smelters, located in Sudbury, Ontario, Canada, began operation in 1890 and by 1960 was pouring 2.6 million tons of SO_2 per year into the atmosphere. By 1970 the environmental damage extended to 72,000 hectares (278 square miles) of injured vegetation, lakes, and soils surrounding the site; within this area 17,000 hectares (66 square miles) were barren. The land was devastated not only by acid deposition but also by the accumulation of toxic metals in the soil, the clear-cutting of forested areas for fuel, and soil erosion caused by wind, water, and frost heave. (The situation improved dramatically in the ensuing decades, but only after construction in 1972 of a tall "superstack" that dispersed emissions farther from the smelting facility, followed by an extensive tree-planting program and other major environmental reclamation efforts, as well as installation of industrial scrubber systems at the facility during the 1990's.)

In urban areas, high concentrations of sulfur corroded metal and accelerated the erosion of stone buildings and monuments. Historical structures such as the Acropolis in Greece suffered serious damage from elevated acidity. During the winter, added emissions from home heating and stagnant weather conditions caused severe air pollution episodes characterized by sulfuric acid fogs and thick, black soot. In 1952 a four-day air-pollution episode in London, England, killed an estimated four thousand people.

After World War II, large coal-burning utilities in Western Europe and the United States built their plants with particulate control devices and stacks higher than 100 meters (328 feet) to improve the local air quality. (By contrast, huge industrial facilities throughout Eastern Europe and the Soviet Union operated without air-pollution controls for most of the twentieth century.) The tall stacks increased the dispersion and transport of air pollutants from tens to hundreds of kilometers. While this measure eased localized impacts, it simultaneously made the problem more widespread. Worldwide emissions of SO_2 increased; in the United States emissions climbed from 18 million tons in 1940 to a peak of 28 million tons in 1970. Acid deposition evolved into an interstate and even an international problem.

In major cities, exhaust from automobiles combined with power plant and industrial emissions to create a choking, acrid smog of ozone mixed with nitric and organic acids formed by photochemical processes. The rapid deterioration of air quality in cities, with attendant health and environmental consequences, spurred the passage of environmental laws such as the U.S. Clean Air Act (CAA) of 1963, which was amended and expanded in 1970, 1977, and 1990.

Each amendment to the CAA brought new requirements for air-pollution controls.

EFFECTS ON AQUATIC ECOSYSTEMS

Landscapes rich in limestone or acid-buffering soils are less sensitive to acid deposition. Regions that are both sensitive and exposed to acid deposition include the eastern United States, southeastern Canada, the northern tip of South America, southern Sweden and Norway, central and Eastern Europe, the United Kingdom, southeastern China, northeastern India, Thailand, and the Republic of Korea (South Korea). Within these regions, acid rain disrupts aquatic ecosystems and contributes to forest decline.

A strong correlation has been found between fish extinction and lake and stream acidity. Researchers have also found that the diversity of not only fish but also phytoplankton, zooplankton, invertebrates, and amphibian species diminishes by more than 50 percent as surface water pH drops from 6.0 to 5.0. Below pH 5.6, aluminum released from bottom sediments or leached from the surrounding soils interferes with gas and ion exchange in fish gills and can be toxic to aquatic life. At pH 5.0, most fish eggs cannot hatch; below pH 4.0, no fish survive.

In southern Norway, the virtual extinction of salmon in all the larger salmon rivers is attributed to acid rain. In 2008 it was found that twelve salmon stocks were endangered and another eighteen had been wiped out altogether. Records and long-term monitoring show that the decline of fish populations began during the early twentieth century, with dramatic losses during the 1950's. Between 1950 and 1990, fish mortality in the region became more widespread. In the three decades following 1980, pollution-control measures in Norway and the rest of Europe cut acid precipitation over Norway by roughly half. Aquatic animal and plant populations and ecosystems continue to recover from the damage.

In the United States the New York State Department of Environmental Conservation reported in 2008 that 26 percent of the state's lakes in the Adirondack Mountains were unable to neutralize incoming acids to concentrations that fish could tolerate. During certain times of year, up to 70 percent of the lakes had the potential to become intolerably

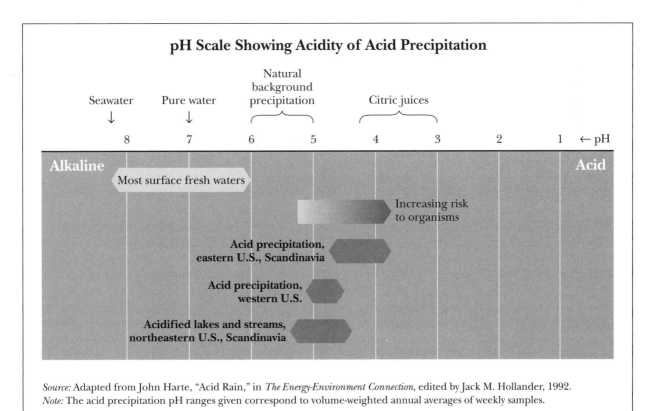

pH Scale Showing Acidity of Acid Precipitation

Source: Adapted from John Harte, "Acid Rain," in *The Energy-Environment Connection*, edited by Jack M. Hollander, 1992.
Note: The acid precipitation pH ranges given correspond to volume-weighted annual averages of weekly samples.

acidic. Of forty-eight Adirondack lakes surveyed, six-teen had aluminum concentrations above levels that juvenile fish could withstand. Ecosystems in this re-gion have been affected by plumes of air pollutants carried by prevailing winds from the Ohio River Val-ley. Like their Norwegian counterparts, the Adiron-dack lakes have low acid-neutralizing capacity. Fish declines that began during the early twentieth cen-tury and continued through the 1980's corresponded to reductions in pH. Fish kills often followed spring snowmelt, which filled the waterways with acid accu-mulated in winter precipitation.

While the CAA and its amendments have done much to reduce the acidity of the precipitation in the Adirondacks, and thus of the lakes, recovery takes time. Lake chemistry that is acidic year-round may take twenty-five to one hundred years to return to lev-els that aquatic life can tolerate, while seasonally acidic lake waters may take a few years or several de-cades. Once a lake regains its ability to neutralize acid, wildlife in the lake's ecosystem require additional time to become reestablished.

EFFECTS ON FORESTS AND CITIES

In areas exposed to acid rain, dead and dying trees stand as symbols of environmental change. In Ger-many the term *Waldsterben*, or forest death, was coined to describe the rapid declines of Norway spruce, Scotch pine, and silver fir trees during the early 1980's, followed by beech and oak trees during the late 1980's, especially at high elevations in the Black and Bavarian forests. At higher altitudes, clouds fre-quently shroud mountain peaks, bathing the forest canopy in a mist of heavy metals and sulfuric and nitric acids. Under drought conditions, invisible plumes of ozone from sources hundreds of kilome-ters distant intercepted the mountain slopes. Several forests within the United States have been likewise af-fected by both ozone and acidic deposition, including the pine forests in California's southern Sierra Ne-vada and high-elevation spruce-fir forests in the Northeast.

Intensive field and laboratory investigations of for-est decline in North America and Europe have yielded contradictory results regarding the link be-tween dead trees and acid deposition. Some labora-tory experiments have found that acid rain has no ef-fect, or even a fertilizing effect, on trees. Symptoms of tree stress include changes in foliage color, size, and shape; destruction of fine roots and associated fungi;

and stunted growth. Many researchers attribute these symptoms and forest decline to complex interactions among a variety of stressors, including acid precipita-tion, ozone, excessive nitrogen deposition, land man-agement practices, climate change, drought, and pes-tilence. Acid rain generally does not kill trees directly; rather, it weakens them by damaging their leaves, im-pairing their leaves' functionality, stripping nutrients from the soil, and releasing toxic substances from the soil.

Ambient air concentrations of SO_2 and NO_x are typically higher in major cities owing to the high den-sity of emission sources in these locations. The result-ing haze may be carried by winds hundreds of kilome-ters from where it was generated, obscuring visibility even in remote wilderness areas. The acids formed ac-celerate the weathering of exposed stone, brick, con-crete, glass, metal, and paint. For example, the calcite in limestone and marble reacts with water and sulfuric acid to form gypsum (calcium sulfate). The gypsum washes off stone with rain or, if eaves protect the stone, accumulates as a soot-darkened crust. Acid-induced weathering has obscured the details of elabo-rate carvings on medieval cathedrals, ancient Greek columns, and Mayan ruins at alarming rates.

PREVENTION EFFORTS

Discovery of the connection between sulfur emis-sions in continental Europe and lake acidification in Scandinavian countries ultimately led to the 1979 Convention on Long-Range Transboundary Air Pol-lution. This legally binding international agreement addresses air-pollution issues on a broad regional basis. Subsequent protocols to the convention deal specifically with sulfur emissions (1985 and 1994 pro-tocols), NO_x emissions (1988 protocol), and acidifica-tion abatement (1999 protocol).

In the United States the Acidic Deposition Control Program, Title IV of the CAA amendments of 1990, di-rects the Environmental Protection Agency (EPA) to reduce the adverse effects of acid rain, specifically through a reduction in emissions of SO_2 and NO_x. The program sets an annual cap on SO_2 emissions from power plants and establishes allowable NO_x emission rates based on boiler type. Between 1990 and 2006, the program achieved a reduction of SO_2 emissions by more than 6.3 million tons from 1990 levels. The program, in combination with other ef-forts to reduce emissions, also cut NO_x emission by roughly 3 million tons, making 2006 emissions less than half what would have been expected without the program.

The National Acid Precipitation Assessment Program coordinates interagency acid deposition monitoring and research and assesses the cost, benefits, and effectiveness of acid deposition control strategies. Acid deposition reduction schemes in the United States target large electric utilities, which as of 2006 were responsible for about 67 percent of the country's SO_2 and 19 percent of the NO_x emissions from anthropogenic sources. Utilities participate in a novel market-based emission allowance trading and banking system that permits great flexibility in controlling SO_2 emissions. For example, utilities may choose to remove sulfur from coal by cleaning it, to burn a cleaner fuel such as natural gas, or to install a gas desulfurization system to reduce emissions. They may also buy or sell emissions allowances. U.S. research efforts cost-shared between government and industry, such as the Clean Coal Power Initiative and similar programs before it, have developed technologies—for example, the catalytic conversion of NO_x to inert nitrogen—that can radically decrease emissions of acid gases from coal-fired power plants.

Between 1990 and 2006 annual atmospheric concentrations of SO_2 in the United States decreased by 53 percent, while nitrogen dioxide concentrations fell by 30 percent. Emissions of SO_2 dropped during this period by 38 percent; NO_x emissions declined 29 percent, with most of the decrease occurring after 1998. These reductions resulted in significant decreases in acid rain. Between 1985 and 2002, nitrate deposition declined in the New England states; however, in the western states, increasing oil and gas production and other factors caused nitrate deposition to rise. In the eastern states, sulfate concentrations in the air generally declined, affected regions decreased in size, and the magnitude of the highest concentrations dropped. During the periods from 1989 to 1991 and 2004 to 2006, sulfate deposition in the Northeast and the Midwest decreased by more than 30 percent. As a result, surface water quality improved.

Noreen D. Poor
Updated by Karen N. Kähler

FURTHER READING

Brimblecombe, Peter, et al., eds. *Acid Rain: Deposition to Recovery.* New York: Springer, 2007.

Finlayson-Pitts, Barbara J., and James N. Pitts. *Chemistry of the Upper and Lower Atmosphere: Theory, Experi-*

ments, and Applications. San Diego, Calif.: Academic Press, 2000.

Hill, Marquita K. "Acidic Deposition." In *Understanding Environmental Pollution.* 3d ed. New York: Cambridge University Press, 2010.

Lane, Carter N., ed. *Acid Rain: Overview and Abstracts.* New York: Nova Science, 2003.

Little, Charles E. *The Dying of the Trees: The Pandemic in America's Forests.* New York: Viking Press, 1995.

McGee, Elaine. *Acid Rain and Our Nation's Capital: A Guide to Effects on Buildings and Monuments.* Washington, D.C.: U.S. Geological Survey, 1995.

Smith, Robert Angus. *Air and Rain: The Beginnings of a Chemical Climatology.* 1872. Reprint. Whitefish, Mont.: Kessinger, 2007.

Visgilio, Gerald R., and Diana M. Whitelaw, eds. *Acid in the Environment: Lessons Learned and Future Prospects.* New York: Springer, 2007.

SEE ALSO: Air pollution; Automobile emissions; Clean Air Act and amendments; Coal-fired power plants; Convention on Long-Range Transboundary Air Pollution; Fish kills; London smog disaster; Nitrogen oxides; Power plants; Sudbury, Ontario, emissions; Sulfur oxides.

Acid mine drainage

CATEGORY: Water and water pollution
DEFINITION: The flow of acidified waters from mining operations and mine wastes
SIGNIFICANCE: Acid mine drainage can pollute groundwater, surface water, and soils, producing adverse effects on plants and animals.

During mining, rock is broken and crushed, exposing fresh rock surfaces and minerals. Pyrite, or iron sulfide, is a common mineral encountered in metallic ore deposits. Rainwater, groundwater, or surface water that runs over the pyrite leaches out sulfur, which reacts with the water and oxygen to form sulfuric acid. In addition, if pyrite is present in the mining waste materials that are discarded at a mine site, some species of bacteria can directly oxidize the sulfur in the waste rock and tailings, forming sulfuric acid. In either case, the resulting sulfuric acid may run into groundwater and streams downhill from the mine or mine tailings.

An official with the Pennsylvania Department of Environmental Protection's Bureau of Abandoned Mine Reclamation shows acid mine drainage flowing from an entrance to an old mine in Fallston in 2005. The mine had been abandoned since the 1960's. (AP/Wide World Photos)

Acid mine drainage (AMD) pollutes groundwater and adjacent streams and may eventually seep into other streams, lakes, and reservoirs to pollute the surface water. Groundwater problems are particularly troublesome because the reclamation of polluted groundwater is very difficult and expensive. Furthermore, AMD dissolves other minerals and heavy metals from surrounding rocks, producing lead, arsenic, mercury, and cyanide, which further degrade the water quality. Through this process, AMD has contributed to the pollution of many lakes.

AMD runoff can be devastating to the surrounding ecosystem. Physical changes and damage to the land, soil, and water from AMD directly and indirectly affect the biological environment. Mine water immediately adjacent to mines that are rich in sulfide minerals may be as much as 100,000 to 1,000,000 times more acidic than normal stream water. AMD water poisons and leaches nutrients from the soil so that few, if any, plants can survive. Animals that eat those plants, as well as microorganisms in the leached soils, may also die. AMD is also lethal for many water-dwelling animals, including plankton, fish, and snails. Furthermore, people can be poisoned by drinking water that has been contaminated with heavy metals produced by AMD, and some people have developed skin cancer as a result of drinking groundwater contaminated with arsenic generated by AMD leaching.

Alterations in groundwater and surface-water availability and quality caused by AMD have also had indirect impacts on the environment by causing changes in nutrient cycling, total biomass, species diversity, and ecosystem stability. Additionally, the deposition of iron as a slimy orange precipitate produces an unsightly coating on rocks and shorelines.

AMD was so severe in the Tar Creek area of Oklahoma that the U.S. Environmental Protection Agency designated the area as the nation's foremost hazardous waste site in 1982. The largest complex of toxic waste sites in the United States was produced by the mines and smelters in Butte and Anaconda, Montana, with much of the pollution attributed to direct and indirect effects of AMD. AMD is also a widespread problem in many coal fields in the eastern United States.

Alvin K. Benson

FURTHER READING

Bell, F. G. *Basic Environmental and Engineering Geology.* Boca Raton, Fla.: CRC Press, 2007.
Younger, Paul L., Steven A. Banwart, and Robert S. Hedin. *Mine Water: Hydrology, Pollution, Remediation.* Norwell, Mass.: Kluwer Academic, 2002.

SEE ALSO: Groundwater pollution; Heavy metals; Water pollution.

Adams, Ansel

CATEGORIES: Activism and advocacy; preservation and wilderness issues
IDENTIFICATION: American photographer and environmental activist
BORN: February 20, 1902; San Francisco, California
DIED: April 22, 1984; Carmel, California
SIGNIFICANCE: Through his spectacular photographs and his advocacy, Adams helped to increase Americans' awareness of the beauty of the nation's wilderness areas and the importance of preserving that beauty.

Ansel Adams was the only son of Charles and Olive Bray Adams. Although a gifted child, he detested public schools and was educated primarily by tutors, graduating from a private school in 1917. His early interests centered on music and, after he acquired his first camera, photography. A visit to Yosemite National Park in California in 1916 sparked his interest in nature photography, and he made frequent return visits to Yosemite throughout his lifetime.

Adams was able to support himself with music and photography during his early working years. As his interest in photography increased, however, it became the dominant factor in his life. He traveled extensively throughout the American West, photographing landscapes. His first published photograph appeared in the *Sierra Club Bulletin* in 1927, and the first of his numerous photographic collections was published in the same year. In 1928 he married Virginia Rose Best in Yosemite and held his first one-person exhibit in San Francisco. The Adamses had two children: Michael, born in 1933, and Anne, born in 1935.

In 1934 Adams was elected to the board of directors of the Sierra Club, a position he held until 1971. He was a cofounder of Group f/64, an organization dedicated to the use of photography to emphasize and preserve the natural beauty of the American West.

Adams's photographs, almost exclusively in black and white, are renowned for their sharp contrast, detail, use of light and shadow, and ability to capture the beauty of their natural settings. During his lifetime, Adams often exhibited his work in museums and at universities that recognized his artistic talent. In addition to this artistry, he developed several innovative techniques for developing photographs to enhance their contrast and embellish their appearance. He published several how-to books on photography and taught workshops on how to utilize his techniques. In addition, he established his own studio to exhibit and sell his works.

Largely self-taught, Adams is recognized as one of the leaders in nature photography. He not only photographed nature but also realized the need for environmental activism and lobbied extensively for conservation measures. As early as 1936 he approached the U.S. Congress to promote the establishment of additional national parks in the western states. He was a member of President Lyndon B. Johnson's environmental task force and later met with both President Gerald Ford and President Ronald Reagan to discuss environmental issues.

Adams's work extended beyond nature photography to include architectural studies, portraits, and commercial photography. In 1943 he photographed the Manzanar Relocation Center, where Japanese Americans were interned by the U.S. government

Photographer and environmental activist Ansel Adams. (Joe Munroe/LANDOV)

during World War II. In 1960 he was commissioned to photograph scenes from all nine campuses of the University of California. In his later years he received several honorary doctoral degrees and other awards, including one named in his honor by the Wilderness Society. During these years he traveled widely, giving lecturers and exhibiting his works.

Gordon A. Parker

FURTHER READING

Adams, Ansel. *Ansel Adams: Four Hundred Photographs.* Edited by Andrea G. Stillman. Boston: Little, Brown, 2007.

Alinder, Mary Street. *Ansel Adams: A Biography.* New York: Henry Holt, 1996.

Lichtenstein, Therese. *Master of Light: Ansel Adams and His Influences.* New York: New Line Books, 2006.

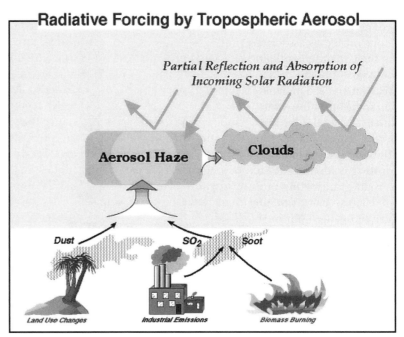

Aerosols can both reflect and absorb solar radiation, making their role in the greenhouse effect particularly complex. (NOAA)

SEE ALSO: National forests; National parks; Sierra Club; Wilderness Society; Yosemite Valley.

Aerosols

CATEGORY: Atmosphere and air pollution

DEFINITION: Aggregations of small particles, both liquid and solid, and the atmospheric gases in which they are suspended

SIGNIFICANCE: Atmospheric aerosols are responsible for the diminishment of environmental air quality throughout much of the world.

An aerosol is a multiphasic system consisting of tiny liquid and solid particles and the gas in which they are suspended. In unpolluted areas such as New Zealand, aerosols contain impurities from natural sources; acidity comes from carbonic acid (H_2CO_3). In central Europe and other industrialized areas throughout the world, fossil-fuel combustion contributes large amounts of oxides of sulfur (SO_x) and oxides of nitrogen (NO_x) to the atmosphere, leading to the formation of sulfuric acid and nitric acid aerosols. In these polluted areas, acidity levels are much higher in fog than in rain, and dry deposition of sulfuric and nitric acid particulates from impactions of acid fog may be more damaging to buildings and the environment than acid rain. Forest canopies tend to scavenge acid aerosols; on conifer-covered mountains, cloud droplets are the major source of acid deposition. Dry deposition on canopies rapidly affects root systems and increases soil acidification over the long term. Soil acidification has been linked to a number of adverse effects on vegetation, especially tree dieback and forest decline in Europe.

Aerosols in industrialized areas often contain heavy metals—including chromium (Cr), iron (Fe), copper (Cu), cadmium (Cd), cobalt (Co), nickel (Ni), and lead (Pb)—latex, surfactants, and asbestos. When tetraethyl lead was used as a gasoline additive, inhalation of lead aerosols contributed a substantial fraction of the body burden of lead in urban dwellers. Recognition of this hazard led to a ban on leaded gasoline. Dry deposition of heavy metals in aerosols can have adverse effects on ecosystems; $CdSO_4$ and $CuSO_4$ are known to reduce root elongation in trees. Fluorides released by heavy industry contribute to tree dieback.

The adverse impact of aerosols on air quality has been noted by many writers since the time of John Evelyn, who, on January 24, 1684, recorded a marked decrease in atmospheric visibility and increased respiratory problems associated with London smog. During the late nineteenth and early twentieth centuries, the lethal effects of aerosols were evident in the greatly increased mortality that occurred during London smog episodes. A historic episode in 1952, during which smog in London reduced visibility to zero, caused an estimated four thousand excess deaths, doubling the normal death rate for children and for adults ages forty-five to sixty-four. This episode led to legislation regarding emissions controls, and by the late 1960's, the health effects related to coal burning were reduced to minimal levels.

Urban aerosols—with their mix of latex, soot, hydrocarbons, SO_x and NO_x, and other pollutants—have been implicated in the growing prevalence of asthma. In the United States, the National Institutes of Health (NIH) estimated that asthma prevalence rose 34 percent between 1983 and 1993. About 4.8 million children in the United States were estimated to suffer from asthma in 1993. Asthma deaths have rapidly increased in the United States, rising from 0.9 per 100,000 people in 1976 to 1.5 per 100,000 people in 1986. Deaths from asthma among African Americans of all ages rose from 1.5 per 100,000 in 1976 to 2.8 per 100,000 in 1986. African Americans between the ages of fifteen and twenty-four had an asthma death rate of 8.2 per one million in 1980; by 1993, the rate had increased to 18.8 per one million.

Aerosols also increase the absorption and scattering of light in the atmosphere, reducing visibility. When smoke particle concentrations exceed 80-100 milligrams per cubic meter (mg/m^3), visibility falls below 1 kilometer (0.62 miles). Smoke palls can travel considerable distances; for example, during the spring of 1998, agricultural burning in the Yucatán in Mexico created a pall that markedly diminished visibility in Dallas, Texas.

Particulates in aerosols serve as condensation nuclei and may enhance fog formation. Visibility reductions by hygroscopic (moisture-retaining) air pollutants are noticeable at relative humidities of about 50 percent. A number of multicar chain-reaction accidents have occurred downwind of industrial plants when hygroscopic plumes have passed over cooling ponds adjacent to highways, creating abrupt reductions in visibility. When drivers of cars and trucks moving at high speeds suddenly become engulfed in an area with visibility of only a few feet, they often respond erratically, causing accidents.

Extremely low air temperatures can lead to the formation of ice fogs—known as "arctic haze"—over cities; such fogs, which are entirely anthropogenic (caused by humans), are fed by moisture and by particles given off by combustion. They are composed of minute ice crystals that substantially reduce visibility to the point where air travel is restricted. Ice fogs last for several days at a time and are frequent during winters in Fairbanks, Alaska, and many Canadian cities. The Siberian city of Irkutsk reports an average of 103 days of fog yearly, all during winter.

Anita Baker-Blocker

FURTHER READING

Colbeck, Ian, ed. *Environmental Chemistry of Aerosols.* Ames, Iowa: Blackwell, 2008.

Cotton, William R., and Roger A. Pielke, Sr. *Human Impacts on Weather and Climate.* 2d ed. New York: Cambridge University Press, 2007.

Katsouyanni, K., et al. "Short-Term Effects of Ambient Sulphur Dioxide and Particulate Matter on Mortality in Twelve European Cities: Results from Time Series Data from the APHEA Project." *British Medical Journal* 314 (June 7, 1997).

SEE ALSO: Acid deposition and acid rain; Air pollution; London smog disaster.

Africa

CATEGORIES: Places; animals and endangered species; ecology and ecosystems; forests and plants; preservation and wilderness issues

SIGNIFICANCE: The ecosystems of Africa range from the desolate expanses of the Sahara to the dense confines of equatorial rain forests. Threats to the plant and animal life found across the continent have placed Africa in the forefront of global environmentalist campaigns, some of which aim not only to save menaced species but also to conserve other natural resources, including soils and minerals.

The image of the African continent containing primarily dense tropical jungles holds true for limited regions only—those near the equator and subject to heavy seasonal rainfall. In fact, Africa has several different environmental regions. These correspond

Habitats and Selected Vertebrates of Africa and Madagascar

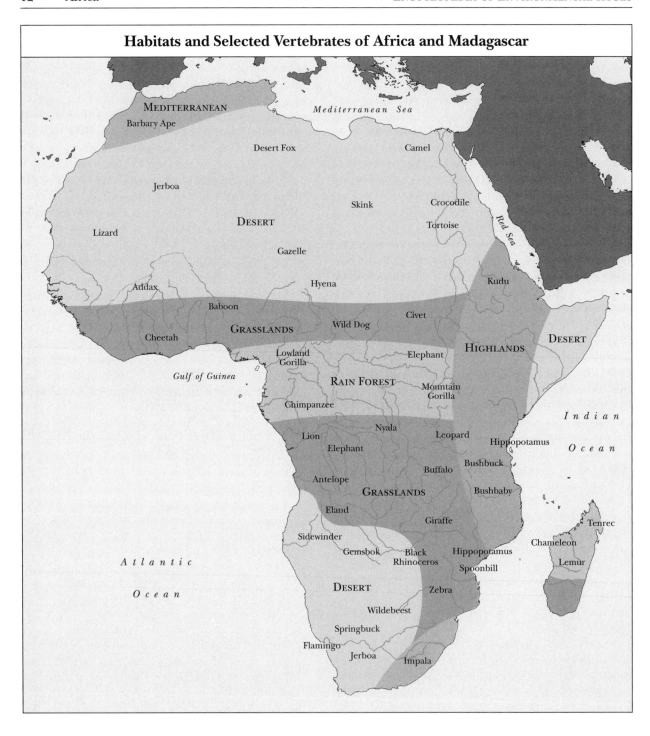

mainly, but not only, to relative latitudinal position north or south of the equator. Geographers generally separate the continent into five zones: North, South, East, West, and Central. The environmental conditions that are found in the zones vary widely.

GENERAL ENVIRONMENTAL CHARACTERISTICS

In comparison to other continents, Africa possesses relatively few major mountain ranges. The most important tectonic upliftings are in South Africa—notably the Drakensberg range, which is 1,000 kilometers (620 miles) long, with elevations reaching in places nearly 3,500 meters (11,500 feet)—and in the northwestern and northeastern areas of the continent (the Aurès and Atlas mountains in Tunisia and Morocco, and the Simyen and Bale mountains in Ethiopia). The geology of the Ethiopian ranges, which boast peaks such as Ras Dashen in the Simyens, more than 4,500 meters (14,800 feet) high, and many volcanic peaks, is closely connected with the tectonic phenomena that created the Great Rift Valley running north to south from the Red Sea to Malawi in Southern Africa.

Whereas these long ranges contribute to extensive regional environmental patterns—combining patterns of flora and fauna that determine the livelihoods of entire sections of Africa—other quite different mountains (mostly volcanic "monoliths") lend themselves to more localized environmental, and therefore human cultural, settings. Prototypical examples would be Mount Kenya and Mount Kilimanjaro in Tanzania (the latter being the highest mountain in Africa—reaching almost 6,000 meters, or 19,700 feet) and Mount Cameroon near the Gulf of Guinea coast in West Africa. Mount Kenya became the

site of one of Africa's earliest national parks, founded in 1949. Each of these famous mountains is associated with indigenous legends tying them to local religious beliefs. High levels of rainfall close to such prominent peaks often combine with fertile volcanic soil to create not only lush natural vegetation but also fairly high levels of domestic and commercial agricultural productivity (both Kenya and Tanzania, for example, produce famous coffee strains).

Africa contains the most extensive and arid expanse in the world: the Sahara. This desert region, which covers most of North Africa, is bordered on the south by the Sahel, a region that forms a broad continental dividing line between North Africa (where basic environmental characteristics are shared among the five modern countries of Morocco, Algeria, Tunisia, Libya, and Egypt) and sub-Saharan Africa. Much of North Africa (except for Egypt's Nile River Valley, which runs south to north between desolate desert littorals, reaching to the Red Sea in the east and to Libya's Fezzan desert region to the west) is characterized by rain-fed coastal plains bordered by the higher altitudes of the Aures and Atlas mountain chains. Once the Sahara begins, however, vegetation is very

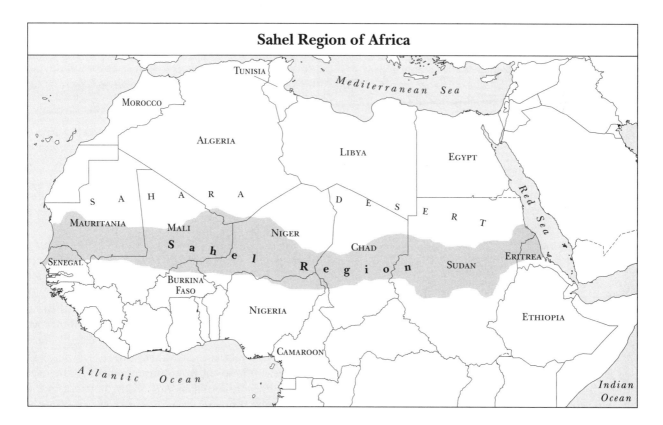

scant or nonexistent, except in oasis areas, where subsurface water sources are available. The estimated average annual rainfall for the Sahara is approximately 25 millimeters (about 1 inch).

Another major desert in Africa is the Kalahari, which, in combination with the contiguous Namib Desert, covers much of the land area of South Africa from the Orange River into modern Angola and Namibia. The Kalahari is not nearly as arid as the Sahara, having widely dispersed thorny trees and bushes.

The greatest territorial expanses of Africa are characterized by semiarid savanna conditions; their ecology includes extensive grasslands with widely dispersed trees. Savannas, which receive quite limited and seasonally specific rainfall, are typically located between desert expanses and tropical zones near the equator. African savannas are found in the immediate sub-Saharan zones of Senegal, Mali, Niger, Chad, and Sudan, and in Central and Southern Africa stretching eastward to the Indian Ocean. Because of their extensive grasslands, these zones form (or formed) the habitats of many of Africa's famous wild animal herds (especially in the Serengeti of Tanzania) and areas of Zambia, Zimbabwe, and South Africa. Rains totaling as much as 635 millimeters (25 inches) come from May into the summer, but then disappear by early fall.

Clearly, the major factor contributing to dense vegetation in Africa is rainfall, as the label "tropical rain forest" suggests. Regions close to the equator (including in Guinea, southern Côte d'Ivoire, and Ghana, but especially in Cameroon, Equatorial Guinea, Gabon, and the Congo) receive the highest rainfall, and their environments support extensive multitiered forests that are habitats for many animal species, including primates such as gorillas and chimpanzees. The extensive and complicated watershed of the Congo basin produces vegetation necessary to support both human and animal life and is the source of the Congo River. Another very long African river, the Niger, has its sources in the West African tropical zone of Guinea.

Africa's major inland lakes are found in different environmental settings. What was once the continent's largest lake, Lake Chad, is located in the middle of the dry interior of what was formerly part of French West Africa and French Equatorial Africa. This lake is probably the remnant of an ancient inland sea. Although it is still fed by the Chari River and its tributary the Logone (which carry water from equatorial Africa), its shores have receded considerably. This is part of the increasing desiccation of the African Sahel, which has caused major environmental problems for inhabitants of the region.

By contrast, many of the main lakes of East Africa—including Lake Tanganyika and Lake Malawi, also known as Lake Nyasa—are formed by water runoff from the high elevations surrounding the Great Rift Valley. The more centrally located Lake Victoria, also known as Victoria Nyanza, is an exception. Lake Malawi is the largest East African lake (covering nearly 70,000 square kilometers, or 27,000 square miles), and Lake Tanganyika (covering almost 34,000 square kilometers, or 13,000 square miles) is the longest and second-deepest lake in the world (some 680 kilometers, or 420 miles, long, with a maximum depth of almost 1,500 meters, or 5,000 feet). Although all major and lesser East African lakes provide natural habitat for cichlid fish populations, Lake Malawi has a greater variety of such fish than any other place in the world, including many species that are unknown anywhere else.

In addition to the obvious environmental features of Africa's vast landmass and its lakes and rivers, the continent's natural resources are tied to its Mediterranean Sea, Atlantic Ocean, and Indian Ocean coastal waters and estuaries. Beyond their value for commercially viable fishing (and fishing for domestic consumption) some coastal regions of Africa offer features that began to attract particular attention during the late decades of the twentieth century.

Environmental Issues

Among the environmental issues affecting Africa are concerns about declining numbers and even extinctions of indigenous plant and animal species. The list of menaced African species ranges from the cichlid tropical fish populations in East Africa's freshwater lakes through the African elephant.

Various factors have caused a sharp decline in what was once a thriving habitat for more than five hundred species of cichlids in Lake Victoria. It is estimated that almost half of the cichlids have disappeared as the result of the decision to introduce Nile perch into the lake waters to increase local fishing harvests for food and commercial sale. The nonnative perch proved to be predatory and fed off the local cichlid populations.

There are many other examples of rising concerns among environmentalists who study declining numbers of animal species that were once very numerous in different areas of Africa. Those that have drawn the greatest attention are the large mammals, including mountain gorillas, elephants, and members of the "great cat" family (lions, leopards, and cheetahs). Questionable human actions—including poaching for food or trophies—are generally behind the threats to these species.

By the end of the twentieth century, organized conservation efforts were on the rise in Africa. The obvious importance of saving large game species led, for example, to "safari tourism" in protected parks. Among the best known of these parks is the Selous Game Reserve in Tanzania, which covers more than 54,000 square kilometers (21,000 square miles) and is home to an incredibly diverse array of animals, including elephants, black rhinoceroses, giraffes, and crocodiles.

Some environmental activists have attempted to create programs to promote the conservation of both animal and plant species. Some, such as Conserve Africa, a nonprofit organization based in the United Kingdom, support environmental conservation efforts that can benefit not only flora and fauna but also local African populations that depend directly on the environment for their livelihoods. The primary goal of Conserve Africa is to help develop skills among Africans that will benefit both the local environment and the Africans' material welfare. Examples include the development of apiculture (beekeeping), the training of Africans in the careful selection of tree and medicinal plant species for cultivation, and the promotion of ecotourism.

The African Conservation Foundation (ACF) coordinates the sharing of information among conservation groups working in Africa and posts alerts concerning alarming developments. A prime example of an ACF alert is one stating the organization's opposition to illicit mining (particularly in the violence-torn eastern Congo) of what the ACF calls the "Three T's"—tin, tantalum, and tungsten (all minerals in great demand because of their use in developed countries' high-technology industries).

In some cases, efforts to protect the environment in Africa are complicated by various political factors. For example, pressures toward economic development conflict with concerns about the environment in decisions regarding offshore petroleum drilling, particularly along the coasts of West Africa (extending from the Niger Delta through Cameroon, Equatorial Guinea, and Gabon). The potential threat that such drilling poses for marine life and vegetation, particularly in environmentally unique coastal mangrove ecosystems, may rank as high as inland deforestation and mineral extraction concerns as major African environmental issues in the twenty-first century.

Some areas of Africa face environmental concerns stemming from natural causes presumed to be tied to global warming. This is particularly the case in the Sahel region to the south of the Sahara. Beginning during the mid-1970's and intermittently since then, drought conditions have caused major crop shortages as well as widespread losses in animal herds that provide the only means of survival for many Sahel populations, as well as for inhabitants of other arid regions of Africa. As human migration out of such stricken zones rises, environmental problems are often seen in the areas receiving the migrants; these problems include urban sprawl, shortages of safe water supplies, and the spread of communicable animal and human diseases.

Byron Cannon

FURTHER READING

Doyle, Shane. *Crisis and Decline in Bunyoro*. London: British Institute in Eastern Africa, 2006.

Goldblatt, Peter, ed. *Biological Relationships Between Africa and South America*. New Haven, Conn.: Yale University Press, 1993.

Maddox, Gregory H. *Sub-Saharan Africa: An Environmental History*. Santa Barbara, Calif.: ABC-CLIO, 2006.

Shorrocks, Bryan. *The Biology of African Savannahs*. New York: Oxford University Press, 2007.

United Nations Environment Programme. *Africa: Environment Outlook*. Nairobi: Author, 2006.

SEE ALSO: Congo River basin; Elephants, African; Extinctions and species loss; Kalahari Desert; Mountain gorillas; Nile River; Poaching; Rain forests; Savannas.

Africanized bees. *See* Killer bees

Agenda 21

CATEGORIES: Treaties, laws, and court cases; ecology and ecosystems; resources and resource management

IDENTIFICATION: Plan of action adopted by 178 nations at the United Nations Conference on Environment and Development in Rio de Janeiro, Brazil

DATE: Adopted on June 13, 1992

SIGNIFICANCE: Agenda 21 represents an important step by the international community in planning for the changes necessary to protect the environment, including reducing human consumption patterns and promoting sustainable development. It outlines the techniques that should be implemented to balance the needs of humans with the preservation of natural resources.

The 1992 United Nations Conference on Environment and Development, widely known as the Earth Summit, produced five documents: Agenda 21, the Rio Declaration on Environment and Development, the United Nations Framework Convention on Climate Change, and the United Nations Convention on Biological Diversity. Agenda 21 challenges governments to adopt practices that promote sustainable development. It encourages national governments to work with nongovernmental organizations (NGOs), local governments, and businesses to create change and to reduce wasteful actions that harm the earth. Although designed to be implemented by national governments, Agenda 21 also includes directives for the regional and local levels; these are commonly referred to as Local Agenda 21, or LA21, programs.

Agenda 21 addresses both current problems and foreseeable problems of the future. One theme is improvement of the economic situation for those living in poverty. Agenda 21 notes that conditions can improve if the less fortunate have access to the resources necessary to lead sustainable lives. As the world's less industrialized countries strive to achieve their maximum potential, they must approach development in ways that do not harm the environment. Agenda 21 proposes that since the more industrialized countries have produced more pollution than others, these countries should provide the funding necessary to help underdeveloped countries grow in a sustainable manner. The plan recognizes the roles that nontraditional groups—such as indigenous peoples, women, industries, young people, and NGOs—can play in such development.

Another theme of Agenda 21 is the need for proper management of natural resources. Along with acknowledging that progress must be made, the plan emphasizes the need to combat desertification and deforestation, to care for the atmosphere and the oceans, to protect biodiversity, and to encourage sustainable agriculture practices. It also seeks to prevent disputes over the high seas and to discourage practices that deplete the oceans' fish stocks. Agenda 21 calls for increasingly high standards of energy efficiency and encourages the development and use of clean technologies. It promotes the preservation of forests, the planting of new trees, and the elimination of slash-and-burn agriculture methods that destroy forests. Noting the scarcity of clean drinking water for many around the world, Agenda 21 challenges nations to provide safe water and sanitation for all citizens by the year 2025.

Kathryn A. Cochran

A Global Partnership

The opening paragraph of the Preamble to Agenda 21 presents an unusually stark statement of the challenges facing humanity at the beginning of the twenty-first century and the need for international cooperation to meet those challenges.

Humanity stands at a defining moment in history. We are confronted with a perpetuation of disparities between and within nations, a worsening of poverty, hunger, ill health and illiteracy, and the continuing deterioration of the ecosystems on which we depend for our well-being. However, integration of environment and development concerns and greater attention to them will lead to the fulfilment of basic needs, improved living standards for all, better protected and managed ecosystems and a safer, more prosperous future. No nation can achieve this on its own; but together we can—in a global partnership for sustainable development.

FURTHER READING

Porter, Gareth, Janet Welsh Brown, and Pamela Chasek. "The Emergence of Global Environmental Politics." In *Global Environmental Politics.* Boulder, Colo.: Westview Press, 2000.

Sitarz, Daniel, ed. *Agenda 21: The Earth Summit Strategy to Save Our Planet.* Boulder, Colo.: EarthPress, 1993.

Speth, James Gustave, and Peter M. Haas. "From Stockholm to Johannesburg: First Attempt at Global Environmental Governance." In *Global Environmental Governance*. Washington, D.C.: Island Press, 2006.

SEE ALSO: Convention on Biological Diversity; Earth Summit; Environmental law, international; Johannesburg Declaration on Sustainable Development; Rio Declaration on Environment and Development; Sustainable development; United Nations Commission on Sustainable Development; United Nations Environment Programme; United Nations Framework Convention on Climate Change.

Agent Orange

CATEGORIES: Pollutants and toxins; human health and the environment

DEFINITION: Powerful defoliant used extensively by the U.S. military during the Vietnam War

SIGNIFICANCE: The spraying of Agent Orange resulted in the deforestation of large sections of Southeast Asia, and exposure to the herbicide has been linked to the development of serious health problems in both military personnel and civilians.

Between 1962 and 1971, some 19 million gallons of herbicides were sprayed by the U.S. military over South Vietnam and Laos by airplane, helicopter, boat, truck, and manual sprayers in an effort to reduce ground cover for enemy troops and to destroy enemy crops. About 11 million gallons of this total was sprayed in the form of Agent Orange, a fifty-fifty mixture of the herbicides 2,4-D and 2,4,5-T combined with a kerosene-diesel fuel for dispersal. The principal components of both 2,4-D and 2,4,5-T decompose within weeks after application; however, 2,4,5-T contains between 0.05 and 50 parts per million of dioxin, primarily 2,3,7,8-tetrachlorodibenzo-para-dioxin (TCDD), which is among the deadliest chemicals known and which has a half-life of decades. Approximately 368 pounds of dioxin were released during the spraying, with profound effects on both the ecology of the region and the subsequent health of U.S. military personnel, the residents of Vietnam and Laos, and their offspring.

Agent Orange constituted about 60 percent of total volume of herbicides sprayed by U.S. personnel in the region; other substances used included dinoxol, trinoxol, bromacil, diquat, tandex, monuran, diuron, and dalapon as well as compounds known by such code names as Agent White, Agent Blue, Agent Purple, Agent Green, Agent Pink, and Agent Orange II ("Super Orange"). In addition to 2,4-D and 2,4,5-T, various mixes included picloram (a growth regulator similar to 2,4-D and 2,4,5-T) and cacodylic acid, an arsenic-containing organic that dehydrated and killed plants. Unlike 2,4-D and 2,4,5-T, which acted on plant surfaces, mobile picloram penetrated soil to be absorbed by roots and induce systemic effects.

MILITARY BACKGROUND

The U.S. military developed weapon herbicides during World War II at Fort Detrick in Frederick, Maryland, and considered using them against Japanese food plots on Pacific islands. The British used herbicides in Malaya in the 1950's to destroy rebel food plots. Domestic tests of Agent Orange were conducted at Camp Drum, New York, and Eglin Air Force Base, Florida, in the 1950's. The first field tests were conducted in South Vietnam in 1961 and in Thailand in 1964 and 1965.

"Operation Ranch Hand" was the code name given to the U.S. Air Force program of herbicide application during the Vietnam War. The program involved a total of thirty-six aircraft that sprayed roughly 10 percent of the area of South Vietnam. The spraying and defoliation targeted jungles, inland forests, camp edges, roads, trails, railroads, and canals to make enemy movement conspicuous and easier to attack. Because of the possible risk to the crops of the United States' South Vietnamese allies, spraying of enemy food plots was not as extensive. In addition, the Army Chemical Corps conducted truck, helicopter, and manual spraying (particularly around base camps and transportation routes); the U.S. Navy, using small riverboats, sprayed edges of rivers and canals; and Special Forces troops conducted covert spraying operations.

ENVIRONMENTAL DAMAGE IN VIETNAM

Many of the trees in the tropical mangrove forests near the southernmost coasts of South Vietnam were killed by a single spraying. It is estimated that it will take up to a century for the mangrove forests to recover without reseeding. Since these and adjacent waters served as breeding and nursery grounds for

U.S. Air Force planes spray Agent Orange over dense vegetation in South Vietnam in 1966. (AP/Wide World Photos)

wildlife, the area's ecology was catastrophically affected.

Inland forests were less sensitive to Agent Orange. They usually recovered after single or double sprayings with only temporary foliage loss. However, three or more sprayings eventually induced tree death and converted forests to grasslands. Wildfires burned seeds and seedlings of native trees, further delaying recovery. Erosion washed sediment into deltas near river outlets, compounding environmental damage. Since approximately one-third of sprayed land was sprayed more than once, and 52,600 hectares (130,000 acres) were sprayed more than four times, damage to Vietnam's ecology was extensive.

Elevated concentrations of toxins have continued to affect Vietnam's population and environment. Because of the persistent effects of dioxin, which is fat-soluble and bioaccumulates up the food chain, health

problems related to the spraying have affected not only Vietnamese alive during the war but also their offspring. Malformations and birth defects are common in Vietnamese children, and other maladies are suspected to be linked to dioxin exposure.

It took more than two decades for a serious effort at characterizing and alleviating these problems to be begun, because it took years after the war for the problems to become recognized. In addition, Vietnam had other pressing economic, military, and rebuilding concerns. In the 1990's, the Vietnamese National Committee for the Investigation of the Consequences of the Chemicals Used During the Vietnam War, also known as the 10-80 Committee, solicited international help. In an effort to determine existing dioxin levels, from 1994 to 1998, Hatfield Consultants of West Vancouver, British Columbia, sampled soils, crops, fish, poultry, livestock, and human blood in the A Luoi Valley in central Vietnam near the old Ho Chi Minh Trail. The investigators also performed satellite characterization of topography changes in the Ma Da region of southern Vietnam caused by defoliation and sediment washing.

The Vietnamese Red Cross established a fund for Agent Orange victims in 1998; in April, 1998, Vietnam's prime minister ordered the first nationwide survey of Agent Orange-related problems. Vietnam also established several "peace villages," each of which could accommodate up to five hundred patients with health problems related to Agent Orange.

HEALTH EFFECTS IN AMERICAN VIETNAM VETERANS

Agent Orange and dioxin are now known to have caused comparable health problems in American veterans of the Vietnam War. Health problems linked to exposure include the cancers soft-tissue sarcoma, non-Hodgkin's lymphoma, and Hodgkin's disease and the skin diseases chloracne and porphyria cutanea tarda. There is a definite correlation between Agent Orange exposure and respiratory cancer, prostate cancer, multiple myeloma, acute and subacute peripheral neuropathy, and spina bifida (abnormal spine development in children of veterans). Other suspected health effects include immune system disorders, reproductive difficulties and cancers, diabetes, endocrine and hormone imbalances, cancer in offspring, and malformations and defects (there is much stronger evidence linking birth defects to dioxin in Vietnamese). It appears that many of these health problems (including spina bifida, birth defects, immune system problems, and cancer propensity) can be passed on to the children and even the grandchildren of those initially exposed.

These issues were controversial for years after the war, but as evidence accumulated, the Department of Veterans Affairs (VA) eventually addressed the issue. Free medical examinations and care were offered to Vietnam veterans with suspected Agent Orange-induced problems in 1978. By 1981, the VA had established a program providing follow-up hospital care to veterans with any health problem of which the cause was unclear. The VA now provides monthly compensation for those with the ten diseases for which there is a proven cause-and-effect or positive correlation. Also, compensation, health care, and vocational rehabilitation are provided to children of veterans with spina bifida. The VA now presumes that all military personnel who served in the Vietnam theater were exposed to Agent Orange.

Other groups of people—including farmers, foresters, ranchers, chemical industry workers, incinerator workers, and paper mill workers—are often exposed to trace dioxin and related chemicals such as polychlorinated biphenyls (PCBs) at levels in excess of those experienced by the typical Vietnam veteran. Workers in these areas occasionally exhibit increased frequencies of diseases (for example, prostate cancer is especially common among farmers). Trace dioxins are commonly produced by the burning of chlorine-containing organics and other chemical processes. It is likely that nearly all persons in the industrialized world have some amount of dioxin in their bodies. However, because of many epidemiological and exposure variables, symptoms vary widely, and it is difficult to establish conclusive links between symptoms and particular chemicals or activities.

Defenders of the spraying program in Vietnam point out that the use of Agent Orange and related herbicides was highly successful in meeting the U.S. military's goals and, hence, likely saved the lives of many American servicemen. Moreover, although the resulting environmental and health effects might have been less severe if another chemical had been used rather than the dioxin-containing 2,4,5-T, dioxin's long-term effects were not known at the time.

The long-term consequences of the use of Agent Orange in Vietnam thus drive home the lesson that unanticipated outcomes often result from the employment of new technologies.

Robert D. Engelken

FURTHER READING

Buckingham, William A., Jr. *Operation Ranch Hand: The Air Force and Herbicides in Southeast Asia, 1961-1971.* Washington, D.C.: Office of Air Force History, U.S. Air Force, 1982.

Griffiths, Philip Jones. *Agent Orange: Collateral Damage in Vietnam.* London: Trolley, 2004.

Institute of Medicine. *Veterans and Agent Orange: Health Effects of Herbicides Used in Vietnam—Update 2008.* Washington, D.C.: National Academy Press, 2009.

Young, Alvin Lee. *The History, Use, Disposition, and Environmental Fate of Agent Orange.* New York: Springer, 2009.

SEE ALSO: Birth defects, environmental; Dioxin; Pesticides and herbicides.

Agricultural chemicals

CATEGORIES: Pollutants and toxins; agriculture and food

DEFINITION: Chemicals utilized by the agriculture industry to improve crop yield or the quality of produce

SIGNIFICANCE: The chemical fertilizers and pesticides used by the agriculture industry have the potential to stay in soil and water for long periods, creating unintended environmental impacts.

To grow, plants require sunshine, water, carbon dioxide from the atmosphere, and mineral nutrients from the soil. Mineral nutrients may be subdivided into macronutrients (calcium, magnesium, sulfur, nitrogen, potassium, and phosphorus) and micronutrients (iron, copper, zinc, boron, manganese, chloride, and molybdenum). Plant growth, and thus crop yields, is reduced if any one of these nutrients is not present in sufficient amounts. Micronutrients are required in small quantities, and deficiencies occur infrequently; therefore, the majority of agricultural fertilizers contain only macronutrients. Magne-

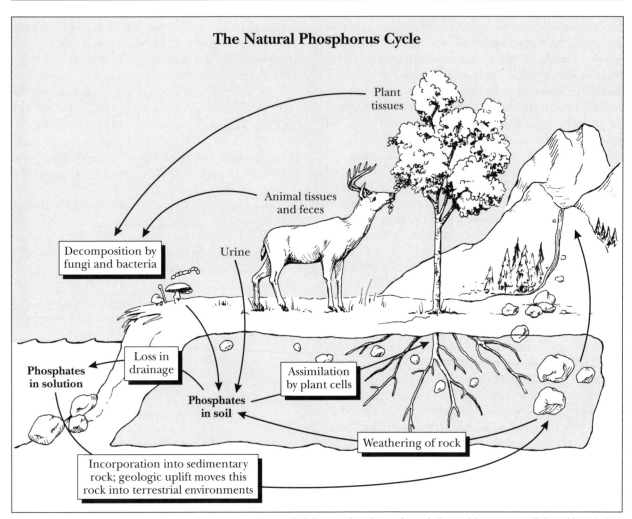

The Natural Phosphorus Cycle

Plant tissues

Animal tissues and feces

Decomposition by fungi and bacteria

Urine

Loss in drainage

Phosphates in solution

Assimilation by plant cells

Phosphates in soil

Weathering of rock

Incorporation into sedimentary rock; geologic uplift moves this rock into terrestrial environments

The biogeochemical phosphorus cycle is the movement of the essential element phosphorus through the earth's ecosystems. Released largely from eroding rocks, as well as from dead plant and animal tissues by decomposers such as bacteria and fungi, phosphorus migrates into the soil, where it is picked up by plant cells and is assimilated into plant tissues. The plant tissues are then eaten by animals and released back into the soil via urination, defecation, and decomposition of dead animals. In marine and freshwater aquatic environments, phosphorus is a large component of shells, from which it sediments back into rock and can return to the land environment as a result of seismic uplift.

sium and calcium are utilized in large quantities, but most agricultural soils contain an abundance of these two elements, either derived from parent material or added as lime. Most soils also contain sufficient amounts of sulfur from the weathering of sulfur-containing minerals, the presence of sulfur in other fertilizers, and atmospheric pollutants. The remaining three macronutrients (nitrogen, potassium, and phosphorus) are readily depleted; these are referred to as fertilizer elements, as they must be added to most soils on a regular basis. Mixed fertilizers contain two or more nutrients. For example, a fertilizer labeled

10-10-10 contains 10 percent nitrogen, 10 percent phosphorus, and 10 percent potassium.

FERTILIZERS AND ENVIRONMENTAL CONCERNS

The application of fertilizer to agricultural soil is by no means new. Farmers have been applying manures to improve plant growth for more than four thousand years. For the most part, this practice had little environmental impact. Since the development of chemical fertilizers in the late nineteenth century, however, fertilizer usage has increased tremendously. During the second half of the twentieth century, the amount of fertilizer applied to the soil increased more than 450 percent. While this increase more than doubled the worldwide crop production, it also generated some environmental problems.

The production of fertilizer requires the use of a variety of natural resources, and some people have argued that the increased production of fertilizers has required the use of energy and mineral reserves that could have been used elsewhere. For every crop, there is a point at which the yield may continue to increase with the application of additional nutrients, but the increase will not offset the additional cost of the fertilizer. The economically feasible practice, therefore, is to apply the appropriate amount of fertilizer that produces maximum profit rather than maximum yield. Unfortunately, many farmers still tend to overfertilize, which wastes money and contributes to environmental degradation. Excessive fertilization can result in adverse soil reactions that damage plant roots or produce undesired growth patterns; it can actually decrease yields, as some micronutrients, if applied in excessive amounts, are toxic to plants and will dramatically reduce their growth.

The most serious environmental problem associated with fertilizers is their contribution to water pollution. Excess fertilizer elements, particularly nitrogen and phosphorus, are carried from farm fields and cattle feedlots by water runoff and are eventually deposited in rivers and lakes, where they contribute to the pollution of aquatic ecosystems. High levels of plant nutrients in streams and lakes can result in increased growth of phytoplankton, a condition known as eutrophication. During the summer months, eutrophication can deplete oxygen levels in lower layers of ponds and lakes. Excess nutrients can also leach through the soil and contaminate underground water supplies. In areas where intense farming occurs, nitrate concentrations are often above recommended safe levels. Water that contains excessive amounts of plant nutrients poses health problems if consumed by humans and livestock, and it can be fatal if ingested by newborns.

PESTICIDES

Pesticides are chemicals designed to kill unwanted organisms that interfere, either directly or indirectly, with human activities. The major types of pesticides in common use are those designed to kill insects (insecticides), nematodes (nematocides), fungi (fungicides), weeds (herbicides), and rodents (rodenticides). Herbicides and insecticides make up the majority of the pesticides applied in the environment.

The application of chemicals to control pests is not new. Humans have been using sulfur and heavy metal compounds as insecticides for more than two thousand years. Residues of toxic metals such as arsenic, lead, and mercury continue to be accumulated in plants that are grown on soil where these materials have been used. The commercial introduction of dichloro-diphenyl-trichloroethane (DDT) in 1942 opened the door for the synthesis of a host of synthetic organic compounds to be used as pesticides.

Chlorinated hydrocarbons such as DDT were eventually banned or severely restricted in the United States because of their low biodegradability and persistence in the environment. They were replaced by organophosphates, which biodegrade rapidly but are generally much more toxic to humans and other animals than chlorinated hydrocarbons. In addition, they are water-soluble and, therefore, more likely to contaminate water supplies. Carbamates have also been used in place of the chlorinated hydrocarbons, and while these compounds biodegrade rapidly and are less toxic to humans than organophosphates, they are also less effective in the killing of insects.

A large number of herbicides have also been developed. These chemicals are generally classified as one of three types: contact herbicides, systemic herbicides, and soil sterilants. Contact herbicides kill when they come in contact with the leaf surface. Systemic herbicides circulate throughout the plant after being absorbed and cause abnormal growth. Soil sterilants kill microorganisms necessary for plant growth and act as systemic herbicides.

PESTICIDES AND ENVIRONMENTAL CONCERNS

Approximately fifty-five thousand different pesticides are available in the United States, and Americans apply about 500 million kilograms (1.1 billion pounds) of pesticides each year. Fungicides account for 12 percent of all pesticides used by farmers, while insecticides account for 19 percent and herbicides account for 69 percent. Billions of pounds of pesticides are applied each year throughout the world. While most of these chemicals are applied in developed countries, the amount of pesticides used in underdeveloped countries is rapidly increasing.

There is no doubt that pesticides have had a beneficial impact on the lives of humans by increasing food production and reducing food costs. Alongside these benefits, however, the use of pesticides has caused some environmental problems, including the development of pesticide resistance. When a new pesticide is first utilized, it is effective in reducing the number of target pests, but within a few years, a number of species develop genetic resistance to the chemical and are no longer controlled with it, necessitating the development of another pesticide to replace the one that is no longer effective. The pests in turn develop resistance to the newer pesticides, and so on. As a result, many synthetic chemicals have been introduced into the environment, yet the pest problem is as great as it ever was.

Another problem with pesticides is that they do not go away after they have done their job; they can remain in the environment for varying lengths of time. For example, chlorinated hydrocarbons can persist in the environment for up to fifteen years. This can be beneficial from an economic standpoint because the pesticide has to be applied less frequently, but it can be detrimental from an environmental standpoint. When many pesticides are degraded, many of their breakdown products, which are often toxic to other organisms, may also persist in the environment for long periods of time.

A third problem with pesticides is their tendency to become more concentrated as they move up the food chain. A pesticide may not affect species at the base of the chain, but it may be toxic to organisms that feed at the apex because the concentration of the chemical increases at each higher level of the food chain. Many algal species, for example, can be sprayed with an insecticide without any apparent effect to the algae. However, the chemical can be detrimental to organisms such as birds that eat fish that have eaten insects that have fed on the algae.

A fourth problem is broad-spectrum poisoning. Few, if any, chemical pesticides are selective. They kill a wide range of organisms rather than just the target pests. Many insects are beneficial to the environment, and more damage than good may be done by the use of insecticides that kill these organisms along with the harmful insects.

Many pesticides, particularly insecticides, are also toxic to humans. Thousands of people, many of them children, have died as the result of direct exposure to high concentrations of these chemicals. In addition, many workers in pesticide factories have been poisoned through job-related contact with the chemicals. Numerous agricultural laborers, particularly in coun-tries that have no laws in place specifying stringent guidelines for the handling of pesticides, have also died as a result of direct exposure to these chemicals. Some pesticides have been classified as carcinogens by the U.S. Environmental Protection Agency.

D. R. Gossett

FURTHER READING

Akinyemi, Okoro M. *Agricultural Production: Organic and Conventional Systems.* Enfield, N.H.: Science Publishers, 2007.

Altieri, Miguel A. *The Scientific Basis of Alternative Agriculture.* Boulder, Colo.: Westview Press, 1987.

Carson, Rachel. *Silent Spring.* 40th anniversary ed. Boston: Houghton Mifflin, 2002.

Miller, G. Tyler, Jr., and Scott Spoolman. "Food, Soil, and Pest Management." In *Living in the Environment: Principles, Connections, and Solutions.* 16th ed. Belmont, Calif.: Brooks/Cole, 2009.

Montgomery, David R. *Dirt: The Erosion of Civilizations.* Berkeley: University of California Press, 2007.

Ohkawa, H., H. Miyagawa, and P. W. Lee, eds. *Pesticide Chemistry: Crop Protection, Public Health, Environmental Safety.* New York: Wiley-VCH, 2007.

SEE ALSO: Agricultural revolution; Dichloro-diphenyl-trichloroethane; Integrated pest management; Intensive farming; Pesticides and herbicides.

Agricultural revolution

CATEGORY: Agriculture and food

DEFINITION: Advances in agricultural technology during the twentieth century

SIGNIFICANCE: The agricultural revolution led to a dramatic increase in the worldwide production of food and fiber crops, but this increase did not occur without negative environmental consequences.

Early agricultural centers were located near large rivers that helped maintain soil fertility through the deposition of new topsoil during each annual flooding cycle. Agriculture, however, eventually moved into other regions that lacked the annual flooding of large rivers, and humans began to utilize a technique known as slash-and-burn agriculture, in which they cleared the land by burning the existing vegetation

With the agricultural revolution, farmers in developed nations became increasingly reliant on monoculture and on the use of heavy machinery such as combines. (©Marian Mocanu/Dreamstime.com)

and then used the ashes to fertilize their crops. This type of agriculture is still practiced in many developing nations and is one reason the tropical rain forests are disappearing at such a fast rate. During the nineteenth century, the Industrial Revolution led the way for the development of many different types of agricultural machinery, which resulted in the mechanization of most farms and ranches in industrialized nations.

The Green Revolution—as advances in agricultural science during the twentieth century came to be known—resulted in the development of new, higher-yielding varieties of numerous crops, particularly the seed grains that supply most of the calories necessary for maintenance of the world's growing population. While higher-yielding crops, along with improved farming methods, resulted in tremendous increases in the world's food supply, they also led to an increased reliance on monoculture, the practice of growing only one crop over a vast area of land. This practice decreased the genetic variability of many agricultural plants, increased the need for commercial fertilizers, and produced an increased susceptibility to damage from a host of biotic and abiotic factors.

The agricultural revolution has resulted in the development of an agricultural unit that requires relatively few employees, is highly mechanized, devotes large amounts of land to the production of a single crop, and is highly reliant on agricultural chemicals such as fertilizers and pesticides. While this has led to tremendous increases in agricultural productivity, it has also had significant impact on the environment. During the past one hundred years, the world has undergone a continual loss of good topsoil. Even under ideal conditions, the process of soil formation is very slow. Many agricultural techniques lead to the removal of trees and shrubs that provide windbreaks or to the depletion of soil fertility, which reduces the plant cover on the fields. The result of these practices has been the exposure of the soil to increased erosion from wind and moving water, to the extent that as much as one-third of the world's croplands are losing topsoil more quickly than it can be replaced.

Agriculture represents the largest single user of global water. Approximately 73 percent of all water withdrawn from freshwater supplies is used to irrigate crops. Some irrigation practices are actually detrimental. Overwatering can waterlog the soil, and irrigation of crops in dry climates can result in salinization of the soil, which occurs when irrigation water rapidly evaporates from the soil, leaving behind the mineral salts that were dissolved in the water. The salts accumulate and become detrimental to plant growth. It has been estimated that as much as one-third of the world's agricultural soils have been damaged by salinization. In addition, debate exists as to whether the increased usage of water for agriculture has decreased the supply of potable water fit for other human uses.

The nutrients most often depleted from agricultural soils are nitrogen, phosphorus, and potassium, and these nutrients must be applied to the soil regularly in order to maintain fertility. During the second half of the twentieth century, the amount of fertilizer applied to the soil increased more than 450 percent, causing environmental problems in some areas. Fertilizer elements, particularly nitrogen and phosphorus, are carried away by water runoff and are eventually deposited in rivers and lakes, where they contribute to the pollution of aquatic ecosystems. In addition, nitrates can accumulate in underground water supplies.

Agriculture is highly dependent on the use of pesticides to kill organisms such as insects, nematodes, weeds, and fungi that directly or indirectly interfere with crop production. The use of pesticides has dramatically improved crop yields, primarily because pesticides are designed to kill pests before significant damage can occur to the crop. Pesticides often kill nonpests as well, however, and evidence suggests that indiscriminate use of these chemicals can have detrimental effects on wildlife, the structure and function of ecosystems, and even human health. In addition, the overuse of pesticides can lead to the development of resistance in the target species, which can result in a resurgence of the very pest the pesticide was designed to control.

Modern agriculture also consumes large amounts of energy. The operation of farm machinery requires large supplies of liquid fossil fuels, and the production of fertilizers, pesticides, and other agricultural chemicals is another energy cost associated with agriculture. Energy used in food processing, distribution, storage, and cooking after a crop leaves the farm may be five times as much as the energy used to produce the crop. Most of the foods consumed in the United States require more calories of energy to produce, process, and distribute to the market than they provide when they are eaten.

D. R. Gossett

FURTHER READING

Carson, Rachel. *Silent Spring*. 40th anniversary ed. Boston: Houghton Mifflin, 2002.

Conkin, Paul K. *A Revolution Down on the Farm: The Transformation of American Agriculture Since 1929.* Lexington: University Press of Kentucky, 2008.

Federico, Giovanni. *Feeding the World: An Economic History of Agriculture, 1800-2000.* Princeton, N.J.: Princeton University Press, 2005.

Miller, G. Tyler, Jr., and Scott Spoolman. *Environmental Science: Problems, Concepts, and Solutions.* 13th ed. Belmont, Calif.: Brooks/Cole, 2010.

Pierce, Christine, and Donald VanDeVeer. *People, Penguins, and Plastic Trees.* Florence, Ky.: Wadsworth, 1995.

SEE ALSO: Agricultural chemicals; Green Revolution; Pesticides and herbicides.

Air-conditioning

CATEGORY: Atmosphere and air pollution

DEFINITION: Intentional modification of indoor air temperature, humidity, or other characteristics for the purpose of comfort, health, or protection of items sensitive to temperature or humidity

SIGNIFICANCE: Air-conditioning benefits humankind in many ways, but it also has some negative environmental impacts. The refrigerants long used in air conditioners have proven to have adverse effects on ozone in the earth's stratosphere, and the use of air-conditioning accounts for a high percentage of energy consumption, particularly in regions with warm climates.

Since ancient times, human beings have tried to make the air inside buildings more comfortable. The Romans circulated cool water through buildings. Other cultures created fans or encouraged breezes to blow across pools of evaporating water to cool buildings. Cooling by evaporation of water, however, can lower temperatures only by a few degrees and is not effective when the air is humid.

Most modern air conditioners are similar to that first proposed by Willis Carrier in 1902. Air conditioners work in much the same way as do electric refrigerators. Both kinds of devices make use of the thermodynamics of fluids. A fluid with a boiling point somewhat below room temperature is selected, and a compressor is used to compress the fluid. Under sufficiently high pressure, the fluid liquefies. This process produces heat, so, in a part of the system outside the area to be cooled, fans blow air over tubes containing the warm liquid to cool it. The liquid is then circulated into the part of the system within the area to be cooled, where it passes through tubes in which the pressure drops, allowing the fluid to evaporate. The evaporating fluid then cools to a temperature near its boiling point. A fan blows air over the cold tubes, cooling the air. Many early air conditioners used fluids such as ammonia, methyl chloride, or even propane. The problem with these materials is that they are health hazards, explosion or fire hazards, or both.

In 1928, Thomas Midgley, Jr., developed chlorofluorocarbons (CFCs), a family of related chemicals that were nontoxic and nonexplosive and that had boiling temperatures and other properties making them suitable for use in air conditioners and refrigerators. It was unknown at the time that these nontoxic CFCs, when released into the atmosphere, break down ozone in the earth's stratosphere. Because ozone in the stratosphere blocks ultraviolet light from the sun, maintenance of the ozone layer is important to life on earth. Without the ozone layer, the earth's surface would receive a lethal dosage of ultraviolet light. Some damage has already been done to stratospheric ozone by CFCs released from leaking air-conditioning systems. In addition to depleting stratospheric ozone, CFCs are greenhouse gases, which means that they contribute to global warming.

The international agreement known as the Montreal Protocol calls for the gradual reduction of the production of CFCs and other chemicals known to harm the ozone layer, and since the adoption of the protocol in 1987 the use of these substances has been greatly reduced around the world. Some non-CFC refrigerants have been developed that are less damaging to the environment; however, these newer refrigerants have slightly different properties and cannot simply be substituted for CFCs in older air-conditioning systems without modifications to the systems, some of which can be expensive.

In regions with warm climates, air-conditioning accounts for the largest use of electricity in buildings. With electricity becoming increasingly expensive, many manufacturers of air conditioners emphasize energy efficiency in their marketing. The standard reference for air-conditioner energy efficiency is the seasonal energy-efficiency ratio (SEER), which is the thermal energy removed (measured in British thermal units) divided by the energy used (measured in watt-hours). The larger the SEER, the more energy-efficient the air conditioner.

Raymond D. Benge, Jr.

FURTHER READING

Achermann, Marsha E. *Cool Comfort: America's Romance with Air-Conditioning.* Washington, D.C.: Smithsonian Institution Press, 2002.
Miller, Rex, and Mark R. Miller. *Air Conditioning and Refrigeration.* New York: McGraw-Hill, 2006.

SEE ALSO: Chlorofluorocarbons; Energy-efficiency labeling; Freon; Greenhouse effect; Greenhouse gases; Montreal Protocol; Ozone layer.

Air pollution

CATEGORIES: Atmosphere and air pollution; weather and climate

DEFINITION: Atmospheric presence of materials or energy in quantities sufficient to harm living organisms, affect weather changes, or damage human-made materials and structures

SIGNIFICANCE: Air pollution's effects are profound and far-reaching. Acute pollution episodes can cause human fatalities, while ongoing pollution at lower concentrations results in cumulative damage to health. Pollutants can be transported long distances from their sources, where they can affect populations in neighboring cities, states, or nations. Pollutants such as ozone-depleting substances and greenhouse gases can have global impacts.

Earth's atmosphere is a mixture of two types of gases: The gases of one type are present in concentrations that are constant over long periods of time, and the gases of the other type are present in variable concentrations. Among the former are nitrogen (N_2), which makes up approximately 78 percent of the atmosphere, and oxygen (O_2), which constitutes about 21 percent of the atmosphere. Along with these two are argon (Ar), almost 1 percent of the atmosphere, and trace amounts of neon (Ne), helium (He), krypton (Kr), hydrogen (H_2), and xenon (Xe), none of which seems to have any major effect on the atmosphere. Among the gases that vary in concentration are carbon dioxide (CO_2), water vapor (H_2O), methane (CH_4), carbon monoxide (CO), ozone (O_3), ammonia (NH_3), hydrogen sulfide (H_2S), and several oxides of nitrogen and sulfur. Water vapor has the highest degree of variability and has a significant effect on the atmosphere because of its ability to change phase readily, absorbing or emitting energy as it does so. The trace components of the atmosphere are continually redistributed by the complex circulation patterns of the air known as winds.

In addition to its chemical makeup, the earth's atmosphere is characterized by the various physical effects acting on it or taking place within it. The most important of these are solar radiation and thermal energy. The sun acts as an almost perfect black-body radiator at an effective temperature of 3,316 degrees Celsius (6,000 degrees Fahrenheit). Some of the incoming solar radiation is absorbed by atmospheric gases such as oxygen, ozone, carbon dioxide, and water vapor, allowing about 80 to 85 percent to reach the ground under clear sky conditions. Cloud cover ensures that only about 50 percent of the solar radiation reaches the planet's surface on the average.

This incident solar radiation is absorbed by the atmosphere and the earth's surface and reradiated at longer wavelengths, mostly infrared. The amount of this radiation that reaches space is affected by atmospheric concentrations of carbon dioxide and water, both of which absorb some infrared radiation. This absorption produces the greenhouse effect, an important process that keeps the lower atmosphere at a higher temperature than the upper atmosphere, supporting life on the planet.

The atmosphere is a chemical system that is not in equilibrium, mainly because of the activities of living organisms. The respiration of many organisms produces carbon dioxide, which, by means of photosynthesis, produces oxygen. The huge amount of oxygen in the atmosphere is almost entirely the result of photosynthesis. Methane, the main hydrocarbon in the atmosphere, is produced by microbial degradation of organic matter in marshes, paddy fields, and the digestive systems of animals. Microorganisms that de-

grade nitrogen compounds in animal urine create ammonia. Forests are a large source of more complex hydrocarbons such as alkanes, alkenes, and esters (which produce the odors of flowers and fruits).

These natural processes have resulted in an atmosphere that has, since the formation of the planet, reached a steady-state composition. Since the nineteenth century's Industrial Revolution, however, human activities have significantly changed the amounts of some of the compounds found naturally in the atmosphere and have introduced many new compounds. Some of these substances are toxic to plant and animal life. In addition, some anthropogenic (human-caused) compounds have the potential to affect seriously the delicate balance that exists among the earth, its atmosphere, and living organisms.

ACID RAIN

The acidity of precipitation is usually expressed in terms of its pH values, where pH represents the concentration of hydrogen ions present. The pH scale commonly in use extends from 0 to 14, with the pH value 7 representing neutral solutions, values greater than 7 basic solutions, and values less than 7 acidic solutions. The pH scale is logarithmic, so that an increase of one pH unit corresponds to a tenfold increase in hydrogen ion concentration. Unpolluted precipitation has a pH value of around 5.6. This natural acidity results mainly from the atmospheric presence of carbon dioxide, which forms carbonic acid, and of chlorine from sea salt, which forms hydrochloric acid. Other natural contributions that affect pH, regionally or globally, are ammonia, soil particles, and volcanic emissions of sulfur dioxide and hydrogen sulfide. Testing of precipitation that predates the Industrial Revolution, which has been preserved in glaciers, generally has found a pH of more than 5, sometimes as high as 6.

Precipitation with a pH value of less than 5.6 is generally referred to as acid rain. Its average pH values commonly range from 4 to 5; as of 2000, the most acidic rainfalls in the United States had a pH of around 4.3. Individual storms may be much more acidic, such as a rainfall in West Virginia in 1978 that had an unofficial pH of 2.0. The major causes of acid rain are human-generated sulfur dioxide and nitrogen dioxide. Ice cores covering the period between 1869 and 1984 have shown that by the mid-1980's precipitation in Greenland had experienced a threefold increase in sulfate ion concentrations since 1900, and nitrate concentrations had undergone a twofold increase since 1955. These increases are believed to be the result of sulfur oxide and nitrogen oxide emissions carried from North American and Eurasian sources.

The fact that acid rain is largely a regional phenomenon gives clues to the sources of the major contributors. Much acid precipitation results from the combustion of fossil fuels, especially high-sulfur coal. The sulfur is oxidized by burning into sulfur dioxide (SO_2), which is released into the atmosphere with the main combustion products, carbon dioxide and water. In the atmosphere, sulfur dioxide reacts with water to produce sulfuric acid (H_2SO_4). This in turn falls to the earth's surface as both wet and dry acidic deposition. High atmospheric concentrations of sulfur dioxide and soot particles produce a grayish haze known as "London smog," a particularly severe occurrence of which affected London, England, in December, 1952. During a four-day period, many people in London experienced respiratory difficulty, and thousands of deaths from respiratory causes directly paralleled measured average smoke and sulfur dioxide concentrations.

Nitrogen and oxygen in the air do not react at any significant rate but readily combine to form nitrogen oxides (NO_x) in the high-temperature combustion processes found in power plants, smelters, steel mills, and internal combustion engines. In the atmosphere, gaseous nitrogen oxides produce the brownish haze often seen over cities such as Los Angeles and Denver. The nitrogen oxides go through various reactions in the atmosphere, some of which result in the production of nitric acid (HNO_3), which eventually reaches the earth's surface.

Among the effects of these acids are the corrosion of human-made objects such as metallic structures, stone buildings, statues, and automotive paints and other coatings. Lakes and other surface waters are affected by acid rain not only through direct alteration of their pH but also by metals such as aluminum, manganese, zinc, nickel, lead, mercury, and cadmium, which are leached from surrounding soils by the overly acidified precipitation. These alterations in water chemistry can kill aquatic organisms outright or affect their health and the viability of their offspring. Acid rain also harms trees and other plants, weakening them by damaging their leaves, stripping soils of necessary nutrients, and releasing substances from soils that are toxic to plant life.

Percentage Change in U.S. Emissions
(millions of tons per year)

	1980 vs. 2008
Carbon monoxide	−56
Lead	−99
Nitrogen oxides	−40
Volatile organic compounds	−47
Direct particulate matter (10-micron-diameter)	−68
Direct particulate matter (2.5-micron-diameter)	—
Sulfur dioxide	−56

Source: Data from U.S. Environmental Protection Agency, *Air Quality Trends,* 2009.

In the United States a program to reduce acid rain that began in 1995 has made significant progress. While the problem of acid rain has not been eliminated, by 2009 the program had reduced annual sulfur dioxide emissions from electricity-generating units by 67 percent compared with 1980 levels and 64 percent compared with 1990 levels. From 1995 to 2009 annual emissions of nitrogen oxides from these units dropped by 67 percent.

OZONE

Ozone is a form of oxygen found in small quantities throughout the earth's atmosphere. In the troposphere (the lowest layer of the atmosphere), ozone is of interest for a number of reasons. First, it plays an important role in the control of photochemistry, a group of processes in which compounds produced in the reduced state from natural or anthropogenic sources are oxidized to chemically inert materials such as carbon dioxide or to materials that can be precipitated from the atmosphere, such as nitric acid. Photochemical reactions in the troposphere provide the chief cleansing mechanism by which some materials are removed from the atmosphere. The importance of ozone to this process arises from its dissociation by ultraviolet radiation to produce reactive atomic oxygen. Some of the atomic oxygen reacts with water to produce hydroxyl radicals, which are responsible for the oxidation of most trace gases.

Ozone in the troposphere is also an important pollutant. It is implicated in the breakdown of natural polymers such as rubber, cotton, leather, cellulose, some paints, plastics, nylon, and fabric dyes. Because ozone is a very strong oxidant, it is a potential irritant to the lungs of humans and animals. Finally, tropospheric ozone, because of its oxidizing ability, is involved in global climate control because of its ability to influence concentrations of such greenhouse gases as carbon dioxide and methane. In addition, ozone is itself a greenhouse gas.

In the stratosphere, ozone provides an essential umbrella that partially shields the earth's surface from dangerous ultraviolet radiation. In the upper atmosphere, ozone formation involves oxygen and ultraviolet radiation. The reaction is $O_2 + h\upsilon \rightarrow O + O$, where $h\upsilon$ is ultraviolet energy. The oxygen atoms then react with oxygen molecules to produce ozone according to the reaction $O_2 + O \rightarrow O_3$. This photochemical process by which ozone is produced is balanced by the photochemical process that destroys it: $O_3 + h\upsilon \rightarrow O_2 + O$. Both processes involve the absorption of ultraviolet radiation, and the dynamic chemical equilibrium that exists between them removes a portion of the ultraviolet energy as it travels toward the earth's surface.

Knowledge of the equilibrium chemistry between oxygen and ozone allowed prediction of the equilibrium concentration of ozone in the upper atmosphere. Since the early 1970's, measurements of stratospheric ozone concentrations over the Antarctic continent have pointed to concentrations much lower (sometimes by as much as 50 percent) than expected. Reductions in ozone concentrations have also been seen over the Arctic and at midlatitudes.

The reason for this ozone depletion was ultimately traced to human activity, notably the widespread use of a group of compounds called chlorofluorocarbons (CFCs). These chemicals were especially effective in air-conditioning and refrigeration systems and as blowing agents in plastic-forming processes, solvents in the electronics industry, and propellants in aerosol spray cans. CFCs were initially believed to be free of side effects, and as a result large amounts of them were expelled into the atmosphere. The very inertness that makes CFCs so useful for industrial and consumer applications, however, became a great disadvantage as the compounds made their long journey to the stratosphere.

Once in the stratosphere, CFC molecules absorb ultraviolet radiation and break down, yielding chlorine atoms. These chlorine atoms catalyze the conversion of ozone to oxygen. Because the chlorine atom is a catalyst in the process, it is released to continue its destructive activity for many years to come. Halons—bromofluorocarbon compounds with excellent fire-suppressant properties—are similarly destructive to ozone, as their molecules break down in the presence of ultraviolet radiation to produce bromine atoms that destroy ozone.

The Montreal Protocol, an international environmental agreement that has undergone several amendments since it was adopted in 1987, has phased out CFCs, halons, and other ozone-depleting substances, with related ozone-depleting compounds to be phased out in future. While overall concentrations of these substances are on the decline in the stratosphere, their long residence time there means that ozone levels are unlikely to rebound to pre-1980 levels any sooner than the mid-twenty-first century.

GLOBAL CLIMATE CHANGE

When energy in the form of electromagnetic radiation strikes a molecule, the energy may be reflected, absorbed, or transmitted. Solar energy striking the earth's surface is absorbed and heats the land and water, which in turn radiate energy in the form of infrared radiation back toward space. Eventually an equilibrium state is reached in which the amount of energy absorbed by the earth equals the amount radiated. In the absence of an atmosphere, the equilibrium temperature of the earth would be about −21 degrees Celsius (−5.8 degrees Fahrenheit).

The atmosphere contains gases that transmit ultraviolet and visible radiation but absorb infrared wavelengths. Therefore, the infrared energy radiated by earth toward space is trapped in the air layer, increasing its temperature and that of the earth's surface. The equilibrium temperature of the earth because of this phenomenon is about 12 degrees Celsius (53.6 degrees Fahrenheit), 33 degrees Celsius (59.4 degrees Fahrenheit) warmer than it would be without those gases. These gases do for the earth what glass walls and roofs do for the temperature of a greenhouse; therefore, they are known as greenhouse gases.

Any molecule with two or more atoms that has no center of symmetry is a potential greenhouse gas. Important greenhouse gases in the earth's atmosphere include carbon dioxide, methane, nitrous oxide, ozone, and CFCs. This collection of gases absorbs radiation across the infrared range of wavelengths so that there are no windows for reflected infrared radiation to escape back into space. Since preindustrial times, atmospheric concentrations of greenhouse gases have been rising because of human activity. Between 1970 and 2004, total anthropogenic greenhouse gas emissions increased by approximately 70 percent.

Water vapor is the greatest contributor to the greenhouse effect, but its concentration is generally considered to be unaffected by human activities. After water the most important of the greenhouse gases is carbon dioxide. Carbon dioxide from fossil-fuel use alone accounted for 56.6 percent of the world's anthropogenic greenhouse gas emissions in 2004. The concentration of carbon dioxide in the atmosphere increased by about 80 percent between 1970 and 2004, primarily from the burning of fossil fuels. Among the products resulting from any hydrocarbon combustion are water and carbon dioxide. All processes that depend on energy from coal, oil, or natural gas are contributing to the total amount of greenhouse gases in the atmosphere.

Whether weather phenomena such as frequent serious storms, El Niño conditions, droughts, and floods are directly related to the greenhouse effect is a topic of heated debate among scientists and nonscientists. Cores covering thousands of years of accumulation taken from the Antarctic ice pack have been examined for clues about concentrations of atmospheric gases and average temperatures. An almost direct correlation has been found between carbon dioxide concentration and surface temperature. The historical evidence seems to point to potentially serious consequences if humankind does not quickly develop and implement the use of forms of energy that do not contribute to carbon dioxide emissions.

OTHER POLLUTANTS

In addition to gases, air contains suspended particulate matter. The particles are collections of molecules, sometimes similar, sometimes different. The constituents of particulate matter differ over time and space. In urban areas, particulate matter often contains sulfuric acid and other sulfates, carbon, or higher molecular weight hydrocarbons that result from incomplete combustion of fossil fuels. Particulate matter and sulfur dioxide are common pollutants found in urban smog. Over time, suspended particles tend to increase mass by combining or acting as nuclei on which vapors condense. Eventually these fall to the ground or are washed out by precipitation.

The greatest concern over particulate matter in the atmosphere is the fact that often the particles are small enough to be inhaled and retained in the respiratory system. Vegetation is affected when particles coat leaves and thus reduce plants' absorption of carbon dioxide and suppress photosynthesis and hence plant growth.

Major Air Pollutants

Pollution of the earth's atmosphere comes from many sources. Some sources are natural, such as volcanoes and lightning-caused forest fires, but most sources of pollution are by-products of industrial society, such as that of Donora, Pennsylvania. Each of the following eight major forms of air pollution has an impact on the atmosphere. Often two or more forms of pollution have a combined impact that exceeds the impact of the two acting separately.

1. **Suspended particulate matter:** This is a mixture of solid particles and aerosols suspended in the air. These particles can have a harmful impact on human respiratory functions.

2. **Carbon monoxide (CO):** An invisible, colorless gas that is highly poisonous to air-breathing animals.

3. **Nitrogen oxides:** These include several forms of nitrogen-oxygen compounds that are converted to nitric acid in the atmosphere and are a major source of acid deposition.

4. **Sulfur oxides, mainly sulfur dioxide:** This sulfur-oxygen compound is converted to sulfuric acid in the atmosphere and is another source of acid deposition.

5. **Volatile organic compounds (VOCs):** These include such materials as gasoline and organic cleaning solvents, which evaporate and enter the air in a vapor state. VOCs are a major source of ozone formation in the lower atmosphere.

6. **Ozone and other petrochemical oxidants:** Ground-level ozone is highly toxic to animals and plants. Ozone in the upper atmosphere, however, helps to shield living creatures from ultraviolet radiation.

7. **Lead and other heavy metals:** Generated by various industrial processes, lead is harmful to human health even at very low concentrations.

8. **Air toxics and radon:** Examples include cancer-causing agents, such as radioactive materials and asbestos. Radon is a radioactive gas produced by natural processes in the earth.

Particulate matter adheres to painted surfaces and buildings, reducing the lifetimes of materials and coatings and often causing corrosion, especially in moist atmospheres.

Other pollutants in the atmosphere include radioactive materials, carbon monoxide, lead, and hydrocarbons. Radioactive materials result from natural processes (including the decay of materials such as uranium) and from human nuclear technology. Radioactive nuclides produce ionizing radiation, which has the potential for long-term effects on cells, including cell death, genetic mutations, and malignant tumor formation.

Carbon monoxide results from the incomplete combustion of hydrocarbons. It is an unstable compound that quickly oxidizes to carbon dioxide. It is absorbed through the lungs and forms a complex with hemoglobin that is more tightly bound than oxygen. In this way carbon monoxide prevents oxygen from reaching individual cells, eventually resulting in death.

Metals such as beryllium, cadmium, chromium, lead, manganese, mercury, nickel, and vanadium may also be found in the air. Lead in particular is widely dispersed throughout the environment, mainly because of its use as an additive in gasolines. Most countries have taken major steps to ban leaded gasolines, but residual concentrations still affect humans, especially children in urban areas. Another problematic metal, mercury, can enter the atmosphere through power plant emissions. Many coal deposits contain mercury, which is released when the coal is burned. Airborne mercury can travel far from its source before it settles into water or onto land, where it can enter the food web and bioaccumulate within living organisms.

Hydrocarbons and their derivatives may be found as solids, liquids, or gases. Although some are the results of natural processes, most are by-products of combustion processes. Some of the hydrocarbons are toxic even in small concentrations, but the major contribution of hydrocarbons is their involvement in atmospheric photochemistry.

Grace A. Banks
Updated by Karen N. Kähler

FURTHER READING

Godish, Thad. *Air Quality.* 4th ed. Boca Raton, Fla.: Lewis, 2004.

Hilgenkamp, Kathryn. "Air." In *Environmental Health: Ecological Perspectives.* Sudbury, Mass.: Jones and Bartlett, 2006.

Hill, Marquita K. "Air Pollution." In *Understanding Environmental Pollution.* 3d ed. New York: Cambridge University Press, 2010.

Jacobson, Mark Z. *Atmospheric Pollution: History, Science, and Regulation.* New York: Cambridge University Press, 2002.

McKinney, Michael L., Robert M. Schoch, and Logan Yonavjak. "Air Pollution: Local and Regional." In *Environmental Science: Systems and Solutions.* 4th ed. Sudbury, Mass.: Jones and Bartlett, 2007.

Seinfeld, John H., and Spyros N. Pandis. *Atmospheric Chemistry and Physics: From Air Pollution to Climate Change.* 2d ed. Hoboken, N.J.: John Wiley & Sons, 2006.

Sokhi, Ranjeet S., ed. *World Atlas of Atmospheric Pollution.* London: Anthem Press, 2008.

U.S. Environmental Protection Agency. *Our Nation's Air: Status and Trends Through 2008.* Research Triangle Park, N.C.: Author, 2010.

Vallero, Daniel. *Fundamentals of Air Pollution.* 4th ed. Boston: Elsevier, 2008.

SEE ALSO: Acid deposition and acid rain; Air-pollution policy; Air Quality Index; Automobile emissions; Clean Air Act and amendments; Coal-fired power plants; Greenhouse effect; Greenhouse gases; London smog disaster; Ozone layer; Smog.

Air-pollution policy

CATEGORY: Atmosphere and air pollution

DEFINITION: High-level governmental plan of action for establishing and maintaining acceptable air quality and regulating individual air pollutants

SIGNIFICANCE: Laws and regulatory agencies establish air-pollution policy to control human-generated pollutants that can have negative impacts on human life and health, ecosystems, and global processes such as stratospheric ozone replenishment. Regulatory policy typically seeks to control pollutants by setting ambient air-quality standards, limiting allowable emissions, and requiring the use of specific pollution-control technologies.

The Clean Air Act, passed by the U.S. Congress in 1963, laid the foundation for what some consider to be the most progressive, wide-reaching, and complicated environmental cleanup legislation in the world. When the Clean Air Act and other early fed-

Milestones in Air-Pollution Policy

Year	Event
1963	The Clean Air Act sets aside $95 million to reduce air pollution in the United States.
1970	The Environmental Protection Agency is established to enforce environmental legislation.
1970	Clean Air Act amendments establish stricter air-quality standards.
1977	Additional Clean Air Act amendments extend compliance deadlines established by the 1970 amendments and allow the EPA to bring civil lawsuits against companies that do not meet air-quality standards.
1979	The United Nations sponsors the Convention on Long-Range Transboundary Air Pollution, which is designed to reduce acid rain and air pollution.
1987	The Montreal Protocol is signed by twenty-four nations pledging to reduce the output of ozone-depleting chlorofluorocarbons.
1990	Clean Air Act amendments increase regulations on emissions that cause acid rain and ozone depletion and also establish a system of pollution permits.
1997	The Environmental Protection Agency issues updated air-quality standards.
1998	California institutes tougher emission control standards for new cars; other states follow with similar laws.
2003	Proposed Clear Skies Act is designed to amend the Clean Air Act with a cap-and-trade system.
2005	EPA's Clean Air Interstate Rule (CAIR) begins a cap-and-trade program to keep air pollution generated in one state from rendering other states noncompliant with air-quality standards.
2008	A federal appeals court rules that CAIR exceeds the EPA's regulatory authority but later orders temporary reinstatement.
2009	The EPA officially finds that the greenhouse gases methane, carbon monoxide, nitrous oxide, hydrofluorocarbons, perfluorocarbons, and sulfur hexafluoride constitute a threat to the public health and welfare.
2010	The American Lung Association reports that about 58 percent of Americans endure unhealthy air-pollution levels.
2010	The EPA replaces CAIR with the Transport Rule, requiring eastern states to decrease power plant emissions severely by 2014.

eral, state, and local clean air laws proved to be relatively ineffective, several sweeping amendments to the laws were enacted.

The groundbreaking 1970 amendments to the Clean Air Act resulted in emissions standards for automobiles and new industries in addition to establishing air-quality standards for urban areas. Devised through an exceptionally cooperative bipartisan effort, the 1970 amendments were proclaimed by President Richard M. Nixon to be a "historic piece of legislation" that put the United States "far down the road" toward achieving cleaner air. The amendments established specific maximum concentration levels for several hazardous substances, and the individual states

were charged with developing comprehensive plans to implement and maintain these standards.

Tightly controlled scientific methodology was used for the first time to assess and determine acceptable levels for public and environmental health for six "priority air pollutants": carbon monoxide, sulfur dioxide, nitrogen dioxide, respirable particulate matter, ground-level ozone, and lead. Emission standards for air-pollution sources such as automobiles, factories, and power plants were established that also limited the discharge of air pollutants in geographical areas where air quality was already acceptable, thus preventing its deterioration.

The major Clean Air Act amendments of 1970 also

stimulated many states to pass regional and local air-pollution legislation, with some areas eventually passing laws that later proved to be even more stringent than federally established guidelines. During this period, the newly created U.S. Environmental Protection Agency (EPA) began strongly suggesting the tightening of rules regulating the amount of lead that could be added to gasoline, a significant source of lead poisoning in urban children and young adults, thus laying the groundwork for the future elimination of all leaded gasolines. Many sectors of the business community challenged the wording of some of the 1970 amendments, arguing that the language was vague and required clarification, particularly regarding the deterioration of air quality in areas that were already meeting federal standards.

1977 AMENDMENTS

The 1977 amendments to the Clean Air Act were stimulated by growing public and government awareness of the necessity for further clarification of standards and the increased knowledge that came from a decade of scientific pollution-control research. Industrial areas that were in violation of air-quality standards, called nonattainment areas, were allowed to expand their factories or build new ones only if the new sources achieved the lowest possible emission rates. Additionally, other sources of pollution under the same ownership in the same state were required to comply with pollution-control provisions, and unavoidable emissions had to be offset by pollution reductions by other companies within the same region. These emissions-offset policies forced new industries within geographical regions to make formal requests that existing local companies reduce their pollution production; such situations often resulted in new companies paying the considerable expense of new emissions-control devices for existing companies.

Protection of air quality in regions that were already meeting federal standards sparked congressional debate, as many environmentalists asserted that existing air-quality standards gave some industries a theoretical license to pollute the air up to permitted levels. Rules for the "prevention of significant deterioration" within areas that already met clean air standards were set for sulfur oxides and particulates in 1977, and many individual experts and organizations lobbied for the inclusion of other pollutants, such as ozone, the chief component of smog.

A final major change mandated by the 1977 amendments was the strengthening of the authority of the EPA to enforce laws by allowing the agency to use civil lawsuits in addition to the criminal lawsuits that were previously required. Civil lawsuits have the advantage of not carrying the burden-of-proof requirements needed for criminal convictions; this legal dilemma previously motivated violating companies to take part in lengthy legal battles, as the legal costs were lower than the costs of purchasing and maintaining the necessary pollution-control devices. The EPA was also empowered to levy noncompliance penalties without having to file lawsuits, using the argument that violators have an unfair business advantage over competitors that are currently complying with established legislation. Additionally, several "right-to-know" laws went into effect beginning in 1985 that required manufacturing plant managers to make health and safety information regarding toxic materials available to current and prospective employees, business partners, and sponsors.

1990 AMENDMENTS

In 1990 the Clean Air Act was further amended to address inadequacies in previous amendments, with major changes including the establishment of standards and attainment deadlines for 190 toxic chemicals. The amendments were approved through the same kind of bipartisan effort as the one that resulted in the 1970 amendments, prompting President George H. W. Bush to state that the new legislation moved society much closer toward the clean air environment that "every American expects and deserves."

The 1990 amendments established a market-based measure for pollution taxes on toxic chemical emissions, thus enhancing the incentive for businesses to comply as quickly as possible. Emissions standards were tightened for automobiles, and mileage standards for new vehicles were raised; these provisions attacked the pollution problem at its center by prompting numerous significant steps toward improved fuel efficiency. Notable results of these measures included significant reductions in vehicular emissions of sulfur dioxide and nitrogen oxide (50 percent), carbon monoxide (70 percent), and other harmful substances (20 percent).

The 1990 amendments also established market-based incentives to reduce nitrogen and sulfur oxides because of their role in the growing controversy regarding acid deposition within rainwater. The EPA was empowered to create tradable permits that stipulated permissible emissions levels for nitrogen and sulfur oxides. The permits were issued to U.S. companies that had emission rates lower than those set by current requirements for the improvement of air quality. This landmark legislation enabled companies that implemented innovative and cost-effective means to reduce air pollution to sell their unused credits to other companies.

Other significant legislation passed within or assisted by the 1990 amendments included the beginning of the phasing out of numerous ozone-depleting chemicals and the implementation of strategies that would help sustain the environment. Discovery of a seasonal "ozone hole" over Antarctica in 1985 had sparked international concern regarding the state of the earth's ozone layer and its ability to continue to shield the planet's surface from harmful ultraviolet radiation. Many businesses complained that the considerable additional expenses associated with implementing these new laws created unnecessary burdens for industry that in many cases outweighed the potential environmental benefits. In some cases, this economic pressure merely transferred environmental problems elsewhere, with many businesses choosing to operate outside the United States, in countries with less stringent environmental requirements.

Another important clause in the 1990 amendments required the EPA to regulate emissions coming from solid waste incinerators, including incinerators used for disposal of medical waste. Medical waste incinerators are among the largest sources of airborne dioxin and mercury, which are widely believed to contribute to serious health problems.

Subsequent Developments

President Bill Clinton continued tightening acceptable levels of smog and soot in the United States but did begin allowing flexible methods for reaching these improved goals over a ten-year period. This marked a significant change from the earlier administration of President Ronald Reagan, which proposed a relaxation of environmental standards to favor industrial and technological interests. Clinton is credited with associating the problem of controlling fossil-fuel emissions with the threat of global warming, an issue that would undergo considerable debate within the United Nations and elsewhere for years to come. A 1990 amendment requiring the use of gasoline containing 2 percent oxygen by weight in regions classified as being in severe or extreme nonattainment for

the federal ozone standard was followed by several state-level requirements.

The clean air changes of 1990 led the California Air Resources Board (CARB) to introduce the country's most stringent vehicle emissions quality controls to date later that year. Under the state's ambitious program, 2 percent of all new cars sold in California in 1998 were to have pollution-control devices that released no environmentally harmful emissions at all, and the figure was required to rise to 10 percent by 2003. These monumental state laws also dictated that the hydrocarbon emissions of all new cars sold in California be at least 70 percent less than those sold in 1993 by the year 2003. Thirteen northeastern states later passed similar, but somewhat less rigorous, laws; only New York retained the 2 percent goal.

Given 1990's technology, the CARB standards were in effect a mandate for the automobile industry to develop battery-powered electric vehicles. Automakers failed to mass-produce cost-effective, high-performance, battery-powered electric vehicles soon enough to meet regulatory requirements, however. When California adjusted its requirements in response to this slow progress, the result was a lowering of pressure on the auto industry to meet regulatory demands through innovation. By the early years of the twenty-first century, zero-emission vehicles had yet to become an industry standard, but low-emission vehicles such as hybrids had become commonplace, bringing the industry closer to realizing those goals.

During President George W. Bush's administration, the Clear Skies Act of 2003 sought to amend the Clean Air Act with a cap-and-trade system. The controversial bill was never enacted, but measures from it were included in the EPA's 2005 Clean Air Interstate Rule (CAIR), a cap-and-trade program to keep air pollution generated in one state from rendering another state noncompliant with air-quality standards.

In July, 2008, a federal appeals court ruled that CAIR exceeded the EPA's regulatory authority, but five months later the court ordered a temporary reinstatement until the EPA could develop a satisfactory replacement rule. A proposed replacement, known as the Transport Rule, was issued in July, 2010. The Transport Rule is intended to improve air quality in the eastern United States through a decrease in power plant emissions in thirty-one states and the District of Columbia. It requires that, by 2014, power plants reduce their sulfur dioxide emissions by 71 percent and their nitrogen oxides emissions by 52 percent. Each state must meet firm emissions requirements by 2014, which leaves room for only limited trading of pollution credits. The EPA estimates that the rule, if implemented, could prevent 14,000 to 36,000 premature deaths per year, as well as hundreds of thousands of cases of upper respiratory illness.

In late 2009 the EPA issued findings that current and projected atmospheric concentrations of the greenhouse gases (GHGs) methane, carbon monoxide, nitrous oxide, hydrofluorocarbons, perfluorocarbons, and sulfur hexafluoride constitute a threat to the public health and welfare. While this EPA action did not impose regulatory requirements, it paved the way for the agency to finalize GHG emissions standards for new motor vehicles, which contribute to atmospheric GHG concentrations. The EPA also proposed GHG emissions thresholds that would define whether Clean Air Act permits for such emissions are required. These thresholds would target the nation's largest stationary sources, such as power plants, refineries, and cement production facilities, but would not affect small businesses and farms. This proposed rule became the focus of a host of legal challenges.

INDOOR AIR QUALITY

Indoor air pollution began to receive serious public attention in the United States in the wake of the 1970's energy crisis. Interest in conserving energy drove changes in the construction of new buildings, and the retrofitting of old ones, aimed at retaining desired indoor temperatures. Making structures more airtight meant that air contaminants were also retained indoors. By the late 1980's, enough cases had emerged of people experiencing discomfort or various debilitating health problems—chronic respiratory issues, sinus infections, sore throats, headaches, and more—as a result of time spent in particular buildings that the phenomenon had been dubbed sick building syndrome (SBS). The U.S. Occupational Safety and Health Administration has estimated that 30 to 70 million American workers are affected by SBS, although the vast majority do not suffer serious health problems as a result of exposure.

With most Americans spending more than 90 percent of their lives working, learning, and spending leisure time indoors, indoor air quality has the potential to have profound impacts on the population's health. Despite this fact, no comprehensive federal legislation has addressed indoor air quality in the United States; rather, various federal and state regulatory standards address indoor air quality by focusing on specific pollutants, activities, and types of structures.

Common indoor air pollutants include radon (a natural breakdown product of uranium in soil or rock), tobacco smoke, asbestos, formaldehyde, biological contaminants (such as mold and mildew), combustion products (such as carbon monoxide), cleansers and other household products, and pesticides. A number of federal and state laws address asbestos, including the 1976 Toxic Substances Control Act (TSCA), which gives the EPA broad authority to control the production, distribution, and disposal of potentially hazardous chemicals. Under TSCA, federal standards have also been established for the amount of formaldehyde (a chemical present in many adhesives, resins, and solvents) allowable in composite wood-based products

Other legislation, such as the 1976 Consumer Product Safety Act, has granted federal and state authority over consumer products that are potentially dangerous to public health and the environment, with many products that generate indoor air pollution falling under that jurisdiction. For example, carbon monoxide and other hazardous combustion products can be emitted by stoves, and formaldehyde can outgas from plywood and textiles.

Developing countries have different problems related to indoor air pollution than do the United States and other developed nations. In comparison with developed nations, developing nations generally have fewer and less restrictive environmental regulations in place, and indoor air pollution in these nations is less likely to be caused by the airtightness of buildings than by the use of substances indoors that can be harmful. For example, wood, dung, and crop residues are primary sources of cooking and heating fuels in many developing nations, and these can generate unhealthful air pollutants when burned. The United Nations is involved in various efforts to address the problem of indoor air pollution in developing nations.

AIR QUALITY

In July, 1997, the EPA issued updated air-quality standards following the most complete scientific review process in the history of the organization. Based on the findings of this review, which was conducted by hundreds of internationally recognized scientists, industry experts, and public health officials, major steps were taken toward the improvement of environmental and public health through the revision of ozone

(continued on page 36)

National Ambient Air Quality Standards for Criteria Pollutants

POLLUTANT	AVERAGING TIME	POLLUTANT LEVEL	EFFECTS ON HEALTH
Carbon monoxide: colorless, odorless, tasteless gas; it is primarily the result of incomplete combustion; in urban areas the major sources are motor vehicle emissions and wood burning.	1-hour 8-hour	35 ppm 9 ppm	The body is deprived of oxygen; central nervous system affected; decreased exercise capacity; headaches; individuals suffering from angina, other cardiovascular disease; those with pulmonary disease, anemic persons, pregnant women and their unborn children are especially susceptible.
Ozone: highly reactive gas, the main component of smog.	1-hour 8-hour	0.120 ppm 0.080 ppm	Impaired mechanical function of the lungs; may induce respiratory symptoms in individuals with asthma, emphysema, or reduced lung function; decreased athletic performance; headache; potentially reduced immune system capacity; irritant to mucous membranes of eyes and throat.
Particulate matter < 10 microns (PM10): tiny particles of solid or semisolid material found in the atmosphere.	24-hour Annual arithmetic mean	$150 \, \mu g/m^3$ $50 \, \mu g/m^3$	Reduced lung function; aggravation of respiratory ailments; long-term risk of increased cancer rates or development of respiratory problems.
Particulate matter < 2.5 microns (PM2.5): fine particles of solid or semisolid material found in the atmosphere.	24-hour Annual arithmetic mean	$65 \, \mu g/m^3$ $15 \, \mu g/m^3$	Same as PM10 above.
Lead: attached to inhalable particulate matter; primary source is motor vehicles that burn unleaded gasoline and re-entrainment of contaminated soil.	Calendar quarter	$1.5 \, \mu g/m^3$	Impaired production of hemoglobin; intestinal cramps; peripheral nerve paralysis; anemia; severe fatigue.
Sulfur dioxide: colorless gas with a pungent odor.	3-hour 24-hour Annual arithmetic mean	0.5 ppm 0.14 ppm 0.03 ppm	Aggravation of respiratory tract and impairment of pulmonary functions; increased risk of asthma attacks.
Nitrogen dioxide: gas contributing to photochemical smog production and emitted from combustion sources.	Annual arithmetic mean	0.053 ppm	Increased respiratory problems; mild symptomatic effects in asthmatics; increased susceptibility to respiratory infections.

Notes: ppm equals parts per million and $\mu g/m^3$ equals micrograms per cubic meter.
Source: United States Environmental Protection Agency (EPA); URL http://www.epa.gov.

standards for the first time in twenty years. In addition, annual exposure standards for fine particulate matter were introduced. (Short-term standards for coarse and fine particulates had been in place for a decade. Short-term standards that applied specifically to fine particulates were not introduced until 2006.) The EPA's 1997 study concluded that many previously imposed standards were not resulting in enough protection for the environment and public health. Data indicated that repeated exposure to pollutants at levels previously considered to be acceptable could cause permanent lung damage in children and in adults who regularly exercise and work outdoors in many urban environments.

The EPA regularly reviews national air-quality standards for the Clean Air Act's six priority air pollutants. Between 1990 and 2008 the Clean Air Act and its supporting legislation enabled national emissions reductions of 78 percent for lead, 14 percent for ozone, 68 percent for carbon monoxide, 35 percent for nitrogen dioxide, 59 percent for sulfur dioxide, and 31 percent for respirable particulates. According to one study, between 1980 and 2000 the reduction in particle pollution alone increased life expectancy in fifty-one cities in the United States by an average of five months. Thanks to control programs for chemical plants, dry cleaners, coke ovens, incinerators, and mobile sources, total emissions of toxic air pollutants decreased by approximately 40 percent between 1990 and 2005. Haze and acid precipitation were also on the decline.

Despite the progress that has been made, air pollution remains a critical environmental risk in the United States. In 2008 thirty-one areas in the United States failed to meet ambient air-quality standards for ozone, eighteen areas failed to meet standards for particulates, and two failed to meet the standard for lead. In 2010 the American Lung Association reported that approximately 58 percent of the nation's population was continuing to experience unhealthy air-pollution levels. According to the U.S. Centers for Disease Control and Prevention, chronic lower respiratory diseases were the fourth leading cause of death in the United States in 2007; the asthma death rate for children under nineteen years old increased by nearly 80 percent between 1980 and 2001.

On the international level, a 1987 United Nations conference held in Canada saw twenty-four nations agree to guidelines established to protect the ozone layer through the Montreal Protocol on Substances That Deplete the Ozone Layer. The Montreal Protocol, which by 2010 had been ratified by 196 countries and amended several times, provides a framework for the phaseout of certain ozone-depleting compounds and includes a mechanism through which developed nations can aid developing countries in making this transition. The United States has also accepted several of the protocols of the 1979 Geneva Convention on Long-Range Transboundary Air Pollution, an international agreement that addresses the impacts of air-pollution migration across political boundaries. These protocols include those concerning reduction strategies for emissions of nitrogen oxides, cadmium, lead, mercury, sulfur, volatile organic compounds, and ammonia.

Daniel G. Graetzer
Updated by Karen N. Kähler

FURTHER READING

Bailey, Christopher J. *Congress and Air Pollution: Environmental Politics in the US.* New York: Manchester University Press, 1998.

Ferrey, Steven. "Air Quality Regulation." In *Environmental Law: Examples and Explanations.* 5th ed. New York: Aspen, 2010.

Godish, Thad. "Regulation and Public Policy." In *Air Quality.* 4th ed. Boca Raton, Fla.: Lewis, 2004.

Kessel, Anthony. *Air, the Environment, and Public Health.* New York: Cambridge University Press, 2006.

Melnick, R. Shep. *Regulation and the Courts: The Case of the Clean Air Act.* Washington, D.C.: Brookings Institution Press, 1983.

Rushefsky, Mark E. "Environmental Policy: Challenges and Opportunities." In *Public Policy in the United States: At the Dawn of the Twenty-first Century.* 4th ed. Armonk, N.Y.: M. E. Sharpe, 2008.

U.S. Environmental Protection Agency. *The Plain English Guide to the Clean Air Act.* Research Triangle Park, N.C.: Office of Air Quality Planning and Standards, 2007.

SEE ALSO: Air pollution; Automobile emissions; Clean Air Act and amendments; Convention on Long-Range Transboundary Air Pollution; Environmental law, international; Environmental law, U.S.; Environmental Protection Agency; Indoor air pollution; Montreal Protocol; Pollution permit trading; Sick building syndrome.

Air Quality Index

CATEGORY: Atmosphere and air pollution
DEFINITION: The U.S. Environmental Protection Agency's tool for indicating the health risks posed by ambient air quality in given areas at particular times
SIGNIFICANCE: With air pollution a growing problem the world over, the Air Quality Index and equivalent tools in other nations have become an indispensable part of regional and federal governments' efforts to convey information on ambient air quality to the general public.

Nations around the world employ a number of variations on the Air Quality Index (AQI) developed by the U.S. Environmental Protection Agency (EPA), but the basic methodology used to assess air quality is the same. The concentrations of various air pollutants in a testing area are measured regularly, and, based on this information, the area is assigned a number. This number fits into a color-coded rating system in which different tiers correspond to the severity of the health threat the air quality poses; the higher the number, the more severe the threat.

Although a host of air pollutants have the potential to cause adverse health effects, only a handful of criteria contaminants are generally used in the assessment of basic air quality. Individual criteria differ from place to place, but the most common pollutants monitored are suspended particulate matter, airborne lead, ground-level ozone, nitrogen dioxide, sulfur dioxide, and carbon monoxide. These hazardous compounds are capable of causing severe respiratory irritation, heart and circulatory problems, and other negative health effects.

Monitoring sites are typically limited to cities and towns, where pollution levels and population densities are both high. Different levels of pollution may fall into different tiers based on the stringency of a particular agency's approach to measuring air quality. For example, Hong Kong's Air Pollution Index has come under heavy criticism for its relatively lax standards, as the air quality it rates as safe sometimes contains pollutants at levels several times higher than those considered acceptable by the World Health Organization.

Daniel J. Connell

SEE ALSO: Air pollution; Air-pollution policy; Airborne particulates; Carbon monoxide; Nitrogen oxides; Smog; Sulfur oxides.

Airborne particulates

CATEGORY: Atmosphere and air pollution
DEFINITION: Tiny particles found in the air
SIGNIFICANCE: Some airborne particulates, such as dust, dirt, soot, and smoke, are large enough to be visible to the naked eye, while other forms are so small that they require electron microscopes for detection. The inhalation of microscopic particles can have serious adverse effects on human respiratory and cardiovascular health.

Airborne particulate matter (PM) represents a complex mixture of organic and inorganic substances and varies in size, composition, and origin. Some particles, known as primary particles, are emitted directly from sources such as construction sites, unpaved roads, fields, smokestacks, and fires. Secondary particles are formed by reactions of gases, such as sulfur dioxide and nitrogen oxides, that are emitted from power plants, industrial plants, and automobiles. Secondary particles make up most of the fine-particle pollution in the United States.

Particle pollution contains microscopic solids or liquid droplets that are small enough to travel deep into the lungs and cause serious health problems. Breathing such pollution can lead to respiratory symptoms such as coughing and difficult breathing; it can decrease lung function and aggravate existing asthma. Also associated with exposure to particle pollution are chronic obstructive pulmonary disease and emphysema, chronic bronchitis, irregular heartbeat, nonfatal heart attacks, and premature death in people with heart or lung disease.

The size of airborne particulates is directly linked to their potential for causing health problems. The U.S. Environmental Protection Agency (EPA) has established air-quality standards concerning two sizes (or fractions) of particles: PM10 and PM2.5. PM10 particles are those with a diameter of 10 micrometers or smaller (10 micrometers is equal to 0.004 inch, or one-seventh the width of a human hair); they include both coarse and fine particles. PM10 particles smaller than 10 micrometers can settle in the bronchi and lungs and cause health problems. PM2.5 particles are

2.5 micrometers in diameter or smaller. Particles in this fraction tend to penetrate further, reaching the gas exchange regions of the lung, and even smaller particles (0.1 micrometer or smaller) may pass through the lungs into the bloodstream and affect other organs, particularly the cardiovascular system. These particles can also adsorb harmful gases or other components (such as iron, carcinogens, or ozone) and release them within lung cells. Particles emitted from modern diesel engines are typically 0.1 micrometer or smaller. PM2.5 inhalation can lead to high plaque deposits in the arteries, causing vascular inflammation and atherosclerosis (hardening of the arteries that reduces elasticity and can lead to heart attacks).

The federal Clean Air Act requires the EPA to review the latest scientific information every five years and promulgate the National Ambient Air Quality Standards for six pollutants, among them PM. U.S. air-quality standards for PM were first established in 1971 and were not significantly revised until 1987, when the EPA changed the indicator of the standards specifically to regulate PM10 levels. Ten years later, the agency set a separate standard for PM2.5 particles based on new research findings regarding their link to serious health problems. The 1997 standards also retained but slightly revised the PM10 standards, which were intended to regulate inhalable coarse particles ranging from 2.5 to 10 micrometers in diameter. The EPA revised the air-quality standards for airborne particle pollution in 2006, lowering the acceptable level of PM2.5 over a 24-hour period from 65 micrograms per cubic meter to 35 micrograms per cubic meter (1 cubic meter is roughly equivalent to 35 cubic feet). It retained the 24-hour PM10 standard of 150 micrograms per cubic meter.

Bernard Jacobson

FURTHER READING

Hilgenkamp, Kathryn. "Air." In *Environmental Health: Ecological Perspectives*. Sudbury, Mass.: Jones and Bartlett, 2006.

National Research Council. *Research Priorities for Airborne Particulate Matter, IV: Continuing Research Progress*. Washington, D.C.: National Academies Press, 2004.

Peters, Annette, and C. Arden Pope III. "Cardiopulmonary Mortality and Air Pollution." *The Lancet* 360 (October 19, 2002): 1184-1185.

What Is Haze?

The Environmental Protection Agency's Office of Air Quality Planning and Standards defines "haze" as follows:

Haze is caused when sunlight encounters tiny pollution particles in the air. Some light is absorbed by particles. Other light is scattered away before it reaches an observer. More pollutants mean more absorption and scattering of light, which reduce the clarity and color of what we see. Some types of particles, such as sulfates, scatter more light, particularly during humid conditions.

Where does haze-forming pollution come from?

Air pollutants come from a variety of natural and manmade sources. Natural sources can include windblown dust, and soot from wildfires. Manmade sources can include motor vehicles, electric utility and industrial fuel burning, and manufacturing operations. Some haze-causing particles are directly emitted to the air. Others are formed when gases emitted to the air form particles as they are carried many miles from the source of the pollutants.

What else can these pollutants do to you and the environment?

Some of the pollutants which form haze have also been linked to serious health problems and environmental damage. Exposure to very small particles in the air have been linked with increased respiratory illness, decreased lung function, and even premature death. In addition, particles such as nitrates and sulfates contribute to acid rain formation which makes lakes, rivers, and streams unsuitable for many fish, and erodes buildings, historical monuments, and paint on cars.

SEE ALSO: Air pollution; Air-pollution policy; Air Quality Index; Carbon monoxide; Nitrogen oxides; Smog; Sulfur oxides.

Alar

CATEGORY: Agriculture and food
DEFINITION: Brand name for the chemical daminozide, used as a plant growth regulator
SIGNIFICANCE: The use of Alar to improve the quality and appearance of apples became controversial in the late 1980's, when a debate arose over the chemical's carcinogenic properties.

Alar is a growth-regulating chemical manufactured by Uniroyal Chemical Company. In the late 1960's, apple farmers began using the product to improve the quality and appearance of their fruit. The use of Alar by apple growers to preserve crispness as the fruit was sent to market was a common practice for more than twenty years. In 1989, however, a controversy erupted regarding the potential harmful effects of the chemical, and Alar was accused—erroneously, according to many experts—of being the most potent carcinogen in the food supply.

The first questions about the chemical were raised in the 1970's by Dr. Bela Toth of the University of Nebraska, whose research appeared to indicate that Alar created tumors in mice but not in rats. Reports of Toth's discovery did not stress the fact that the rodents were fed massive amounts of the chemical, far exceeding the maximum tolerated dose used in cancer testing.

At the time, the U.S. Environmental Protection Agency (EPA) disregarded the study. In 1983, however, under attack by environmental and media groups critical of the environmental policies of Ronald Reagan's presidential administration, the EPA began its questioning of Alar. Steve Schatzow, a lawyer, was appointed to lead the Office of Pesticide Programs (OPP), and in conjunction with other organizations, including the Natural Resources Defense Council (NRDC) and the American Council on Science and Health (ACSH), the OPP began the fight against Alar, proclaiming it to be the most potent cancer-causing substance in the food industry.

The effect of OPP's announcement was dramatic. Consumers poured apple juice down drains, stores pulled apple products from their shelves, and farmers suffered losses estimated in the hundreds of millions of dollars. The anti-Alar campaign became more aggressive when such celebrities as *Sixty Minutes* newsman Ed Bradley, consumer activist Ralph Nader, and actor Meryl Streep—who set up a group called Mothers and Others for Pesticide Limits—expressed fears about the chemical.

Despite the anti-Alar fanfare, subsequent studies on the consumption of traces of the chemical, whether in apple juice or any other form of apples, proved to be negative. After 1990, no mainstream, peer-reviewed research demonstrated any linkage between the chemical or its breakdown product, UMDH, and cancer. For example, Dr. Jose R. P. Cabral, an investigator with the International Agency for Research on Cancer, declared that Alar is safe to use and that his group's experiments had not found tumors in rodents that had consumed reasonable quantities of the chemical. In the wake of such findings, the U.S. Food and Drug Administration issued a statement affirming that eating apples that have been treated with Alar poses no health threat.

Soraya Ghayourmanesh

FURTHER READING

Blay-Palmer, Alison. *Food Fears.* Burlington, Vt.: Ashgate, 2008.

Ohkawa, H., H. Miyagawa, and P. W. Lee, eds. *Pesticide Chemistry: Crop Protection, Public Health, Environmental Safety.* New York: Wiley-VCH, 2007.

SEE ALSO: Agricultural chemicals; Carcinogens; Pesticides and herbicides.

Alaska Highway

CATEGORIES: Places; land and land use

IDENTIFICATION: Important transportation and tourist corridor through northwest Canada and Alaska

DATE: Completed in 1943

SIGNIFICANCE: The presence of the Alaska Highway has caused environmental change in the regions through which it passes, in conjunction with the land-use development associated with transportation and tourism.

In the early 1940's, during World War II, fear that Japan would invade Alaska prompted the construction of the Alaska Highway for military use. Begun in 1942 and completed in 1943, the Alaska Highway was an all-weather gravel road that connected Dawson Creek in British Columbia, Canada, with Fairbanks, Alaska. The highway was initially a military road only and, as such, was designed for military transport; little attention was given to environmental considerations in deference to the war effort. The road had to be constructed for ease and speed of movement of military goods, and engineering efforts were aimed at meeting these needs. Grades, road cuts, and stream crossings were made quickly, without much consideration of the environmental impact of constructing a highway through some 2,400 kilometers (1,500 miles) of previously untraversed wilderness.

For a few years following World War II, civilian travel on the Alaska Highway, also known as the Alcan

Highway, was limited and carefully controlled. With the removal of travel restrictions in 1947, however, traffic increased, and environmental degradation inevitably followed. While early travelers had to attend to most of their needs themselves, the flow of traffic proved an incentive for the provision of services. Gas stations, restaurants, and lodging facilities all grew to meet the needs of the traveling public on the Alaska Highway. In turn, the presence of a well-maintained highway and available public facilities fostered additional growth, and the new facilities themselves created a need for waste removal, storage areas, and underground fuel tanks. Interest in and use of the highway—which passes through wilderness areas, historically important settings such as the Klondike, and active mining areas—have continued to grow.

Regular improvements have been made to the highway. It has been upgraded from a gravel to a paved surface, curves have been straightened, and grades have been improved. These changes themselves have imposed additional imprints on the adjacent environment. Successive efforts to widen and straighten the highway have increased the direct physical impact on the surrounding landscape.

Jerry E. Green

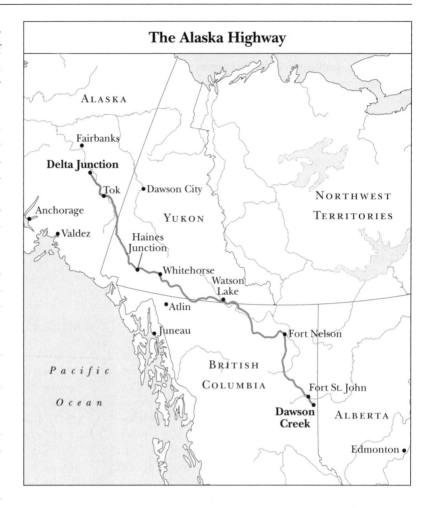

The Alaska Highway

FURTHER READING

Brown, Tricia. *The World-Famous Alaska Highway: A Guide to the Alcan and Other Wilderness Roads of the North.* 2d ed. Anchorage: Alaska Northwest Books, 2005.

Nelson, Daniel. *Northern Landscapes: The Struggle for Wilderness Alaska.* Washington, D.C.: Resources for the Future, 2004.

SEE ALSO: Conservation; Road systems and freeways; Trans-Alaska Pipeline.

Alaska National Interest Lands Conservation Act

CATEGORIES: Treaties, laws, and court cases; land and land use; preservation and wilderness issues

THE LAW: U.S. law intended to resolve conflicts regarding landownership in Alaska among Native Alaskans and the federal and state governments

DATE: Enacted on December 2, 1980

SIGNIFICANCE: The Alaska National Interest Lands Conservation Act more than tripled the amount of land protected as wilderness in the United States, but it did not lead to the resolution of all conflicts over Alaska land use.

The 1959 Alaska Statehood Act allowed the state government to choose 41.5 million hectares (102.5 million acres) of federal land from that part of the public domain not reserved for parks or other designated use. The land-selection process proceeded slowly, complicated by such issues as Native Alaskan claims, which were not clarified in the statehood bill. The discovery of oil fields in northern Alaska prompted passage of the 1971 Alaska Native Claims Settlement Act (ANCSA), which extinguished Native Alaskan land claims and allowed for construction of the Trans-Alaska Pipeline. To secure support for the ANCSA from conservation groups, the U.S. Congress included a provision in the bill that allowed the secretary of the interior to withdraw up to 32.4 million hectares (80 million acres) from the public domain for the establishment of conservation units, including national parks and wilderness areas. Congress would make the final determination regarding the final disposition of withdrawn lands. In order to provide a forum for cooperative planning on the issue, Congress created the Joint Federal-State Land Use Planning Commission.

An intense political battle ensued. Federal agencies, including the Forest Service and the National Park Service, competed for the authority to manage the withdrawn lands. Environmentalists favored agencies that would limit development, while Alaska state officials threw their support behind agencies that might prove willing to allow multiple use and the exploitation of natural resources for economic gain. After years of contentious debate, in 1979 the House of Representatives passed a bill establishing 51.4 million hectares (127 million acres) of conservation units, 26.3 million hectares (65 million acres) of which were designated wilderness areas. The Senate version of the bill significantly reduced the amount of land, and the legislation did not become law. The following year, the election of Ronald Reagan to the U.S. presidency and a Republican majority to the Senate convinced House members that they had no alternative but to accept the Senate version of the bill, which passed in 1980 and became known as the Alaska National Interest Lands Conservation Act (ANILCA).

ANILCA set aside 42.1 million hectares (104 million acres) of land as conservation units, with 23 million hectares (57 million acres) of wilderness. The National Park Service received 17.8 million hectares (44 million acres), the Fish and Wildlife Service 20.2 million hectares (50 million acres), and the Forest Service 1 million hectares (2.5 million acres). The Bureau of Land Management, the federal agency most amenable to development, received a patchwork of marginal lands that no other agency desired.

The act more than tripled the amount of land protected as wilderness in the United States, and its provisions placed 75 percent of national park land in Alaska. It did not solve the conflicts over Alaska land use, however. The state contested some land withdrawals. The act allowed for the construction of pipelines and roadways through conservation units. Moreover, ANILCA contained no measures that protected the habitats of migratory animals. Finally, the bill included provisions that allowed for the future exploitation of mineral resources, including oil, should both the Congress and the president deem it necessary.

Thomas Clarkin

FURTHER READING

Haycox, Stephen W. *Frigid Embrace: Politics, Economics, and Environment in Alaska.* Corvallis: Oregon State University Press, 2002.

Nelson, Daniel. *Northern Landscapes: The Struggle for Wilderness Alaska.* Washington, D.C.: Resources for the Future, 2004.

Ross, Ken. *Pioneering Conservation in Alaska.* Boulder: University Press of Colorado, 2006.

SEE ALSO: Bureau of Land Management, U.S.; Forest Service, U.S.; National parks; Trans-Alaska Pipeline; Wilderness areas.

Alliance of Small Island States

CATEGORIES: Organizations and agencies; weather and climate

IDENTIFICATION: Lobbying group that represents small island and low-lying coastal nations

DATE: Founded in 1990

SIGNIFICANCE: Members of the Alliance of Small Island States worry that global warming caused by human activity will lead to a rise in sea levels that could obliterate much of their land and cause increasingly extreme tropical cyclones. The group quickly developed into a powerful force for the promotion of international treaties such as the 1997 Kyoto Protocol, which limits emissions of greenhouse gases into the atmosphere.

The Alliance of Small Island States (AOSIS) was founded at the Second World Climate Conference, which was held from October 29 to November 7, 1990, in Geneva, Switzerland. AOSIS is an association

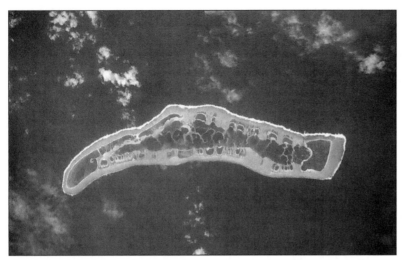

Millennium Island, part of Kiribati, a member of the Alliance of Small Island States, as seen from the International Space Station. The island, no part of which is more than 6 meters (20 feet) above the ocean, has been identified by the United Nations as being at significant risk from rises in sea level. (NASA)

of its member states' ambassadors to the United Nations; it does not have a charter, a budget, or its own secretariat. By 2010 AOSIS had thirty-nine member nations and four observers.

Because global warming is expected to raise seawater levels as glaciers melt at both of the earth's poles, the member nations of AOSIS fear future inundation by rising seas. Moreover, they fear that the rising salt water could contaminate their freshwater resources and erode their coasts, and that global warming could bleach their coral reefs and cause increasingly severe storms that could wreck their lands and their coastlines.

Most of the founding AOSIS member nations are small islands or coastal states in the Caribbean Sea, the Indian Ocean, and the Pacific Ocean. Given that their own emissions of greenhouse gases have generally been very low, owing to the small size and developing nature of their economies, the member states expressed unhappiness at having to carry the burden of global warming caused by others—namely, the developed and large developing nations. This led to AOSIS's consistent demand for money transfers from the countries emitting the majority of greenhouse gases to the AOSIS nations, to help them mitigate the expected effects of global warming on their endangered islands and coastlines.

From 1991 to 1994, under its first chair, Surinameborn African American activist Robert Van Lierop, who represented the Pacific Island nation of Vanuatu

at the United Nations from 1981 to 1994, AOSIS worked proactively to shape global climate policy at the United Nations. AOSIS members were initial signatories to the United Nations Framework Convention on Climate Change (UNFCCC) in June, 1992. Next, AOSIS proposed and helped to draft the text of the 1997 Kyoto Protocol, which committed those nations that signed it to controlling, limiting, and reducing their greenhouse gas emissions.

AOSIS also works very closely with the Small Island Developing States group at the United Nations and has been active at each of the annual Conferences of the Parties of the UNFCCC. At the fifteenth Conference of the Parties in Copenhagen, Denmark, on December 10, 2009, AOSIS submitted an ambitious proposal to limit human-made global warming to absolutely no more than 1.5 degrees Celsius (2.7 degrees Fahrenheit) instead of the previous limit of 2 degrees Celsius (3.6 degrees Fahrenheit). However, in an atmosphere of increased skepticism about the science behind the models of human-caused global warming and a lingering global economic crisis, other nations did not take up this proposal. Some scientists argued that at that time, many of the problems of the small island and low-level coastal states had been caused by coastal mismanagement and population growth, and that these were a greater threat than future global warming.

R. C. Lutz

FURTHER READING

Ashe, John W., Robert Lierop, and Anilla Cherian. "The Role of the Alliance of Small Island States (AOSIS) in the Negotiation of the United Nations Framework Convention on Climate Change." *Natural Resources Forum* 23, no. 3 (1999): 209-220.

Barnett, Jon, and John Campbell. *Climate Change and Small Island States: Power, Knowledge, and the South Pacific.* Washington, D.C.: Earthscan, 2010.

Kasa, Sjur, Anne Gullberg, and Gørild Heggelund. "The Group of 77 in the International Climate Negotiations: Recent Developments and Future Directions." *International Environmental Agreements: Politics, Law, and Economics* 8, no. 2 (2008): 113-127.

Risse, Mathias. "The Right to Relocation: Disappearing Island Nations and Common Ownership of the Earth." *Ethics and International Affairs* 23, no. 3 (2009): 281-300.

SEE ALSO: Climate change and oceans; Climate change skeptics; Climate models; Climatology; Coral reefs; Glacial melting; Global warming; Intergovernmental Panel on Climate Change; Kyoto Protocol; Maldive Islands; Pacific Islands; Sea-level changes; United Nations Framework Convention on Climate Change.

Alternative energy sources

CATEGORY: Energy and energy use

DEFINITION: Sources of energy other than the dominant fossil and mineral fuels

SIGNIFICANCE: Energy sources that offer alternatives to the burning of fossil fuels such as coal and petroleum are urgently needed to address rising demand for energy in ways that will not contribute to air pollution and climate change. The ideal alternative energy source is renewable or inexhaustible and causes no lasting environmental damage.

Both the extraction and the burning of fossil fuels have caused severe and growing damage to the environment, contributing to such problems as air pollution, the release of greenhouse gases (which retain heat and contribute to climate change), and sulfuric acid in rainfall. Nuclear energy sources are very limited in supply and expensive, require extreme amounts of processing, and produce long-lasting radioactive waste. In the long term, energy release from nuclear fusion has been proposed as a limitless supply of power, but industrial-scale production of fusion power continues to pose large and uncertain obstacles and hazards.

SOLAR POWER

The sun powers winds, ocean currents, rain, and all biomass growth on the earth's surface. Because the availability and extraction means for each of these secondary sources of solar power are diverse, each forms a different field of alternative energy technology. Where solar power is extracted and converted to energy directly, the capture can be by means of flat-plate receivers that collect at the incident intensity but can operate in diffuse light, or by means of concentrators that can achieve intensities of several hun-

Banks of solar panels at an electricity-generating facility in Pittsfield, Massachusetts. (AP/Wide World Photos)

dred suns but work poorly in diffuse light.

In solar photovoltaic power (PV) technology, solar radiation is directly converted to useful power through PV cell arrays, which require semiconductor mass-production plants. PV cell technologies have evolved from using single-crystal silicon to using thinner polycrystalline silicon, gallium arsenide, thin-film amorphous silicon, cadmium telluride, and copper indium selenide. The needed materials are believed to be abundant enough to meet projected global growth. The process of purifying silicon requires large inputs of energy, however, and it generates toxic chemical waste. Regeneration of the energy required to manufacture a solar cell requires about three years of productive cell operation.

Solar cell technology continues to evolve. Broadband solar cell technologies have the potential to make cells sensitive to as much as 80 percent of the energy in the solar spectrum, up from about 60 percent. High-intensity solar cells could enable operation at several hundred times the intensity of sunlight, reducing the cell area required when used with concentrator mirrors and enabling high thermal efficiency.

Direct solar conversion is another option. Laboratory tests have shown 39 percent conversion from broadband sunlight to infrared laser beams using neodymium-chromium fiber lasers. Direct conversion of broadband sunlight to alternating-current electricity or beamed power through the use of optical antennae is projected to achieve 80 to 90 percent conversion. Such technologies offer hope for broadband solar power to be converted to narrowband power in space and then beamed to the earth by satellites.

Another way of harnessing solar power is through solar thermal technology. Solar concentrators are used with focal-point towers to achieve temperatures of thousands of kelvins and high thermal efficiency, limited by containment materials. The resulting high-

Construction of an offshore wind turbine, part of a wind farm off the coast of Denmark. (©Yobidaba/Dreamstime.com)

temperature electrolysis of water vapor generates hydrogen and oxygen in an efficient manner, and this technology has demonstrated direct solar decomposition of carbon dioxide (CO_2) to carbon monoxide (CO) and oxygen.

WIND POWER

Winds are driven by temperature and pressure gradients, ultimately caused by solar heating. Wind energy is typically extracted through the operation of turbines. Power extraction is proportional to the cube of wind speed, but wind-generated forces are proportional to the square of wind speed. Wind turbines thus can operate safely only within a limited range of wind speed, and most of the power generation occurs during periods of moderately strong winds. Turbine efficiency is strongly dependent on turbine size and is limited by material strength. The largest wind turbines exceed 5 megawatts in capacity. Denmark, the Netherlands, and India have established large wind turbine farms on flat coastal land, and Germany and the United Kingdom have opted for large offshore wind farms. In the United States, wind farms are found in the Dakotas, Minnesota, and California, as well as on Colorado and New Mexico mountain slopes and off the coasts of Texas and Massachusetts.

Because of wind fluctuations and the cubic power relation, wind power is highly unsteady, and means must be established for storing and diverting the power generated before it is connected to a power grid. In addition, offshore and coastal wind farms must plan for severe storms. Smaller wind turbines are sometimes used for power generation on farms and even for some private homes in open areas, but these tend to be inefficient and have high installation costs per unit power. They are mainly useful for pumping irrigation water or for charging small electrical devices.

Environmentalists have raised some concerns about large wind turbines. The machinery on wind farms causes objectionable noise levels, and many assert that the wind turbine towers themselves constitute a form of visual pollution. Disturbances to wildlife, particularly deaths and injuries among bird populations, are another area of concern. In addition, the construction of wind farms often requires the building of roads through previously pristine areas to enable transportation of the turbines' large components.

Hydroelectric Power and Tidal Power

Large dams provide height differences that enable the extraction of power from flowing water using turbines. Hydroelectric power generated by dams forms a substantial percentage of the power resources in several nations with rivers and mountains. However, the building of large dams raises numerous technical, social, and public policy issues, as damming rivers may displace human inhabitants from fertile lands and may result in the flooding of pristine ecosystems, sometimes the habitats of endangered species. Increased incidence of earthquakes has also been associated with the existence of very large dams.

In some of the world's remote communities, micro hydroelectric (or micro hydel) plants provide power,

generating electricity in the 1-30 megawatt range. Very small-scale systems, known as pico hydel, extract a few kilowatts from small streams; these can provide viable energy sources for individual homes and small villages, but the extraction technology has to be refined to bring down the cost per unit of power.

Although tidal power is abundant along coastlines, the harnessing of that power has been slow to gain acceptance, in part because of the difficulties of building plants that can survive ocean storms. Tidal power is extracted in two principal ways. In one method, semipermeable barrages are built across estuaries with high tidal ranges, and the water collected in the barrages is emptied through turbines to generate power. In the second, offshore tidal streams and currents are harnessed through the use of underwater equivalents of wind turbines.

Tidal power plants typically use pistons that are driven up and down by alternating water levels or the action of waves on turbines. A rule of thumb is that a tidal range of 7 meters (23 feet) is required to produce enough hydraulic head for economical operation. One drawback is that the 12.5-hour cycle of tidal operation is out of synchronization with daily peak electricity demand times, and hence some local means of storing the power generated is desirable. In many cases, impellers or pistons are used to pump water to high levels for use when power demand is higher.

Biomass Power

Biomass, which consists of any material that is derived from plant life, is composed primarily of hydrocarbons and water, so it offers several ways of usage in power generation. Combustion of biomass is considered to be carbon-neutral in regard to greenhouse gas emissions, but it may generate smoke particles and other pollution.

One large use of biomass is in the conversion of corn, sugarcane, and other grasses to ethyl alcohol (ethanol) to supplement fossil petroleum fuels. This use is controversial because the energy costs associated with producing and refining ethanol are said to be greater than the savings gained by using such fuel. It is argued that subsidies and other public policies and rising energy prices entice farmers to devote land to the production of ethanol crops, thus triggering shortages and increases in food prices, which hurt the poorest people the most. Brazil has advanced profitable and sustainable use of ethanol extracted from sugarcane to replace a substantial portion of the nation's transportation fossil-fuel use.

Jatropha plants, as well as certain algae that grow on water surfaces, offer sources of biodiesel fuel. Biodiesel from jatropha is used to power operations on several segments of India's railways, and vegetable oil from peanuts and groundnuts, and even from coconuts, has been used in test flights of aircraft ranging from strategic bombers to jetliners.

Biogas and Geothermal Energy

Hydrocarbon gases from decaying vegetation form large underground deposits that have been exploited as sources of energy for many years. Technology similar to that used in extracting energy from these natural deposits, which are not considered a renewable energy source, can be used to tap the smaller but widely distributed emissions of methane-rich waste gases from compost pits and landfills. Creating the necessary infrastructure to capture these gases over large areas poses a difficult engineering challenge, however. In addition, care must be taken to avoid the release of methane from these deposits into the atmosphere, as methane is considered to be twenty times as harmful as carbon dioxide as a greenhouse gas.

Geothermal energy comes from heat released by radioactive decay inside the earth's core, perhaps augmented by gravitational pressure. Where such heat is released gradually through vents in the earth's surface, rather than in volcanic eruptions, it forms an abundant and steady, reliable, long-term source of thermal power. Hot springs and geothermal steam generation are used on a large scale in Iceland, and geothermal power is used in some American communities and military bases.

Narayanan M. Komerath and Padma P. Komerath

Further Reading

Charlier, R. H., and C. W. Finkl. *Ocean Energy: Tide and Tidal Power.* London: Springer, 2009.

Edwards, Brian K. *The Economics of Hydroelectric Power.* Northampton, Mass.: Edward Elgar, 2003.

Klass, Donald L. *Biomass for Renewable Energy, Fuels, and Chemicals.* San Diego, Calif.: Academic Press, 1998.

Pollan, Michael. *The Omnivore's Dilemma: A Natural History of Four Meals.* New York: Penguin Press, 2007.

Traynor, Ann J., and Reed J. Jensen. "Direct Solar Reduction of CO_2 to Fuel: First Prototype Results." *Industrial and Engineering Chemistry Research* 41, no. 8 (2002): 1935-1939.

Vaitheeswaran, Vijay. *Power to the People: How the Coming Energy Revolution Will Transform an Industry, Change Our Lives, and Maybe Even Save the Planet.* New York: Farrar, Straus and Giroux, 2003.

Walker, John F., and Nicholas Jenkins. *Wind Energy Technology.* New York: John Wiley & Sons, 1997.

Wenisch, A., R. Kromp, and D. Reinberger. *Science or Fiction: Is There a Future for Nuclear?* Vienna: Austrian Ecology Institute, 2007.

SEE ALSO: Alternative fuels; Biomass conversion; Ethanol; Geothermal energy; Hydroelectricity; Hydrogen economy; Nuclear fusion; Photovoltaic cells; Renewable energy; Solar energy; Tidal energy; Wind energy.

Alternative fuels

CATEGORY: Energy and energy use

DEFINITION: Materials or substances that can be substituted for commonly used fossil fuels

SIGNIFICANCE: The development of alternatives to fossil fuels (gasoline, diesel, natural gas, and coal) has been spurred by growing awareness of the environmental damage associated with the burning of fossil fuels, as well as by the knowledge that at some time in the future the earth's supply of fossil fuels will be exhausted.

With the exception of nuclear-powered seagoing vessels, most vehicles are powered by internal combustion engines that use either gasoline or diesel fuel. Gasoline and diesel release significant amounts of greenhouse gases into the atmosphere when burned; these gases include water vapor, carbon dioxide, ozone, nitrous oxide, and methane. These gases absorb and emit radiation in the infrared range; thus they increase the earth's temperature. In addition to the fact that the internal combustion engine burns an environment-polluting fossil fuel, it also is an inefficient method for transferring the energy stored in the fuel into propulsion. Most of the stored energy is lost in heat, which escapes through the exhaust pipe. In addition, the pistons within the engine accelerate up, stop, accelerate down, and stop with each revolution. This rapid cycle of acceleration and deceleration wastes energy. Many of the alternative fuels available are used to power internal combustion engines and thus have the same limitations as fossil fuels in this regard. Using an alternative fuel such as stored electricity does not have these limitations because it does not produce heat and it propels an electric motor, which rotates (no starting and stopping with each cycle).

Comparisons of the costs and levels of pollutant production of nonfossil fuel sources must take into account the costs associated with production of the fuels. An example is the use of corn for the production of ethanol. Raising the crop requires energy for production, such as fuel for tractors. The corn must then be fermented (yeast converts the sugar in the corn into ethanol), and the fermented product must be distilled (boiled to release the alcohol), which requires fuel to heat the still. The process increases the cost of ethanol and also produces pollution. The electricity that charges an electric vehicle may have been produced by a fossil-fuel source such as a diesel generator. Another problem with alternative fuels lies in the difficulty consumers may have in replenishing their supplies. Facilities distributing gasoline and diesel are prevalent throughout most developed nations; in contrast, sources of alternative fuels such as hydrogen and ammonia are not readily available. The ideal alternative fuel is one that is nonpolluting, cheap to produce, and easy to replenish.

BIOFUELS

Biofuels are derived from plant sources such as corn, sugarcane, and sugar beets; in some cases, they are blended with a fossil fuel, usually gasoline. Alcohol, methanol, butanol, biodiesel, biogas, and wood gas are all examples of biofuels.

Alcohol was initially used as a fuel in the Ford Model T automobile, which was first produced in 1908. The carburetor (a device that mixes fuel with air prior to entry into the engine) of the Model T could be adjusted to burn gasoline, ethanol, or a mixture. Many modern-day vehicles can run on a mixture of 10-15 percent ethanol and gasoline (E10, or gasohol, is 10 percent alcohol). The fuel known as E85 is a mixture of 85 percent ethanol and 15 percent gasoline; this fuel can be used only in flexible-fuel vehicles (FFVs).

FFVs are designed to run on gasoline, E85, or any other gasoline-ethanol mixture. A disadvantage of ethanol is that it has approximately 34 percent less energy per volume than gasoline. Because ethanol has a high octane rating, ethanol-only engines may have relatively high compression ratios, which increases efficiency. In developed nations such as the United States, ethanol blends are available in many areas. Critics of ethanol as an alternative fuel note that it requires a large amount of agricultural land, which is diverted from producing crops used for food; also, the use of crops such as corn for ethanol production drives up food prices.

Methanol can be used as an alternative fuel, but automakers have not yet produced any vehicles that can run on it. Butanol is more similar to gasoline than

A groundskeeper fills a mowing/snow-removal machine with biodiesel fuel at Sinclair Community College in Dayton, Ohio. The fuel is made on campus from used cooking oil that comes from the college's kitchens. (AP/Wide World Photos)

Reagan Signs the Alternative Motor Fuels Act

President Ronald Reagan made the following statement before signing the Alternative Motor Fuels Act of 1988:

Well, Members of Congress and distinguished guests, good afternoon. We're here today to sign into law an investment in America's future: the Alternative Motor Fuels Act of 1988. This bill is a landmark in the quest for alternative forms of energy. And believe me, when you're my age you just love hearing about alternative sources of energy.

I'm particularly proud this afternoon because I remember more than 4 years ago, at a Cabinet meeting in January 1984, and I asked Vice President George Bush to launch a thorough investigation of alternative energy and see what he could find—not pie-in-the-sky demonstration projects but real-world possibilities and realistic options that would help keep our air clean and our nation less dependent on foreign oil.

That's what's so exciting about the bill before us today: The forms of energy encouraged by this bill are already in use. Methanol, for example, is used in the Indianapolis 500 and in other race cars because it simultaneously enhances performance and safety. And cars that run on methanol have the potential to reduce emissions by an amazing 50 percent and improve efficiency. For areas like southern California, that could be a Godsend. A few months ago, Vice President Bush dedicated the first methanol pump on Wilshire Boulevard in Los Angeles. And this bill gives American automobile companies a real incentive to start building cars powered by alternative fuels by adjusting the federally mandated average fuel economy ratings to reflect the gasoline saved by these vehicles.

This legislation also opens up new markets for natural gas and coal, our two most plentiful energy resources in this country. The success of these projects could improve employment and the economies in the hard-pressed oil- and gas-producing areas of the country. This bill takes advantage of existing government programs and mechanisms to assist alternative fuels. Most important, it's not intended to create massive new bureaucracies or new taxpayer subsidies.

ethanol and can be used in some engines designed for use with gasoline without modification.

Biodiesel can be manufactured from vegetable oils and animal fats, including recycled restaurant grease. It is slightly more expensive than diesel; however, it is a safe, biodegradable fuel that produces fewer pollutants than diesel. Diesel engines are more efficient than gasoline engines (44 percent versus 25-30 percent efficiency); thus they have better fuel economy than gasoline engines. Some diesel engines can run on 100 percent biodiesel with only minor modifications. Biodiesel can be combined with regular diesel in various concentrations (for example, B2 is 2 percent biodiesel, B5 is 5 percent biodiesel, and B20 is 20 percent biodiesel).

Biogas is produced by the biological breakdown of organic materials—for example, rotting vegetables,

plant wastes, and manure produce biogas—and the energy produced varies depending on the source. Biogas can replace compressed natural gas for fueling internal combustion engines. Wood gas is another biofuel that can power an internal combustion engine. It is produced by the incomplete burning of sawdust, wood chips, coal, charcoal, or rubber. Depending on the source, the gas produced varies in energy content and contaminants. Contaminants in wood gas can foul an engine.

ELECTRIC VEHICLES

At the beginning of the twentieth century, automobiles powered by steam, gasoline, and electricity were available. Electric vehicles were popular into the early 1920's, but then the automotive industry became dominated by gasoline-powered vehicles. The decline in electric-powered vehicles occurred for several reasons: Road improvement allowed travel over longer distances, and the range of electric vehicles was limited; fossil fuels became cheap and plentiful; the electric starter replaced the hand crank on gasoline engines, which greatly simplified starting such engines; and mass production of automobiles by Henry Ford's company made gasoline-powered vehicles much less expensive than electric-powered vehicles ($650 versus $1,750 average price at that time). By the end of the twentieth century, a growing emphasis on environmentally friendly energy sources encouraged the reemergence of electric vehicles and the development of hybrid vehicles powered by both gasoline (or diesel) and electricity.

A hybrid vehicle contains an electric motor that can both propel the vehicle and recharge the battery. Hybrid vehicles have achieved greater popularity than electric-only vehicles, as electric-only vehicles continue to have some of the same basic problems as earlier electric cars: limited range and higher cost than gasoline-powered or hybrid vehicles. Public recharging facilities for electric vehicles remain few and far between; furthermore, recharging takes time. The latest electric and hybrid vehicles use lithium-ion batteries rather than the lead-acid batteries used by earlier versions (and still used in gasoline and diesel vehicles). Lithium-ion batteries are much lighter than lead-acid batteries and can be molded into a variety of shapes to fit available areas. One criticism of electric vehicles, including hybrids, is that many are small and lightweight and thus less safe for passengers, in the case of collisions, than are larger gasoline-powered vehicles.

Other Fuels Derived from Nonfossil Sources

Ammonia has been evaluated for use as an alternative fuel. It can run in either a spark-ignited engine (that is, a gasoline engine) or a diesel engine in which the fuel-air mixture ignites upon compression in the cylinder. Modern gasoline and diesel engines can be readily converted to run on ammonia. Although ammonia is a toxic substance, it is considered no more dangerous than gasoline or liquefied petroleum gas (LPG). Ammonia can be produced by electrical energy and has half the density of gasoline or diesel; thus it can be placed in a vehicle fuel tank in sufficient quantities to allow the vehicle to travel reasonable distances. Another advantage of ammonia is that it produces no harmful emissions; upon combustion, it produces nitrogen and water.

Compressed-air engines are piston engines that use compressed air as fuel. Air-engine-powered vehicles have been produced that have a range comparable to gasoline-powered vehicles. Compressed air is much less expensive than fossil fuels. Ambient heat (normal heat in the environment) naturally warms the cold compressed air upon the air's release from the storage tank, increasing its efficiency. The only exhaust is cold air, which can be used to cool the interior of the vehicle.

Hydrogen vehicles can be powered by the combustion of hydrogen in the engine much as the typical gasoline engine operates. Fuel cell conversion is another method of using hydrogen; in this type of vehicle, the hydrogen is converted to electricity. The most efficient use of hydrogen to power motor vehicles involves the use of fuel cells and electric motors. Hydrogen reacts with oxygen inside the fuel cells, which produces electricity to power the motors. With either method no harmful emissions are produced, as the spent hydrogen produces only water. Hydrogen is much more expensive than fossil fuels, and it contains

significantly less energy on a per-volume basis, meaning that the vehicle's range is reduced. Experimental fuel cell vehicles have been produced, but such vehicles remain far too expensive for the average consumer.

Liquid nitrogen (LN_2) contains stored energy. Energy is used to liquefy air, then LN_2 is produced by evaporation. When LN_2 warms, nitrogen gas is produced; this gas can power a piston or turbine engine. Nitrogen-powered vehicles have been produced that have ranges comparable to gasoline-powered vehicles; these vehicles can be refueled in a matter of minutes. Nitrogen is an inert gas and makes up about 80 percent of air. It is virtually nonpolluting. Furthermore, it produces more energy than does compressed air.

Oxyhydrogen is a mixture of hydrogen and oxygen gases, usually in a 2:1 ratio, the same proportion as water. Oxyhydrogen can fuel internal combustion engines, and, as in hydrogen-fueled engines, no harmful emissions are produced.

Steam was a common method of propulsion for vehicles during the early twentieth century, but, like electricity, it fell into disfavor with the advent of the electric starter, cheap gasoline, and mass production of Ford automobiles. A disadvantage of steam-powered vehicles is the time required to produce the steam. A steam engine is an external combustion engine—that is, the power is produced outside rather than inside the engine. Steam engines are less energy-efficient than gasoline engines. Fuel for steam engines can be derived from fossil fuels or from nonfossil fuel sources.

Alternative Fossil Fuels

Some fossil fuels are less polluting than gasoline or diesel, and some are in plentiful supply. Natural gas vehicles use compressed natural gas (CNG) or, less commonly, liquefied natural gas (LNG). Internal combustion engines can be readily converted to burn natural gas. Natural gas is 60-90 percent less polluting than gasoline or diesel and produces 30-40 percent less greenhouse gases. Furthermore, it is less expensive than gasoline. Limitations of natural gas vehicles include a lack of available fueling stations and limited space for fuel, given that natural gas must be stored in cylinders, which are commonly located in the vehicle's trunk.

Liquefied petroleum gas is suitable for fueling internal combustion engines. Like natural gas, LPG is less polluting than gasoline, with 20 percent less carbon dioxide emissions; it is also less expensive. LPG is added to a vehicle's fuel tank through the use of a specialized filling apparatus; a limitation to LPG is the lack of fueling stations.

Robin L. Wulffson

FURTHER READING

DeGunther, Rik. *Alternative Energy for Dummies.* Hoboken, N.J.: John Wiley & Sons, 2009.

Gibilisco, Stan. *Alternative Energy Demystified.* New York: McGraw-Hill, 2007.

Hordeski, Michael F. *Alternative Fuels: The Future of Hydrogen.* 2d ed. Lilburn, Ga.: Fairmont Press, 2008.

Lee, Sunggyu, James G. Speight, and Sudarshan K. Loyalka. *Handbook of Alternative Fuel Technologies.* Boca Raton, Fla.: CRC Press, 2007.

Nersesian, Roy L. *Energy for the Twenty-first Century: A Comprehensive Guide to Conventional and Alternative Sources.* New York: M. E. Sharpe, 2007.

SEE ALSO: Air pollution; Alternative energy sources; Alternatively fueled vehicles; Automobile emissions; Carbon dioxide; Carbon monoxide; Electric vehicles; Fossil fuels; Hybrid vehicles; Methane; Synthetic fuels.

Alternative grains

CATEGORY: Agriculture and food

DEFINITION: Grains cultivated for food as alternatives to traditional high-yield grain crops

SIGNIFICANCE: Various minor cereals, new cereals, and pseudocereals can be cultivated in climates that are not conducive to the growing of other high-yield grain crops, providing alternative food sources and avoiding the environmental damage caused by the fertilizers and pesticides used in large-scale intensive farming.

More than one-half of the calories consumed daily by the world's human population come from grains. Most of these grains are produced by plants of the grass family, Poaceae. Major cereal plants domesticated many centuries ago include rice (*Oryza sativa*), wheat (*Triticum aestivum*), and maize or corn (*Zea mays*). Other important grain crops, also plants of the grass family, include barley (originating in Asia), millets and sorghum (originating in Africa), and oats and rye (originating in Europe).

Since the early twentieth century, the scientific principles of genetics have been applied to improvements of crop plants, with some of the most notable improvements occurring between 1940 and 1970. As a result of irrigation, improved genetic varieties, and the use of large amounts of fertilizers and pesticides, yields of major crops greatly increased. Norman Borlaug received the Nobel Peace Prize in 1970 for his contributions to these developments, which came to be called the Green Revolution. However, it soon became apparent that the Green Revolution was not the boon first envisioned. For maximum yield, large-scale farming involving huge investments of capital is required. Also, environmentalists became concerned over the resulting erosion and the environmental damage caused by the use of large amounts of fertilizer and pesticides.

Various alternatives to such farming have been proposed. In the case of grain crops, several approaches offer promise, including more widespread use of minor cereals, especially those tolerant of unfavorable growing conditions; development of new cereal plants through hybridization or other genetic manipulations; and utilization of pseudocereals, nongrass crop plants that produce fruits (grains) similar to those of cereal plants.

MINOR CEREALS AND NEW CEREALS

Most sorghum (*Sorghum bicolor*) grown in the United States is used for silage (milo) or molasses (sweet sorghum). In Africa and India, various grain sorghums are grown in regions where rainfall is insufficient for most other grain crops. Well adapted to hot, dry climates, these grains are used to make a pancakelike bread. "Millet" refers to several grasses that are useful cereal plants because they also tolerate drought well. In Africa the most important are pearl millet (*Pennisetum glaucum*) and finger millet (*Eleusine coracana*). Grains of both species can be stored for long periods and are used to make breads and other foods. Other, perhaps less important, grain plants also called millet include foxtail millet (*Setaria italica*), native to India but now grown in China; proso millet (*Pamicum milaeceum*), native to China but grown in Russia and central Asia; sanwa millet (*Echinochloa frumentacea*), cultivated in East Asia; and teff (*Eragrostis teff*), an important food and forage plant of Ethiopia. Such grain sorghums and millets have the potential to grow in areas with hot, dry climates far beyond the regions where they are now being utilized.

In a distinct category is wild rice (*Zizania aquatica*). Native to the Great Lakes region of the United States and Canada, it has been, and still is, harvested by American Indians. Like the common (but unrelated) rice, it grows in flooded fields. Attempts to cultivate wild rice since the 1950's have been somewhat successful as the result of the development of nonshattering varieties. However, it remains an expensive, gourmet item.

Two cereal plants have promise because of the high protein content of their grains. Wild oat (*Avena sterilis*) is a disease-resistant plant with large grains. Job's tears (*Coix lachryma-jobi*), native to Asia, is now planted throughout the Tropics. Research on these and related species continues.

Although all important cereal plants have been improved, either by genetic engineering or by more conventional genetic techniques, the most notable new alternative grain plant is triticale (*Triticosecale* sp.). The first human-made cereal, it is the result of crossing wheat with rye. The sterile hybrid from such a cross was made fertile through the doubling of its chromosomes; thus triticale varieties produce viable seeds. Triticale combines the superior traits of each its parents: the cold tolerance of rye and the higher yield of wheat. The protein content of triticale compares favorably with that of wheat, and its quality, as measured by lysine content, is higher. However, flour made from triticale is inferior for making bread unless mixed with wheat flour.

PSEUDOCEREALS

Pseudocereals are plants that are not in the grass family but that produce nutritious hard, grainlike fruits that can be stored, processed, and prepared for food much like grains. They belong to several plant families. Many grow under conditions not suitable for the major cereal crops. Buckwheat (*Fagopyrum esculentum*), of the buckwheat family, Polygonaceae, probably originated in China. It tolerates cool conditions and is adapted to short growing seasons, thus permitting it to be grown in the temperate regions of North America and Europe. In the United States it is often associated with pancakes, but it is used in larger quantities for livestock feed. In Eastern Europe, the milled grain is used for soups.

Quinoa (*Chenopodium quinoa*), of the goosefoot family, Chenopodiaceae, has been cultivated by Indians of the Andes mountains for centuries. The leafy annual produces grainlike fruits (actually achenes) with a high protein content and exceptional quality (high in lysine and other essential amino acids). After its bitter saponins have been removed, quinoa can be cooked and eaten like rice or made into a flour. Quinoa has been cultivated in the Rocky Mountains of Colorado since the 1980's and has become a gourmet food in the United States.

Most amaranth (*Amaranthus* sp.) plants are New World weeds. They belong to the amaranth family, Amaranthaceae. A few species were used by Aztecs and other native peoples, but their use was banned by the Spanish. Since the late 1970's, plant breeders have targeted several species for improvement. The results are highly nutritious grains rich in lysine that are suitable for making flour. Research in Pennsylvania and California has resulted in improved varieties.

Thomas E. Hemmerly

FURTHER READING

Carver, Brett F., ed. *Wheat: Science and Trade*. Ames, Iowa: Wiley-Blackwell, 2009.

Levetin, Estelle, and Karen McMahon. *Plants and Society*. 5th ed. Boston: McGraw-Hill Higher Education, 2008.

Schery, Robert W. *Plants for Man*. 2d ed. Englewood Cliffs, N.J.: Prentice Hall, 1972.

Simpson, Beryl B., and Molly C. Ogorzaly. *Economic Botany: Plants in Our World*. 3d ed. New York: McGraw-Hill, 2001.

SEE ALSO: Food and Agriculture Organization; Genetically modified foods; Green Revolution; High-yield wheat; Sustainable agriculture.

Alternatively fueled vehicles

CATEGORIES: Energy and energy use; atmosphere and air pollution

DEFINITION: Vehicles fueled, wholly or partially, by energy derived from sources other than petroleum

SIGNIFICANCE: The development of vehicles that operate efficiently using little or no petroleum-based fuel is an important part of efforts to address the problems of declining petroleum resources and the pollution caused by emissions from petroleum-based fuels.

Almost all of the fuel used for transportation in the United States is derived from petroleum. In California, a huge consumer of fuel to run its millions of vehicles, a mere one-fourth of 1 percent of those vehicles used alternative fuel sources in 2010. As the world's oil fields become depleted and petroleum reserves shrink, and as awareness grows regarding the harm to the environment caused by the burning of fossil fuels, the pressure to develop alternative fuels and vehicles that can operate using them has become intense. Since the late twentieth century, significant progress has been made in this area: Between 1970 and 2010, advances in automotive technologies decreased toxic vehicle emissions by an estimated 90 percent. These advances included the development of vehicles powered by non-petroleum-based fuels.

Auto manufacturers long resisted producing vehicles that would not be dependent on petroleum-based energy sources; for many years, they rejected alternative technologies in favor of the entrenched gasoline-burning internal combustion engine. Significant gasoline shortages in 1973 and 1979, however, clearly demonstrated the need for mass production of vehicles that could be powered by renewable resources. Whereas automobile companies and the petroleum industry had previously discouraged technologies related to the development of practical electric vehicles, many such corporations came to recognize the need to explore and encourage the production of alternative fuel sources.

Hybrid and Electric Vehicles

By the late twentieth century, automobile manufacturers, first in Japan but later in the United States and Europe, turned their attention to producing reasonably priced, fuel-efficient, nonpolluting vehicles. The most popular of such vehicles are hybrid vehicles; by 2010, nearly twenty versions of such vehicles were available commercially.

Hybrids such as Toyota's Prius and Honda's Insight are powered by gasoline-fueled internal combustion engines in combination with electric motors. These motors, which receive their electricity from batteries, provide power for the automobiles, but when they run down, small internal combustion engines take over. The batteries in hybrid vehicles are recharged by the friction created every time the brakes are applied. The ranges of such vehicles—that is, the distances that they can be driven between refuelings—are comparable to those of conventional vehicles.

Plug-in electric vehicles have a much more limited range than do hybrids; most can be driven only about 160 kilometers (100 miles) before they require recharging. Such vehicles are useful for service within limited areas, where they can be driven for short periods and recharged when not in use. Recharging a plug-in electric vehicle's batteries fully can take four to eight hours. Anticipated improvements in electric vehicle technology include drastic shortening of the time required to recharge batteries and enhancement of battery life to extend the vehicles' range.

Other Fuels for Internal Combustion Engines

The conventional internal combustion engines found in most vehicles can often be run on nonpetroleum fuels or on mixtures of petroleum-based and other fuels. One frequently used alternative fuel is ethanol, an alcohol derived from plants that can be mixed with conventional gasoline. Such mixtures usually consist of 5 percent ethanol and 95 percent conventional gasoline, however, so the gasoline savings are negligible.

Another nonpetroleum fuel is hydrogen, which can be used to fuel slightly modified existing vehicles. As of 2010, more than one hundred buses fueled by hydrogen were operating in the United States, Canada, Mexico, Brazil, Japan, Egypt, Iceland, and India. Hydrogen has the advantage of producing water as its sole emission, so it does not contribute to air pollution. The major problem with using hydrogen is that it is not readily available to consumers. Proponents of hydrogen as a motor vehicle fuel assert that successful solution of the problem of distribution might revolutionize how vehicles are fueled.

Although hydrogen is among the most plentiful elements in the universe, it is not available merely for the taking. Because it bonds with other elements, such as oxygen, it is a carrier of energy rather than a source of energy like petroleum or coal. To power a vehicle, hydrogen must pass through onboard fuel cells in which a chemical reaction reduces water into its component parts, hydrogen and oxygen. This hydrogen, available in a free state, is converted to electricity and used to fuel the vehicle. When hydrogen is used in its gaseous state, leakage is a problem. Hydrogen can be liquefied and can also exist in powder form. Its most efficient use in fueling vehicles is usually in its gaseous or liquefied form.

Liquefied natural gas, which is less expensive to produce than gasoline, is used to fuel some fleets of

Stephanie White explains how her hydrogen fuel cell-powered car is refueled—one cable goes to a data port at the rear, and the filler hose goes into the side—during a stop on the Hydrogen Road Tour 2009, a nine-day road rally of cars powered by hydrogen fuel. (AP/Wide World Photos)

taxicabs and other commercial vehicles. Liquefied petroleum gas, or propane, is also used to provide the power for many commercial fleets. It is less expensive and less polluting than regular gasoline.

R. Baird Shuman

FURTHER READING

Chan, C. C., and K. T. Chau. *Modern Electric Vehicle Technology.* New York: Oxford University Press, 2001.

Ehsani, Mehrdad, et al. *Modern Electric, Hybrid Electric, and Fuel Cell Vehicles: Fundamentals, Theory, and Design.* Boca Raton, Fla: CRC Press, 2005.

Erjavec, Jack, and Jeff Arias. *Hybrid, Electric, and Fuel-Cell Vehicles.* Clifton Park, N.J.: Thomson Delmar Learning, 2007.

Halderman, James D., and Tony Martin. *Hybrid and Alternative Fuel Vehicles.* Upper Saddle River, N.J.: Pearson/Prentice Hall, 2009.

Hanselman, Duane C. *Brushless Permanent-Magnet Motor Design.* New York: McGraw-Hill, 1994.

King, Nicole Bezic, ed. *Renewable Energy Resources.* North Mankato, Minn.: Smart Apple Media, 2004.

Lee, Sunggyu, James G. Speight, and Sudarshan K. Loyalka. *Handbook of Alternative Fuel Technologies.* Boca Raton, Fla.: CRC Press, 2007.

SEE ALSO: Air pollution; Alternative fuels; Automobile emissions; Catalytic converters; Electric vehicles; Ethanol; Fossil fuels; Hybrid vehicles; Internal combustion engines; Smog; Synthetic fuels.

Amazon River basin

CATEGORIES: Places; ecology and ecosystems; forests and plants

IDENTIFICATION: River basin in South America bounded by the Guiana Highlands to the north and the Brazilian Highlands to the south

SIGNIFICANCE: The Amazon River basin contains the world's largest rain forest, in which one-tenth of the known species in the world are found. Because of the unique biodiversity and geography of the region, it has been the subject of mass efforts by international conservation groups and an ongoing topic in discussions regarding global environmental protection and sustainability. Environmentalists and human rights advocates argue that the destruction of habitat in the basin will result in losses not only in biodiversity but also in human lives, as many indigenous peoples still live in the forest and rigidly continue their ways of life, refusing to integrate into modern society.

The Amazon River basin is best known for two characteristics: its high biodiversity and its enormous vegetative productivity. These characteristics give the basin a special significance to the biological world. The rain forest itself produces an estimated 20 percent of the earth's oxygen. The Amazon River, which is some 6,400 kilometers (4,000 miles) long, has the largest drainage basin in the world, accounting for nearly one-fifth of total global river flow. Further, this basin harbors the largest concentration of animal and plant species on the planet. In addition to the self-evident reasons to protect the nonhuman species of the basin from extinction, the basin holds great economic potential for the discovery of important medicines and plant-based products.

Many people live in the Amazon basin, including farmers and loggers with South American national ties and groups of indigenous peoples. Most of the area's original indigenous population was wiped out by disease and conflict during and after the European invasion; from an estimated pre-Columbian population of some millions, the numbers have been reduced to roughly two hundred thousand. Many of these natives have cultures and languages that are still alive and vibrant, but they remain under threat as their land and resources are coveted by outsiders, both South American nationals and prospectors from abroad.

ECOLOGICAL DIVERSITY

Much of the Amazon basin remains unexplored and therefore certainly uncataloged, and only estimates exist for total species numbers. However, more than four hundred species of mammals, twelve hundred species of birds, three thousand species of fish, four hundred species of amphibians, three hundred species of reptiles, two and one-half million species of insects, and forty thousand species of plants have been described or projected. The density, inaccessibility, and harshness of the terrain all make species cataloging difficult, but efforts are ongoing, and the numbers continue to grow.

Flagship species include the Brazilian tapir, a large forest ungulate with a prehensile snout; the jaguar, a cousin of the leopard; the golden lion tamarin, a small and charismatic monkey with striking orange hair; the scarlet macaw, a multicolored large parrot; the quetzal, a brilliantly green frugivorous (fruit-eating) bird; the poison dart frog, the toxins of which are used by some indigenous groups to tip the darts they shoot through blowguns; the anaconda, a very large constricting snake that can grow to up to 9 meters (30 feet) long; and the howler monkey, which makes vocalizations that can be heard through the forest for up to 5 kilometers (3 miles). Nearly all these species are under threat of extinction, but a plenitude of other lesser known species suffer the same threat, and many of those unknown to science must be under similar circumstances. The dangers to Amazon species include habitat destruction, poaching and the pet trade, hunting for food and medicine, and the impending yet unclear effects of climate change.

CULTURAL DIVERSITY

The human population of the Amazon River basin prior to the arrival of Europeans in the sixteenth century was somewhere on the order of five million, with around five thousand tribes. Nearly five hundred years later and after exploitation, disease, and violence, about half a million individuals remain living in the basin in traditional ways, with approximately five hundred tribes, seventy-five of which are uncontacted and live in voluntary isolation. Great linguistic diversity exists in the basin, where more than thirty language families are recognized. With the advent of industrial logging and oil extraction, conflicts between prospectors and indigenous Amazonians began to rise, and many still go unreported in extremely remote areas.

(continued on page 56)

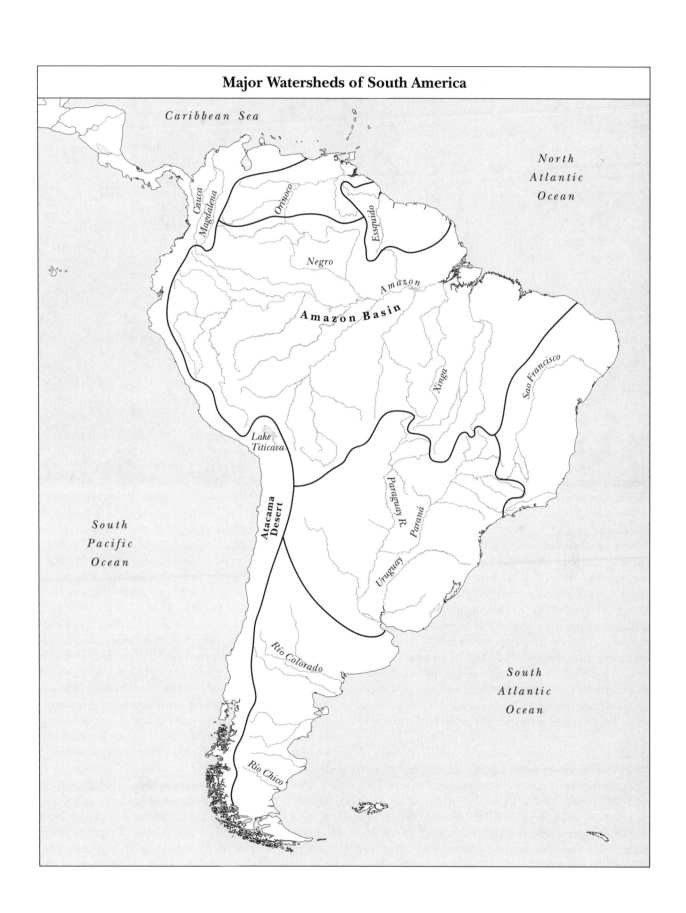

Major Watersheds of South America

Caribbean Sea

North
Atlantic
Ocean

Cauca

Magdalena

Orinoco

Essquido

Negro

Amazon

Amazon Basin

Xingu

Sao Francisco

Lake
Titicaca

Atacama
Desert

Paraguay R.

Paraná

Uruguay

South
Pacific
Ocean

South
Atlantic
Ocean

Río Colorado

Río Chico

The Amazon River basin is known for its high biodiversity and its enormous vegetative productivity. (©Paura/Dreamstime.com)

Groups such as the U.S.-based Amazon Watch fight for the legal rights of the indigenous peoples to own their land and to dictate who may and may not enter and extract resources. This movement is separate from the ecological conservation movement and is better characterized as an environmental justice movement, but the two necessarily go hand in hand as the indigenous peoples depend on forest resources to continue their ways of life. Certainly, not all indigenous Amazonians live in the forest; many have moved to cities and have chosen modern lifestyles. People with indigenous blood make up significant portions of the populations of most South American countries.

HABITAT DESTRUCTION

In the Amazon River basin, habitat destruction usually takes the form of deforestation. Between 1991 and 2000, the total area of forest lost rose from 415,000 to 587,000 square kilometers (160,000 to 227,000 square miles), nearly six times the size of Portugal. Although the logging and oil industries are responsible for much of the degradation of the land-

scape, a large percentage is the result of the conversion of forest to pasture and farmland by local people. Most of the formerly forested land is used for livestock pasture, as grass grows well even in poor Amazonian soils and raising livestock has relatively high returns for minimal effort. Long-term, large-scale farming is usually ineffective because of the paucity of nutrients in the soil, and therefore slash-and-burn agriculture is widely practiced. This involves logging all the trees in an area, setting a fire and letting it burn through the remaining tree and other plant growth, then planting crops in the resulting ash, which provides temporary nutrients. The ability to grow crops on the land lasts for only a short time, so a slash-and-burn farmer must move to a new spot every few years, thus accelerating deforestation.

Another important contributor to deforestation has been the Trans-Amazonian Highway, which cuts straight through the rain forest from Brazil to Peru and opens up much of the surrounding regions to farming and resource extraction. It has been projected that if the rates of deforestation seen in the early twenty-first century continue, in two decades the extent of the Amazon rain forest will have decreased by 40 percent. Many innovative projects have been developed to oppose unsustainable use of the forest, one of the most notable of which is the ecovillage of Gaviotas in Colombia, which was founded in 1971 as part of an effort to build and develop a sustainable model for a modern Amazonian village.

Jamie Michael Kass

Further Reading

Benjamin, Alison. "More than Half of the Amazon Will Be Lost by 2030, Report Warns." *Guardian*, December 6, 2007.

London, Mark, and Brian Kelly. *The Last Forest: The Amazon in the Age of Globalization.* New York: Random House, 2007.

Sheil, Douglas, and Sven Wunder. "The Value of Tropical Forest to Local Communities: Complications, Caveats, and Cautions." *Conservation Ecology* 6, no. 2 (2002): 9-24.

Slater, Candace. *Entangled Edens: Visions of the Amazon.* Berkeley: University of California Press, 2002.

Smith, Nigel J. H. *The Amazon River Forest: A Natural History of Plants, Animals, and People.* New York: Oxford University Press, 1999.

See also: Biodiversity; Logging and clear-cutting; Rain forests; Rainforest Action Network; Slash-and-burn agriculture; South America; Sustainable forestry.

American alligator

Category: Animals and endangered species

Definition: Crocodilian species native to the southeastern United States

Significance: Once an endangered species, the American alligator made a striking comeback in the last few decades of the twentieth century, thanks to conservation efforts.

The American alligator (*Alligator mississippensis*) is a surviving member of the crocodilians—a family of reptiles that roamed the earth along with dinosaurs

Conservation efforts in the late twentieth century brought the American alligator back from endangered status. (©Vanessagifford/ Dreamstime.com)

230 million years ago. Ranging between 2.5 and 4.6 meters (8 to 15 feet) in length and weighing up to 450 kilograms (1,000 pounds), it is the largest reptile in North America.

The American alligator inhabits coastal areas, swamps, ponds, and marshes of the southeastern United States, from North Carolina to Florida and along the Gulf Coast to Texas. Spanish explorers were the first Europeans to come across the American alligator, which they named *el lagarto* ("the lizard"), an expression Americans later turned into "alligator." Various descriptions of alligators by eighteenth century trappers, explorers, adventurers, and naturalists quickly earned the reptile a place in legend and folklore. Americans and Europeans alike became fascinated with false, but entertaining, portrayals of the reptile as an almost mythological, smoke-breathing dragon.

Although white southerners and Native Americans hunted the alligator for centuries, the reptile faced no serious, widespread threat as a species until the 1870's, when a worldwide demand arose for its soft hides, which were turned into belts, hats, shoes, and handbags. Their value grew in the following decades, and by the 1940's alligator populations were so dangerously reduced that southern states outlawed the hunting and trapping of the reptiles. These actions, however, stimulated illegal poaching during the 1960's and drove the alligator to the brink of extinction.

In 1973 the U.S. government put the alligator—along with its cousin, the American crocodile (*Crocodylus acutus*)—on the endangered species list and banned the trafficking of its hides. The protection worked so well that alligators are no longer regarded as endangered. Open hunting of alligators is still illegal, though Florida and Louisiana allow permitted hunts to control alligator populations and protect fur-bearing animals.

New problems, however, confront the American alligator. Land development continues to destroy its natural habitat, but human activity also creates new artificial living spaces for alligators in canals and drainage ditches. These new environs often put alligators in close proximity to humans. As a result, alligators commonly appear in the swimming pools or yards of private homes, on docks, at highway toll booths, and even at schools and shopping centers, and their encounters with humans are on the rise. Though alligators feed mainly on fish, snails, crabs, amphibians, and small mammals, they also occasionally consume dogs, cats, and even calves. Attacks on humans, especially small children, do occur, though they are rare.

Pollution also threatens alligators. In 1996, researchers from the University of Florida reported serious reproductive problems among alligators in Lake Apopka, Florida's third-largest lake. The researchers suspected that the fertility problems were the result of a 1980 chemical spill and agricultural pesticide runoff into the lake. According to one University of Florida study, Lake Apopka lost 90 percent of its alligator population in a recent twenty-year period. More research is needed to determine whether the Lake Apopka situation is an isolated case or an indicator of a wider problem that could again put alligators in grave danger.

John M. Dunn

FURTHER READING

Fujisaki, Ikuko, et al. "Possible Generational Effects of Habitat Degradation on Alligator Reproduction." *Journal of Wildlife Management* 71 (September, 2007): 2284-2289.

Lockwood, C. C. *The Alligator Book.* Baton Rouge: Louisiana State University Press, 2002.

SEE ALSO: Agricultural chemicals; Endangered Species Act; Endangered species and species protection policy; Hunting; Poaching.

American bison

CATEGORY: Animals and endangered species

DEFINITION: Large, oxenlike animal native to North America

SIGNIFICANCE: The American bison was nearly exterminated during the late nineteenth century, but efforts to protect the species—first undertaken in 1870—became one of the first North American environmental success stories.

The American bison (often commonly called the American buffalo) is a large, oxenlike animal with a large head, well-developed front quarters, pronounced hump, and thin rear quarters. The ancestors of the American bison originated in the colder climates of Eurasia and made their way from Siberia to Alaska across the Bering land bridge during the ice

A herd of American bison in the plains near Delta Junction, Alaska. (©iStockphoto.com)

ages three hundred thousand to six hundred thousand years ago. They underwent several changes to evolve into the modern species about five thousand years ago and to become one of the most abundant land animals of all time. By 1500, some thirty million were in North America, mostly on the plains between the Mississippi River and the Rocky Mountains, and between what are now Canada and Mexico. The species owed its success to its abilities to subsist on the grasses of the plains, to survive often bitterly cold and snowy winters, and to outrun all predators, including human beings.

After 1500, with the arrival of the horse in North America, humans could better keep up with these fleet and unpredictable animals, and by the late nineteenth century, three new technologies contributed to their wholesale slaughter. The first was an effective weapon, the Sharps 50-caliber rifle, which permitted long-range killing. The second was the transcontinental railroad (completed in 1869), which provided cheap transportation of the bisons' heavy hides to market. The third was the development of a new tanning process that made the bison hides malleable to processing into leather, which was used primarily for

industrial machine belts. In addition, the extermination of bison was promoted as a way of controlling the Native American tribes that depended on bison meat and hides for survival. By 1883, the American bison was no longer viable as a commercially exploitable species and was nearly biologically extinct.

As early as 1870, attempts were made to protect and preserve the American bison, but few federal or state laws were adopted or enforced. Although 200 bison were living in Yellowstone National Park when it was established in 1872, their numbers had dwindled to 25 from poaching by 1894. At that time, the Lacey Yellowstone Protection Act was enacted, finally providing public protection to the surviving wild bison. Meanwhile, a few people motivated by their own interest in the animals and by plans to sell them to zoos and others developed herds in Texas, Kansas, Nebraska, South Dakota, and Montana.

During the early twentieth century, efforts to provide public protection for bison were successful. Foremost in this endeavor was the American Bison Society, one of the first environmental organizations in the United States, founded in 1905 and consisting primarily of eastern industrialists. The society succeeded in getting the U.S. Congress to set aside four additional areas for bison preservation: the National Bison Range in Montana, Wichita Mountain in Oklahoma, Fort Niobrara in Nebraska, and Wind Cave National Park in South Dakota. The society then purchased the animals to stock these preserves primarily from private holdings, although the Bronx Zoo donated 15 animals to Wichita Mountain in 1907. A census in 1929 counted 3,385 bison, and the species had been rescued from extinction.

As of 2010 the population of American bison in the United States and Canada was estimated at 500,000. More than 90 percent were in private holdings, farmed like livestock and subject to genetic selection to make them more useful to humans and less unpredictable. Such breeding practices have the effect of reducing the animals' wild features, and some commentators have noted that it would be unfortunate for the species to survive physical extinction only to face genetic obliteration. Those bison managed in the public domain have the chance to remain the majestic, wild animals that long dominated North America. Because most public preserves are small, a Great Plains Park has been proposed (perhaps on both sides of the U.S.-Canadian border) to provide habitat for large, nomadic herds of bison such as those that once roamed the middle of the continent.

James L. Robinson

FURTHER READING

Hodgson, Bryan. "Buffalo Back Home on the Range." *National Geographic,* November, 1994, 64-89.

Lott, Dale F. *American Bison: A Natural History.* Berkeley: University of California Press, 2002.

Zontek, Ken. *Buffalo Nation: American Indian Efforts to Restore the Bison.* Lincoln: University of Nebraska Press, 2007.

SEE ALSO: Captive breeding; Conservation; Extinctions and species loss; Grazing and grasslands; North America; Preservation; Range management; Resource depletion; Wildlife management; Yellowstone National Park.

Amoco Cadiz oil spill

CATEGORIES: Disasters; water and water pollution

THE EVENT: Grounding of the tanker *Amoco Cadiz* off the coast of Brittany, France, resulting in the spilling of its cargo of crude oil into the sea

DATE: March 16, 1978

SIGNIFICANCE: The *Amoco Cadiz* oil spill resulted in the deaths of thousands of fish and seabirds and the pollution of coastal waters important to France for the harvest of marine life and for the tourism industry.

The four-year-old very large crude carrier (VLCC) *Amoco Cadiz* had been built in Spain. At 331 meters (1,086 feet) long and 68.6 meters (225 feet) wide, and with a draft of 19.8 meters (65 feet), it was one of the largest ships afloat. The vessel was American owned (by Standard Oil of Indiana), Liberian flagged, and crewed by Italians.

The *Amoco Cadiz* was in the final stages of its voyage from the Persian Gulf to Rotterdam in the Netherlands with 223,000 tons of mixed Kuwaiti and Iraqi crude oil. The vessel was northbound along the coast of Brittany, France, at about 10:00 A.M. when a steering failure occurred. The ship was about 24 kilometers (15 miles) off the French coast. Within two hours, the tugboat *Pacific* was alongside the *Amoco Cadiz* connecting a towline. The tug ran out about 914 meters (3,000 feet) of steel towing wire in an attempt to keep the large tanker off the rocks, but after only two hours of towing, the towline broke. During this time, shipboard engineers had attempted to fix the tanker's damaged rudder, but the system was beyond repair. The ship was 10.5 kilometers (6.5 miles) off the coast of France.

By 9:00 P.M., the tug had reattached a towline to the stern of the *Amoco Cadiz*, but shortly thereafter the large tanker grounded on the Roches de Portsall. It immediately began leaking its cargo of crude oil over the coast of Brittany. This area accounts for almost 40 percent of the marine life and 7 percent of the oysters harvested in France. Within three days, the oil slick covered almost 80 kilometers (50 miles) to the north of the ship and 32 kilometers (20 miles) to the south. The French government implemented its oil-spill cleanup plan. Within six days of the spill, all of the vessel's tanks were open to the sea, and the slick measured 129 kilometers (80 miles) by 29 kilometers (18

Aerial view of the Amoco Cadiz *as it sinks off the coast of Brittany, France.* (AP/Wide World Photos)

miles). The French were unable to pump oil off the grounded vessel because of both poor weather and poor charts of the area.

Ten days after the ship grounded, the highest tide of the period occurred, and beach cleaning began in earnest. By the beginning of April, the French had mustered almost six thousand military personnel, three thousand civilians, twenty-eight boats, and more than one thousand vehicles for the cleanup operation. By the end of May, 206,000 tons of material had been cleaned off the shores of Brittany. Only 25,000 tons of this was actually oil—the rest was sand, rock, seaweed, and other plant life. Those involved in rescue efforts found some ten thousand dead fish and twenty-two thousand dead seabirds.

The cleanup effort was declared a success even though many criticized the French government for a slow response and fragmented efforts. The bulk of the oil cleanup was attributed to the sea itself. The relatively deep water and fast current along the shore helped the sea to disperse and dissipate the 223,000 tons of crude oil that had spilled into it over a six-day period.

Robert J. Stewart

FURTHER READING

Clark, R. B. *Marine Pollution.* 5th ed. New York: Oxford University Press, 2001.

Fairhall, David, and Philip Jordan. *The Wreck of the Amoco Cadiz.* New York: Stein and Day, 1980.

Fingas, Merv. *The Basics of Oil Spill Cleanup.* 2d ed. Boca Raton, Fla.: CRC Press, 2001.

SEE ALSO: *Argo Merchant* oil spill; *Braer* oil spill; *Exxon Valdez* oil spill; Oil spills; *Sea Empress* oil spill; Tobago oil spill; *Torrey Canyon* oil spill.

Amory, Cleveland

CATEGORIES: Activism and advocacy; animals and endangered species

IDENTIFICATION: American author and animal rights activist

BORN: September 2, 1917; Boston, Massachusetts

DIED: October 14, 1998; New York, New York

SIGNIFICANCE: Amory's decades of activism for animal rights and animal protection saved thousands of animals from extermination and helped bring the issue of cruelty to animals into the public spotlight.

As an author, Cleveland Amory is perhaps best known for his best-selling books featuring his cat, Polar Bear: *The Cat Who Came for Christmas* (1987), *The Cat and the Curmudgeon* (1990), and *The Best Cat Ever* (1993). He also published a number of social history studies, and in 1974 he published *Man Kind? Our In-*

Animal rights advocate Cleveland Amory. (Getty Images)

credible War on Wildlife, a work that has been credited with influencing the antihunting movement in the United States.

From the time he was a young child, Amory harbored a dream to create a sanctuary for animals where they would be protected from harm and allowed to roam free. To this end, he established the Fund for Animals, a nonprofit organization, in 1967. In 1977, the Fund for Animals initiated its first major rescue of animals when the U.S. Park Service scheduled the extermination, by shooting, of all the wild burros living in the Grand Canyon in Arizona. Over the next two years, Amory's organization orchestrated a helicopter airlift of 577 burros from the floor of the Grand Canyon, some seven thousand feet below its rim. An intensive, nationwide adoption campaign was conducted for the burros; those animals not adopted found homes at the Black Beauty Ranch, an animal sanctuary founded by Amory in Murchison, Texas, in 1979. Two years later, the Fund for Animals rescued another 5,000 burros that had been earmarked for destruction at Death Valley National Monument and the Naval Weapons Center at China Lake, both in California.

In the early 1980's the organization rescued more than sixty wild Spanish Andalusian goats from San Clemente Island off the coast of California after the U.S. Navy had decided to eradicate the island's population of the animals, saying that the move was mandated by the Endangered Species Act (1973) because the goats were eating the native, endangered species of vegetation on the island. The Navy intended to allow hunters to kill the animals for a fee. The Fund for Animals arranged a helicopter airlift of the animals to temporary quarters in San Diego, California, and another adoption campaign was successfully conducted.

Black Beauty Ranch, which was renamed Cleveland Amory Black Beauty Ranch in 2004 to honor its founder, has been home to thousands of domestic and exotic animals rescued from neglectful or abusive situations. Among the animals given sanctuary at the ranch since its founding have been chimpanzees, kangaroos, wild horses, buffalo, and elephants.

Through his work with the Fund for Animals, Amory developed the resources to conduct high-profile rescues of large numbers of animals, and the organization's continuing efforts have received much national and international publicity. This publicity has helped raise the public's consciousness about issues of cruelty and neglect toward animals.

Karen E. Kalumuck

FURTHER READING

Greenwald, Marilyn S. *Cleveland Amory: Media Curmudgeon and Animal Rights Crusader.* Hanover, N.H.: University Press of New England, 2009.

Marshall, Julie Hoffman. *Making Burros Fly: Cleveland Amory, Animal Rescue Pioneer.* Boulder, Colo.: Johnson Books, 2006.

SEE ALSO: Animal rights; Animal rights movement; People for the Ethical Treatment of Animals; Pets and the pet trade; Wild horses and burros.

Amphibians

CATEGORY: Animals and endangered species

DEFINITION: Class of vertebrate animals with thin, permeable skin and generally characterized by an aquatic larval stage and a more or less terrestrial adult stage

SIGNIFICANCE: Amphibian populations have been suffering worldwide declines as a consequence of habitat destruction and alteration, parasites and diseases, and pollution, to which their permeable skin renders them particularly vulnerable.

Amphibians include frogs and toads, salamanders, and caecilians (rarely encountered legless tropical forms). Because nearly all amphibians lead double lives, they are affected by habitat alterations and environmental contaminants in both aquatic environments (as larvae, such as tadpoles) and terrestrial environments (as adults). Also, because amphibians breathe primarily through their skin, they are very sensitive to chemicals and to poor water quality. Like the proverbial canary in the coal mine, amphibians serve as environmental monitors. Consequently, declining amphibian populations are of concern not only because amphibians are important components of many ecosystems but also because the declines may signal the onset of deteriorating conditions that will soon affect other forms of life, including humans.

About 32 percent of the world's amphibian species are known to be threatened or extinct, and insufficient data are available to allow determination of the threat status of another 25 percent. At least 38 amphibian species are known to be extinct and at least another 120 species have not been found in recent years and could be extinct. At least 42 percent of all

A tiny corroboree frog sits on a zookeeper's thumb. The corroboree is among Australia's most endangered amphibians. (Mick Tsikas/Reuters/LANDOV)

amphibian species are declining, suggesting that the number of threatened species will rise in the future.

Most of the declines are attributable to habitat alterations and destruction. Especially in the Tropics, deforestation has affected a number of species, and the cool, wet mountaintop habitats favored by many amphibians are particularly vulnerable to global climate change. Complicating the issue is the rapid worldwide spread of the chytrid fungus, which has devastated amphibian populations. The fungus invades the thin skin of amphibians and kills them by disrupting their ability to regulate the movement of water and oxygen in their bodies. The fungus spreads in water and through direct body contact. A few resistant species, such as bullfrogs, clawed frogs, and cane toads, serve as vectors, transmitting the disease to sensitive species. Chytrid thrives in the same cool, wet conditions favored by most amphibians.

Deformed frogs with missing or extra limbs were brought to the attention of the world when they were discovered by schoolchildren in Minnesota in 1995. At first, chemical pollutants were thought to be the cause, but it became apparent that parasitic flatworms (trematodes) were responsible. However, experts believe that, like chytrid, stress imposed by deteriorating habitats and a changing climate may well be aggravating the problem in a synergistic fashion.

Another environmental concern associated with amphibians is not related to declines but is, in fact, caused by too many frogs in the wrong places. A few very hardy species have been transported by humans to places far from their native ranges. These invasive species upset biotic communities by competing with or eating native species. Large and voracious cane toads, introduced widely to control insect pests in sugarcane fields, have been implicated in the decline or extinction of native frogs in places as far from one another as Florida and Australia. Equally large and voracious bullfrogs, intentionally imported around the world for food, are responsible for declines of native frogs in the western United States, Latin America, and even Europe. Coquís, small tree frogs native to Puerto Rico, pose a threat to unique Hawaiian insects. Cuban tree frogs, introduced into Florida, have hitchhiked on tropical plants shipped around the world, and in their new habitats they eat other frogs and even small lizards.

Robert Powell

FURTHER READING

Collins, James P., and Martha L. Crump. *Extinction in Our Times: Global Amphibian Decline*. New York: Oxford University Press, 2009.

Heatwole, Harold, and John W. Wilkinson, eds. *Amphibian Decline: Diseases, Parasites, Maladies, and Pollution*. Chipping Norton, N.S.W.: Surrey Beatty & Sons, 2009.

Lannoo, Michael, ed. *Amphibian Declines: The Conservation Status of United States Species*. Berkeley: University of California Press, 2005.

Stuart, S. N., et al., eds. *Threatened Amphibians of the World*. Arlington, Va.: Conservation International, 2008.

SEE ALSO: Extinctions and species loss; Indicator species; Introduced and exotic species.

Animal rights

CATEGORIES: Animals and endangered species; philosophy and ethics

DEFINITION: Term for the philosophy that promotes the valuing of animals and the protection of their basic interests, sometimes to the same extent as the interests of humans

SIGNIFICANCE: Supporters of animal rights in the United States have succeeded in influencing public opinion on issues such as the eating of meat and the wearing of fur, and they have secured substantial state and federal legislation aimed at protecting the lives and welfare of animals.

The animal rights philosophy covers a wide range of beliefs, from simple interest in the welfare of companion animals such as cats and dogs to the stance that all animals should be left in nature and beyond the reach of human interference. Among the most common concerns of animal rights activists are the use of animals as a source of food and the use of animals in biomedical research and education. Activists also stage protests against the fur industry, zoos, circuses, the pet industry, and hunting. Some animal rights activists seek allies within certain environmentalist groups that share a similar counterculture philosophy, but many environment-oriented organizations reject the core tenets of animal rights purists, as they believe that the human use of animals serves the perpetuation of natural food chains and ecosystem management.

ANIMAL CRUELTY

All U.S. states have laws against animal abuse to protect most birds and mammals from cruel treatment and abandonment. Most of these laws do not include in their coverage the larger numbers of invertebrates, fish, reptiles, and amphibians, and most states exclude coverage of animals used in science research. Such laws are directed toward human behavior and do not establish "rights" of animals any more than laws concerning land use establish "land rights." States define wildlife as the property of either the landowner or the state.

Animal cruelty includes direct acts of physical abuse—such as beating a horse—and neglect, as when an owner allows an animal to starve to death. Cruelty usually involves intentional, malicious, or knowingly committed acts and excludes instances of accident or ignorance. Many states focus on unjustified infliction of pain in defining cruelty. Therefore animal breeders and those who operate racetracks and other facilities that confine animals usually must meet minimal requirements in providing the animals with exercise, space, light, ventilation, and clean living conditions. The long tradition in U.S. law of punishing animal cruelty reflects widespread public acceptance of animal welfare principles. Many animal rights activists want to expand these principles to challenge the use of animals for food, for research, for education, for fur, and for hunting; many also object to the keeping of animals in zoos.

Until the 1980's, most animal rights organizations in the United States were not activist; rather, they focused on animal welfare and providing shelters. The public was aware of the role that animal research had played in the earlier development of antibiotics, insulin, surgical techniques, and vaccinations, and most adults had recent memories of a nation in which no cures or treatments were available for many major diseases. In the 1980's, however, members of a new generation who had little experience with the effects of

widespread disease began to criticize the U.S. Food and Drug Administration (FDA) protocols for the approval of new drugs based on animal testing.

Two books—*Animal Liberation* (1975), by Peter Singer, and *The Case for Animal Rights* (1983), by Tom Regan—provided an updated philosophy for activists who questioned the human use of animals for medical research, food, fur, and education. New animal rights organizations were formed, and old organizations shifted their missions, including People for the Ethical Treatment of Animals (PETA), the New England Antivivisection Society (NEAVS), the Ethical Science Education Committee, and the underground Animal Liberation Front (ALF).

ANIMALS AS A FOOD SOURCE

Farmers own agricultural animals as property. Meat processing is a well-established industry, and a large majority of the American population traditionally eats meat. Animal rights efforts in this area have therefore concentrated on attacking specific farming practices, such as by portraying the confinement of animals to produce veal, poultry, and eggs as inhumane. They have also promoted the ideas that the eating of meat is unhealthy and that the practices used in raising and processing animals for meat are bad for the environment.

Animal rightists have leveraged nutritional concerns among some members of the medical community into a broad condemnation of consuming meat. Meat, cheese, and milk once formed a major food group as defined by the U.S. Department of Agriculture (USDA), but in 1992 the USDA began to promote a set of dietary guidelines using a "food pyramid" that recommends a smaller amount of meat, showing this group near the pyramid peak. Humans demonstrate an evolutionary history of eating both plant and animal products, in tooth structure, digestive anatomy, and the need for some amino acids found originally in animals. New plant breeds can provide some of these nutrients, however, and a vegetarian diet is now a viable alternative for many people.

Some animal rightists contend that vegetarianism is an ecological necessity.

Animal Rights Milestones

YEAR	EVENT
1977	Peter Singer publishes *Animal Liberation*, which elaborates an animal rights philosophy and gains a substantial following.
1980	People for the Ethical Treatment of Animals (PETA) is founded.
1983	Tom Regan's *The Case for Animal Rights* reinforces the message of Singer's books and adds more people to the movement.
1988	The first public school dissection opt-out law in the United States is passed in California; Florida, New York, Maine, Pennsylvania, and Illinois soon follow.
1992	The U.S. Department of Agriculture begins promoting dietary guidelines recommending that people eat less animal meat.
1993	The First Congress in Alternatives in Animal Usage provides a forum for discussion and debate between animal research advocates and animal rights groups.
2009	The European Union begins phasing out animal testing in the cosmetics industry.
2010	PETA claims worldwide membership of two million.

Their primary argument is based on energy flow. On average, only 10 percent of the chemical energy of one food level is stored in the next food level; for example, one hundred units of grain sustain ten units of cow, which in turn produce one unit of human. Using this logic, the human population could increase tenfold if all humans switched from eating meat to eating grain directly. Critics of this argument point out that most of a cow's diet is grass that is indigestible to humans and is often grown on land not suitable for crops. The majority of meat animals worldwide, from pigs to fish, are not fed on grain, and their feeding may even be critical to the recycling of wastes. Some animal rightists have asserted that a "hamburger connection" exists between tropical deforestation and the beef used by American fast-food chains. Although there is no export of rain-forest beef to the United States, the destruction of South and Central American rain forests to make room for cattle ranches and farms continues at a rapid rate.

BIOMEDICAL RESEARCH

Researchers made extensive use of laboratory animals in the development of germ theory and in the conquest of most major infectious diseases. Nearly all of the scientists who have received the Nobel Prize in Physiology or Medicine since the award was first presented in 1901 have used animals in their work, and, until the 1980's, their research met with widespread public approval. Antivivisection and humane societies spoke with a weak voice and addressed standard animal cruelty issues until the publication of *Animal Liberation* and *The Case for Animal Rights*, which provided a more detailed rationale for their cause. During the 1970's and 1980's, many young people in the United States—the first full generation raised in the absence of major infectious diseases, away from contact with farm animals, and in a climate that was increasingly antiscience—joined animal rights organizations.

Animal rightists have stated four major objections to animal research. First is the assertion that all organisms have an intrinsic right to live free of human interference; some activists refer to the favoring of human needs over animal needs as "speciesism." A second set of concerns centers on the suffering and pain caused to research animals. A third concern is that some researchers use animals that were once pets, acquired through animal shelters. Finally, many animal rightists assert that new technologies have been developed that make direct research on animals

unnecessary; they contend that animal research can be replaced with tissue-culture techniques and computer simulations.

Although more than 85 percent of research animals are rodents, most protests are targeted toward research using dogs and cats (less than 2 percent of research animals) and nonhuman primates (less than 0.5 percent). Activists often complain that the laws concerning requirements for animal care in laboratories are not being enforced. Some even assert the existence of a major criminal industry that steals pets to supply research animals.

Animal rights groups seek to control local animal shelters and restrict the transfer of animals from such shelters to uses in research and education. Activists are involved in ongoing attempts to influence the passage of legislation that would restrict the uses of shelter animals; eliminate cosmetic and other product safety testing on animals; allow private citizens to sue for enforcement of the Animal Welfare Act (AWA); include farm animals, rats, and mice under the AWA; and include the research use of animals under state animal cruelty laws.

Many scientists have responded to the objections of animal rightists by arguing that biological systems are far more complex than simple computer models and that drugs do not always respond in tissue cultures as they would in whole organisms. In regard to the use of laboratory animals, most research organizations now follow the "three R's": replacement, reduction, and refinement. That is, they make efforts to replace some animals in research (with tissue cultures, for example), they attempt to reduce the numbers of animals used, and they make refinements to reduce the pain and distress of lab animals. Animals taken from shelters for research constitute less than 1 percent of the dogs and cats that would be otherwise be euthanized. Research facilities are inspected by the USDA's Animal and Plant Health Inspection Service (APHIS), which enforces AWA criteria. The FDA and the Environmental Protection Agency (EPA) also have established regulations concerning the use of laboratory animals. Advocates of animal testing claim that 95 percent of lab animals are not subjected to pain; the remaining animals, involved in studies of pain itself, are administered pain-relieving drugs or anesthetics as soon as the study permits.

Medical research using animals is defended by the National Association for Biomedical Research (NABR) and the Incurably Ill for Animal Research (IIFAR).

This issue continues to generate a large proportion of the mail directed toward legislators in the United States. In 1989 a backlash following raids by animal rights activist on research facilities led Senator Howell Heflin of Alabama to introduce legislation in the U.S. Senate specifically aimed at protecting animal research facilities.

EDUCATION AND OTHER ISSUES

The first U.S. state law enacted to allow a student to opt out of lessons using animals was passed in California in 1988 in response to a high school student who protested a science class requirement that she perform a dissection on a frog. Although she failed to win her federal court case to establish a right to avoid such course work, the California law made the issue moot in that state. Florida, New York, Pennsylvania, Rhode Island, Virginia, Oregon, New Jersey, Vermont, and Illinois also eventually passed laws that allowed students to opt out of classroom dissection. In other states (including Maine and and New Mexico), state school boards or departments of education have developed policies under which students who object to dissecting animals may do alternative assignments.

Animal rights organizations have staged protests against some of the companies that provide animals to schools for student dissection. Some pamphlets allege that the use of these animals in classrooms has contributed to the decline in amphibians and endangerment of other species. Some activists contend that dissection and animal experimentation desensitize students to the suffering of animals, and they assert that only a small percentage of students need to learn lab skills. They promote the use of plastic models, computer simulations, and other alternatives as effective and economical alternatives for students of biology and anatomy. One organization maintains a toll-free telephone line for use by students who do not wish to participate in dissections and provides materials on the topic written for elementary, secondary, and college levels.

Public schools in the United States are averse to entering into public debates and controversy concerning animal rights. Many scientists and educators, however, have responded to objections to dissection by claiming that alternatives fail to provide the multisensory experiences that make laboratory research meaningful; they assert that it is critical for students to be exposed to genuine lab results if they are to understand the reality base of science. Interest in laboratory work continues to be the main incentive for many students to decide to enter careers in health care. Field biologists claim that nearly all dissection animals come from food harvesting, captive rearing, or animal shelters. They also claim that dissection, rather than desensitizing students, normalizes students' attitudes toward organs, blood, and feces. While not all students will become doctors, all will become patients, and they are benefited by learning about their own internal anatomy through hands-on experience.

Another concern of animal rights activists is the killing of animals for their fur. Although increasingly the furs used in clothing production come from farm-bred animals, such as minks, protests continue to target fur-trapping techniques. Demonstrators who

Members of the Animal Rights Action Network protest fur farming in Dublin, Ireland, in 2010. (AP/Wide World Photos)

splash red paint on expensive fur coats are usually given news coverage. Animal rights activists also object to the hunting of animals. Some have staged protests against hunting, a rural tradition with widespread support in the American West and Midwest, by "shadowing" hunters to drive away game animals.

Many animal rightists also object to the keeping of animals in zoos, often portraying zoos as prisons. This contrasts with the stance taken by many environmentalist organizations that value zoos for maintaining animal populations that would otherwise have gone extinct and sometimes reintroducing species to the wild. Animal rightists often portray circuses as promoting both animal slavery and animal abuse. The training of animals used in circuses, other live shows, and film and television productions may be subject to scrutiny and harassment by activists.

Persons who view themselves as animal rightists vary widely in their philosophies toward pet ownership. Pet owners constitute the major group of contributors to animal rights organizations, some of which have begun a movement away from the use of the word "pet," which they see as demeaning to animals, preferring instead the term "companion animal." The extent to which organizations are willing to advocate pure animal freedom defines the sometimes volatile differences among animal rights groups and animal welfare groups.

An analysis of membership in animal rights organizations reveals that more women than men join such groups, and they tend to have a relatively high level of education; increasing numbers of pet veterinarians are becoming members, but not farm veterinarians. Although the membership of animal rights groups represents less than 0.5 percent of the U.S. population, these organizations have succeeded in securing substantial state and federal legislation and continue to influence public policy.

John Richard Schrock

FURTHER READING

Baird, Robert M., and Stuart E. Rosenbaum, eds. *Animal Experimentation: The Moral Issues.* Amherst, N.Y.: Prometheus Books, 1991.

Beers, Diane L. *For the Prevention of Cruelty: The History and Legacy of Animal Rights Activism in the United States.* Athens: Swallow Press/Ohio University Press, 2006.

Franklin, Julian H. *Animal Rights and Moral Philosophy.* New York: Columbia University Press, 2005.

Leavitt, Emily Stewart. *Animals and Their Legal Rights: A Survey of American Laws from 1641 to 1990.* Rev. ed. Washington, D.C.: Animal Welfare Institute, 1990.

National Academy of Sciences and Institute of Medicine. *Science, Medicine, and Animals.* Washington, D.C.: National Academy Press, 1991.

National Research Council. *Education and Training in the Care and Use of Laboratory Animals: A Guide for Developing Institutional Programs.* Washington, D.C.: National Academy Press, 1991.

Orlans, F. Barbara. *In the Name of Science: Issues in Responsible Animal Experimentation.* New York: Oxford University Press, 1993.

Regan, Tom. *The Case for Animal Rights.* 2d ed. Berkeley: University of California Press, 2004.

Singer, Peter. *Animal Liberation.* 1975. Reprint. New York: HarperPerennial, 2009.

Sunstein, Cass R., and Martha C. Nussbaum, eds. *Animal Rights: Current Debates and New Directions.* New York: Oxford University Press, 2006.

Verhestel, Ernest. *They Threaten Your Health: A Critique of the Antivivisection/Animal Rights Movement.* Tucson, Ariz.: Nutrition Information Center, 1985.

SEE ALSO: Animal rights movement; Animal testing; People for the Ethical Treatment of Animals; Pets and the pet trade; Singer, Peter; Speciesism; Zoos.

Animal rights movement

CATEGORIES: Activism and advocacy; animals and endangered species; philosophy and ethics

IDENTIFICATION: Social movement involving groups and individuals concerned with the basic rights and welfare of animals

SIGNIFICANCE: The animal rights movement has brought about a number of changes in laws as well as in public perceptions of issues involving animals, but debates continue concerning the extent of the rights to which animals are entitled.

People involved in the animal rights movement share philosophical beliefs based on the idea that all animals are entitled to an equal claim on life and liberty and possess the same rights to existence as humans. Animal rightists oppose those who believe that animals exist for human use as objects of study and experimentation, as food, as beasts of burden, or as objects of amusement and recreation.

Major Animal Rights Organizations

Organization	Founded
American Society for the Prevention of Cruelty to Animals	1866
American Anti-Vivisection Society	1883
Animal Defense League	1934
Friends of Animals	1957
Fund for Animals	1967
Actors and Others for Animals	1971
American Fund for Alternatives to Animal Research	1977
People for the Ethical Treatment of Animals	1980
Alliance for Animals	1988

The philosophical concept of animal rights arose during the seventeenth and eighteenth centuries along with the development of biological science. The growing interest in biology gave rise to a sort of sideshow in which living, conscious dogs were cut open so that the animals' internal organs could be displayed to crowds of onlookers. A variety of blood sports were popular as well, including bullbaiting and bearbaiting. In these, a bull or a bear was chained in a ring along with one or more dogs that were trained to attack the larger animals. Dogfighting, in which various terrier breeds were encouraged to attack each other, was also popular.

Birth of Animal Rights

In 1824 the Royal Society for the Prevention of Cruelty to Animals (RSPCA) was founded in Great Britain to enforce new anticruelty laws. However, the laws and their enforcement had little, if any, effect in rural areas, which were far from the watchful eyes of the police or RSPCA agents. On many farms, animals were still kept in filthy conditions and beaten if they balked at hauling overloaded wagons. The slaughter of animals for market was carried out as simply and quickly as possible.

Biomedical research, especially in human anatomy and physiology, advanced rapidly in Europe during the early to mid-nineteenth century. While anatomical study could be satisfied with human corpses, physiologists required living material, and animals became their targets. Although many of the animals used in medical research were rats and mice, dogs were also frequently used. Many of the animals were stolen pets, while others were strays that were found roaming on the streets. The treatment that dogs received in medical laboratories varied, but, for the purposes of good

science, the animals had to be maintained and treated in clean, sanitary, and relatively stress-free environments. Many dogs died at the hands of medical researchers, and opposition to the practice quickly grew. Objections came from members of the general public, who had heard stories of both real and imagined horrors suffered by animals in the experiments.

The British government sought to quiet the complaints by passing the Cruelty to Animals Act of 1876. The act did not prohibit the practice of experimenting on live animals; rather, it set regulatory procedures that had to be followed in the laboratories. Animal rights were no less an issue in the United States at that time. The first documented humane society in the United States was the American Society for the Prevention of Cruelty to Animals (ASPCA), incorporated in 1866. Another pioneer group, the American Antivivisection Society (AAS), was founded in 1883.

The animal rights movement in the United States was relatively quiet until the first Earth Day in 1970, after which it rapidly expanded. Of the more than eighty animal rights and animal welfare organizations in the United States, fifty-seven (70 percent) were founded after 1970. In addition to the older groups, among the organizations are such diverse associations as Actors and Others for Animals, the Coalition for Non-Violent Food, the Animal Political Action Committee, and the American Fund for Alternatives to Animal Research. The list also includes several adversarial and confrontational groups, such as Greenpeace, People for the Ethical Treatment of Animals (PETA), and the Animal Liberation Front (ALF).

Animal Rights Issues

One issue that quickly attracted animal rights proponents was the plight of the whales. For centuries whaling was conducted from sailing ships with handheld harpoons in the manner made famous in Herman Melville's novel *Moby Dick* (1851). Even with such crude equipment and methodology, whalers reduced the whale population in the Atlantic Ocean and turned their attention to the Pacific Ocean. In the late nineteenth century, steam (and, later, diesel) vessels and cannon-fired harpoons increased the whalers' efficiency. The methods of the whalers aroused the ire of many people. Frequently, a harpooned whale was forced to tow the steel "catcher" ship for hours until the animal succumbed to the injuries from the explosive-headed harpoon. Another whaling technique was to harpoon and kill a whale calf. The mother and other adults hovered around the injured or killed calf and were in turn harpooned.

Economics rather than animal rights brought about the formation of the International Whaling Commission (IWC) in 1946. The commission was established to manage the whale stocks. It had no regulatory authority, however, and Norway, Iceland, and Japan continued to hunt whales despite the recommendations of the IWC. These nations have defended their activities as a sustainable use of a natural resource. Opponents view whaling as an archaic activity and a violation of animal rights. In 1972 the U.S. government passed the Marine Mammal Protection Act as a move to protect the whales. The apparent recovery of the Pacific gray whale stocks suggests that the act may have been a step toward achieving the goals of animal rights activists.

Dolphins are also at risk, but from indirect human exploitation. In the eastern tropical Pacific Ocean, yellowfin tuna frequently swim below pods of dolphins, and commercial fishermen learned to set their nets around the schools of dolphins to capture the tuna. As the fish are netted, the trapped dolphins drown. The killing of as many as 132,000 dolphins each year in this manner led to protests by animal rights groups that included a boycott of canned tuna. In response to the public outcry, the U.S. government instituted regulations that require both domestic and foreign fishers to follow practices that release the dolphins from their nets and still retain most of the tuna. Within five years after the regulations were put in place, the accidental catch and kill of dolphins was reduced to 25 percent of what it had been. The reduced tuna catches, however, forced many U.S. fishers out of the industry, leaving a void that was quickly filled by foreign fishers.

The animal rights movement has also brought increased scrutiny to the fur trade and to circuses, zoos, theme parks, and any other activity in which live animals are used. For example, in February, 1995, when the Ringling Brothers and Barnum & Bailey Circus was preparing to visit Richmond, Virginia, the show's management asked the news media not to mention the time of the circus train's arrival, because the circus wanted to move its animals from the rail yard to the show grounds with as little fanfare as possible. The circus had received threats from animal rights groups and sought to avoid any confrontation and possible risks to animals and the public. Although the animal rights movement has brought about a number of changes in laws as well as in public perceptions of animal issues, it is unlikely that ongoing controversies surrounding animal rights will soon be resolved.

Albert C. Jensen

FURTHER READING

Beers, Diane L. *For the Prevention of Cruelty: The History and Legacy of Animal Rights Activism in the United States.* Athens: Swallow Press/Ohio University Press, 2006.

Fellenz, Marc R. *The Moral Menagerie: Philosophy and Animal Rights.* Urbana: University of Illinois Press, 2007.

Franklin, Julian H. *Animal Rights and Moral Philosophy.* New York: Columbia University Press, 2005.

Garner, Robert. *The Political Theory of Animal Rights.* New York: Manchester University Press, 2005.

Harnack, Andrew, ed. *Animal Rights: Opposing Viewpoints.* Farmington Hills, Mich.: Greenhaven Press, 1996.

Sunstein, Cass R., and Martha C. Nussbaum, eds. *Animal Rights: Current Debates and New Directions.* New York: Oxford University Press, 2006.

Whisker, James B. *The Right to Hunt.* 2d ed. Bellevue, Wash.: Merril Press, 1999.

SEE ALSO: Animal rights; Animal testing; Greenpeace; People for the Ethical Treatment of Animals; Singer, Peter; Speciesism; Zoos.

Animal testing

CATEGORIES: Animals and endangered species; philosophy and ethics

DEFINITION: Use of nonhuman animals for research purposes, particularly medical research

SIGNIFICANCE: Animal testing is an integral component of modern science, product testing, and education. Most significant developments in medicine directly or indirectly rely on animal testing. Public debate about the moral and legal status of animals in society has resulted in numerous regulations on animal testing, yet it continues to be a controversial subject.

Animal Welfare Act

The main provisions of the Animal Welfare Act of 1966 follow:

An Act,

To authorize the Secretary of Agriculture to regulate the transportation, sale, and handling of dogs, cats, and certain other animals intended to be used for purposes of research or experimentation, and for other purposes.

Be it enacted by the Senate and House of Representatives of the United States of America in Congress assembled. That, in order to protect the owners of dogs and cats from theft of such pets, to prevent the sale or use of dogs and cats which have been stolen, and to insure that certain animals intended for use in research facilities are provided humane care and treatment, it is essential to regulate the transportation, purchase, sale, housing, care, handling, and treatment of such animals by persons or organizations engaged in using them for research or experimental purposes or in transporting, buying, or selling them for such use.

SEC. 2. When used in this Act— . . .

(d) The term "dog" means any live dog (*Canis familiaris*);

(e) The term "cat" means any live cat (*Felis catus*);

(f) The term "research facility" means any school, institution, organization, or person that uses or intends to use dogs or cats in research, tests, or experiments, and that (1) purchases or transports dogs or cats in commerce, or (2) receives funds under a grant, award, loan, or contract from a department, agency, or instrumentality of the United States for the purpose of carrying out research, tests, or experiments;

(g) The term "dealer" means any person who for compensation or profit delivers for transportation, or transports, except as a common carrier, buys, or sells dogs or cats in commerce for research purposes;

(h) The term "animal" means live dogs, cats, monkeys (nonhuman primate mammals), guinea pigs, hamsters, and rabbits.

A nimal testing, also known as animal experimentation, animal research, in vivo testing, and vivisection, is used to advance pure and applied research. Behavior, development, evolution, and genetics research are all forms of pure research involving animals. Applied research includes medical research, defense research, and toxicology studies for drugs, food additives, pesticides, and cosmetics. The use of animals in medical education and training is typically considered to be a form of animal testing as well.

RATIONALE, SCOPE, AND REGULATIONS

The underlying rationale for the use of animal testing is that living organisms provide interactive, dynamic systems that scientists can observe and manipulate in order to understand normal and pathological functioning as well as the effectiveness of medical interventions. The vast majority of animal testing is for human benefit and relies on the physiological and anatomical similarities between humans and other animals. The term "animal model" refers to the use of live animals to study particular biological processes with the end of extrapolating that information to other animals, particularly humans.

Many species are used in animal testing. Nematode worms and fruit flies are commonly used invertebrates. Zebra fish and mice are commonly used vertebrates. While it is difficult to ascertain the exact number of animals used in research, it is estimated that between 50 million and 100 million vertebrates are used annually worldwide. Approximately 90 percent of the animals used in research are mice and rats.

A chief moral concern raised in regard to animal testing is the pain and suffering it involves. Sentience, or subjective awareness, particularly of pain and pleasure, is common to all vertebrates. Evidence for sentience in invertebrates is generally absent, however, and for this reason research on invertebrates is largely unregulated. Cephalopods (the class of animals that includes octopi and squid) are notable exceptions and are covered by regulations in several countries owing to evidence for their sentience.

A societal consensus exists that animal testing for the advancement of science and medicine is justified, provided that there are no alternatives, the use of animals is kept to a minimum, and that animal pain and distress is minimized. Supporters of animal testing commonly cite the number of major medical advances that have resulted from the practice.

Animal testing is heavily regulated in many countries. Regulations have changed significantly since the mid-twentieth century, and they differ in the numbers of species covered, the kinds of animal welfare protections offered, and the regulatory approaches taken. In the United States, animal testing is governed by two federal statutes: the Animal Welfare Act and Regulations of 1966 (AWAR) and the Health Research Extension Act of 1985, the provisions of which are carried out in the Public Health Service Policy on Humane Care and Use of Laboratory Animals (PHS Policy). AWAR establishes the minimum acceptable standards of care and treatment for certain animals in research, testing, experimentation, exhibition purposes, and use as pets. AWAR covers all warm-blooded animals yet specifically excludes birds, mice, and rats bred for research purposes, as well as animals used for food, fiber, or many forms of agricultural research. PHS Policy, which applies to all research funded by the National Institutes of Health, applies to all live vertebrates used for research purposes.

Opposition and Alternatives

Opposition to animal testing is diverse. Disagreement with the practice is based on both scientific and ethical grounds, and it varies both according to species and according to purpose. Research using primates, monkeys, cats, and dogs is particularly controversial. Cosmetics testing on animals is controversial because many consider the benefit of yet another cosmetic product to be of dubious value when weighed against animals' interests. Although cosmetics testing on animals remains legal in the United States, it has been banned in several countries, including the United Kingdom, and the European Union began phasing it out in 2009. Some object to the use of animals in education, asserting that such use encourages the view that animals are objects to be manipulated rather than beings deserving of compassion or respect.

Those opposed to animal testing on scientific grounds cite the unreliability of predicting effects in humans based on animal models. Some argue that animal testing is not cost-effective; they assert that, given the substantial costs of conducting animal tests, which often last years and cost millions of dollars, the goal of improving human health would be more fully and efficiently realized through a reallocation of funding to implement existing medical technologies more widely. Some argue that much animal testing is immoral because the animal suffering caused is greater than the expected benefits to humans. The stronger animal rights view is that each animal has inherent moral worth, which prohibits humans from using them as experimental subjects for any reason.

First articulated by scientists William M. Russell and Rex L. Burch, the "three R's"—replacement, reduction, and refinement—are influential guiding principles for the humane use of animals in research. Replacement involves seeking to increase alternatives to animal testing that generate the desired research data without the use of sentient animals. Examples of replacement include the use of computer models, epidemiological data, tissue cultures, isolated organs, and nonsentient animals. Reduction is the effort to obtain comparable data using fewer animals or to obtain more data using the same number of animals. Refinement involves favoring research protocols that alleviate or minimize animal pain and distress through the use of analgesics, veterinary care, improved living quarters, and enrichment. Further development and increased implementation of alternatives to and refinement of animal testing is an area of common ground between animal advocates and animal researchers.

Joel P. MacClellan

Further Reading

Carbone, Larry. *What Animals Want: Expertise and Advocacy in Laboratory Animal Welfare Policy.* New York: Oxford University Press, 2004.

Cothran, Helen, ed. *Animal Experimentation: Opposing Viewpoints.* San Diego, Calif.: Greenhaven Press, 2002.

Guerrini, Anita. *Experimenting with Humans and Animals: From Galen to Animal Rights.* Baltimore: The Johns Hopkins University Press, 2003.

Monamy, Vaughan. *Animal Experimentation: A Guide to the Issues.* 2d ed. New York: Cambridge University Press, 2009.

Paul, Ellen Frankel, and Jeffrey Paul, eds. *Why Animal Experimentation Matters: The Use of Animals in Medical Research.* New Brunswick, N.J.: Transaction, 2001.

See also: Animal rights; Animal rights movement; Anthropocentrism; People for the Ethical Treatment of Animals; Singer, Peter; Speciesism.

Antarctic and Southern Ocean Coalition

CATEGORIES: Organizations and agencies; preservation and wilderness issues; ecology and ecosystems

IDENTIFICATION: Global coalition of nongovernmental organizations with a common commitment to preserve the lands of the Antarctic region and the southern oceans in perpetuity as a wilderness area

DATE: Founded in 1978

SIGNIFICANCE: The Antarctic and Southern Ocean Coalition is the only nongovernmental organization that serves to advocate for environmental protection and reform for the Antarctic region. It plays an important role in support of the region by monitoring the Atlantic Treaty System.

The Antarctic and Southern Ocean Coalition (ASOC) was established by the Friends of the Earth International, the World Wildlife Fund office of New Zealand, and other environmental organizations to monitor the Atlantic Treaty System, to implement its environmental protocols, and to provide expert witness in Antarctic affairs. ASOC seeks to advance the preservation and protection of the fragile ecosystems of the Antarctic continent and its surrounding waters. It reports regularly to key international organizations serving mandates to protect the Antarctic environment and its ecosystems from illegal hunting and fishing practices and to advocate for sustainable resource management, including waste disposal and habitat protection.

World Wildlife Fund founder Sir Peter Scott visited Antarctica in 1966 in the wake of successive international expeditions to the South Pole in the twentieth century, notably during the International Geophysical Year of 1957-1958. Prior to that time the southern Antarctic region and its complex relationship to global ecosystems were poorly understood. Despite regularly coordinated International Polar Year efforts, data systems were not sufficient to bring into clear focus the singular contributions of

the southern oceans and icy landmasses of the Antarctic continent to the health of planetary biodiversity and the cosmography of global temperatures and climate. The critical roles of the Antarctic Convergence and the Antarctic Circumpolar Current in maintaining powerful hydraulic and chemical processes within the Pacific, Atlantic, and Indian oceans are of particular interest to international climatologists concerned with the possible consequences of global warming on sea levels and acidification.

Several international protocols are in place to preserve the Antarctic region in perpetuity as a pristine wilderness and global commons for international research. As an outcome of the International Geophysical Year of 1957-1958, twelve nations established research stations in the Antarctic, and on December 1, 1959, they signed the Antarctic Treaty proposed by the United States. By 2010, forty-seven international participants had agreed to honor the treaty's commitment to collaborative research and environmental protection while proscribing military operations on the continent. Twenty-eight nations serve as consultative parties with authority to make decisions on behalf of the treaty.

ASOC works closely with the Antarctic Treaty Secretariat and attends meetings sponsored by other international organizations, including the International Maritime Organization and the International

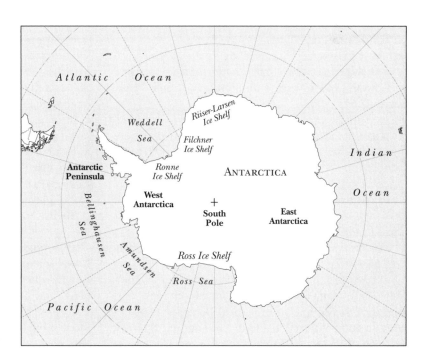

Whaling Commission, as well as the bodies overseeing the Convention on the Conservation of Antarctic Marine Living Resources and the Agreement on the Conservation of Albatrosses and Petrels. Major ASOC campaigns have included the Antarctic Krill Conservation Project, the Southern Ocean Whale Sanctuary, and protection for the Ross Sea.

As an outcome of the 2007-2008 International Polar Year, the Scientific Committee on Antarctic Research (SCAR) developed a network of databases cataloging the extraordinary biodiversity of the South Pole region. The brainchild of Claude De Broyer and Bruno Danis at the Royal Belgian Institute of Natural Sciences, the Marine Biodiversity Information Network (SCAR-MarBIN) is a collaborative effort by hundreds of scientists worldwide to establish the Register of Antarctic Marine Species, the first complete online resource of its kind. This system collates data from more than one hundred international databases to create detailed mappings of the delicate ecology of Antarctica. These images provide valuable information for scientists seeking to understand and analyze the biological equilibrium of the region and the implications of global climate change on life-forms unique to its lands and waters.

Victoria M. Breting-García

FURTHER READING

Bargagli, R. *Antarctic Ecosystems: Environmental Contamination, Climate Change, and Human Impact.* New York: Springer, 2005.

Joyner, Christopher C. *Governing the Frozen Commons: The Antarctic Regime and Environmental Protection.* Columbia: University of South Carolina Press, 1998.

Steig, Eric J., et al. "Warming of the Antarctic Ice-Sheet Surface Since the 1957 International Geophysical Year." *Nature* 457 (January 22, 2009): 459-462.

Triggs, Gillian D., ed. *The Antarctic Treaty Regime: Law, Environment, and Resources.* 1987. Reprint. New York: Cambridge University Press, 2009.

Turner, John, et al., eds. *Antarctic Climate Change and the Environment.* Cambridge, England: Scientific Committee on Antarctic Research, 2009.

SEE ALSO: Antarctic Treaty; Climate change and oceans; Convention on Biological Diversity; Convention on International Trade in Endangered Species; Friends of the Earth International; Glacial melting; Intergovernmental Panel on Climate Change; International Convention for the Regulation of Whaling; Ocean currents; Ozone layer; Sea-level changes; United Nations Framework Convention on Climate Change.

Antarctic Treaty

CATEGORIES: Treaties, laws, and court cases; preservation and wilderness issues; ecology and ecosystems

THE TREATY: International agreement in which signatory nations have committed to setting aside the Antarctic region for scientific and peaceful pursuits

DATE: Opened for signature on December 1, 1959

SIGNIFICANCE: The Antarctic Treaty preserves the continent of Antarctica for peaceful use and protects it from environmental exploitation and degradation; the agreement has in turn led to further international agreements concerning the protection of wildlife.

Antarctica, the large body of land and ice that surrounds the South Pole, is among the earth's most unique and wild places. The only continent with no indigenous human population, Antarctica exceeds 12 million square kilometers (5 million square miles) in size, almost 1.5 times larger than the continental United States. An ice layer averaging more than 1.6 kilometers (1 mile) in thickness covers approximately 95 percent of the land area. Although very few terrestrial species are found on the continent, the surrounding waters are rich in marine life and support large populations of marine mammals, birds, fish, and smaller creatures, some of which are found nowhere else in the world. Antarctica and its surrounding waters play a key but not yet fully understood role in the planet's weather and climate cycles.

In the early twentieth century, seven nations asserted territorial claims on Antarctica, and these claims persisted unresolved for decades. International scientific cooperation among twelve countries during the 1957 International Geophysical Year (IGY) led to the establishment of sixty research stations on the continent. As the IGY drew to a close, the scientific community argued that Antarctica should remain open for continuing scientific investigation

The Antarctic Treaty

Articles I to III and V of the Antarctic Treaty lay out the major goals and objectives of the agreement and the positive obligations of its signatories.

ARTICLE I

1. Antarctica shall be used for peaceful purposes only. There shall be prohibited, inter alia, any measure of a military nature, such as the establishment of military bases and fortifications, the carrying out of military manoeuvres, as well as the testing of any type of weapon.

2. The present Treaty shall not prevent the use of military personnel or equipment for scientific research or for any other peaceful purpose.

ARTICLE II

Freedom of scientific investigation in Antarctica and cooperation toward that end, as applied during the International Geophysical Year, shall continue, subject to the provisions of the present Treaty.

ARTICLE III

1. In order to promote international cooperation in scientific investigation in Antarctica, as provided for in Article II of the present Treaty, the Contracting Parties agree that, to the greatest extent feasible and practicable:

    a. information regarding plans for scientific programs in Antarctica shall be exchanged to permit maximum economy of and efficiency of operations;

    b. scientific personnel shall be exchanged in Antarctica between expeditions and stations;

    c. scientific observations and results from Antarctica shall be exchanged and made freely available.

2. In implementing this Article, every encouragement shall be given to the establishment of cooperative working relations with those Specialized Agencies of the United Nations and other technical organizations having a scientific or technical interest in Antarctica. . . .

ARTICLE V

1. Any nuclear explosions in Antarctica and the disposal there of radioactive waste material shall be prohibited.

2. In the event of the conclusion of international agreements concerning the use of nuclear energy, including nuclear explosions and the disposal of radioactive waste material, to which all of the Contracting Parties whose representatives are entitled to participate in the meetings provided for under Article IX are parties, the rules established under such agreements shall apply in Antarctica.

and should be unfettered by national rivalries over territory. This led to the negotiation of the Antarctic Treaty, which entered into force in 1961.

The treaty prohibits military activity on the continent and promotes scientific cooperation among the parties. Signatory nations agree to freeze existing territorial claims and make no new ones. Initially, twelve nations signed and became consultative parties, agreeing to hold regular consultative meetings to discuss implementation of the treaty. Sixteen more countries have since been granted consultative party status. Twenty additional nations have acceded to the treaty but are not full parties and participate as observers only. The Antarctic Treaty was also the first major arms-control agreement among the nuclear weapons states and provided a model for several subsequent agreements, including the Limited Test Ban Treaty (1963). The Antarctic Treaty specifically prohibits nuclear test explosions in Antarctica.

Consultative meetings have resulted in more than two hundred implementation recommendations to national governments and have led to two additional treaties: the 1972 Convention for the Conservation of Antarctic Seals and the 1982 Convention on the Conservation of Antarctic Marine Living Resources (CCAMLR), which addresses fishery management. An unusual feature of the latter is that the applicable territory is defined by ecosystem criteria rather than by political boundaries. The Antarctic Treaty and all related agreements and recommendations form what is known as the Antarctic Treaty System.

Mineral discoveries in Antarctica in the 1970's led some nations and private companies to contemplate plans to exploit mineral resources there, but environmental advocates and the scientific community, sharing a concern about potential impacts, led a long fight to prevent such activities. In 1991 the historic Protocol on Environmental Protection was finally adopted; this agreement bans mineral and oil exploration in Antarctica for a minimum of fifty years. Annexes to the protocol contain legally binding provisions regarding environmental assessments, protection of indigenous plants and animals, waste disposal, marine pollution, and designation of protected areas. The protocol entered into force in January, 1998, after ratification by all consultative parties. Environmental issues that still posed challenges to the protection of Antarctica in the late 1990's included threats of overexploitation of Antarctic fisheries, impacts of expanding tourism, and the need for an agreement among the treaty parties regarding liability for environmental damages.

In 1978 several environmental organizations interested in Antarctica formed the Antarctic and Southern Ocean Coalition (ASOC), which by 2010 included more than one hundred member organizations from fifty countries. ASOC has been accorded status as an expert observer to the Antarctic Treaty System and represents member group interests. Environmentalists who have endorsed the concept of Antarctica as a "world park" lobbied for and strongly support the environmental protocol.

Phillip A. Greenberg

FURTHER READING

Cioc, Mark. *The Game of Conservation: International Treaties to Protect the World's Migratory Animals*. Athens: Ohio University Press, 2009.

Triggs, Gillian D., ed. *The Antarctic Treaty Regime: Law, Environment, and Resources*. 1987. Reprint. New York: Cambridge University Press, 2009.

SEE ALSO: Antarctic and Southern Ocean Coalition; Limited Test Ban Treaty.

Anthropocentrism

CATEGORY: Philosophy and ethics

DEFINITION: The view that human beings are of central importance in the universe

SIGNIFICANCE: Some environmentalists believe that anthropocentric attitudes are largely responsible for human actions that have led to environmental calamities such as air and water pollution, species extinction, and global climate change.

For anthropocentrists human lives have greater value than do the lives of any other species. Anthropocentrists often point out that humans are the only beings that possess certain capacities. They note that, unlike other animals, humans are typically intelligent, self-aware, autonomous, language users, and moral agents; humans engage in play and make art, among other complex cognitive tasks. For anthropocentrists only humans are intrinsically valu-

able. The rest of nature (including all plant and animal species) has only instrumental value—that is, nature serves only as a means to human ends. From the anthropocentric point of view, biodiversity should be preserved only if it is in the interest of humans to preserve it—any duties that human beings have to preserve biodiversity are owed to other humans, not to any other species.

Anthropocentrism is deeply rooted in most human cultures, but this viewpoint has come under increasing challenges by environmental activists, animal rights advocates, and others. Among the major arguments against anthropocentrism is that it is invidiously perfectionist—that is, logically, those humans who do not display all the characteristics that anthropocentrists assert are uniquely human (intelligence, self-awareness, autonomy, and so on) should be viewed as less valuable than those who do. Some environmental philosophers believe that humans must eradicate both anthropocentrism and the related viewpoint of speciesism, replacing them with deep ecology, biocentrism, sentientism (the view that all sentient beings have moral worth), and ecocentrism. The search for a convincing nonanthropocentric foundation for what is valuable in nature is at the center of environmental philosophy.

Julia Tanner

SEE ALSO: Biocentrism; Biodiversity; Deep ecology; Ecocentrism; Environmental ethics; Speciesism.

Antibiotics

CATEGORY: Human health and the environment

DEFINITION: Molecules that inhibit the growth of or kill bacteria

SIGNIFICANCE: Antibiotics are used routinely in health care, industry, and agriculture to treat and prevent bacterial infections. Antibiotics released into the environment as waste can generate bacteria that are resistant to the antibiotics through the process of selective pressure. These antibiotic-resistant bacteria can enter the food chain and can indirectly or directly promote the generation of antibiotic-resistant bacteria pathogens.

Antibiotics are defined as bacteriostatic or bactericidal substances that can inhibit the growth of or kill bacteria, respectively. Antibiotics can range from

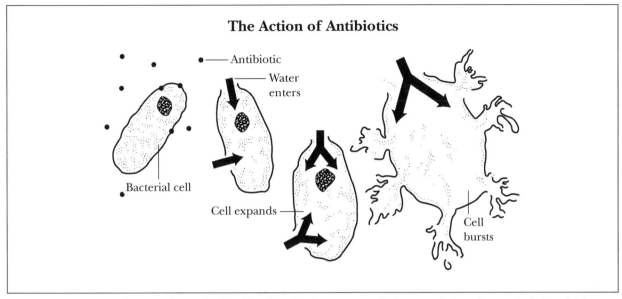

The Action of Antibiotics

An antibiotic destroys a bacterium by causing its cell walls to deteriorate; water will then enter the bacterium unchecked until it bursts.

those isolated as natural compounds, such as aminoglycosides, to those that are completely synthetic, such as quinolones. The first modern antibiotic was penicillin, discovered by scientist Alexander Fleming in 1928; Fleming was later awarded the Nobel Prize for his contributions to medical science. Antibiotics work by inhibiting essential biochemical processes in cells, leading to the inhibition of growth or death of the target bacterium. For example, the bactericidal beta-lactam antibiotics such as penicillin work by inhibiting transpeptidase enzymes, which play an important role in synthesis of the protective peptidoglycan layer of the bacterial cell envelope. The bacteriostatic agent chloramphenicol inhibits protein synthesis by binding to the 50S subunit of the ribosome and interfering with peptide bond formation.

While various in vitro microbiological techniques may indicate whether or not a particular antibiotic is bacteriostatic or bactericidal, the antibiotic's clinical usefulness is dependent on other factors. These include the antibiotic's pharmacokinetic and pharmacodynamic characteristics as well as the host's state of health and immune system defense mechanisms. Antibiotics can generally be categorized as either narrow-spectrum or broad-spectrum. Narrow-spectrum antibiotics target certain families of bacteria, whereas broad-spectrum antibiotics can treat a wide variety of pathogens and also affect environmental bacteria.

Antibiotics have their origin in nature as molecules that are secreted by one type of bacteria to inhibit thegrowth of a competitor. It is believed that antibiotic-producing bacteria are the origin of antibiotic resistance genes, since these organisms must have the capacity to resist the effects of the antibiotics they produce.

ANTIBIOTICS AND THE ENVIRONMENT

Antibiotics are widely used by humans in clinical settings and to treat or prevent bacterial infections in companion and food animals. The Animal Health Institute and the Union of Concerned Scientists have estimated that in the United States alone, several million pounds of antibiotics are consumed per year in the treatment of humans and livestock. Antibiotics, especially nonbiodegradable antibiotics, can enter aquatic and terrestrial environments through avenues such as farms, wastewater, and animal manure containing these drugs. For example, a Norwegian study that investigated the release of pharmaceuticals in the wastewater effluent from two Oslo hospitals found that detectable levels of several antibiotics were present in both hospitals' wastewater. Furthermore, the same study showed that influent and sludge samples from a wastewater treatment works that serves the region contained detectable amounts of most of the antibiotics found in the hospital wastewater. Effluent samples from the treatment works contained reduced amounts of antibiotics; however, the levels of the antibiotic trimethoprim were higher in the effluent than in the influent samples. High levels of the antibiotic ciprofloxacin were found in the sludge samples. Antibiotics released into the environment have the potential to generate harmful and possibly deadly antibiotic-resistant bacteria.

ANTIBIOTIC-RESISTANT BACTERIA

Antibiotic resistance in a bacterium can be the result of a random mutation in the target molecule or the transfer of antibiotic resistance genes from one bacterium to another. A random mutation can result in the positive selection of bacterial mutants resistant to the antibiotic. Another route whereby bacteria can gain resistance is through horizontal gene transfer (HGT). During HGT, a bacterium can gain an antibiotic resistance gene through transformation, transduction, or conjugation. In most bacteria, the major route of transmission of antibiotic resistance genes is through the conjugation of plasmids, which are extra chromosomal pieces of deoxyribonucleic acid (DNA) that are capable of autonomous replication. It is important to note that HGT can occur almost anywhere that bacteria exist, including the environment and the human gastrointestinal tract.

One example of an antibiotic resistance gene is the *bla* gene, which codes for the TEM-1 beta-lactamase enzyme. Beta-lactamases such as TEM-1 can cleave the beta-lactam ring of beta-lactam antibiotics such as ampicillin. This cleavage results in the inactivation of the antibiotic. In an effort to increase the efficacy of some antibiotic drugs, some manufacturers add to antibiotic preparations other compounds that play a role in inactivating the biochemical mechanism in the resistant bacteria that degrades the antibiotic. For example, the drug Augmentin is a mixture of amoxicillin (a beta-lactam-containing antibiotic) and potassium clavulanate. The clavulanate is a beta-lactamase inhibitor—that is, it inhibits the beta-lactamase that would cleave amoxicillin and render it ineffective.

Any type of resistance to antibiotics that is gained by a bacterial population as a result of selective pressure by the population's natural habitat is called intrinsic resistance. Acquired resistance by bacteria, in contrast, is brought about as a result of antibiotic use by humans; such acquired resistance has been involved in the development of antibiotic-resistant human pathogens.

DISSEMINATION OF ANTIBIOTIC-RESISTANT BACTERIA

The dissemination of antimicrobial resistance is not restricted by geographic, phylogenetic, or ecological boundaries, especially given the global economy of the modern world. The use of antibiotics in one ecological niche, such as in agriculture or aquaculture, may give rise to antimicrobial resistance in other ecological niches, such as the human environment. The spread of resistance can be indirect, such as through the horizontal transfer of resistance genes through plasmids, or direct, where antibiotic-resistant bacteria are transferred by direct contact with humans. An example of indirect transfer comes from a study that showed that a plasmid encoding for cephalothin resistance was transferable to *Escherichia coli* (*E. coli*), a bacteria in the human gut, from *Vibrio* strains found in a shrimp pond. Direct transfer can include the intake of harmful antibiotic-resistant bacteria from contaminated drinking water, food animal sources (such as chickens and turkeys), and vegetables.

Infections caused by antibiotic-resistant bacteria present a special challenge to physicians, especially in health care settings. Poor hygiene practices in hospitals (such as failure to decontaminate surgical or respiratory equipment), nursing homes, and dialysis centers have led to the spread and promotion of multidrug-resistant (MDR) bacteria, which usually have resistance to three or more antibiotics. One study on burn patients who had been hospitalized for a long period of time in an intensive care unit showed that the patients carried significantly more MDR bacteria than they had on admission. Methicillin-resistant *Staphylococcus aureus* (or MRSA) is resistant to a wide variety of beta-lactam antibiotics and has become a highly adaptable, drug-resistant pathogen that kills about seventeen thousand hospital patients per year. Antibiotic resistance increases morbidity and mortality from infectious diseases as well as the costs of treating such diseases. With the lack of many new antibiotics on the horizon, antibiotic resistance poses a serious public health dilemma.

Sandra S. Szegedi

FURTHER READING

Heuer, Ole E., et al. "Human Health Consequences of Use of Antimicrobial Agents in Aquaculture." *Clinical Infectious Diseases* 49 (2009): 1248-1253.

Keen, Edward F., III, et al. "Prevalence of Multidrug-Resistant Organisms Recovered at a Military Burn Center." *Burns: Journal of the International Society for Burn Injuries* (January, 2010).

Martinez, Jose L. "The Role of Natural Environments

in the Evolution of Resistance Traits in Pathogenic Bacteria." *Proceedings of the Royal Society B* 276 (2009): 2521-2530.

Rice, Louis B. "The Clinical Consequences of Antimicrobial Resistance." *Current Opinion in Microbiology* 12 (2009): 476-481.

Sachs, Jessica Snyder. *Good Germs, Bad Germs: Health and Survival in a Bacterial World.* New York: Hill & Wang, 2007.

Thomas, Kevin V., et al. "Source to Sink Tracking of Selected Human Pharmaceuticals from Two Oslo City Hospitals and a Wastewater Treatment Works." *Journal of Environmental Monitoring* 9 (2007): 1410-1418.

Walsh, Christopher. *Antibiotics: Actions, Origins, Resistance.* Washington, D.C.: ASM Press, 2003.

SEE ALSO: Bacterial resistance; Chloramphenicol; Horizontal gene transfer; Waste management; Waste treatment.

Antienvironmentalism

CATEGORIES: Philosophy and ethics; activism and advocacy

DEFINITION: Philosophy that holds that human beings' immediate economic and lifestyle needs are more important than concerns about the fate of other species and the general environment

SIGNIFICANCE: The long-standing debates that continue between antienvironmentalists and environmentalists have important influence on both legislators and policy makers as they address environment-related issues.

In the early twentieth century, environmentalism in the United States was largely fostered by wealthy sportsmen who saw the need to protect the outdoors in order to maintain satisfactory areas for their pursuits of hunting, fishing, and camping. The movement got a populist boost in 1962 from the publication of Rachel Carson's *Silent Spring*, which presents an easily understood account of the dangers of toxic substances in the environment. For the first time, the American public began to demand that laws be enacted to protect the environment and clean up land, water, and air that had already been polluted.

GROWTH OF THE ENVIRONMENTAL MOVEMENT

For several years the environmental movement gathered strength as the public voted into office politicians with environmental orientations. Public outcry surged against polluting companies, leading to boycotts of products. Grassroots, citizen-led efforts such as recycling programs and litter patrols gained support as the public became more educated and concerned about environmental issues. Among the issues that pitted environmentalists against the government and industry were toxic waste incineration, habitat destruction by logging and mining companies, and use of public lands, including national parks.

Two oil crises during the 1970's served to focus awareness on energy conservation and the need to develop alternatives to energy derived from fossil fuels. Many feared that oil supplies were dwindling, while others wished to end U.S. reliance on oil-exporting nations in the Middle East. One important result was a general reduction in the size of motor vehicles. This,a long with other technological advances, helped lead to the development of cars that were more fuel-efficient. The research required to accomplish these changes, however, was very costly to automakers.

The 1970's were also characterized by landmark legislation that imposed strict limits on pollution output and resource use and also provided for the remediation of polluted land and water. Large fines were imposed for violations of the new laws, which were enforced by the newly formed U.S. Environmental Protection Agency (EPA). One of the most important and pivotal developments was the passage of the Comprehensive Environmental Response, Compensation, and Liability Act, or Superfund, which provided vast sums of public money for the cleanup of designated industrial and military waste dumps and other degraded sites. Signed into law by the U.S. Congress in 1980, Superfund's provisions allowed the government to bring lawsuits against the responsible parties, requiring them to help pay cleanup costs. In order to avoid fines, many industries were forced to develop and implement costly waste-processing technologies.

The political and economic situation began to change in the late 1970's as industry mounted a counteroffensive against environmental laws. Businesses contended with the burgeoning number of environmental regulations by finding and exploiting loopholes in legislation. A growing number of industries used stalling tactics and countersuits to delay or eliminate the need to implement required changes. Meanwhile, in the western United States, a coalition of loggers, miners, cattle ranchers, farmers, and developers demanded that the federal government transfer control of large tracts of federally owned land to individual states. Members of the so-called Sagebrush Rebellion felt that state ownership would give them more power to exploit the natural resources on the land.

ANTIENVIRONMENTALIST BACKLASH

A severe backlash against environmentalism began to occur when Ronald Reagan replaced Jimmy Carter as president of the United States in early 1981. Many environmental laws and regulations, which the new presidential administration viewed as barriers to economic progress, were weakened or abolished. A large number of federal judges who had started their careers during the 1960's retired, and they were re-placed by politically conservative judges who began to interpret existing laws in favor of industry. The office of the EPA was weakened, and funding for environmental enforcement and remediation was slashed. Secretary of the Interior James Watt, who had been a leader of the Sagebrush Rebellion, promoted legislation to open previously protected areas to mining and oil exploration. The general public, experiencing growth and prosperity for the first time in many years, began to favor short-term economic gains and turned a blind eye to news of the weakening environmental movement.

The late 1980's saw the birth of the wise-use movement, which appeals to the pragmatic and optimistic aspects of human nature by asserting that some optimal balance of resource use and restoration is practicable and that technology, given time and funding, will develop workable solutions to existing environmental problems. This position assumes that human beings can understand the complex ecosystems involved well enough to know what these balances should be. Advocates of wise use believe that all public lands, including national parks, should be opened to mining and drilling. Like the Sagebrush Rebels, they also promote the strengthening of the rights of states and property owners to exploit resources with minimal federal regulation.

According to the tenets of wise use, the harvesting of timber from ancient forests would be followed by the planting of an equivalent acreage of saplings. Logging would be timed according to growth rates, and technology would produce fast-growing varieties of trees that would furnish adequate ecosystems for wildlife in the new forests. Environmentalists, in contrast, argue that ancient forests represent complex, irreplaceable ecosystems that cannot be substituted with new forests planted by logging companies. Similar disagreement exists regarding coastal wetlands, which provide vital habitat to numerous species and contain a high degree of biodiversity. Wetlands are

---

Equating Environmentalism with Marxism

Llewellyn H. Rockwell begins "An Anti-Environmental Manifesto" by comparing environmentalism with Marxism-Leninism:

The last Stalinist, Alexander Cockburn, has gone from attacking Gorbachev (for selling out Brezhnev) to defending Mother Earth. His new book, *The Fate of the Forest*, is both statist and pantheist.

Cockburn, a man who supposedly cares about peasants and workers, instead decries their cutting down the Brazilian rainforests to farm and ranch. People are supposed to live in indentured mildewtude so no tree is touched.

But Cockburn is part of a trend. All over Europe and the U.S., Marxists are joining the environmental movement. And no wonder: environmentalism is also a coercive utopianism—one as impossible to achieve as socialism, and just as destructive in the attempt. . . .

Today we face an ideology every bit as pitiless and messianic as Marxism. And like socialism a hundred years ago, it holds the moral high ground. Not as the brotherhood of man, since we live in post-Christian times, but as the brotherhood of bugs. Like socialism, environmentalism combines an atheistic religion with virulent statism. But it ups the ante. Marxism at least professed a concern with human beings; environmentalism harks back to a godless, manless, and mindless Garden of Eden.

If these people were merely wacky cultists, who bought acres of wilderness and lived on it as primitives, we would not be threatened. But they seek to use the state, and even a world state, to achieve their vision.

And like Marx and Lenin, they are heirs to Jean Jacques Rousseau. His paeans to statism, egalitarianism, and totalitarian democracy have shaped the Left for 200 years, and as a nature worshipper and exalter of the primitive, he was also the father of environmentalism.

During the Reign of Terror, Rousseauians constituted what Isabel Paterson called "humanitarians with the guillotine." We face something worse: plantitarians with the pistol.

frequently located in areas that are desired by real estate developers wishing to build vacation homes or resorts. Environmental protection laws based on the tenets of wise use mandate that destroyed wetlands must be replaced by new wetlands of equivalent size. Again, environmentalists worry that too many threatened species would be lost in the process of destroying and replacing wetland areas.

SCIENTIFIC CONTROVERSY

Another significant trend in antienvironmentalism toward the end of the twentieth century involved public confusion over scientific debates concerning such topics as the ozone hole and global warming. Industry and government scientists often questioned and condemned dire predictions advanced by other scientists. The government frequently responded by requesting additional research before requiring vast, expensive reductions of known pollutants. Rather than believe the frightening scenarios painted by some scientists, many people sided with scientists who questioned the validity of these and other threats to the global environment.

Mainstream environmental organizations that had evolved from small groups of fervent individuals were now led by full-time professional lobbyists based in Washington, D.C. Details of new legislation were negotiated among industry, government, and environmental leaders. National environmental organizations grew increasingly cumbersome and expensive to run. Many began accepting large donations from the same industries they were trying to monitor, which created serious conflicts of interest. Top industry executives became members of the boards of directors for environmental organizations. At the same time, these corporations also made large donations to elected officials and thus gained access and influence in government. As the 1990's progressed, membership in large environmental groups began to decline as "donor fatigue" set in, questions persisted about the true urgency of various environmental issues, and cynicism arose about the possibility of environmental progress under such circumstances.

A relaxation of concern about environmental problems came about in the late 1990's in the wake of encouraging news about improvements of environmental indicators such as air-pollution levels of certain gases in the aftermath of implementation of cleaner energy production. For example, atmospheric levels of sulfur dioxide, which leads to acid rain, decreased in the United States and Europe after cleaner coal- and oil-burning technologies were implemented. The air-quality goals of many cities were met through a combination of fuel efficiency and "scrubber" smokestacks.

A controversial strategy advanced by a coalition of industry, government, and environmental leaders involves tradable pollution permits. According to the plan, the government assigns utilities a certain number of pollution units per year. An especially clean-running plant will not need all of its units and will be able to sell them to plants that exceed their allotments. This system has been criticized by many environmentalists, who contend that some utilities are able to buy their way out of the need to reduce pollution. The position thought to be antienvironmentalist in this context would maintain that the plan is a realistic method of controlling overall levels of pollution without putting older utilities out of business while they endeavor to upgrade their performance.

The revelation that environmental degradation can, in certain cases, be reversed over a fairly short span of time led to the argument on the part of antienvironmentalists that nature is surprisingly resilient, and therefore environmental protection does not need to be so stringent, costly, and regressive. Environmentalists counter such arguments with a call to remain vigilant and to include the health of the environment in national and global visions of the future.

Wendy Halpin Hallows

FURTHER READING

Dowie, Mark. *Losing Ground: American Environmentalism at the Close of the Twentieth Century.* Cambridge, Mass.: MIT Press, 1995.

Easterbrook, Gregg. *A Moment on the Earth: The Coming Age of Environmental Optimism.* New York: Penguin Books, 1995.

Helvarg, David. *The War Against the Greens: The "Wise-Use" Movement, the New Right, and the Browning of America.* Rev. ed. Boulder, Colo.: Johnson Books, 2004.

Hirt, Paul W. *A Conspiracy of Optimism: Management of the National Forests Since World War Two.* Lincoln: University of Nebraska Press, 1994.

Jacques, Peter J. *Environmental Skepticism: Ecology, Power, and Public Life.* Burlington, Vt.: Ashgate, 2009.

McKibben, Bill. *The End of Nature.* 1989. Reprint. New York: Random House, 2006.

Shabecoff, Philip. *A Fierce Green Fire: The American Environmental Movement.* Rev. ed. Washington, D.C.: Island Press, 2003.

Young, John. *Sustaining the Earth: The Story of the Environmental Movement—Its Past Efforts and Future Challenges.* Cambridge, Mass.: Harvard University Press, 1990.

SEE ALSO: Climate change skeptics; Global warming; Pollution permit trading; Public opinion and the environment; Sagebrush Rebellion; Watt, James; Wise-use movement.

Antinuclear movement

CATEGORIES: Activism and advocacy; nuclear power and radiation

IDENTIFICATION: Social movement comprising a loose collection of organizations and individuals opposed to nuclear weapons and nuclear power

SIGNIFICANCE: The antinuclear movement, which emerged during the 1950's and gained momentum during the 1960's, initially focused its attention on the proliferation of nuclear weapons and the threat of nuclear warfare. It later broadened its scope to include opposition to nuclear power plants and nuclear waste facilities. Antinuclear activists have had some success in raising public awareness of the dangers of nuclear proliferation and in using safety regulations and environmental law to slow the development of nuclear power, particularly in the United States.

Antinuclear activists generally believe that if nuclear weapons are available, these weapons will eventually be used; they therefore seek to rid the earth of all such armaments. By organizing seminars, rallies, protests, and public outreach campaigns, activists seek to stimulate public debate on issues previously left to insiders, and they help shape the political climate. For example, activists from the environmental organization Greenpeace drew international attention to French resumption of nuclear weapons testing in 1995. The outrage that this publicity fueled may have contributed to the French government's decision to conduct fewer tests than planned.

Antinuclear activists have often been successful in capturing favorable media coverage and mobilizing hundreds of thousands of citizens to protests. However, translating this success into practical results has often proved elusive. At a conference on abolishing nuclear weapons held at Boston College in October, 1997, American Friends Service Committee staffer David McCauley offered a possible explanation. He suggested that the movement is fundamentally anarchistic and that distrust of people in power stifles activists' ability to cooperate with established political practices.

NUCLEAR FREEZE AND DISARMAMENT GOALS

In the multilateral Nuclear Non-Proliferation Treaty (NNPT), which entered into force in 1970, five recognized nuclear weapons states—the United States, the United Kingdom, France, China, and the Soviet Union—agreed not to transfer nuclear weapons to states without such weapons. With the treaty, signatory nations committed to work toward nuclear disarmament, yet the arms race continued. Antinuclear activists claimed that arms negotiators were allowing themselves to be defeated by complexities and suggested a mutual and verifiable freeze by the United States and the Soviet Union on the testing, production, and deployment of all nuclear weapons. The campaign attracted sympathetic media coverage and gained momentum, so that by the late 1970's hundreds of thousands of people were attending massive rallies in the United States and Europe to protest the planned deployment of the ground-launched cruise missile (GLCM) and the Pershing II missile.

Leaders of the North Atlantic Treaty Organization (NATO), however, dared not ignore the complexities inherent in nuclear negotiations. In 1976 the Soviet Union began updating its forces by deploying SS-20 missiles. NATO viewed this as a destabilization of the balance of power because the SS-20's were mobile, more accurate, equipped with three warheads each, and able to fly 1,500 to 3,000 kilometers (930 to 1,860 miles) farther than the missiles they replaced. After all attempts at diplomatic negotiations with the Soviets failed, NATO responded in 1983 by deploying Pershing IIs and GLCMs, which were more accurate than the SS-20's but had less than one-half their range.

NATO and the Soviet Union eventually agreed to intrusive, on-site verification procedures. In late 1987, the two nations signed the Intermediate-Range Nuclear Forces Treaty, which completely eliminated an entire class of weapons: SS-20's, Pershing IIs, GLCMs, and all other ground-launched nuclear missiles with ranges between 500 and 5,500 kilometers (310 and 3,420 miles), along with their launchers, support facilities, and bases. The treaty also banned flight testing and production of these missiles. The antinuclear movement deserves credit for helping educate the public and adding to the pressure that pushed politicians to the bargaining table. However, based on history, it seems unlikely that the treaty would have come about had NATO not deployed its own weapons.

In 1995, the year the NNPT was extended indefinitely, an antinuclear initiative called Abolition 2000 was proposed. Within a few years, more than twelve hundred nongovernmental organizations on six continents had voiced support for it; by 2010 it had grown to a network of more than two thousand organizations in more than ninety countries. Abolition 2000 called for movement toward "clean, safe, renewable forms of energy production that do not provide the materials for weapons of mass destruction and do not poison the environment for thousands of centuries."

It also called for negotiations on a nuclear weapons abolition convention that would establish a timetable for phasing out all nuclear weapons and include provisions for effective verification and enforcement.

At its 2003 annual meeting, Abolition 2000 launched a collaborative effort with Mayors for Peace, an international program established in 1982 in which the heads of city governments around the globe work to encourage city-by-city support of the abolition of nuclear weapons. By August 6, 2010 (the sixty-fifth anniversary of the first use of the atomic bomb), 4,069 cities in 144 countries and regions around the world had become members of the Conference of Mayors for Peace and were calling for complete nuclear disarmament by 2020.

Nuclear Power in the United States

Many members of the antinuclear movement are also against the development and expansion of nuclear power generation. They believe that nuclear technology is too dangerous and see reactor accidents

A "ban the bomb" rally is held in London, England, in April, 1961. The beginnings of the antinuclear movement focused on the elimination of nuclear weapons. (AP/Wide World Photos)

caused by natural disasters, equipment failures, or human errors as inevitable. They do not believe that radioactive waste can be safely disposed of, and they see such waste as an unfair burden to pass on to future generations. Furthermore, many argue that terrorists or nations without their own nuclear arsenals could divert reactor materials to make nuclear weapons.

During the 1960's the antinuclear movement became concerned with the possible effects of low-level radiation from nuclear power plants. Scientists are divided as to whether low levels of radiation can cause increased incidence of cancer in communities located near nuclear power plants; whereas some studies point to significant risk, others report considerable evidence that the human body is able to repair damage caused by sufficiently low levels of radiation.

The antinuclear movement has been somewhat successful in using safety regulations and environmental law to effect change. By lobbying for stricter regulations and suing to force nuclear plants to follow safety regulations that they previously had been allowed to bypass, activists have contributed to making nuclear plants safer. These actions have also added to the costs of nuclear power plants and to delays in plant licensing. Using the courts to delay construction of nuclear plants has been a powerful tool for the movement. In 1967 construction time for a nuclear power plant in the United States averaged 5.5 years; by 1980 it had reached 12 years.

The price of electricity from nuclear plants has increased over time with lengthening construction time, the addition of safety features, and the tendency of nuclear power companies to try new designs instead of settling on a standard design. In 1976 the price of electricity from coal-fired and nuclear plants was nearly the same, but by 1990 nuclear power was twice as expensive as coal power in the United States. One goal of the antinuclear movement was reached in the United States: With economics against them, planners ceased construction of new nuclear power plants.

The U.S. Nuclear Regulatory Commission (NRC) eventually combined the construction and operating licensing procedures to minimize delays. Also, the industry significantly reduced operating costs. These factors, combined with mounting concerns about the finite nature of fossil-fuel supplies and the potential of the burning of such fuels to affect global climate, made nuclear power a more attractive prospect by the early twenty-first century. Some previously antinuclear environmentalists reconsidered their stance on nuclear power and came out in favor of it as an alternative to fuels that produce greenhouse gas emissions.

NUCLEAR POWER IN ASIA AND EUROPE

According to the International Atomic Energy Agency, in 2008 nuclear power provided 19.7 percent of the electricity in the United States, 24.9 percent in Japan, 28.8 percent in Germany, 42.0 percent in Sweden, and 76.2 percent in France. Countries that must import much of their fuel, such as Japan and France, find nuclear power particularly attractive.

Antinuclear sentiment led Sweden to announce a phaseout of nuclear power beginning in 1998. The shutdown was delayed, however, for several reasons, including lack of replacement power, fears by workers and industry that the shutdown would lead to higher energy prices and exacerbate unemployment, and a lawsuit by a nuclear plant's owners seeking indemnification. In 2009, citing climate concerns, the Swedish government officially abolished its phaseout scheme. Some 52 percent of the Swedes who responded to a 2010 survey favored keeping nuclear power and replacing old reactors with new ones.

As part of the price of forming a ruling coalition with Germany's Social Democratic Party, the environmentalist Green Party extracted a promise that Germany's nuclear power reactors would eventually be eliminated. Shutdown plans were made official in 2002. Germany's government faced problems similar to Sweden's in trying to phase out nuclear power and felt additional pronuclear pressure from England and France, which held contracts worth $6.5 billion to reprocess German nuclear fuel. By 2008 concerns about power availability, costs, and carbon dioxide emission reduction goals had prompted Germany to explore how to keep its existing nuclear power plants online, even as thousands of protesters turned out in opposition to the shipment of nuclear reactor waste from France to a German storage site in Gorleben.

SCIENTIFIC GROUPS

A few scientific groups have been especially influential in the area of nuclear arms control. Many scientists who developed the atomic bomb were against leaving decisions about its uses to a few elite government and military officials. Immediately after World War II, some of them formed the Atomic Scientists of Chicago and began publication of the *Bulletin of the Atomic Scientists*. Another group formed the Federation of Atomic Scientists. The efforts of these scientists contributed to the founding in 1946 of the Atomic Energy Commission, a civilian agency that took control of the materials, facilities, production, research, and information relating to nuclear fission from the military.

In response to the escalating arms race, in 1955 the physicist Albert Einstein and the philosopher, mathematician, and social critic Bertrand Russell published a manifesto in which they called upon scientists to assemble and appraise the perils of nuclear weapons. Cyrus Eaton, a wealthy industrialist and admirer of Russell, invited twenty-two scientists from both sides of the Iron Curtain to a conference in July, 1957, that was held in Eaton's summer home in the small village of Pugwash, Nova Scotia. The scientists were able to function as icebreakers between governments; in fact, some were government advisers. More Pugwash conferences followed, providing invaluable contacts, networks, and facts for those involved in arms control. The antinuclear movement in general and scientific organizations in particular were instrumental in bringing about several treaties, including the Limited Test Ban Treaty of 1963, in which signatory nations agreed to end aboveground nuclear testing.

The Union of Concerned Scientists (UCS) was formed in 1969 by faculty and students at the Massachusetts Institute of Technology who felt that too much emphasis was being placed on research with military applications and not enough on research that could address environmental and social concerns. The UCS subsequently grew to become a coalition of scientists and citizens across the United States and expanded its focus to include renewable energy and other environmental issues. The UCS combated the establishment of an antiballistic missile defense system in the United States and played a key role in defeating a scheme to rotate two hundred MX missiles among 4,600 protective silos. The plan would have cost $37 billion and would have swallowed vast tracts of the western desert of the United States.

The UCS has also waged campaigns calling for U.S. support of the 1996 Comprehensive Nuclear-Test-Ban Treaty, which prohibits all nuclear explosions (the United States signed the treaty in 1996 but has not ratified it). The UCS has worked in support of other nuclear disarmament treaties as well, including a series of agreements between the United States and the Soviet Union (later the Russian Federation) limiting the number of nuclear warheads and delivery vehicles that these nations may deploy: the 1991 Strategic Arms Reduction Treaty (START), the 2002 Strategic Offensive Reductions Treaty (SORT), and the 2010 New START.

ALTERNATIVE ENERGY SOURCES

One of the great challenges facing the antinuclear movement is that of finding alternative sources of energy. Many nonnuclear power plants have serious environmental effects of their own. By some estimates, cardiopulmonary illness linked to air pollution accounts for tens of thousands of deaths in the United States each year. Power plants are major contributors to such pollution, with older, coal-fired plants being among the worst offenders. Burning coal not only produces copious amounts of carbon dioxide but also releases particulates, sulfur compounds, lead, arsenic, mercury, naturally occurring radioactive elements, and other harmful elements. Pollution-control equipment increases the construction and operating costs of new coal-fired plants. Further, running this equipment expends energy, and the equipment can only reduce pollution, not eliminate it.

Many segments of the antinuclear movement actively support the development of alternative energy sources, such as solar, geothermal, wind, and biomass power. The United States leads the world in the use of alternative energy sources, but by 2008 the U.S. Department of Energy reported that renewable energy sources (including hydropower) accounted for only 9 percent of the electricity generated in the nation.

Charles W. Rogers
Updated by Karen N. Kähler

FURTHER READING

Bodansky, David. *Nuclear Energy: Principles, Practices, and Prospects.* 2d ed. New York: Springer, 2004.

Cooke, Stephanie. *In Mortal Hands: A Cautionary History of the Nuclear Age.* New York: Bloomsbury, 2009.

Cortright, David, and Raimo Väyrynen. *Towards Nuclear Zero.* New York: Routledge, 2010.

Evangelista, Matthew. *Unarmed Forces: The Transnational Movement to End the Cold War.* 1999. Reprint. Ithaca, N.Y.: Cornell University Press, 2002.

Giugni, Marco. *Social Protest and Policy Change: Ecology, Antinuclear, and Peace Movements in Comparative Perspective.* Lanham, Md.: Rowman & Littlefield, 2004.

Gusterson, Hugh. *Nuclear Rites: A Weapons Laboratory at the End of the Cold War.* Berkeley: University of California Press, 1996.

Peterson, Christian. *Ronald Reagan and Antinuclear Movements in the United States and Western Europe, 1981-1987*. Lewiston, N.Y.: Edwin Mellen Press, 2003.

Price, Jerome. *The Antinuclear Movement*. Rev. ed. Boston: Twayne Publishers, 1990.

Wittner, Lawrence S. *Confronting the Bomb: A Short History of the World Nuclear Disarmament Movement*. Stanford, Calif.: Stanford University Press, 2009.

_____. *Toward Nuclear Abolition: A History of the World Nuclear Disarmament Movement, 1971 to the Present*. Stanford, Calif.: Stanford University Press, 2003.

SEE ALSO: Coal-fired power plants; Commoner, Barry; Comprehensive Nuclear-Test-Ban Treaty; Green movement and Green parties; Greenpeace; Intergenerational justice; Limited Test Ban Treaty; Nuclear accidents; Nuclear and radioactive waste; Nuclear power; Nuclear Regulatory Commission; Nuclear testing; Nuclear weapons production; SANE; Union of Concerned Scientists.

Antiquities Act

CATEGORIES: Treaties, laws, and court cases; preservation and wilderness issues

THE LAW: U.S. federal law empowering the president to set aside as national monuments public lands deemed to be of historic or scientific importance

DATE: Enacted on June 8, 1906

SIGNIFICANCE: The Antiquities Act was one of the earliest pieces of U.S. federal legislation aimed at historic and cultural preservation. Since passage of the act, U.S. presidents have used it more than one hundred times to protect natural phenomena as well as human-built sites of cultural interest.

The original impetus for the Antiquities Act of 1906 was public reaction to the widespread looting of prehistoric Native American ruins in the Southwest during the late nineteenth century. One of the individuals seeking some form of federal protection for these sites was anthropologist Edgar Lee Hewett, and Congressman John F. Lacey, a Republican from Iowa, joined in the effort during the early part of the century. President Theodore Roosevelt signed the resulting legislation into law in June of 1906, and three months later he used the act for the first time to establish the Devils Tower National Monument in Wyoming.

All subsequent U.S. presidents have exercised the powers given them by the law to establish a wide variety of protected sites, from natural phenomena such as the Grand Canyon in Arizona and the Muir Woods in California to historic sites such as the Statue of Liberty and the birthplace of George Washington. The environmental significance of the act was expanded during the presidency of Jimmy Carter when it was used (in 1978) to establish a series of fifteen national monuments preserving a total of 22.7 million hectares (56 million acres) of wilderness land in Alaska, protecting the land from oil drilling and other forms of commercial development.

Scott Wright

SEE ALSO: Carter, Jimmy; Grand Canyon; National parks; Preservation; Roosevelt, Theodore; Wilderness areas; Wildlife refuges.

Appropriate technology

CATEGORY: Resources and resource management

DEFINITION: Technology that is developed with the needs of the intended users and potential impacts on the environment taken into account

SIGNIFICANCE: The rise of the appropriate technology movement during the 1970's was a response to a growing appreciation of the idea that the people most qualified to find technological solutions to their societies' needs are the people who live in the societies and understand how various technologies will affect their environment.

The beginning of what would become the appropriate technology (AT) movement is commonly credited to Mohandas Gandhi, who helped lead India to independence from Great Britain during the 1930's and 1940's. Gandhi advocated the use of small, local, and mostly community-based technological solutions as a way to help people become self-reliant and independent. He rejected the common belief that modern, industrialized technological development always amounts to progress. Instead, he thought that technology is best produced in a decentralized manner and best used in ways that benefit the people as much as possible. Gandhi believed in a radical revolution of production; he rejected the factory model of industrialization, which values the product and production over the individual worker.

Intermediate Technology

Gandhi's ideas had a great influence on E. F. Schumacher, who is best known for his 1973 book *Small Is Beautiful: Economics As If People Mattered* and for coining the phrase "intermediate technology" (IT) and later creating the Intermediate Technology Development Group (ITDG). Intermediate technologies, in Schumacher's usage, are technologies and tools that are more effective and expensive than traditional or indigenous methods and techniques but ultimately much cheaper than technologies of the developed world. Intermediate technologies, whether hard or soft, can presumably be acquired and used by poor people and can lead to increased productivity with minimal social and environmental costs. They are usually labor-intensive, rather than capital-intensive, and make use of local skills and materials. They are conducive to decentralization and empowerment of the people, compatible with ecological considerations, and better able to serve the interests of the poor than more advanced technologies. More important, they do not contribute to making the people who use them dependent and subservient to machines over which they have no control.

Schumacher's formulation of an alternative approach came at a time of world energy crisis and a time of growing disenchantment with the broad social, political, and ecological implications of advanced technology; in particular, awareness was growing of the undesirable consequences of the wholesale transfer of such technologies to the less developed countries (LDCs). This may explain why, despite its problematic nature, it had a very substantial impact on the thinking of administrators and planners within aid agencies, as well as officials within the LDCs themselves. Later years witnessed a rapid expansion in the scale and scope of activities undertaken by ITDG (which in 2005 was renamed Practical Action) and the creation of similar organizations in many parts of the world.

Beginnings of AT

Around the time that Schumacher's concept of intermediate technology was catching on, the closely related concept of appropriate technology became enshrined in the operating principles of such agencies as the United Nations Environment Programme and the United Nations Industrial Development Organization, as well as other development organizations. The AT movement had taken off.

The concept of AT is not solely applicable to the nonmodern (as opposed to the modern) sectors of LDCs. In the developed countries, technological and engineering solutions that have minimal impacts on the environment and society are referred to as AT as well, in the sense that they are considered to be sustainable. Although Schumacher's proposal had an unquestionable impact, critics have been keen to point out that his construal of IT or AT excludes from consideration a whole range of issues relating to the interplay of political, social, and economic factors with technology. They have objected to his characterization of the dual economies of LDCs, in which the modern and nonmodern sectors are regarded as separate entities, such that activities designed to alleviate poverty in the latter are unaffected by the nature of the former, or of the forces giving rise to an apparently dual economy in the first place. Some observers have also questioned the underlying conception of technology at work in Schumacher's proposal, namely, that technology can somehow be regarded as a wholly independent variable in the development process, determining social, economic, and political relations, but in no sense being determined by them.

Some have even undermined Schumacher's position by turning it against itself. If the choice of industry is governed, according to Schumacher, by powerful forces that include entrepreneurial interests, why is it that these forces cease to operate somehow when the question of technology arises? Clearly, they cannot. By assuming otherwise, Schumacher is led into the contradictory position of arguing that radical social change in the form of alleviation of mass poverty in the LDCs can, by implication, be brought about without prior change or alteration of other elements in the political, economic, and social contexts within which it arises.

Resolving Contradictions

Despite efforts made over the years to reformulate Schumacher's view, proponents of AT have never really managed to escape fully from the contradiction of his original argument. The problem is both determined by and reflected in the way in which the concept of AT has taken over from its intermediate predecessor, assuming in the process a rather rigid set of connotations. It is difficult to locate the precise point in time at which AT replaced IT, and some would even argue that a valid distinction can still be drawn between the concepts. In general usage, however, the two terms have come to be virtually interchangeable, and the root of the problem may be traceable back to Schumacher himself and the clear impression he gives that IT is *the* appropriate solution for LDCs to pursue.

All these problems notwithstanding, some observers have argued, a renewed and strengthened effort can and must be made to reconceptualize the notion of appropriate, progressive, and sustainable technology and set it on new foundations. Proponents of this view assert that, given the morally outrageous and avoidable poverty plaguing the world, not to mention the dwindling supplies of some natural resources and the looming ecological catastrophe caused by runaway industrialization and technological explosion, such a task is a moral imperative. They argue that greater attention needs to be paid to questions of how technology is created, adapted, and modified in specific physical, social, cultural, political, and economic contexts, and by whom.

Nader N. Chokr

FURTHER READING

Darrow, Ken, and Mike Saxenian. *Appropriate Technology Sourcebook: A Guide to Practical Books for Village and Small Community Technology.* Rev. ed. Stanford, Calif.: Appropriate Technology Project, Volunteers in Asia Press, 1993.

Howes, Michael. "Appropriate Technology: A Critical Evaluation of the Concept and the Movement." *Development and Change* 10 (1979): 115-124.

Jequier, Nicholas. *Appropriate Technology: Problems and Promises.* Paris: OECD, 1976.

Schumacher, E. F. *Small Is Beautiful: Economics As If People Mattered.* 1973. Reprint. Point Roberts, Wash.: Hartley & Marks, 1999.

SEE ALSO: Green Revolution; Johannesburg Declaration on Sustainable Development; Naess, Arne; Recycling; Schumacher, E. F.; Sustainable development; United Nations Environment Programme; World Summit on Sustainable Development.

Aquaculture

CATEGORY: Agriculture and food

DEFINITION: Production of marine and freshwater food sources in controlled, farmlike environments in ponds, canals, lakes, and confined coastal areas

SIGNIFICANCE: Although aquaculture provides a good source of protein for the demands of a growing world population, as well as an economic endeavor for some, like many forms of farming it is not without its negative environmental impacts. Residues from fish wastes, pollution, coastal erosion, and impacts on adjacent species are some of the challenges to sustainable forms of aquaculture.

Aquaculture is one of the fastest-growing forms of food production in the world. It includes the processes of propagating, raising, and processing marine and freshwater food sources such as fish, shellfish, and even kelp. In some cases by-products of the production of these commodities produce fertilizers and feeds for other animals. Ornamental products such as cultured pearls are also produced within aquacultural systems.

The practice of aquaculture is not new. Fish farming has been a part of the cultures of Pacific islanders and Southeast Asian peoples for centuries. Early forms of aquaculture were quite rudimentary compared with modern techniques, however; they provided foods and other products on a limited scale, mostly for local villages. In the twenty-first century, sophisticated aquaculture systems control the complete life cycles of the fish and other animals under cultivation. In many of these modern facilities, automated feeding and harvesting systems alleviate the need for labor-intensive practices, reducing the costs of production. In some cases, however, because of the nature of the commodities, intensive human labor is still required. Oysters, for example, are harvested from beds and then must be opened by hand, a process known as shucking, so that the integrity of the shells is maintained for the seeding of a future crop.

One of the elements that has spurred the development of aquaculture is the decline in production in fisheries in the open oceans owing to overfishing. According to the United Nations Food and Agriculture Organization (FAO), all major fishing regions in the world are now being fished intensely. The needs of the growing world population for protein and water pollution affecting natural sources of seafoods have also contributed to the growth of aquaculture.

INDUSTRY GROWTH

Commercial aquaculture ventures have been established in nations throughout the world. The majority of these operations utilize naturally existing water sources in coastal regions, canals, or rivers; others are built at sites where traditional agriculture could not take place. An ideal aquaculture site has certain conditions that are favorable to the healthful management of the stocks of fish and other animals produced. These include clean water, appropriate water temperatures and levels of dissolved gases, good site topography, and food sources.

In the United States aquaculture operations are found in coastal regions and in dry interior areas, where ponds or rivers are used. Clams and oysters are harvested from beds in the marine environment of Washington State's Puget Sound, for example, while in the eastern interior of the state, the Columbia River provides a freshwater source for the raising of salmon, trout, and steelhead, all for commercial uses. In dry west Texas, gravel pits filled with saline waters were some of the first ponds used to cultivate shrimp, an industry that now supplies millions of pounds annually from ponds filled with natural saline water. The state of Virginia has developed a large aquaculture industry, growing clams and oysters in hatcheries.

Crawfish, tilapia, trout, shrimp, cod, clams, oysters, frogs, and catfish are just some of the products found in aquacultural operations in the United States. According to the U.S. National Oceanic and Atmospheric Administration, shellfish account for two-thirds of the total output of the U.S. aquaculture industry; salmon accounts for 25 percent, and shrimp for about 10 percent. Catfish is one of the most popular farmed fish in the United States.

The monetary value of worldwide aquaculture production is some $70 billion per year. FAO has estimated that almost half of the fish eaten in the world, nearly 45 billon tons of fish, is produced in aquacultural environments rather than caught in the open seas. According to FAO, the top ten aquaculture-producing nations are, in descending order, China, India, Vietnam, Thailand, Indonesia,

Bangladesh, Japan, Chile, Norway, and the United States. Even with its tenth-place ranking, the U.S. aquaculture industry contributes about $1 billion annually to the nation's domestic economy.

ENVIRONMENTAL CONCERNS

Aquaculture operations have a number of negative environmental impacts. Among these is that they generally require the installation of a considerable amount of technology, and thus disturb existing habitats, and they use large amounts of energy. Feed must be supplied to penned fish, and waste from the fish must somehow be managed. Diseases must be controlled, not only within the fish and other seafood stocks but also with respect to the possible spread of disease to adjacent native fish and other organisms outside the aquaculture pens or ponds. A particularly controversial element of aquaculture involves the genetic modification of many forms of fish stocks to enhance their ability to grow quickly under confined

Farm-raised salmon move across a conveyor belt as they are brought aboard a harvesting boat in an aquaculture operation near Eastport, Maine. (AP/ Wide World Photos)

conditions. Critics of these practices question the safety of eating such fish and have also voiced concerns that if genetically modified species escape their aquaculture pens they may introduce unforeseeable problems into natural fish populations.

In many aquacultural settings, self-contained ecosystems are set up to supply nutrients for fish stocks. In some cases sewage and even commercial fertilizers have been used in such ecosystems to stimulate the production of phytoplankton, which in turn is eaten by zooplankton and bottom dwellers. These are then eaten by fish stocks within the pen or pond. Unless a consistent flow of water can be maintained through the pen area where the fish stock is raised, the bottom can become silted up with waste. In confined estuaries this condition can have negative effects on fish outside the aquacultural operation. It has been suggested that the decay of fecal matter at the bottom of aquaculture pens and in areas where concentrated fish-rearing operations are conducted can contribute to the depletion of oxygen levels in the water near the bottom. This is especially the case if water currents are not efficient enough to purge the buildup of waste at the bottoms of these ponds and pens.

Shrimp, a major commodity grown aquaculturally, is a good example of some other environmental concerns. Aquaculture operations sometimes use antibiotics to control the diseases that shrimp are prone to acquire when they are being raised in confined areas. During the early part of 2000, the European Union banned shipments of shrimp containing chloramphenicol, an antibiotic used on shrimp that has been found to be dangerous for human health. In addition, fishing communities and coastal residents have been critical of shrimp aquaculture, asserting that it reduces fish catches because it has negative impacts on coastal habitats and because it uses wild fish as food for the shrimp being raised.

Additional concerns about the negative environmental impacts of aquaculture have to do with the use of pesticides and other chemicals in the management of unwanted organisms in aquacultural waters. To control infestations of fish lice, for example, aquaculture operations often disperse pesticides into the water. Other chemicals are often used to cut down on algae growth on the nets and floats used in pen operations.

Within food webs or trophic levels, energy flows from lower to higher levels. If a larger fish eats a smaller fish, not all of the energy available from the

U.S. Aquaculture Production, 2006

	THOUSANDS OF POUNDS	METRIC TONS	THOUSANDS OF DOLLARS
Finfish			
Baitfish	—	—	38,018
Catfish	566,131	256,795	498,820
Salmon	20,726	9,401	37,439
Striped bass	11,925	5,409	30,063
Tilapia	18,738	8,500	32,263
Trout	61,534	27,912	67,745
Shellfish			
Clams	12,564	5,699	72,783
Crawfish	80,000	36,288	96,000
Mussels	962	436	4,990
Oysters	13,711	6,219	92,602
Shrimp	8,037	3,646	18,684
Miscellaneous	—	—	254,738
Totals	**794,328**	**360,305**	**1,244,145**

Source: Data from the National Oceanic and Atmospheric Administration, National Marine Fisheries Association.

Note: Miscellaneous includes ornamental and tropical fish, alligators, algae, aquatic plants, eels, scallops, crabs, and others.

smaller fish is passed on to the larger. This transference of energy is known as ecological efficiency. Aquaculture can exhibit very low levels of ecological efficiency. For example, farm-raised salmon are reared on food made from other fish, which are caught and ground up before being fed to the salmon as pellets or in meal form. It has been estimated that it takes 10 grams (0.35 ounce) of feed to produce 1 gram (0.035 ounce) of salmon. Considering the pressure already existing on wild fish stocks, it has been argued that this low level of ecological efficiency is a considerable detriment for the economy of aquaculture. In contrast, some kinds of aquaculture operations, such as catfish farms, are relatively ecologically efficient, and catfish, which are herbivorous, require no fish-based foods.

Aquaculture operations are also sometimes negatively affected by environmental conditions that come from outside their confines. Along coastlines and in low-lying areas, for example, pollution from agricultural irrigation runoff can cause fish kills such as those that have taken place in aquaculture installations in Indonesia, Malaysia, and the Philippines. Washington State enacted a law to control the flow of manure effluent from dairy farms into the Nooksack River. Prior to the control of this pollution, coastal clam beds managed by the Lummi Nation were decimated; this aquacultural operation recovered after passage of the law.

M. Marian Mustoe

FURTHER READING

Jahncke, Michael L., et al., eds. *Public, Animal, and Environmental Aquaculture Health Issues.* Hoboken, N.J.: John Wiley & Sons, 2002.

Mathias, Jack A., Anthony T. Charles, and Hu Baotong, eds. *Integrated Fish Farming.* Boca Raton, Fla.: CRC Press 1997.

Miller, G. Tyler, Jr., and Scott Spoolman. "Food, Soil, and Pest Management." In *Living in the Environment: Principles, Connections, and Solutions.* 16th ed. Belmont, Calif.: Brooks/Cole, 2009.

Pilay, T. V. R., and M. N. Kutty. *Aquaculture: Principles and Practices.* 2d ed. Ames, Iowa: Blackwell, 2005.

Stickney, Robert R. *Aquaculture: An Introductory Text.* 2d ed. Cambridge, Mass.: CABI, 2009.

SEE ALSO: Antibiotics; Asia; Chloramphenicol; Commercial fishing; Fish and Wildlife Service, U.S.; Fisheries; Introduced and exotic species; Marine debris; Water pollution.

Aquaponics

CATEGORY: Agriculture and food

DEFINITION: Combination of aquaculture and hydroponics that employs engineered aquatic ecosystems to grow edible plants and animals

SIGNIFICANCE: Aquaponics is a method of food production that has few of the negative impacts on the environment seen with traditional agriculture. Aquaponics can efficiently produce food for local consumption without the use of pesticides or fertilizers, without degrading soils, and without altering land use on a major scale.

The production of food to meet the needs of the world's rapidly growing population has many effects on the environment. At least 35 percent of ice-free land is used to grow food for human consumption, and agricultural practices account for 70 percent of water use, 25 percent of global warming potential, 85 percent of the six million tons of pesticides used annually, and nearly all of the 170 million tons of nitrogen, 50 million tons of phosphate, and 50 million tons of potash used annually. It has been estimated that in the United States, food travels about 3,219 kilometers (2,000 miles) before it reaches the consumer's plate, and about half of the food produced gets discarded before it can be consumed. High-production agricultural practices contribute to habitat loss and degradation, water pollution, salinization of soils, and the overdrawing of groundwater. Aquaponics, in contrast, is a technology that can efficiently produce food for local consumption without the use of pesticides or fertilizers, without degrading soils, and without altering land use on a major scale.

Aquaponics combines the time-tested practices of aquatic animal farming (aquaculture) and the growing of plants without soil (hydroponics) in ways that recirculate water and capitalize on the beneficial ecosystem services of both practices. In aquaponic systems, water from the aquaculture subsystem, which contains waste from the animals, is pumped into the hydroponics subsystem, where the nutrient-rich water fertilizes edible vegetation and the vegetation purifies the water, which then flows back to the aquaponics subsystem.

All aquaponic systems include basic components such as tanks, pumps, and vegetable beds, but the design configurations of these components are limited only by the creativity of their makers. Some aquaponic systems are as much art as they are food-generating systems. The materials used to construct aquaponic systems are often salvaged from discarded materials; this tendency to employ recycling further reduces the environmental and economic impacts of aquaponics. For example, aquaponic systems are often built from inexpensive repurposed barrels, tubs, pools, pumps, and plumbing. Whether new or recycled items are used to make aquaponic systems, those operating the systems must take care to ensure that toxins and other harmful substances are not allowed to leach into the systems.

Because aquaponics uses little water and space, it can be a source of food production practically anywhere in the world, even in places where traditional farming is impractical. Aquaponic systems range in size from tabletop 37.8-liter (10-gallon) fish tanks with small beds of vegetables to large industrial greenhouses with tanks holding more than 37.8 kiloliters (10,000 gallons) of water and floating racks of lettuce.

Aquaponics is a central component in the operation of Growing Power, a nonprofit organization located in Milwaukee, Wisconsin, that is dedicated to growing fresh, healthy food with community involvement. Growing Power, which was founded by former professional basketball player Will Allen, uses a variety of sustainable agricultural practices, including aquaponics, to grow food year-round in greenhouses. In a 1.2-hectare (3-acre) greenhouse, Growing Power uses aquaponics to raise approximately one million pounds of produce and ten thousand fish every year. Growing Power has demonstrated that food can be grown year-round even in an environment as cold as Milwaukee.

Greg Cronin

FURTHER READING

Bridgewood, Les. *Hydroponics: Soilless Gardening Explained.* Marlborough, Wiltshire, England: Crowood Press, 2003.

Nelson, Rebecca L. "Ten Aquaponic Systems Around the World." *Aquaponics Journal,* no. 46 (2007).

Roberto, Keith. *How-to Hydroponics.* 4th ed. Farmingdale, N.Y.: Futuregarden Press, 2005.

Stickney, Robert R. *Aquaculture: An Introductory Text.* 2d ed. Cambridge, Mass.: CABI, 2009.

SEE ALSO: Appropriate technology; Aquaculture; Biofertilizers; Community gardens; Composting; Eat local movement; Organic gardening and farming; Pesticides and herbicides; Rainwater harvesting; Sustainable agriculture.

Aqueducts

CATEGORY: Water and water pollution

DEFINITION: Artificial waterways constructed to move water from one area to another

SIGNIFICANCE: Human beings' redirection of water from one area to another through aqueducts can have numerous environmental effects. Ecosystems on both ends of an aqueduct are influenced by the reduction or increase in available water, and the building of aqueducts often involves the disturbance of what was formerly pristine land.

Sufficient water is an absolute necessity for all forms of life on earth (both plants and animals); this fact dictates that life-forms in areas that receive insufficient precipitation or that are not near lakes or rivers must either migrate to more favored regions to survive or develop techniques to bring water in from more favored areas. Certain xerophytic plants (such as cactus) and animals (such as the kangaroo rat and the jackrabbit) manage to survive on minimal amounts of water, but they are exceptions; most species need abundant supplies of water. Over time, human societies have either adopted the migration alternative or developed techniques to import water into drier areas by building conveyance devices. The artificial waterways known as aqueducts date back several thousand years; they have allowed human settlement and agriculture to flourish in several regions.

ANCIENT WATER DELIVERY SYSTEMS

The Minoan civilization, with its capital of Knossus on the Greek island of Crete in the Mediterranean Sea, represents the earliest known record of the development of water-supply infrastructure. Aqueducts were used to bring water into the city about five thousand years ago, until a major earthquake sometime around 1450 B.C.E. destroyed the region. The next major development in the building of aqueducts occurred circa 1000 B.C.E., when underground tunnels called *qanats* were used in the Middle East and North Africa to bring water from upland sources to villages and local farms. Modern-day Iran has the largest number of *qanats* in the region (more than twenty-two thousand), with lengths that vary from 40 to 45 kilometers (25 to 28 miles) and depths approaching 122 meters (400 feet).

The ancient Romans were renowned for their extensive system of elevated stone aqueducts beginning about 312 B.C.E. By 300 B.C.E., Rome had fourteen major aqueducts in service, bringing to the city about 150 million liters (roughly 40 million gallons) of water per day. Major water-supply systems were also built in those parts of Europe that were within the Roman Empire at that time, including what are now Italy, France, Spain, the Netherlands, and England. In Segovia, Spain, an aqueduct that was presumably built in the first century C.E. by the Romans is still being used.

In North America in the ninth century C.E. the Hohokam built canals that were 9-18 meters (30-60 feet) wide to transfer water from the Salt River, near present-day Phoenix, Arizona, to local farms for irrigation. In the tenth century the Anasazi developed a similar type of irrigation scheme in what is now southwest Colorado. During the fifteenth century, the Aztecs used rock aqueducts to supply both drinking water and water for irrigation to the area of present-day Mexico City.

Modern Aqueduct Systems

As might be expected, modern water-delivery systems use some of the same techniques that were used successfully in ancient times. For example, modern systems, like ancient ones, make use of gravity to move water along as much as possible. Increased populations and the accompanying increased demand for water, however, require modern designers of aqueducts to employ additional techniques. Modern aqueducts must often transport water over great distances from the sources to the consuming populations, and conflicts sometimes arise when water is withdrawn from one area to serve another. Electricity or turbines are used to power huge pumps that can move water long distances. In some areas, the use of underground conduits is necessary to avoid interference with street traffic.

In 1825 the completion of the Erie Canal between Buffalo, New York, and New York City encouraged the export of agricultural products from the American Midwest to Europe and led to an increase in population for New York City. Since the Hudson River was too salty to use as a supply source, the city started work on a reservoir and aqueduct in the Croton River watershed in Westchester and Putnam counties. The project was finished in 1842 with a capacity of 341,000 cubic meters (90 million gallons) per day, an amount that the planners thought would be sufficient for a future city population of one million. New York continued to grow, however, necessitating expansion during the early twentieth century into the Catskill and Delaware watersheds and the construction of six new reservoirs and connecting aqueducts. The overall system now serves some nine million people with an average daily consumption of 4.5 million cubic meters (1.2 billion gallons) per day.

In contrast to New York, which receives average annual precipitation of 1,067 millimeters (42 inches), the Los Angeles metropolitan area receives about 381 millimeters (15 inches) of precipitation per year. The first project to bring in outside water to Los Angeles began in 1907 with the construction of an aqueduct 375 kilometers (200 miles) long, from the Owens Valley in east-central California, a project that was met initially with stiff resistance from Owens Valley residents. The next major undertaking for Los Angeles was the Colorado River Aqueduct Project, which was started in 1928 under the sponsorship of thirteen cities in Southern California that formed the Metropolitan Water District (MWD). By 2010 the MWD consisted of twenty-six water districts and cities and was responsible for supplying drinking water to an estimated population of eighteen million, with an average delivery of 7 million cubic meters (1.8 billion gallons) of water per day.

Other major projects to transfer water in arid to semiarid regions of the United States include the Central Arizona Project, which was authorized in 1968 by the U.S. Bureau of Reclamation. Construction started in 1973 to convey water from the Colorado River at Lake Havasu on the Arizona-California border to central and southern Arizona. The delivery system is 540 kilometers (336 miles) long and includes fourteen pumping plants and three tunnels. This is one prominent instance where huge amounts of power are required to overcome gravity, as the elevations at Lake Havasu, Phoenix, and Tucson are 136 meters (447 feet), 458 meters (1,503 feet), and 875 meters (2,870 feet), respectively.

Robert M. Hordon

Further Reading

Cech, Thomas V. *Principles of Water Resources: History, Development, Management, and Policy.* 3d ed. New York: John Wiley & Sons, 2010.

Hillel, Daniel. *Rivers of Eden: The Struggle for Water and the Quest for Peace in the Middle East.* New York: Oxford University Press, 1994.

Koeppel, Gerald T. *Water for Gotham: A History.* Princeton, N.J.: Princeton University Press, 2000.

Powell, James L. *Dead Pool: Lake Powell, Global Warming, and the Future of Water in the West.* Berkeley: University of California Press, 2008.

Strahler, Alan. *Introducing Physical Geography.* 5th ed. Hoboken, N.J.: John Wiley & Sons, 2011.

See also: Colorado River; Department of the Interior, U.S.; Drinking water; Glen Canyon Dam; Hoover Dam; Irrigation; Los Angeles Aqueduct; Price-Anderson Act; Reclamation Act; Water rights; Water use; Watershed management.

Aquifers

CATEGORY: Water and water pollution

DEFINITION: Water-bearing geological formations that can store and transmit significant amounts of groundwater to wells and springs

SIGNIFICANCE: Aquifers are important because groundwater supplies a substantial amount of the water available in many localities. The contamination of aquifers is thus a matter of concern, and so a variety of aquifer restoration techniques have been developed.

All rocks found on or below the earth's surface can be categorized as either aquifers or confining beds. An aquifer is a rock unit that is sufficiently permeable to allow the transportation of water in usable amounts to a well or spring. (In geologic usage, the term "rock" also includes unconsolidated sediments such as sand, silt, and clay.) A confining bed is a rock unit that has such low hydraulic conductivity (or poor permeability) that it restricts the flow of groundwater into or out of nearby aquifers.

There are two major types of groundwater occurrence in aquifers. The first type includes those aquifers that are only partially filled with water. In those cases, the upper surface (or water table) of the saturated zone rises or declines in response to variations in precipitation, evaporation, and pumping from wells. The water in these formations is then classified as unconfined, and such aquifers are called unconfined or water-table aquifers. The second type occurs when water completely fills an aquifer that is located beneath a confining bed. In this case, the water is classified as confined, and the aquifers are called confined or artesian aquifers. In some fractured rock formations, such as those that occur in the west-central portions of New Jersey and eastern Pennsylvania, local geologic conditions result in semiconfined aquifers, which, as the name indicates, have hydrogeologic characteristics of both unconfined and confined aquifers.

Wells that are drilled into water-table aquifers are simply called water-table wells. The water level in these wells indicates the depth below the earth's surface of the water table, which is the top of the saturated zone. Wells that are drilled into confined aquifers are called artesian wells. The water level in artesian wells is generally located at a height above the top of the confined aquifer but not necessarily above the land surface. Flowing artesian wells occur when the water level stands above the land surface. The water level in tightly cased wells in artesian aquifers is called the potentiometric surface of the aquifer.

Water flows very slowly in aquifers, from recharge areas in interstream zones at higher elevations along watershed boundaries to discharge areas along streams and adjacent floodplains at lower elevations. Aquifers thus function as pipelines filled with various types of earth material. Darcy's law governing groundwater flow was developed in 1856 by Henry Darcy, a French engineer. In brief, Darcy's law states that the amount of water moving through an aquifer per unit of time is dependent on the hydraulic conductivity (or permeability) of the aquifer, the cross-sectional area (which is at a right angle to the direction of flow), and the hydraulic gradient. The hydraulic conductivity depends on the size and interconnectedness of the pores and fractures in an aquifer. It ranges through an astonishing twelve orders of magnitude. Very few other physical parameters exhibit such a wide range of values. For example, the hydraulic conductivity ranges from an extremely low 10^7 to 10^8 meters per day in unfractured igneous rock such as diabase and basalt to as much as 10^3 to 10^4 meters per day in cavernous limestone and coarse gravel. Typical low-permeability earth materials include unfractured shale, clay, and glacial till. High-permeability earth materials include lava flows, coarse sand, and gravel.

In addition to this wide range of values, hydraulic conductivity varies widely in place and directionality within the same

A standard aquifer system, featuring the flow times of different paths. (USGS)

aquifer. Aquifers are isotropic if the hydraulic conductivity is about the same in all directions and anisotropic if the hydraulic conductivity is different in different directions. As a result of all of these factors, groundwater yield is extremely variable both within the same aquifer and from one aquifer to another when they are composed of different rocks.

Because groundwater flows slowly in comparison with surface water, any contaminant that gets into the groundwater could be around for a long time, perhaps hundreds or thousands of years. It is thus simpler and much more cost-effective to prevent groundwater contamination than it is to try to correct a problem that has been in existence for years.

Restoration of a contaminated aquifer may be accomplished, albeit at a price, through one or more of the following procedures: inground treatment or containment, aboveground treatment, or removal or isolation of the source of contamination. The first approach involves natural treatment based on physical, chemical, or biological means, such as adding nutrients to existing subsurface bacteria to help them break down hazardous organic compounds into nonhazardous materials. The second approach uses engineered systems such as pumping wells or subsurface structures, which create hydraulic gradients that make the contaminated water stay in a specified location, facilitating removal for later treatment. Regardless of the restoration method selected, the source that is continuing to contaminate the aquifer must be removed, isolated, or treated.

Robert M. Hordon

FURTHER READING

Ahmed, Shakeel, R. Jayakumar, and Abdin Salih. *Groundwater Dynamics in Hard Rock Aquifers: Sustainable Management and Optimal Monitoring Network Design*. New York: Springer, 2008.

Fetter, Charles W. *Applied Hydrogeology*. 4th ed. Upper Saddle River, N.J.: Prentice Hall, 2001.

Kuo, Jeff. *Practical Design Calculations for Groundwater and Soil Remediation*. Boca Raton, Fla.: CRC Press, 1998.

Nonner, Johannes C. *Introduction to Hydrogeology*. 2d ed. Boca Raton, Fla.: CRC Press, 2010.

Todd, David K. *Groundwater Hydrology*. 3d ed. Hoboken, N.J.: John Wiley & Sons, 2005.

SEE ALSO: Drinking water; Groundwater pollution; Water pollution; Water quality; Wells.

Aral Sea destruction

CATEGORIES: Ecology and ecosystems; water and water pollution

THE EVENT: The decades-long, human-induced decimation of one of the world's largest freshwater lakes, located in the central Asian republics of Kazakhstan and Uzbekistan

DATE: Begun in the 1950's

SIGNIFICANCE: Increased diversion of water from Aral Sea by the Soviet government beginning during the 1950's unleashed a host of drastic social, economic, and environmental changes in the region. Destruction of the Aral Sea is regarded as one of the greatest human-made catastrophes of the twentieth century.

Prior to 1960, the Aral Sea was a freshwater lake, the fourth largest in the world, covering 67,300 square kilometers (26,000 square miles). For thousands of years the Aral was fed by two rivers, the Amu Darya and the Syr Darya, which annually carried fresh water more than 2,400 kilometers (1,500 miles) across largely desert terrain. The Aral has no natural outlet; its water balance was maintained by extremely high losses to evaporation.

During the 1950's the government of the Soviet Union undertook a massive project to increase greatly the amount of land under cultivation for cotton in Central Asia. To accomplish this, the project diverted large amounts of water from the Amu and Syr into irrigation canals. Construction on the largest of these, the 845-kilometer (525-mile) Karakum Canal, was begun in 1954 and completed in 1975. After the development of the canal system, inflow to the Aral dropped dramatically. The deltaic wetlands of both rivers largely disappeared, and the bed of the Amu was left essentially dry where it finally reached the Aral Sea.

During the late 1980's, the depleted lake separated into two bodies of water, the "Small Aral" to the north and the "Large Aral" to the south. By the mid-1990's, as a result of water diversion, the Aral Sea had been reduced in surface area by more than 50 percent and in volume by more than 70 percent (an amount equal to one and one-half times that of Lake Erie). In addition, the water level had dropped by more than 18 meters (60 feet). Towns that were once ports at the water's edge became landlocked many miles from the sea, leaving stranded fishing vessels to rust in the sand.

Continuing water loss caused the Large Aral to form separate east and west basins in 2003. By 2007, the Aral Sea was down to 10 percent of its original surface area. By 2009, the eastern basin of the Large Aral had largely dried up. What was once lake bed had become the Aralkum—the Aral Desert.

CONSEQUENCES

The shrinking of the Aral Sea has had a number of serious social, economic, and environmental consequences. The Aral was once the site of a major fishing industry with twenty-four native species of fish. All disappeared, victims of water salinity that had increased to three times greater than that of the oceans. A saltwater species, the Black Sea flounder, was introduced during the 1970's, but by 2003 elevated salinity in the southern lakes had killed it off. The island of Muynak, which once lay within the delta of the Amu Darya, was eventually more than 150 kilometers (93 miles) from the sea. The thriving cannery industry that existed in Muynak persisted for a few years by processing frozen fish imported from the Baltic Sea and the North At-

lantic, but with the breakup of the Soviet Union in 1991, other former Soviet republics ceased shipping their fish to Muynak, and the industry died.

The region surrounding the Aral Sea has become the site of one of the world's most dramatic examples of environmentally induced disease. The vast area of exposed former lake bed—54,000 square kilometers (20,850 square miles) as of April, 2008—is covered with precipitated natural salts. Deposits of toxic substances such as pesticides and fertilizers from agricultural runoff are also found on the lake bed. Each year millions of tons of dust, dried salts, and pollutants are picked up by the winds and carried as massive dust storms. This wind-borne dust has caused high incidences of emphysema, pulmonary tuberculosis, and chronic bronchitis among persons living in the region. Arthritic diseases have increased sixtyfold. Other health problems that have shown dramatic increases in the region are kidney disease, thyroid dysfunction, and throat, intestinal, and liver cancers. The region around the Aral Sea has the highest maternal mortality (120 in every 100,000 mothers) and infant

In the bed of what was once the Aral Sea, a Kazakh villager pulls water from a well sunk into the sandy ground. (Shamil Zhumatov/ Reuters/LANDOV)

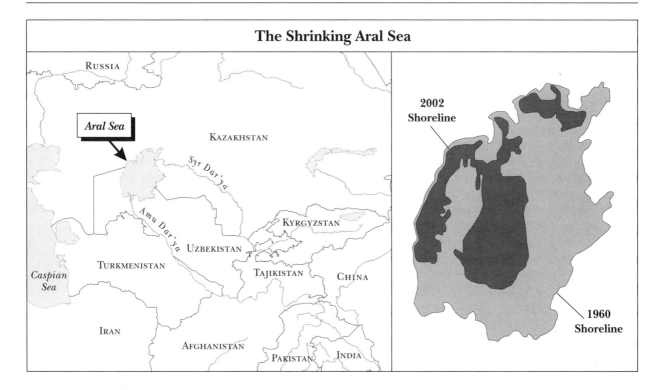

The Shrinking Aral Sea

mortality (75 per 1,000 live births) levels in the former Soviet Union. It is estimated that up to 99 percent of the women of the Karakalpakstan region suffer from anemia. Pesticides are found in mothers' milk, and high levels of heavy metals, salts, and pesticides occur in the drinking water.

Vozrozhdeniye Island may constitute an additional health hazard. Formerly a remote location in the center of the lake, the island was the site of a secret bioweapons facility between 1948 and 1992. Pathogens such as anthrax, tularemia, brucellosis, plague, typhus, smallpox, and botulinum toxin were tested there on laboratory animals and livestock. When receding water levels connected the abandoned island base to the mainland in 2001, the possibility of civilian exposure to remaining biohazardous materials increased. A 2002 joint remediation effort between the United States and Uzbekistan decontaminated ten anthrax burial sites and dismantled the abandoned laboratory complex. It is unclear whether viable pathogenic agents remain in Vozrozhdeniye's soil.

Although the diversion of river water for irrigation initially resulted in a tripling of cotton production, the wind-deposited salts from the drying lake have gradually accumulated in the soils, causing a decline in productivity. Cattle have died in large numbers

from eating salt-poisoned vegetation. In addition to containing high levels of salt, the soils are waterlogged, the result of poor drainage.

High evaporation rates over the Aral Sea had, in the past, formed a column of humid air that had deflected and diverted the hot, dry desert winds in the summer and the cold Siberian winds in the winter. This no longer occurs. Summer temperatures in the region are hotter and winter temperatures are colder. Both seasons show marked decreases in precipitation, and the summer growing season has shortened. The water evaporated from the Aral no longer falls as precipitation on the mountains to the south, the source of the river water; thus the local hydrologic cycle has been disrupted.

MITIGATION MEASURES

A Soviet plan to restore the Aral Sea with waters from distant Siberian rivers was never carried out because of economic limitations and environmental concerns. With the dissolution of the Soviet Union, the newly independent states of Kazakhstan and Uzbekistan were left to compete for the lake's resources. During the early 1990's, all five of the states that occupy the Aral Sea watershed—Kazakhstan, Uzbekistan, Turkmenistan, Tajikistan, and Kyrgyzstan—agreed to contribute to the improvement of the situation in the Aral Sea basin. As these states had conflicting goals, however, and funding was insufficient, no unified water strategy was developed, and environmental concerns continued to have a lower priority than short-term economic gains.

Kazakhstan attempted a partial restoration of the Small Aral. With funding from the World Bank, the Syr Darya was reengineered to improve the river's flow, and a dam was constructed between the Small and Large Arals. After the dam's completion in 2005, the water level in the Small Aral rose by 2 meters (6.5 feet) over an eight-month period. The lake's salinity was sufficiently reduced to support fish populations, making commercial fishing possible once again. Bird, reptile, and plant species also began to reappear. A sluice in the dam is periodically opened to allow water to flow into the depleted lakes to the south. In addition, the Small Aral restoration has affected the region's microclimate, and rain has become more plentiful.

The Large Aral to the south lies in the poorer republic of Uzbekistan. While groundwater has been found to contribute more to the lake's recharge than was previously believed, the Large Aral would still require more inflow to thrive. Uzbekistan's economy depends heavily on cotton, however, and the republic has been more eager to divert water from the Amu Darya for crop irrigation than for Aral restoration. An international consortium of Uzbek, Russian, Korean, Chinese, and Malaysian interests was formed in 2005 to explore and develop oil and gas fields in the exposed lake bed, a venture that is unlikely to have positive environmental impacts.

Donald J. Thompson
Updated by Karen N. Kähler

FURTHER READING

Ellis, William S. "The Aral: A Soviet Sea Lies Dying." *National Geographic*, February, 1990, 73-93.

Ferguson, Robert W. *The Devil and the Disappearing Sea: A True Story About the Aral Sea Catastrophe.* Vancouver: Raincoast Books, 2003.

Glantz, Michael H., ed. *Creeping Environmental Problems and Sustainable Development in the Aral Sea Basin.* New York: Cambridge University Press, 2008.

Hinrichsen, Don. "Requiem for a Dying Sea." *People and the Planet* 4, no. 2 (1995).

Kostianoy, Andrey G., and Aleksey N. Kosarev, eds. *The Aral Sea Environment.* New York: Springer, 2009.

Micklin, Philip P. "Destruction of the Aral Sea." In *Our Changing Planet: The View from Space*, edited by Michael D. King et al. New York: Cambridge University Press, 2007.

Pearce, Fred. *When the Rivers Run Dry: Water, the Defining Crisis of the Twenty-first Century.* Boston: Beacon Press, 2006.

United Nations Environment Programme. *Global International Waters Assessment, Regional Assessment 24: Aral Sea.* Kalmar, Sweden: University of Kalmar, 2005.

Zonn, Igor S., et al. *The Aral Sea Encyclopedia.* New York: Springer, 2009.

SEE ALSO: Airborne particulates; Brucellosis; Cancer clusters; Desertification; Environmental illnesses; Irrigation; Runoff, agricultural; Soil salinization; Water pollution.

Arctic National Wildlife Refuge

CATEGORIES: Places; animals and endangered species
IDENTIFICATION: Large area of land in northeastern Alaska set aside by the U.S. government for the preservation and protection of wildlife
DATES: Established in 1960; expanded in 1980
SIGNIFICANCE: The Arctic National Wildlife Refuge provides a home for a great many species of marine mammals, birds, and terrestrial animals. Its isolation, biodiversity, and protected status make it an important sanctuary for threatened species such as the polar bear. The possibility that oil exploration could be undertaken in a section of the refuge is a topic of ongoing debate.

Even before Alaska gained statehood in 1959, a movement was under way in the National Park Service and among conservationists to protect a small part of northeastern Alaska permanently from development and commercial interests. The Arctic National Wildlife Range was established as a federally protected area in 1960 by the U.S. secretary of the interior, Fred A. Seaton, in order to preserve the uniqueness of the wilderness and its wildlife. In 1980 passage of the Alaska National Interest Lands Conservation Act (ANILCA) expanded the protected area from less than 3.6 million hectares (9 million acres) to approximately 7.3 million hectares (18 million acres) and renamed it the Alaska National Wildlife Refuge (ANWR).

The remoteness and protected status of the refuge have limited the human impact on the environment there. It provides important habitat for marine life, which includes seals and whales as well as fish and seabirds, and terrestrial animals such as caribou, wolves, and the three North American species of bear. The refuge is home to many animals that flourish during the short Arctic summer, such as a variety of insects, and a safe haven where migratory birds can rest while moving south each year. The refuge has been a subject of controversy since the 1970's, and supporters of the area's protected status have fought off many attempts to open it up to oil and gas development.

Geography and Wildlife

The Arctic National Wildlife Refuge, sometimes called the Arctic Refuge, has greater species diversity than any other protected area in the Arctic Circle. The northern boundary of the refuge is the coastline of the Beaufort Sea and has habitats typical of coastal regions, such as river deltas, barrier islands, and salt marshes. These support wildlife that includes migratory seabirds and varieties of fish; in the summer caribou herds travel there to give birth and raise their young. Also during the summer, in addition to the many insects that thrive there, tens of thousands of migratory birds feed and rest on the coastal plain before traveling south. During the winter months, polar bears create birthing dens and hunt seals on the sea ice that grows along the coast.

Further inland, the coastal habitats give way to a plain, dotted with small lakes and braided rivers, that gradually moves upland to the Brooks Range. The Brooks Range region is made up of the foothills, valleys, and mountains north and south of the mountains and provides habitat for, among others, wolves, ducks, and birds of prey such as falcons and eagles. The region of the refuge on the south side of the Brooks Range consists mainly of the boreal forest characteristic of inland Alaska. Year-round residents of this forest include grizzly bears, lynx, wolverines, and moose; caribou herds spend the winter there, and migratory birds breed there during the spring and summer. On this side of the Brooks Range, the rivers that flow south to the Yukon River, the wetlands of the region, and the forest canopy provide a wide variety of habitats and food for many different species, from fish to mammals to birds.

Protection and Controversy

In 1968 the discovery of oil in Alaska's Prudhoe Bay, which is only sixty miles west of the western border of ANWR, began a political controversy that has not abated. When the Alaska National Interest Lands Conservation Act enlarged the refuge in 1980, it also left open the possibility of oil and gas exploration and exploitation on 607,000 hectares (1.5 million acres) of the refuge's coastline and coastal plain, which are areas of great biodiversity and are of vital importance to many species. Section 1002 of ANILCA allowed an opening for oil exploration in ANWR, but only if mandated by Congress and the president; this provision of

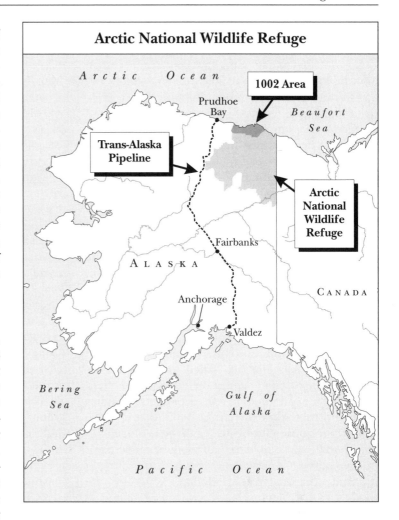

Arctic National Wildlife Refuge

the law was an unhappy compromise that each side vowed to change in its own favor as soon as possible.

Since 1980 many attempts have been made to open what became known as the 1002 Area to exploration and drilling, particularly under Republican presidential administrations and by Republican members of Congress. Alaskan local, state, and national politicians have been among the greatest proponents of drilling in the refuge and have generally been supported by their constituents. Taxes and royalties from oil drilling in Prudhoe Bay provide a large amount of revenue for Alaska annually, and each state resident gets a share in the form of a yearly check; in 2000 the amount of this payment was reportedly about $1,900 per person. The refuge is federal land, however, so the issue must be decided in Washington, D.C., not in Alaska.

During President George W. Bush's two terms in office, moves to open the refuge to oil drilling occurred repeatedly in the Republican-controlled Congress, at first in specific proposals to allow drilling and then, when that approach failed, through attachments to major bills, such as defense-spending and energy bills, or additions to federal budgets. Whether enough oil actually exists in the 1002 Area to make drilling economically worthwhile to the United States and morally acceptable to many Americans remains unknown. What is clear is that the issue of drilling in the Arctic National Wildlife Refuge exemplifies the modern global dilemma between the need for energy and the need for a healthy planet.

Megan E. Watson

FURTHER READING

Kaye, Roger. *Last Great Wilderness: The Campaign to Establish the Arctic National Wildlife Refuge.* Fairbanks: University of Alaska Press, 2006.

Standlea, David M. *Oil, Globalization, and the War for the Arctic Refuge.* Albany: State University of New York Press, 2006.

Vaughn, Jacqueline. *Conflicts over Natural Resources: A Reference Handbook.* Santa Barbara, Calif.: ABC-CLIO, 2007.

SEE ALSO: Alaska National Interest Lands Conservation Act; Biodiversity; Endangered species and species protection policy; Energy Policy and Conservation Act; Indigenous peoples and nature preservation; Marine Mammal Protection Act; Oil drilling; Trans-Alaska Pipeline; Wildlife refuges.

Argo Merchant oil spill

CATEGORIES: Disasters; water and water pollution

THE EVENT: Grounding of the tanker *Argo Merchant* off the coast of Rhode Island, resulting in the spilling of heavy fuel oil into the sea

DATE: December 15, 1976

SIGNIFICANCE: A northwesterly wind helped to prevent extensive environmental damage to the New England shoreline when the *Argo Merchant* grounded.

The *Argo Merchant* was built as the *Arcturus* in Hamburg, Germany, in 1953 and renamed the *Permina Samudia III* in 1968. It was then renamed the *Vari* in 1970 and finally the *Argo Merchant* in 1973. Regardless of the name or the owners, the ship had a long history of accidents. The *Argo Merchant* was not an extremely large vessel. It was 195 meters (641 feet) in length and 25.6 meters (84 feet) wide and had a draft of 10.7 meters (35 feet). It was owned by the Thebes Shipping Company of Greece, was chartered to Texaco, and carried a crew of Greek officers and Filipino crewmen under the Liberian flag.

The vessel departed Puerto la Cruz in Venezuela bound for Salem, Massachusetts, with a cargo of 7.6 million gallons of heavy fuel oil in its thirty cargo tanks. The *Argo Merchant*'s problems began the evening before the grounding, as the onboard gyrocompass had broken, and the officers were unable to determine the ship's position.

At 6:00 A.M. on December 15, 1976, the *Argo Merchant* grounded on the southern end of Fish Rap Shoal in 5.5 meters (18 feet) of water. The ship was 48 kilometers (30 miles) north of the Nantucket lightship station, 24 kilometers (15 miles) outside the normal shipping lanes, and 39 kilometers (24 miles) off its charted track line. When the crew of the *Argo Merchant* called the U.S. Coast Guard to report the grounding, they had no idea of their position; the position they gave turned out to be 48 kilometers (30 miles) from where the tanker was located.

The Coast Guard, however, responded quickly and soon located the grounded vessel. Attempts were made to float the ship and to tow it, but neither method worked. The sea continued to batter the grounded ship, which began to bend and twist on the reef until oil started leaking from the cargo tanks. Within one week the *Argo Merchant* had broken in half and was leaking large amounts of oil. As in the case of the 1967 *Torrey Canyon* oil spill in the English Chan-

The tanker Argo Merchant *floats in the water in two pieces, its stern visible behind the larger bow section, after running aground off the coast of Nantucket.* (©Bettmann/CORBIS)

nel, attempts were made to ignite the oil, but no attempt was made to destroy the oil remaining in the ship's tanks.

Within ten days of the grounding, the oil slick from the *Argo Merchant* extended 160 kilometers (100 miles). Although the weather was rainy and windy, which made oil transfer and salvage operations difficult, the wind blew predominantly from the northwest, and this drove the oil offshore, so that very little oil came ashore in the rich fishing and shellfish areas along the coast of New England. The northwesterly wind did drive the slick across Georges Bank, which is one of the richest fishing areas in the Atlantic Ocean, but the short-term effects of the windblown oil were marginal, and only a limited number of birds and marine mammals were killed. The long-term effects of the oil spill in the water column and the oil's impact

on bottom-dwelling creatures have not been documented.

Robert J. Stewart

FURTHER READING

Clark, R. B. *Marine Pollution.* 5th ed. New York: Oxford University Press, 2001.

Fingas, Merv. *The Basics of Oil Spill Cleanup.* 2d ed. Boca Raton, Fla.: CRC Press, 2001.

Wang, Zhendi, and Scott A. Stout. *Oil Spill Environmental Forensics: Fingerprinting and Source Identification.* Burlington, Mass.: Academic Press, 2007.

SEE ALSO: *Amoco Cadiz* oil spill; *Braer* oil spill; *Exxon Valdez* oil spill; Oil spills; *Sea Empress* oil spill; Tobago oil spill; *Torrey Canyon* oil spill.

Arsenic

CATEGORY: Pollutants and toxins
DEFINITION: Poisonous, crystalline metallic element
SIGNIFICANCE: Ubiquitous in nature, arsenic represents a major health threat to human water supplies and foods, especially in the world's poorest countries.

Arsenic is naturally brittle and relatively soluble in water. It is found activated in the environment owing to a combination of natural processes (weathering, biological activity, and volcanic eruptions) and human activities. In some cases arsenic is naturally present in aquifers. It is also associated with geothermal waters, and its presence has been reported in several areas with hot springs. Human beings have played a significant role in increasing the rate of natural mobilization of arsenic through activities such as mining, combustion of fossil fuels, and the use of arsenic in pesticides, herbicides, crop desiccants, and food additives. Industrial effluents add to the environmental availability of arsenic. Even though human use of arsenic has been significantly reduced since the mid-twentieth century, the impacts of arsenic on the environment are likely to continue for some time.

Arsenic is found in various forms and is a major constituent of more than two hundred different minerals. Some of the elements with which arsenic often forms compounds are oxygen, chlorine, tungsten, tin, molybdenum, cadmium, sulphur, lead, silver, iron, and even gold. Arsenic is widely found in ores such as arsenopyrite.

Contamination of drinking water is one of the most significant ways in which arsenic threatens human health. Commonly four types of arsenic compounds exist in water: arsenite and arsenate, which are inorganic arsenic compounds, and methyl arsenic acid and dimethyl arsenic acid, which are organic forms of arsenic. The amount of arsenic present in drinking water varies among different local water sources, which include surface water (rivers, lakes, reservoirs, and ponds), groundwater (aquifers), and rainwater. Of these, groundwater contains the highest concentrations.

Long-term exposure to this metal results in chronic arsenic poisoning, or arsenicosis. Prolonged exposure to arsenic causes various types of cancer; other common health effects include diabetes mellitus, hypertension, and respiratory diseases. Particu-

larly in the world's poorest nations, people suffering from arsenicosis often withdraw from society, as the disease causes a deformation of the skin that is difficult to differentiate from leprosy. Arsenicosis caused by contaminated drinking water affects the poor more often than it affects the rich.

Because arsenic is carcinogenic, the World Health Organization has set a standard for the amount that is permissible in drinking water that is deemed to be safe. The original standard was 50 micrograms per liter, but the amount was reduced in 1993 to 10 micrograms per liter; in most developing countries that are plagued by the problem of arsenic contamination, however, the higher level is still allowed, in part because of a lack of testing facilities.

Arsenic contamination of water supplies is relatively site-specific and affects a limited number of people who are in proximity to the contaminated water. A number of aquifers across the world have been identified has having arsenic concentrations greater than 50 micrograms per liter; these include aquifers in India, Bangladesh, Taiwan, China, Hungary, Mexico, Chile, Argentina, Thailand, Ghana, Austria, Japan, France, Argentina, New Zealand, and the United States. For a long time, arsenic was not among the list of elements tested by water-quality testing laboratories; thus scientists have not yet identified all of the world's regions that are vulnerable to arsenic contamination of drinking water.

The concentration of arsenic in natural waters is controlled by solid-solution interaction. The importance of oxides in reducing arsenic concentration in natural waters has been well recognized. Clays can also be used to adsorb arsenic, and iron, aluminum, and manganese salts are often used in the removal of arsenic from water. The extent of adsorption is strongly related to arsenic speciation, arsenic concentration, pH, and the concentration of opposing anions.

Sohini Dutt

FURTHER READING

Henke, Kevin R. *Arsenic: Environmental Chemistry, Health Threats, and Waste Treatment.* Hoboken, N.J.: John Wiley & Sons, 2009.

Hill, Marquita K. "Drinking-Water Pollution." In *Understanding Environmental Pollution.* 3d ed. New York: Cambridge University Press, 2010.

Miller, G. Tyler, Jr., and Scott Spoolman. "Water Pollution." In *Living in the Environment: Principles, Con-*

nections, and Solutions. 16th ed. Belmont, Calif.: Brooks/Cole, 2009.

Ravenscroft, Peter, Hugh Brammer, and Keith Richards. *Arsenic Pollution: A Global Synthesis.* Malden, Mass.: Blackwell, 2009.

SEE ALSO: Drinking water; Groundwater pollution; Hazardous and toxic substance regulation; Heavy metals; Lead; Mercury; Water pollution; Water quality.

Asbestos

CATEGORIES: Pollutants and toxins; human health and the environment

DEFINITION: Industrial term for certain silicate minerals that occur in the form of long, thin fibers

SIGNIFICANCE: The adverse health effects of breathing high concentrations of asbestos over prolonged periods have been known since the early 1970's. The federal Clean Air Act of 1963 classified asbestos as a carcinogenic material, and in 1990 the U.S. Environmental Protection Agency established a broad ban on the manufacture, processing, importation, and distribution of asbestos products.

Asbestos-form minerals are natural substances that are common in many types of igneous and metamorphic rocks found over large areas of the earth. Erosion continually releases these fibers into the environment, and most people typically inhale thousands of fibers each day, or more than 100 million over a lifetime. Asbestos fibers also enter the body through drinking water. Drinking-water supplies in the United States typically contain almost 1 million fibers per quart, but water in some areas may have as many as 100 million or more fibers per quart.

PROPERTIES

Many silicate minerals occur in fibrous form, but only six have been commercially produced as asbestos. In order of decreasing commercial importance, these are chrysotile (white asbestos), crocidolite (blue asbestos), amosite (brown asbestos), anthophyllite, tremolite, and actinolite. All these minerals except chrysotile are members of the amphibole group of minerals, which have a chainlike arrangement of atoms. In contrast, chrysotile, as a member of

the serpentine family, has atoms arranged in a sheetlike fashion.

Although the individual properties of these minerals differ greatly from one another, they share several characteristics that make them useful and cost-effective. These include great resistance to heat, flame, and acid attack; high tensile strength and flexibility; low electrical conductivity; resistance to friction; and a fibrous form, which allows them to be used for the manufacture of protective clothing. Asbestos thus was widely used until the 1970's in a great variety of building and industrial products. Such common materials as vinyl floor tiles, appliance insulation, patching and joint compounds, automobile brake pads, hair dryers, and ironing board covers all might have contained asbestos. Most such products now contain one or more of several substitutes for asbestos instead of asbestos itself. However, many of the substitutes may not be hazard-free, a fact that is starting to be recognized by legislators. For example, in 1993 the World Health Organization (WHO) stated that all substitute fibers must be tested to determine their carcinogenicity. Germany now classifies glass, rock, and mineral wools as probable carcinogens.

HEALTH EFFECTS

The U.S. Department of Health and Human Services classifies asbestos as a carcinogen. Studies leading to this determination were mostly based on asbestos workers who had been exposed to extremely high levels of fibers for many years. These studies concluded that the asbestos workers had increased chances of developing two types of cancer: mesothelioma (a cancer of the thin membrane surrounding the lungs) and cancer of the lung tissue itself. These workers were also at increased risk of developing asbestosis, an accumulation of scarlike tissue in the lungs that can cause great difficulty in breathing and permanent disability. None of these diseases develops immediately; all have long latency periods, typically fifteen to forty years. Contrary to common misconception, exposure to asbestos does not cause muscle soreness, headaches, or any other immediate symptoms. The effects of asbestos exposure typically are not noticed for many years.

It is generally agreed that the risk of developing disease after asbestos exposure depends on the number of fibers in the person's body, how long the fibers have been in the body, and whether or not the person is a smoker, since smoking greatly increases the risk of de-

U.S. End Uses of Asbestos, 1977 vs. 2003

END USE	METRIC TONS	
	1977	2003
Cement pipe	145,000	—
Cement sheet	139,500	—
Coatings and compounds	32,500	1,170
Flooring products	140,000	—
Friction products	83,100	—
Insulation: electrical	3,360	—
Insulation: thermal	15,000	—
Packing and gaskets	25,100	—
Paper products	22,100	—
Plastics	7,260	—
Roofing products	57,500	2,800
Textiles	8,800	—
Other	30,200	677

Source: Data from the U.S. Geological Survey.
Note: U.S. mining of asbestos ended in 2002.

veloping disease. There is no agreement on the risks associated with low-level, nonoccupational exposure. The U.S. Environmental Protection Agency (EPA) has concluded that there is no safe level of exposure to asbestos fibers, but the Occupational Safety and Health Administration (OSHA) allows up to 0.1 fiber per cubic centimeter of air during an eight-hour workday.

OTHER CONTROVERSIES

Another area of controversy stems from scientific studies showing that all forms of asbestos are not equally dangerous. Evidence has shown that the amphibole forms of asbestos, and particularly crocidolite, are hazardous, but the serpentine mineral chrysotile—accounting for 95 percent of all asbestos used in the past and 99 percent of current production—is not. For example, one case study involved a school that was located next to a 150,000-ton rock dump containing chrysotile. Thousands of children played on the rocks over a one-hundred-year period, but not a single case of asbestos-related disease developed in any of the children. The difference seems to be in how the human body responds to amphibole compared to chrysotile. The immune system can eliminate chrysotile fibers much more readily than amphibole, and there is also evidence that chrysotile

in the lungs dissolves and is excreted. This remains a controversial area, and the U.S. government still treats all forms of asbestos the same. This is not true of some European governments.

The risk of developing any type of disease from exposure to normal levels of asbestos fibers in outdoor air or the air in closed buildings is extremely low. The calculations of Melvin Benarde in Asbestos: The Hazardous Fiber (1990) show that the risk of dying from nonoccupational exposure to asbestos is one-third the risk of being killed by lightning. The Health Effects Institute made similar calculations in 1991 and found that the risk of dying from asbestos is less than 1 percent the risk of dying from exposure to secondary tobacco smoke.

Gene D. Robinson

FURTHER READING

Bartrip, Peter. *Beyond the Factory Gates: Asbestos and Health in Twentieth Century America.* New York: Continuum, 2006.

Carroll, Stephen, et al. *Asbestos Litigation.* Santa Monica, Calif.: RAND, 2005.

Castleman, Barry. *Asbestos: Medical and Legal Aspects.* 5th ed. New York: Aspen, 2005.

Chatterjee, Kaulir Kisor. "Asbestos." In *Uses of Industrial Minerals, Rocks, and Freshwater.* New York: Nova Science, 2009.

Craighead, John E., and Allen R. Gibbs, ed. *Asbestos and Its Diseases.* New York: Oxford University Press, 2008.

Deffeyes, Kenneth S. "Asbestos." In *Nanoscale: Visualizing an Invisible World.* Illustrations by Stephen E. Deffeyes. Cambridge, Mass.: MIT Press, 2009.

Dodson, Ronald, and Samuel Hammar, eds. *Asbestos: Assessment, Epidemiology, and Health Effects.* Boca Raton, Fla.: CRC Press, 2005.

McCulloch, Jock, and Geoffrey Tweedale. *Defending the Indefensible: The Global Asbestos Industry and Its Fight for Survival.* New York: Oxford University Press, 2008.

Maines, Rachel. *Asbestos and Fire: Technological Trade-Offs and the Body at Risk.* New Brunswick, N.J.: Rutgers University Press, 2005.

SEE ALSO: Asbestosis; Environmental illnesses; Hazardous and toxic substance regulation; Sick building syndrome.

Asbestosis

CATEGORY: Human health and the environment
DEFINITION: Disease of the lungs caused by
  repeated exposure to asbestos
SIGNIFICANCE: The realization that exposure to asbestos could result in serious lung disease led to the banning of this material for many uses, as well as greater awareness of the concept of environmental illness.

The generic term "asbestos" is applied to such minerals as amosite, anthophyllite, chrysotile, and crocidolite. These silicates, first used extensively in the 1940's, have remarkable qualities that initially made them desirable as both thermal and electrical insulators. After the insulating properties of asbestos were recognized, such substances were commonly used in the building trades, notably to insulate pipes and boilers. Some cements and floor tiles contained substantial quantities of asbestos, which was used as a fireproof filler. Forms of asbestos were also used to make blankets designed to smother fires and in the manufacture of safety garments for firefighters. The substance is still used in the manufacture of automotive brake and clutch linings, where its insulating qualities are particularly valued.

Asbestosis, a disease of the lungs, occurs in people who have been exposed to this silicate, but the onset of symptoms may occur years after exposure. Even those whose exposure to asbestos has been moderate may contract the disease, an early symptom of which is perpetual shortness of breath, especially following strenuous physical activity. This symptom is often accompanied by a persistent, hacking cough. Asbestosis is a progressive disease that characteristically begins in the lower lung and spreads to the middle and upper lungs, with disabling, sometimes fatal, results as the air spaces in the lungs narrow. Little aggressive treatment exists for the disease. Asbestosis victims are urged not to smoke and are usually treated with oxygen to improve their breathing.

Direct exposure to asbestos, especially when it occurs over extended periods, often results in asbestosis. The disease has been found in those who have worked in asbestos mines as well as in persons who have worked as pipe fitters, boilermakers, automotive mechanics working with brake and clutch linings, and demolition workers who have razed buildings in which asbestos was used for insulation.

Indirect exposure can also result in asbestosis. The disease has been reported among people who have regularly laundered work clothes that have been directly exposed to asbestos as well as among those who live or work in buildings where asbestos has been used as an insulator. Widespread community exposure has been noted in situations where steel girders in large buildings were sprayed with asbestos as a fire precaution. Cases occurred in a London neighborhood near an asbestos plant among persons who had no direct contact with the plant. South Africa had outbreaks among the general population near its asbestos mines.

After 1975 the use of asbestos declined substantially as it was replaced by other mineral fibers. By the end of the twentieth century, persons working with asbestos in the United States, particularly demolition workers, began receiving special training to help them avoid exposure that could lead to the onset of the disease. Environmental laws in most jurisdictions now prohibit the use of asbestos as an insulating material in new buildings, and many laws protect workers from exposure. Persons who were exposed in the past through work-related activities and currently suffer from the disease are usually eligible to receive workers' compensation payments.

R. Baird Shuman

FURTHER READING

Bartrip, Peter. *Beyond the Factory Gates: Asbestos and Health in Twentieth Century America.* New York: Continuum, 2006.

Castleman, Barry. *Asbestos: Medical and Legal Aspects.* 5th ed. New York: Aspen, 2005.

Dodson, Ronald F., and Samuel P. Hammar, eds. *Asbestos: Risk Assessment, Epidemiology, and Health Effects.* Boca Raton, Fla.: CRC Press, 2006.

SEE ALSO: Asbestos; Environmental health; Environmental illnesses; Hazardous and toxic substance regulation; Sick building syndrome.

Ashio, Japan, copper mine

CATEGORIES: Human health and the environment; water and water pollution

IDENTIFICATION: Copper mine that operated at the headwaters of the Watarase River from the seventeenth century to 1972

SIGNIFICANCE: Development of the Ashio copper mine propelled Japan's industrial revolution and also set the stage for the nation's first conflict over environmental quality. Hazardous runoff from the mine caused tension between agriculture and industry and dramatized the human and environmental costs of industrial pollution.

The Ashio copper mine, located 110 kilometers (68 miles) north of Tokyo, first operated during the seventeenth century, but private ownership spurred development of the mine in 1877. By 1890, it was the largest copper mine in Asia. Production expanded in 1950 because of the Korean War.

As copper production increased, the Ashio mine's impacts on the surrounding area also increased. Located at the headwaters of the Watarase River, the mine caused environmental damage by depositing waste products in the river. By 1880 fish were beginning to die, and people who ate fish from the river became ill. Almost all marine life in the river had died by 1890. Deforestation compounded the pollution problem. The operation of the mine required timber to shore up the mine shafts, for railroad ties, for the construction of buildings, and as fuel for steam engines. This timber was obtained through the deforestation of 104 square kilometers (40 square miles) of surrounding land, which destroyed the watershed at the head of the Watarase River. As a consequence, flooding became a serious problem in the Watarase Valley and the surrounding rice fields.

Although natural flooding had occurred before the development of the mine, such flooding had brought layers of rich silt that contributed to abundant crops. Later floods, however, produced vastly different results: Vegetation did not survive contact with the contaminated floodwaters. Floods became more frequent, more severe, and more damaging because they left poisons in the soil. Soil samples revealed concentrations of sulfuric acid, ammonia, magnesium, iron, arsenic, copper, and chlorine. These substances poisoned the rice fields, and new seeds would not grow. Earthworms, insects, birds, and animals succumbed to the contamination.

Although the Japanese government ordered pollution-control measures, they were ineffective. In 1907 the government forced the evacuation and relocation of the inhabitants of a contaminated village. The collapse of a slag pile in 1958 introduced 2,000 cubic meters (71,000 cubic feet) of slag into the Watarase River, contaminating 6,000 hectares (14,820 acres) of rice fields. The Japanese government set limits on the amount of copper that could be deposited in the river water and the soil, but the damage had already been done. In 1972 the mine was shut down. That year, soil samples from 3 meters (9.8 feet) down still contained excessive amounts of copper as well as significant amounts of lead, zinc, and arsenic. The government ordered the destruction of all rice that had been grown in the area.

Thousands of fishermen, rice farmers, and valley citizens suffered severe economic losses. Serious health problems in the area included a high infant mortality rate, the failure of new mothers to produce milk for their infants, sores on those who worked in the fields, and a high death rate. In 1973 the Japanese Environmental Agency's Pollution Adjustment Committee began to review farmers' claims. The mine was required to admit to being the source of the contamination, and the farmers were awarded a sum equivalent to five million U.S. dollars in 1974. Attempts to reforest the area have failed.

Louise Magoon

FURTHER READING

Tsuru, Shigeto. *The Political Economy of the Environment: The Case of Japan.* Vancouver: University of British Columbia Press, 1999.
Wilkening, Kenneth E. *Acid Rain Science and Politics in Japan: A History of Knowledge and Action Toward Sustainability.* Cambridge, Mass.: MIT Press, 2004.

SEE ALSO: Acid mine drainage; Ducktown, Tennessee; Fish kills; Heavy metals; Mine reclamation.

Asia

CATEGORIES: Places; population issues; ecology and ecosystems; resources and resource management; preservation and wilderness issues

SIGNIFICANCE: Population growth, economic development, rising consumption, and unsustainable exploitation of natural resources have exerted severe stress on the environments of most Asian nations. The region faces the challenge of moving toward environmentally sustainable growth while reversing major environmental damage, including deforestation, habitat destruction, and air, water, and soil pollution.

Great progress in reducing poverty through economic growth has come at the expense of the environment in many Asian countries, particularly from the 1970's onward. Whereas 60 percent of all Asians lived in poverty in 1970, that number was almost halved, to 33 percent, by 2000. However, by 2005 about 670 million Asians still lived on less than the equivalent of one U.S. dollar per day (adjusted for purchasing-power effects), a situation that led to the undernourishment of more than 500 million people out of a total Asian population that reached just over 4 billion in 2010. The continent was subject to ongoing pressures on the environment owing particularly to unsustainable economic growth that depleted and degraded many natural resources.

INTENSIVE AGRICULTURAL DEVELOPMENT

In Asia, the twin pressures of population growth and poverty reduction have often led to severe environmental degradation. A negative process often begins with land conversion, when forests, wetlands, and other natural habitats are cleared for agricultural and industrial use or for human settlement. As Asia became home to about 60 percent of the world's human population by 2010 but encompassed only 30 percent of the earth's land, some of which, such as the Gobi Desert, is inhospitable to human population, the pressures on arable and potentially arable land increased. By 2005, Asian arable land per capita was only 80 percent of the global average.

Primarily through deforestation, cropland in Asia (excluding the Middle East and Asian Russia) increased from 210 million hectares (520 million acres) in 1900 to 453 million hectares (1.12 billion acres) by 1994. From the 1970's onward, high-yield strains of staple crops such as rice and wheat were planted and supported with chemical fertilizers, pesticides, and herbicides. In addition, Asian countries such as the People's Republic of China, India, and the nations of Southeast Asia developed 90 percent of the world's aquaculture, focusing on fish, shrimp, and shellfish. Unsustainably intense agriculture—involving irrigation of as much as 30 percent of the cropland in Asia and heavy use of agrochemicals—eroded soils and depleted and contaminated freshwater sources, threatening freshwater ecosystems and fisheries.

By the early twenty-first century, Asian nations were managing to extract more and more food from their land and coastal areas, both for domestic consumption and for interregional and intercontinental export. However, water pollution, soil erosion, and nutrient depletion led to general land degradation and loss of arable land. It was predicted that if unsustainable levels of agricultural use and development were not halted, the continent would eventually face a food crisis. Some observers advocated the use of genetically modified foods, which would lessen reliance on unsupportable levels of agrochemicals, but this alternative remained controversial. Promising developments, however, included the introduction of new, less environmentally destructive approaches to farming, such as conservation tillage practices and agroecology.

ECONOMIC DEVELOPMENT

Whereas Japan began its industrialization with the Meiji Restoration after 1868, and beginnings of Chinese industry were seen at the end of the nineteenth century, most Asian nations did not begin to see full industrialization and above-world-average economic growth until the 1960's and 1970's. In 2004, Asia led all other global regions in economic growth. From 1995 to 2002, manufacturing in developing Asian nations grew by 40 percent. This rapid industrialization demanded land and energy and led to rapidly increasing emissions of pollutants into air, water, and ground. High levels of air pollution caused respiratory illnesses and acid rain that damaged lakes and forests; degraded soil, rivers, and aquifers; and harmed coastal ecosystems.

Initially, many developing Asian countries placed very little emphasis on technical means of pollution control and enforcement of antipollution laws. Power plants and factories were allowed to discharge their emissions into air and water, sometimes without even the most basic treatment. To fuel its industry and provide power and heat to its people, Asia used coal abundantly; by 2003, coal accounted for 41 percent of the energy consumed in Asia, particularly in China, India, and the Republic of Korea (South Korea). Another 39 percent of the energy came from oil, most of which came from the Middle East. Burning this amount of fossil fuel, often with at most rudimentary pollution-control devices in place and operational in many Asian countries, led to severe air pollution and contributed to greenhouse gas emissions linked to global warming. Experts project that one result of global warming could be a rise in sea levels that would submerge island nations of South Asia and the Pacific and inundate heavily populated Asian coastal regions.

Air pollution became especially severe in Asian cities in China and India and in large urban agglomerations such as Bangkok, Thailand, and Hanoi, Vietnam. By 2006, the world's ten most polluted cities were all Chinese. From the late 1990's onward, China made fighting air pollution a national priority, and significant progress was achieved. Air pollution actually decreased as the economy grew, even though actual levels of reduction (for example, a 0.6 percent reduction of sulfur dioxide emissions in 2007) were still small. Emphasis was placed on increasing China's notoriously inefficient generation and use of energy; the nation's stated goal was to increase energy efficiency by 30 percent from 2006 to 2010. However, China's project devoted to calculating the true economic cost caused by environmental pollution and degradation (its "green gross domestic product") was stopped in 2007 when its negative figures exceeded even worst-case scenarios.

Very ambitious Asian projects to increase the share of cleaner hydroelectric power through construction of massive dams, such as the Three Gorges Dam on the Yangtze River, remained controversial. These megaprojects came with significant environmental problems of their own, including the destruction of river ecosystems, riverbank erosion, and resettlement of human populations on the scale of millions. Critics argued that better alternatives would be a series of smaller dams and increased energy efficiency. Another controversial approach to the reduction of pollution, undertaken in China and India, was the washing of coal to decrease its sulfur content; those opposing this practice noted that it depletes water resources. These cases show that no easy or quick solutions are available for the environmental problems caused by economic growth in Asia; the problems require a more holistic approach aimed toward sustainable economic development instead of rapacious resource depletion and environmental degradation.

Vehicles move slowly in a massive traffic jam in Beijing, China, on September 19, 2010, when the city was choked by more than eighty traffic jams in the morning. In China, rapid urbanization and population growth have led to increased levels of traffic congestion and accompanying air pollution. (AP/Wide World Photos)

URBANIZATION AND CONSUMERISM

In 2010 about half of Asia's people, approximately 2 billion persons, still lived in rural areas and drew their livelihoods from farming, fishing, or forestry. In South Asia, 73 percent of all land was used for agriculture. As Asian economies had developed throughout the region, however, urbanization had increased tremendously. From 1970 to 2000, 560 million people moved from the countryside into Asian cities, increasing urban populations by 260 percent. Urbanization continued to rise in the twenty-first century, with average Asian city growth rates of more than 3 percent annually.

The rapid growth of Asian cities severely taxed the environment. Poor land management led to habitat destruction that threatened many kinds of animals, from elephants to birds. Urban sprawl initiated the conversion of forests, wetlands, and agricultural land. The growing numbers of people in cities caused groundwater depletion and shortages of potable water. Municipal solid and liquid wastes polluted water supplies, and waste incineration polluted the air. Increases in motor vehicle use caused traffic congestion and air pollution. Growing suburbs required more roads as the residents depended on private vehicles in the absence of efficient public transport systems. Poor planning and ineffective law enforcement led to the rapid growth of crowded slum areas, where poor infrastructure failed to meet the needs of inhabitants, leading them to add to environmental degradation with untreated waste disposal that resulted in threats to public health.

Significantly rising living standards for many people in most Asian countries led to greater consumerism, which also affected the environment negatively, as the consumption of energy and natural resources increased along with waste production. All this taxed an environment strained already by intensive agriculture, new industry and manufacturing, and power generation. Many Asian nations were unable or unwilling to devote significant financial resources to addressing environmental degradation seriously and effectively when they were faced with the costs of serving increasing numbers of people desiring better living standards. It was not until the quality of daily life began to deteriorate visibly in Asian cities because of environmental degradation, and health and life expectancies of people began to decline owing to pollution-related illnesses, that previously neglectful governments began to take environmental issues seriously.

ENVIRONMENTAL CHALLENGES AND OPPORTUNITIES

Asia enjoys tremendous biodiversity, but the pressures of population growth and economic development obliterated, degraded, or fragmented significant areas of natural land that had served as habitat for wild animals and plants. Asia's biodiversity became severely threatened as natural areas from lowland forests to grasslands and coastal estuaries, with their rivers and bays, were destroyed or degraded as the result of unregulated urban sprawl, legal and illegal logging, and land clearance for agricultural, industrial, or settlement use, with generally little regard for wildlife, plant life, and ecosystems. In addition, many species of animals were exploited legally and illegally for capture and sale to the pet trade or for use as food or in traditional medicines.

By 2010, approximately 7,500 Asian animal and plant species were considered to be threatened, half of them in Southeast and South Asia. About one-third of Southeast Asia's rain forests were fragmented, and two-thirds of Asia's wildlife habitat had been destroyed. Legal and illegal burning of forests in Indonesia had not only destroyed much wildlife habitat but also, together with industrial pollution, caused the annual emergence, beginning in 2001, of a huge brown cloud of polluted haze over South, Southeast, and East Asia. The severity of this problem led to the founding of the Reducing Emissions from Deforestation and Forest Degradation (REDD) initiative, a program sponsored by the United Nations. It was begun in September, 2008, and was reaffirmed in the Copenhagen Accord of December 18, 2009. The goal of REDD is to reward emissions reductions through financial incentives for regional governments and people.

Conservationists, scientists, and environmental activists all agree that Asia's poor overall record on nature preservation must be tackled with bold and effective conservation mechanisms, with a major shift toward environmental sustainability. Efforts have been made to place wildlife sanctuaries, nature reserves, and other open spaces under effective protection from development, with wildlife protected from poaching. The Asian tiger is just one of the species that has suffered greatly from habitat destruction and poaching; its numbers declined from about one hundred thousand in 1900 to a mere five thousand in 2000. One Asian success story has been China's protection of the giant panda in well-monitored preserves; any person caught poaching pandas is subject to a sentence of life in prison.

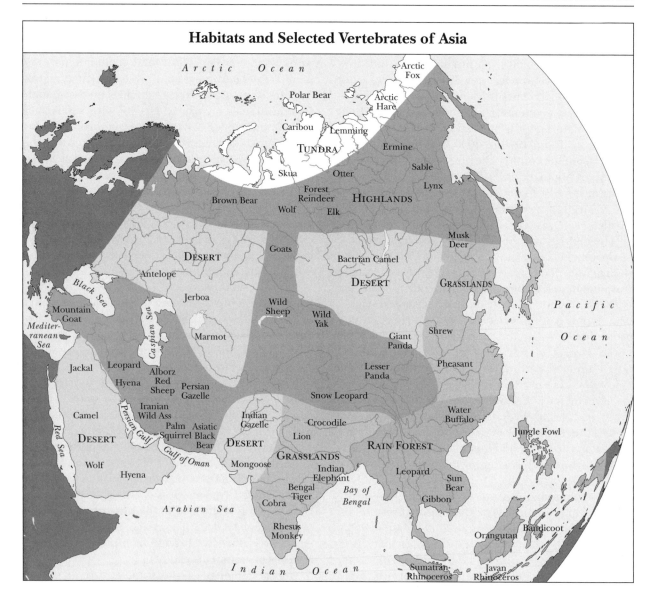

Habitats and Selected Vertebrates of Asia

Asia enjoys biologically rich oceans and coastal regions. Most of these are threatened by unsustainable fishing practices and by high levels of pollution caused by extraordinary population growth and rampant urban, agricultural, and industrial development. The region's coral reefs are in particular danger; 88 percent of these reefs were considered at risk in 2010. The destruction of mangrove forests, more than 50 percent of which were gone by 2000, magnified human losses caused by the tsunami of December, 2004, which killed the most people where protective mangrove forests had been decimated.

The solutions to Asia's significant environmental problems cannot be imposed from the outside; they must come from within the Asian nations. If Asian leaders and policy makers begin to emphasize environmental sustainability as their nations pursue economic development, and if laws protecting the environment are put in place and enforced, the degradation of Asia's environment can be halted and slowly reversed. Such a major turnaround occurred in Japan during the 1970's, after a series of highly publicized human-made environmental disasters, including the mercury poisoning of Minamata Bay, led to a dramatic increase in environmental protection laws and their effective enforcement.

R. C. Lutz

FURTHER READING

Fayzieva, Dilorom, ed. *Environmental Health in Central Asia: The Present and Future.* Boston: WIT Press, 2004.

Hill, Christopher V. *South Asia: An Environmental History.* Santa Barbara, Calif.: ABC-CLIO, 2008.

Hillstrom, Kevin, and Laurie Collier Hillstrom. *Asia: A Continental Overview of Environmental Issues.* Santa Barbara, Calif.: ABC-CLIO, 2003.

Jha, Raghbendra. "Alleviating Environmental Degradation in the Asia Pacific Region: International Co-operation and the Role of Issue Linkage." In *Regional Integration in the Asia Pacific,* edited by Organization for Economic Co-operation and Development. Paris: OECD, 2005.

Kaup, Katherine Palmer. *Understanding Asia Pacific.* Boulder, Colo.: Lynne Rienner, 2007.

SEE ALSO: Air pollution; Aquaculture; Ashio, Japan, copper mine; Bhopal disaster; Coal-fired power plants; Coral reefs; Deforestation; Fossil fuels; Ganges River; Greenhouse gases; Habitat destruction; Minamata Bay mercury poisoning; Population growth; Rain forests; Sustainable development; Water pollution.

Aswan High Dam

CATEGORY: Preservation and wilderness issues
IDENTIFICATION: Dam on the Nile River in Egypt
DATE: Completed in 1970
SIGNIFICANCE: The Aswan High Dam supplies electricity and has helped control seasonal flooding, but it has had several negative environmental effects, including loss of fertility on downriver floodplains, increased erosion, and increased incidence of earthquakes.

The Aswan High Dam was built with the aid of Soviet engineers on the Nile River approximately 960 kilometers (600 miles) south of Cairo, Egypt, between 1960 and 1970 at a cost of US$1 billion. The dam lies 7 kilometers (4.5 miles) south of the city of Aswan and several kilometers from a smaller dam constructed by British engineers between 1898 and 1902. The building of the High Dam followed the signing of the Nile Water Agreement between Egypt and Sudan in November of 1959.

The High Dam is a rock-fill structure with a core of impermeable clay. It measures 3,829 meters (12,562 feet) long, 111 meters (364 feet) high, 980 meters (3,215 feet) wide at the base, and 40 meters (131 feet) wide at the crest. The volume of material contained in the structure, 1.6 million cubic meters (56.5 million cubic feet), would be enough to construct seventeen Great Pyramids. The flow of the river's waters through the dam is via six tunnels, each controlled by a 230-ton gate.

The reservoir impounded by the High Dam is Lake Nasser, named for Egyptian president Gamal Abdel Nasser, who died the year the dam was completed. The portion of the reservoir that lies within Sudan, about 30 percent, is referred to as Lake Nubia. In total, the reservoir measures 499 kilometers (310 miles) in length and has a surface area of approximately 5,996 square kilometers (2,315 square miles) and 9,053 kilometers (5,625 miles) of shoreline. It averages 9.7 kilometers (6 miles) wide, with a maximum width of 16 kilometers (10 miles). Mean water depth is 70 meters (230 feet), maximum depth is 110 meters (360 feet), and the annual vertical fluctuation is 25 meters (82 feet). The reservoir contains enough water to irrigate more than 2.8 million hectares (7 million acres).

ENVIRONMENTAL IMPACTS

The filling of Lake Nasser came with a high cost. Thousands of people were displaced, and natural habitats were significantly altered. A number of ancient temples and monuments, which abound in the region, were to be submerged beneath the rising waters of the reservoir, but some of these were saved—they were cut into large blocks and reassembled at higher locations. High evaporation rates and water loss to infiltration into the underlying permeable Nubian sandstone caused the filling of the reservoir to take much longer than anticipated. A significant amount of water-storage capacity has also been lost as sediment that has been carried in occupies a part of the reservoir's volume.

The benefits derived from the High Dam are principally hydroelectric power generation and the regulation of water flow along the lower Nile for flood control. Twelve turboelectric generators capable of producing 10 billion kilowatt-hours provide 40 percent of Egypt's electrical power. The storage of water in Lake Nasser not only provides flood control but also allows for the irrigation of additional land and the ability to grow multiple crops over the course of a year. Since the filling of the reservoir, a fishing industry has also developed.

The yearly floods of the Nile are the result of late-summer rains that fall in the plateau region to the south in Ethiopia. At peak floods, river volume may increase by as much as sixteen times. More than 100 million tons of soil are carried with the water each year. While the impoundment of water in Lake Nasser has probably saved the lower Nile Valley from disastrous floods and alleviated the effects of regional droughts, the loss of yearly increments of silt—with its associated nutrients—on the floodplain has led to a decline in the floodplain's fertility. Without this natural fertilization, the Egyptians have had to rely on increasing use of artificial fertilizer. The floodwaters also provided a cleansing and draining action for the soil, preventing the accumulation of salts. Further, the floodwaters reduced the numbers of rats and disease-bearing snails. With the decline in flooding, incidences of disease have been on the increase.

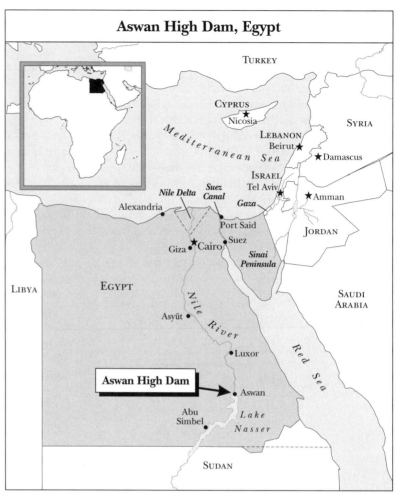

Aswan High Dam, Egypt

Historically, the influx of sediment to the Nile Delta has replenished sediment lost to wave and current erosion at the Delta's margins. Since the construction of the Aswan High Dam, the front of the Delta is being eroded at a rate of 1.8 meters (6 feet) per year. As on the floodplain, the soil of the Delta, a region that has been farmed for more than seven thousand years, also shows evidence of declining fertility.

Since the High Dam traps 98 percent of the Nile sediment, water passing through the dam has an enhanced ability to erode. Consequently, downstream erosion has become a significant problem, scouring the riverbed and undermining riverbanks and bridge piers. In some instances, the increased erosion has also affected the delicate balance of water irrigation systems.

The effects of trapped sediment are not confined to the floodplain and the Nile Delta. Prior to construction of the dam, the river brought sediment and

nutrients into the normally nutrient-poor eastern Mediterranean Sea. This provided for blooms of phytoplankton that formed the base of a food pyramid that included sardines and other commercial varieties of fish. When construction of the dam began, the sardine fishing industry in the Mediterranean declined significantly. From the late 1980's onward, however, a resurgence occurred in sardine fishing, as the filling of Lake Nasser allowed for increased river discharge and nutrient enhancement.

One other environmental consequence of the construction of the High Dam has been the occurrence of earthquakes in the region. These are related to stress that is placed on the earth's crust by the weight of the water impounded in Lake Nasser. A large shock of magnitude 5.6, for example, occurred on November 14, 1981. This was followed by aftershocks for a period of seven months.

Donald J. Thompson

FURTHER READING

Caputo, Robert. "Journey up the Nile." *National Geographic*, May, 1985.

Collins, Robert O. *The Nile*. New Haven, Conn.: Yale University Press, 2002.

Johnson, Irving, and Electra Johnson. "Yankee Cruises the Storied Nile." *National Geographic*, May, 1965.

Smith, Scot E. "The Aswan High Dam at Thirty: An Environmental Impact Assessment." In *Conservation, Ecology, and Management of African Fresh Waters*, edited by Thomas L. Crisman et al. Gainesville: University Press of Florida, 2003.

SEE ALSO: Dams and reservoirs; Erosion and erosion control; Floodplains; Floods; Hydroelectricity; Sedimentation.

Atmospheric inversions

CATEGORIES: Atmosphere and air pollution; weather and climate

DEFINITION: Vertical temperature profiles in which air temperature increases with height in the atmosphere

SIGNIFICANCE: Temperature inversions play an important role in trapping anthropogenic (human-caused) pollutants near the earth's surface, leading to the formation of smog and reduced air quality in many metropolitan areas. Temperature inversions also play an important role in the formation of severe thunderstorms and mixed precipitation (such as freezing rain and sleet).

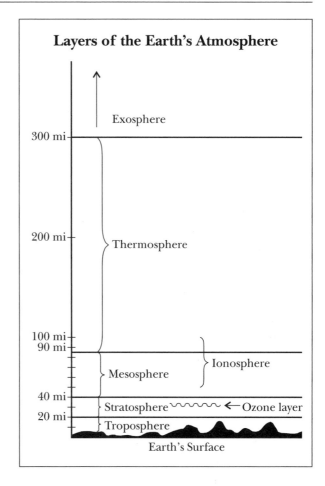

An atmospheric inversion is defined in meteorology as a scenario in which air temperature increases with height. To understand its importance it is necessary to consider the concept in light of the broader structure of the atmosphere. The vertical structure of the atmosphere is characterized by four broad regions; in order of increasing altitude, these are the troposphere, stratosphere, mesosphere, and thermosphere. On average, temperatures decrease with height in the troposphere and mesosphere, and increase with height in the stratosphere and thermosphere. The atmosphere's vertical temperature profile is important because it affects the ability of air to move vertically (that is, to rise and fall) in the atmosphere. Vertical motions are generally permitted in

regions where air temperatures decrease with height, and they are suppressed in regions where air temperatures increase with height. The latter condition, known as a temperature inversion, creates a layer in the atmosphere that has the ability to limit, or cap, vertical mixing. The stratospheric inversion is a prime example; it caps vertical motions associated with storms in the troposphere.

Smaller-scale temperature inversions that occur within the troposphere may have significant impacts on air quality at urban and regional levels because they trap polluted air masses close to the ground. A common example of this is a radiation inversion, which often develops during winter months on days with limited wind and low humidity. These conditions allow the ground to cool quickly, thereby causing the air immediately above the ground to cool more than the air aloft. The warmer air aloft forms a stable layer, or cap, over the atmosphere below it. Radiation inversions are an air-quality concern because they often set up during the evening traffic rush hour and trap associated pollutants close to the surface. In general, radiation inversions affect small geographic regions and dissipate during the early morning, when solar radiation raises temperatures near the surface.

A larger-scale subsidence inversion forms when a high-pressure system causes air to subside and warm adiabatically (without gaining or losing heat) over a broad region. Different rates of subsidence between air near the surface and air aloft allow the air aloft to warm to a greater degree, setting up a large-scale temperature inversion. In contrast to radiation inversions, subsidence inversions have the potential to persist for long periods of time. This scenario is common in many metropolitan areas that are known for poor air quality, particularly Los Angeles, California. In Los Angeles, the impact of temperature inversions is magnified by the fact that the city is surrounded on three sides by mountains, which limit horizontal mixing and promote higher concentrations of air pollutants in the region.

The pollutants that build up in urban areas as the result of temperature inversions have important long-term health implications and, at times, can prove to be serious short-term threats. For example, in 1948 toxic air conditions that resulted from subsidence inversion in the western Pennsylvania town of Donora caused the deaths of twenty residents. A similar incident known as the London smog disaster killed nearly four thousand residents of London, England, in 1952.

Jeffrey C. Brunskill

FURTHER READING

Aguado, Edward, and James E. Burt. *Understanding Weather and Climate.* 5th ed. Upper Saddle River, N.J.: Pearson Education, 2010.

Ahrens, C. Donald. *Meteorology Today: An Introduction to Weather, Climate, and the Environment.* Belmont, Calif.: Brooks/Cole Cengage Learning, 2009.

SEE ALSO: Air pollution; Air-pollution policy; Air Quality Index; Clean Air Act and amendments; Donora, Pennsylvania, temperature inversion; Fossil fuels; London smog disaster; Smog.

Atomic Energy Commission

CATEGORIES: Organizations and agencies; nuclear power and radiation; energy and energy use
IDENTIFICATION: Agency established by the U.S. Congress to promote and monitor the use of nuclear energy
DATES: 1946-1974
SIGNIFICANCE: Because the Atomic Energy Commission was responsible for both the promotion of nuclear power and the safety regulation of nuclear facilities, many critics asserted that the commission was not as diligent as it should be in its regulatory duties.

In 1946, the U.S. Congress passed the Atomic Energy Act, which created the Atomic Energy Commission (AEC) to promote, monitor, and control the development and use of nuclear energy for civilian and military use. Not only was a buildup of nuclear weapons for defense expected, but also the development of nuclear science for peaceful uses. For security reasons, all nuclear production facilities and reactors were owned by the U.S. government, but the information from research was to be controlled by the AEC. The Manhattan Project facilities, where scientists had developed the first atomic bombs during World War II, were taken over by the AEC. In 1954, Congress passed the Atomic Energy Act Amendments, which made possible the development of commercial nuclear power. The AEC was assigned the responsibility of promoting commercial nuclear power at the same time it was charged with developing safety regulations and controls for nuclear power plants.

The work of the AEC encompassed all areas of nuclear science. The AEC studied the production of nuclear weapons, improved and increased nuclear facilities, and oversaw the development of new weapons, such as the thermonuclear device, a fusion bomb, and nuclear propulsion for warships, especially submarines. The study of nuclear reactors led to improved energy-producing reactors and a new type of reactor, the breeder reactor, that produced new fuel while producing energy. Commercial production of nuclear energy became a reality. Regional research centers were established—at the Argonne National Laboratory near Chicago; at Oak Ridge, Tennessee; at Brookhaven, New York; and at the University of California Radiation Laboratory in Berkeley—where scientists could work on projects in nuclear or radiation science. The AEC also contracted with industrial and university researchers to conduct nuclear-related studies in all areas of scientific research, including medical diagnosis using nuclear techniques, disease treatment by radiation, and treatment for radiation exposure. Other areas of study included the effects of radiation, production of new isotopes, uses of new isotopes, and radiation safety. Reactors to produce energy for satellites were developed and sent into space.

Many observers were critical of the AEC's double role as promoter of nuclear power and monitor of radiation safety, saying that this was a conflict of interest. Many asserted that the AEC was promoting nuclear energy at the expense of regulation and that the safety regulations for reactors especially, but also for materials production, location of plants, and environmental protection, were inadequate.

The AEC was abolished in 1974, and its former two roles were split. The promotion of nuclear science in the United States was assigned to the Energy Research and Development Administration; these duties later became the responsibility of the Department of Energy. Responsibility for the safety monitoring of nuclear facilities went to the Nuclear Regulatory Commission.

C. Alton Hassell

FURTHER READING

Cooke, Stephanie. *In Mortal Hands: A Cautionary History of the Nuclear Age.* New York: Bloomsbury, 2009.

Hewlett, Richard G., and Oscar E. Anderson. *A History of the United States Atomic Energy Commission.* 1962. Reprint. Berkeley: University of California Press, 1990.

Walker, J. Samuel. *A Short History of Nuclear Regulation, 1946-1999.* Washington, D.C.: U.S. Nuclear Regulatory Commission, 2000.

SEE ALSO: Breeder reactors; Chalk River nuclear reactor explosion; Hanford Nuclear Reservation; Limited Test Ban Treaty; Nuclear accidents; Nuclear and radioactive waste; Nuclear power; Nuclear Regulatory Commission; Nuclear regulatory policy; Nuclear testing; Nuclear weapons production; Price-Anderson Act; Radioactive pollution and fallout; SL-1 reactor accident; Three Mile Island nuclear accident.

Audubon, John James

CATEGORIES: Activism and advocacy; animals and endangered species

IDENTIFICATION: French American naturalist and wildlife artist

BORN: April 26, 1785; Les Cayes, Saint-Domingue (now Haiti)

DIED: January 27, 1851; New York, New York

SIGNIFICANCE: Through his unique paintings and his writings, Audubon demonstrated ecological relationships among organisms and set new standards for field observation. By illustrating the beauty of birds and animals, he helped to lay the foundation for a national environmental consciousness in the United States.

John James Audubon was born Jean Jacques Audubon, the son of a French naval officer, on April 26, 1785. He grew up in Saint-Domingue (now Haiti) and Nantes, France. At the age of eighteen he was sent to

Naturalist and wildlife artist John James Audubon. (Library of Congress)

live on a farm at Mill Grove, near Philadelphia, Pennsylvania, to escape induction into Napoleon Bonaparte's army. Audubon spent his days in Pennsylvania roaming the woods and observing, collecting, and sketching wildlife. He experimented with bird banding and developed techniques to mount bird specimens so that he could draw them in lifelike poses. During this time he began thinking of himself as an American and began to call himself John James.

As Audubon matured, the hobby of his youth became an obsession. The time he spent outdoors teaching himself to live in the forest and understand its inhabitants cost Audubon dearly in financial terms and tried the patience of his wife, Lucy. However, it gave him unparalleled insights into the lives of his subjects. A perfectionist who periodically tore up drawings he considered less than his best, Audubon developed a distinctive style of illustration, creating drawings and paintings portraying birds and animals in lively interaction with their surroundings.

Audubon established a frontier store and a sawmill in Henderson, Kentucky, but these ventures failed because he so often neglected his business to pursue his art. Forced to declare bankruptcy in 1819, Audubon first thought of giving up painting altogether, then decided to devote himself completely to it. He resolved to publish a portfolio of American birds that would be more complete and more accurately illustrated than any previous work. Lucy taught school to support the family while he worked furiously to produce the necessary sketches and find a publisher. When the work appeared, Audubon suffered mixed reviews from the American scientific and artistic establishments but found a receptive audience in Europe.

In 1826 Audubon made arrangements with an engraver to make folio-sized prints of his illustrations and publish them under the title *The Birds of America.* This large-scale book would present each bird study as a life-size portrait and preserve the fine details of Audubon's watercolors. The hand-colored printing process was laborious and very expensive. To finance the project, Audubon sold subscriptions to a series of separate folios, each containing five engravings. It took him twelve years to complete the 435 plates. Only 176 subscriptions were sold, but this was enough to cover Audubon's expenses. (In the twenty-first century, first-edition copies of *The Birds of America* are prized by collectors, commanding high prices when offered for sale. In December, 2010, a complete first-edition copy was sold at auction for approximately $11.5 million, a record price for a printed book.)

With the appearance of *The Birds of America* in 1838 Audubon established an international reputation. The smaller and less expensive edition of the book, which included excerpts from his *Ornithological Biography* (1839), was a popular and financial success. Finally, Audubon was able to buy an estate on the Hudson River in New York, where he and Lucy mentored many young scholars, artists, and naturalists. One of these students, George Bird Grinnell, founded the first Audubon Society, an organization devoted to promoting bird study and conservation, in 1886. (The present-day National Audubon Society was incorporated in 1905.) As Audubon's energy and eyesight waned, his sons helped him complete paintings for *The Viviparous Quadrupeds of North America* (1849-1854).

With his unique paintings, Audubon demonstrated ecological relationships among organisms by illustrating their food plants, nesting sites, competition, and predators. In his writings, Audubon set new standards for field observation and foresaw the threat of species extinction. Above all, by illustrating the beauty of birds and animals, he promoted the popular study of natural history, helping to lay the foundation for a national environmental consciousness in the United States.

Robert W. Kingsolver

FURTHER READING
Rhodes, Richard. *John James Audubon: The Making of an American.* New York: Alfred A. Knopf, 2004.
Souder, William. *Under a Wild Sky: John James Audubon and the Making of "The Birds of America."* New York: North Point Press, 2004.

SEE ALSO: Conservation; Passenger pigeon extinction; Whooping cranes.

Audubon Society. *See* National Audubon Society

Australia and New Zealand

CATEGORIES: Places; ecology and ecosystems; animals and endangered species; resources and resource management

SIGNIFICANCE: The continent of Australia and the North and South Islands of New Zealand are ancient South Pacific lands of variable climate and extraordinary species extending from the subtropics to the frigid Antarctic. Australia and New Zealand are among the most biologically diverse ecosystems in the world.

The British Commonwealth of Nations is a worldwide association of fifty-four states with an estimated combined population of more than 1.9 billion as of 2005. Australia and New Zealand, both early nineteenth century colonies of the British Commonwealth, have a common Victorian heritage of cultural diffusion, which has had a profound effect on their temperate ecologies and native environments. Of particular significance are the rapid expansion of the nautical, astronomical, geographical, and biological sciences during the eighteenth and nineteenth centuries, advanced by the British Admiralty; capital investment in international trading companies such as the British East India Company; and the rise of professional organizations such as the British Association for the Advancement of Science.

NINETEENTH CENTURY BRITISH COLONIALISM

The famous Pacific voyages of James Cook during the late eighteenth century transformed European understanding of the ocean and its lands. Cook's meticulous observations of the transit of Venus set the stage for subsequent international efforts during the late 1870's. The proliferation of British colonial observatories, numerous expeditionary studies of geomagnetic variation, the international exploration of the Arctic and the Antarctic, and a popular culture that relished global travel to exotic lands gave a distinct intellectual energy to the settling of Australia and New Zealand.

During Anglo-European settlement both countries went through a remarkable series of transformations as British settlers sought to remake their new landscapes in the image of the English countryside, a fitting tribute to the Victorian ideals of progress and civilization. Popular interest in natural history and taxonomy were juxtaposed with early colonial attempts to improve the land with imports of sheep, cattle, pigs, horses, honeybees, flowers and grasses, exotic game animals for sport hunting, tropical species for adaptation in botanical gardens, and familiar birds and mammals—notably the rabbit, the red deer, and the English skylark. These and other unintended invasive transplantations (infectious diseases, cats, dogs, weeds, and rodents included) were made at the expense of native communities and species that were often subject to bounty hunting as a means of control and extermination. Seals and whales are among the species that were hunted to near extinction.

Parkland towns dotted the fringes of both Australia and New Zealand. As successive generations lived and toiled on arable lands, the uniqueness of the terrain and its place in time acquired new significance. Public fondness grew for species unique to the region, such as the kiwi, the platypus, the koala, the kangaroo, and the eucalyptus, and for the novelty of area bushlands. A growing respect for local biodiversity echoed the sentiments of settlers in North America, who saw in the last vestiges of wilderness something sublime that should be preserved as a public good. Amateur studies of natural history were amplified by questions asked by ecologists, creating new systems of knowledge that integrated cultural folkways with empirical science. These insights invited new exchanges with the cultures of the native peoples—the Australian Aborigines and the Maori of New Zealand—creating a foundation for an international statutory framework that protects native land rights.

FOUNDATIONS OF ENVIRONMENTAL AWARENESS

Australia's and New Zealand's contributions to the British war efforts during the early twentieth century helped these nations rise to independent prominence in the global economy. During this transformation both countries experienced remarkable upsurges in population and industry; the resulting environmental burdens in turn stimulated intense public appeals for reform. The world took note of the publication of Rachel Carson's book *Silent Spring* in 1962. In Australia the validity of Carson's warnings was clear in the pollution emitting from coastline commercial and industrial development and in the unprecedented number of predatory crown-of-thorns starfish clustering on the continent's Great Barrier Reef, the world's largest coral reef system. Concerted efforts to understand and to protect the integrity of this coastal

Habitats and Selected Vertebrates of Australia

ecosystem marked a turning point in Australia's self-awareness as the protector of a profound ecological legacy. In 1981 the Great Barrier Reef was registered as a World Heritage Site.

Pollution and climate change continue to endanger thousands of species of whales, dolphins, fish, sea turtles, algae, and shorebirds endemic to this marine environment. In 1975 the government of Australia created the Great Barrier Reef Marine Park, and in 2004 it initiated a massive marine rezoning program—the second-largest marine protection program in the world—to protect nearly 35 percent of the reef ecosystem.

According to a 2006 publication of the Australian Biodiversity Information Services, nearly 10 percent of the world's plant and animal species can be found in Australia and New Zealand. New Zealand's island ecosystem is nearly eighty million years old. Because of its complete isolation, its evolution was unique to land birds; no mammals are recorded until human habitation approximately one thousand years ago.

It is estimated that New Zealand may be home to eighty thousand species of native plants, animals, and fungi.

Victoria M. Breting-García

FURTHER READING

Australian State of the Environment Committee. *Australia: State of the Environment 2006—Independent Report to the Australian Government Minister for the Environment and Heritage.* Canberra : Department of the Environment and Heritage, 2006.

Cosgrove, Denis. *Geography and Vision: Seeing, Imagining, and Representing the World.* New York: Palgrave Macmillan, 2008.

Crosby, Alfred W. *Ecological Imperialism: The Biological Expansion of Europe, 900-1900.* New York: Cambridge University Press, 1986.

Dunlap, Thomas R. *Nature and the English Diaspora: Environment and History in the United States, Canada, Australia, and New Zealand.* New York: Cambridge University Press, 1999.

Finnis, Bill. *Captain James Cook: Seaman and Scientist.* London: Chaucer Press, 2003.

Fitzpatrick, Brian. *The British Empire in Australia: An Economic History, 1834-1939.* 2d ed. 1949. Reprint. Melbourne: Macmillan of Australia, 1969.

Hindmarsh, Richard. *Edging Towards BioUtopia: A New Politics of Reordering Life and the Democratic Challenge.* Crawley: University of Western Australia Press. 2008.

Meredith, David, and Barrie Dyster. *Australia in the Global Economy: Continuity and Change.* New York: Cambridge University Press, 1999.

New Zealand. Ministry for the Environment. *Environment New Zealand 2007.* Wellington: Author, 2008.

Powell, J. M., ed. *The Making of Rural Australia: Environment, Society, and Economy—Geographical Readings.* Melbourne: Sorrett, 1974.

SEE ALSO: Biodiversity; Coral reefs; Ecosystems; Franklin Dam opposition; Great Barrier Reef; Indigenous peoples and nature preservation; Introduced and exotic species; Pacific Islands; Public opinion and the environment; Sea turtles; Seal hunting; Whaling.

Automobile emissions

CATEGORIES: Atmosphere and air pollution; pollutants and toxins

DEFINITION: Combination of vehicular exhaust from incomplete combustion of gasoline or diesel fuel and hydrocarbons that escape these fuels through evaporation

SIGNIFICANCE: Collectively, the emissions produced by the millions of vehicles on the roads of the world's major metropolitan areas have serious impacts on air quality and human health. They also represent a significant contribution to greenhouse gas emissions. The U.S. Environmental Protection Agency has described driving a private car as "probably a typical citizen's most 'polluting' daily activity."

Automobile emissions create ongoing and potentially dangerous environmental problems when gases and particulates are released into the atmosphere at a rate that exceeds the capacity of the atmosphere to dissipate or dispose of them. Motor vehicle emissions are a major component of the smog that blankets such urban areas as Los Angeles, California, and Denver, Colorado. The emissions produced by a motor vehicle consist of exhaust (the by-products of incomplete gasoline or diesel fuel combustion) and fuel that evaporates from the vehicle's fuel tank, engine, and exhaust system during operation, cooldown, and fueling.

Vehicle exhaust contains several problematic compounds. Carbon monoxide, produced by incomplete combustion, reduces the flow of oxygen in the bloodstream. Hydrocarbons, another product of imperfect combustion, are often toxic or carcinogenic. Nitrogen oxides, which are formed when combusting fuel reacts with oxygen in the air, contribute to the formation of acid rain and fine particles that can harm the lungs when breathed in. Together, hydrocarbons and nitrogen oxides react with the heat of sunlight to produce a hazy brown mixture of secondary pollutants. Notable among these pollutants is ground-level ozone, which irritates the eyes and causes damage to the respiratory system. (Nitrogen oxides also readily create ozone by reacting with naturally occurring hydrocarbons produced by trees.) Other secondary pollutants formed by photochemical reactions include the toxic compounds formaldehyde and peroxyacetyl nitrate. Carbon dioxide, a product of complete combustion, is not directly hazardous to human health, but it is chief among the anthropogenic (human-generated) greenhouse gases commonly believed to affect the world's climate.

IMPACTS ON HEALTH AND ENVIRONMENT

Environmental problems associated with automobile emissions include deleterious effects on many forms of agriculture and natural forests, reduction in visibility, and damage to metals, building materials such as stone and concrete, rubber, paint, textiles, and plastics. Automobile emissions cause lung and eye irritation, coughing, chest pain, shallow breathing, and headaches. Automobile-produced air pollution is also a factor contributing to allergies, asthma, emphysema, bronchitis, lung cancer, heart disease, and negative psychological states. Carbon monoxide quickly combines with blood hemoglobin and impairs oxygen delivery to the tissues, particularly in children and the elderly, causing heart and lung problems. Vehicular emissions and other sources of air pollution cost Americans billions of dollars each year in health care and related expenses.

The increased rate and depth of breathing during

Costs of Rural and Urban Air Quality Degradation by Motor Vehicles, 2000

POLLUTANT	IMPACT	RURAL EMISSIONS ($)	URBAN EMISSIONS ($)
Particulate matter	Mortality	12,695	21,558
Particulate matter	Nonfatal illness	3,683	6,232
Sulfur dioxide, nitrogen dioxide, carbon monoxide	Nonfatal illness	0	51
Ozone	Nonfatal illness	28	16
Total		**16,406**	**27,857**

Source: Federal Highway Administration, United States Department of Transportation.
Note: Costs of human illnesses, in millions of 1990 dollars. Costs of crop damage, reduced visibility, and other physical effects on the environment are not included.

physical exertion exposes delicate lung tissues to more polluted air. Research indicates that exercise near a busy freeway may be more harmful than beneficial to the body. At the 1984 Summer Olympics in Los Angeles, the evening rush-hour start of the men's marathon coincided with a stage 2 California health advisory alert, drawing criticism that the organizers of the event were more interested in commercial revenues than in the safety of the athletes and spectators. Later Olympic events were postponed during heavy air-pollution episodes. China faced similar, but more severe, challenges when heavily polluted Beijing hosted the 2008 Summer Olympics. Emergency measures that the city took to improve air quality in time for the games included allowing drivers to use their motor vehicles only every other day, which effectively took more than 1.5 million vehicles off the road daily.

Automobile emissions have been shown to exert their negative effects a considerable distance from the source, depending on atmospheric changes in wind and temperature. Suburbs often exhibit higher levels of pollution than the downtown areas where the emissions are produced. Remote national parks and wilderness areas have had their scenic vistas obscured by haze from distant cities. Fallout of tetraethyl lead from urban automobiles running on leaded gasoline has been observed in oceans and on the Greenland ice sheet. Automobile emissions may also have an impact on a global scale, as they are a significant source of greenhouse gases. Transportation contributed some 13.1 percent of the world's total anthropogenic greenhouse gas emissions in 2004.

EFFORTS TO REDUCE EMISSIONS

Studies in cities such as London, England, have shown that major improvements in air quality can be achieved in less than ten years in urban areas with favorable climatic conditions through the use of more combustion-efficient engines and cleaner-burning fuels . In the United States, the 1970 amendments to the 1963 Clean Air Act (CAA) introduced automobile emissions standards for hydrocarbons, carbon oxides, and nitrogen oxides and ambient air-quality standards for six pollutants—carbon monoxide, sulfur oxides, nitrogen oxides, particulates, ozone, and lead—to protect human health and the environment. Through the CAA and its amendments of 1970, 1977, and 1990, the Environmental Protection Agency (EPA) has established increasingly stringent emissions-control policies for motor vehicles.

Although the EPA sets pollution standards for vehicles, vehicle manufacturers determine how they will meet those standards. Improved engine design, recirculation of exhaust gas to reduce nitrogen oxides, improved evaporative emissions controls, and computerized diagnostic systems have all led to a decline in polluting emissions. One of the most important milestones in the reduction of hydrocarbon and carbon monoxide emissions was the advent of the catalytic converter in 1975. Because lead impedes the catalyst that reduces emissions, unleaded gasoline became widely available at the same time. Ultimately, leaded gasoline was phased out in the United States and in many other countries, with the result that lead concentrations in ambient air have been lowered significantly.

Although the vehicle emissions controls implemented in the United States since 1970 have been effective, much of their success has been offset by the increasing numbers of vehicles on the roads and the greater distances driven, as many Americans travel farther from their homes to reach workplaces, schools, and shopping and recreation centers. Motor vehicle use has also become more widespread in developing countries, where air-quality standards are often more lax than in developed nations.

In the United States, the pollutant associated with vehicle emissions that has been reduced the most has been lead. Ambient concentrations of carbon monoxide and nitrogen oxides have also decreased; however, while they are low in relation to national standards, they remain a matter of concern because of the role they play in producing ozone and particulates and in impairing visibility. Ground-level ozone, toxic hydrocarbons, and particulates continue to be problems—as does carbon dioxide. In May, 2009, the EPA and the U.S. Department of Transportation agreed to establish national standards for greenhouse gas emissions and fuel economy for new cars and trucks sold in the United States.

Daniel G. Graetzer
Updated by Karen N. Kähler

Further Reading

Godish, Thad. *Air Quality.* 4th ed. Boca Raton, Fla.: Lewis, 2004.

Griffin, Roger D. *Principles of Air Quality Management.* 2d ed. Boca Raton, Fla.: CRC Press, 2007.

Hilgenkamp, Kathryn. "Air." In *Environmental Health: Ecological Perspectives.* Sudbury, Mass.: Jones and Bartlett, 2006.

Jacobson, Mark Z. *Atmospheric Pollution: History, Science, and Regulation.* New York: Cambridge University Press, 2002.

McCarthy, Tom. *Auto Mania: Cars, Consumers, and the Environment.* New Haven, Conn.: Yale University Press, 2007.

Rajan, Sudhir Chella. *The Enigma of Automobility: Democratic Politics and Pollution Control.* Pittsburgh: University of Pittsburgh Press, 1996.

U.S. Environmental Protection Agency. *The Plain English Guide to the Clean Air Act.* Research Triangle Park, N.C.: Office of Air Quality Planning and Standards, 2007.

SEE ALSO: Air pollution; Air-pollution policy; Airborne particulates; Carbon monoxide; Catalytic converters; Clean Air Act and amendments; Gasoline and gasoline additives; Hybrid vehicles; Nitrogen oxides; Smog.

B

Bacillus thuringiensis

CATEGORY: Biotechnology and genetic engineering

DEFINITION: Natural microbial pesticide used for biological control of insects

SIGNIFICANCE: *Bacillus thuringiensis* provides an environmentally neutral alternative to chemical pesticides, which can leave pollutants in soil and water.

Chemical control of plant parasites using pesticides, while often effective, has generally been found to create its own problems in areas of environmental pollution. The use of microbial pesticides has often proven as effective at pest management, without the problems of polluting soil and water. In addition,

since many of these organisms are species-specific, targeting only select insects, neither plants nor more desirable insects (such as ladybugs) are at risk.

Bacillus thuringiensis (*B.t.*) is one of several bacteria species that have been used as biopesticides for control of insects. The bacterium produces a proteinaceous parasporal crystal within the cell when it sporulates (that is, forms spores) that is toxic only for a specific insect host. The crystalline protein is harmless in its natural state, becoming toxic only when it is cleaved by specific proteases. Once the bacterium has been ingested by the target insect, the alkaline secretions within the midgut of the insect activate the toxic character of the crystalline body. The toxin binds the intestinal cells, resulting in formation of pores and leakage of nutri-

In Kenya, workers with the International Center of Insect Physiology and Ecology spread the microbial pesticide Bacillus thuringiensis israelensis *to kill mosquitoes.* (©Remi Benali/CORBIS)

ents from the gut, and the insect dies within three to five days. The ability of the bacterium to sporulate also provides a means to disseminate the organism as a spray. Some commercial treatments utilize the crystalline body itself, which is sprayed onto the plants.

Though commercial application of *B.t.* began in the 1960's, widespread use began only decades later with the decline in use of chemical pesticides for protecting food crops and other types of plants. Various strains of *B.t.* have been developed for use against numerous types of insects. The *israelensis* variety is used commercially for control of mosquitos and blackfly larvae. The variety *kurstaki* has been used effectively against both the gypsy moth and cabbage loopers. More recently developed strains of *B.t.* have been shown to be effective in controlling insects that feed on fruit crops such as oranges and grapes. Varieties of *B.t.* are used commercially against some 150 different insect species, including cabbage worms, caterpillars, and other insects that feed on vegetable crops. The targeting of specific insects while using certain strains of *B.t.* also reduces the chance that desirable populations of organisms may be adversely affected.

Attempts have also been made through genetic engineering to introduce the gene that encodes the crystalline body directly into the plant genome. Since the protein is toxic only to insects, such an approach may result in naturally insect-resistant plants without further need for spraying. The procedure has proven effective in controlled experiments using tobacco plants, and more widespread testing in other plants has also been undertaken. The controversial nature of introducing such potentially toxic genes may preclude their use in food crops for some time, but such genetic experiments may prove useful in natural protection of ornamental trees and other plants.

Richard Adler

FURTHER READING

Letourneau, Deborah K., and Beth Elpern Burrows, eds. *Genetically Engineered Organisms: Assessing Environmental and Human Health Effects*. Boca Raton, Fla.: CRC Press, 2002.

Metz, Matthew, ed. *Bacillus Thuringiensis: A Cornerstone of Modern Agriculture*. Binghamton, N.Y.: Haworth Press, 2003.

SEE ALSO: Agricultural chemicals; Biopesticides; Genetically altered bacteria; Integrated pest management; Pesticides and herbicides.

Back-to-the-land movement

CATEGORIES: Philosophy and ethics; activism and advocacy

IDENTIFICATION: Social movement based on the values of self-sufficiency and human harmony with nature

DATES: 1960's-1970's

SIGNIFICANCE: The back-to-the-land movement exemplified a practice that sought to give greater meaning to everyday life through adherence to values of self-sufficiency, simplicity, freedom, and, most important, anticonsumerism.

The expression "back to the land" is commonly used in reference to a North American social and countercultural phenomenon that started during the mid-1960's and continued well into the 1970's. The historical roots of the movement can be traced to Thomas Jefferson's agrarian vision and to the practice of self-reliance espoused by nineteenth century philosophers and essayists Ralph Waldo Emerson and Henry David Thoreau. Those who took part in the so-called back-to-the-land movement migrated from cities to rural areas because they had become increasingly disenchanted with the growing urban and industrial culture; they were attracted to a simple sort of daily life based on a set of values and choices that they saw as being in tune with an agrarian way of life and thus in greater harmony with the natural world.

Among the activities undertaken by those who joined the movement were building homes with natural materials; setting up systems to generate alternative forms of energy, such as solar and wind power; and growing their own food. They also faced choices regarding what types of livelihoods they should pursue and whether to work at home or outside the home; in addition, they somehow had to reconcile their desire to give up the rampant consumerism in society with their need to make a living. Members of the movement were arguably more interested in "making a life" for themselves than they were in "making a living," and many chose to live very simply, at the mercy of nature, and endure some of the inconveniences this may occasionally entail rather than indulge in what they viewed as the rampant and alienating consumerism of North American society at large.

By the end of the 1970's the movement evolved into one that took up the causes of the growing environmental movement, as people became more concerned with sustainable and holistic living. It has been estimated that by this time well over 1 million people had moved from the cities to rural areas. In the 1980's, however, a significant decline was noted in the number of people interested in leaving consumer culture for a simpler life in the countryside. This is explained in part by the booming and widespread prosperity of that decade. According to some observers of these trends, the 1990's saw a return to more environmentally conscious lifestyles among many North Americans, particularly in view of increasing awareness of global ecological crises.

Nader N. Chokr

FURTHER READING

Agnew, Eleanor. *Back from the Land: How Young Americans Went to Nature in the 1970's, and Why They Came Back.* Chicago: Ivan R. Dee, 2004.

Jacob, Jeffrey C. *New Pioneers: The Back-to-the-Land Movement and the Search for a Sustainable Future.* 1997. Reprint. University Park: Pennsylvania State University Press, 2006.

Nearing, Helen, and Scott Nearing. *Living the Good Life: How to Live Sanely and Simply in a Troubled World.* 1954. Reprint. New York: Schocken Books, 1987.

SEE ALSO: Balance of nature; Composting; Deep ecology; Ecology as a concept; Environmentalism; Organic gardening and farming; Overconsumption; Solar energy; Sustainable agriculture; Wind energy.

Bacterial resistance

CATEGORY: Human health and the environment

DEFINITION: The ability of bacteria to withstand antibiotics and other antimicrobial substances

SIGNIFICANCE: Inappropriate use of antimicrobials such as antibiotics, antibacterial soaps, and hand sanitizers has caused some bacteria to develop resistance against the most common antibiotics, and many previously controlled infectious pathogens have become resistant to multiple drugs. Such multiresistant bacteria pose an increasing threat to public health and safety.

Bacteria are the most adaptable living organisms on earth and are found in virtually all environments—from the lowest ocean depths to the highest mountains. Bacteria resist extremes of heat, cold, acidity, alkalinity, heavy metals, and radiation that would kill most other organisms. *Deinococcus radiodurans*, for example, grow within nuclear power reactors, and *Thiobacillus thiooxidans* can grow in toxic acid mine drainage.

The terms "super bacteria" and "superbug" are often used to refer to bacteria that have either intrinsic (naturally occurring) or acquired resistance to multiple antibiotics. For example, two soil organisms—*Pseudomonas aeruginosa* and *Burkholderia* (*Pseudomonas*) *cepacia*—are intrinsically resistant to many antibiotics. Because many of the bacteria that acquire resistance are pathogens that were previously controlled by antibiotics, the development of antibiotic resistance represents a serious public health crisis, particularly for those individuals suffering from compromised immune systems.

HISTORY OF ANTIBIOTIC USE

The history of antimicrobial compounds reaches into the early twentieth century, when German chemist Paul Ehrlich received worldwide fame for discovering Salvarsan, the first relatively specific chemical prophylactic agent against the microorganisms that causes syphilis. Salvarsan had serious undesirable side effects, as arsenic was its active ingredient. In addition, despite advances in antiseptic surgery, secondary infections resulting from hospitalization were a leading cause of death in the early twentieth century. Consequently, when Scottish bacteriologist Alexander Fleming reported his discovery of a soluble antimicrobial compound called penicillin, produced by the fungus *Penicillium*, the news attracted worldwide attention.

Antibiotics such as penicillin are low-molecular-weight compounds excreted by bacteria and fungi. Antibiotic-producing microorganisms most often belong to a group of soil bacteria called actinomycetes. *Streptomyces* are good examples of antibiotic-producing actinomycetes, and most of the commercially important antibiotics are isolated from *Streptomyces*. It is not entirely clear what ecological role the antibiotics play in natural environments. Microorganisms produce the antibiotics in a late, stationary growth phase, rather than during an early, active growth phase, which suggests that their chief function is not to inhibit the growth of competing microorganisms. The results of some studies suggest that, in natural ecosystems, antibiotics are types of signaling molecules that

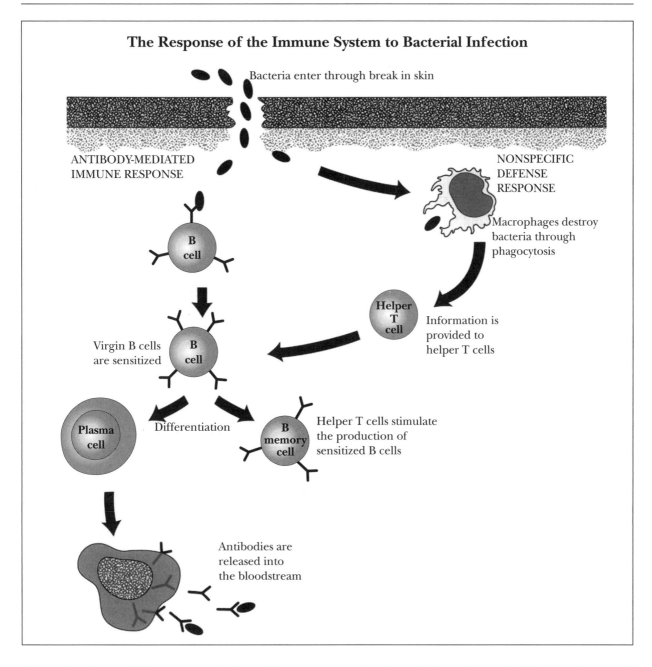

The Response of the Immune System to Bacterial Infection

Bacteria enter through break in skin

ANTIBODY-MEDIATED
IMMUNE RESPONSE

NONSPECIFIC
DEFENSE
RESPONSE

Macrophages destroy
bacteria through
phagocytosis

B
cell

Helper
T
cell

Information is
provided to
helper T cells

Virgin B cells
are sensitized

B
cell

Plasma
cell

Differentiation

B
memory
cell

Helper T cells stimulate
the production of
sensitized B cells

Antibodies are
released into
the bloodstream

play a role in intercellular communication, while antibiotic resistance genes influence metabolism.

Fleming's discovery was not of much clinical importance until two English scientists, Howard Florey and Ernst Chain, took Fleming's fungus and produced purified penicillin just in time for World War II. The success of penicillin as a therapeutic agent with almost miraculous effects on infection prompted other microbiologists to look for naturally occurring antimicrobial compounds. In 1943 Selman Waksman,

an American biochemist born in Ukraine, discovered the antibiotic streptomycin, the first truly effective agent to control *Mycobacterium tuberculosis*, the bacterium that causes tuberculosis. Widespread antibiotic use began shortly after World War II and was regarded as one of the great medical advances in the fight against infectious disease. By the late 1950's and early 1960's, pharmaceutical companies had devoted extensive research and development programs to isolating and producing new antibiotics.

Antibiotics were so effective, and their production ultimately so efficient, that they became inexpensive enough to be routinely prescribed for all types of infections, particularly to treat upper respiratory tract infections. When it was discovered that low levels of antibiotics also promote increased growth in domesticated animals, antibiotics began to be used routinely as feed supplements. Antibiotic use became ubiquitous among humans and livestock, to the degree that downstream from intense urbanization or livestock production, water sources commonly contain detectable levels of nonmetabolized, excreted antibiotics.

DEVELOPMENT OF ANTIBIOTIC RESISTANCE

The widespread use and, ultimately, misuse of antibiotics inevitably caused antibiotic-resistant bacteria to emerge as microorganisms adapted to this new selective pressure. The prevalence of antimicrobial consumer products such as antibacterial soaps and hand sanitizers resulted in additional selective pressure. Eventually, many strains of pathogenic organisms developed on which antibiotics have little or no effect. Relatively recently discovered pathogens, such as the *Helicobacter pylori* bacterium associated with peptic ulcers, have rapidly developed resistance to the antibiotics used to treat them.

Streptococcal infections are a major bacterial cause of morbidity and mortality. In the mid-1970's *Streptococcus pneumonia* was uniformly susceptible to penicillin. However, penicillin-resistant streptococci strains were being isolated as early as 1967. A study in Denver, Colorado, showed that penicillin-resistant *S. pneumonia* strains increased from 1 percent of the isolates in 1980 to 13 percent of the isolates in 1995. One-half of the resistant strains were also resistant to another antibiotic, cephalosporin. It was apparent from the Denver study that a high correlation existed between antibiotic resistance in an individual and whether a member of the individual's family attended day care. Children attending day care are frequently exposed to preventive antibiotics, and these may have assisted in selecting for resistant bacteria. By 1996 penicillin-resistant *S. pneumoniae* represented between 33 and 58 percent of the clinical isolates around the world. In another U.S. study, conducted during the 1990's in middle Tennessee, non-penicillin-susceptible *S. pneumoniae* isolates increased over a nine-year period from 10 percent to 70 percent.

The disease tuberculosis (TB) was once the lead-ing cause of death in young adults in industrialized countries. The disease was so common and feared that it was known as the White Plague. Before 1990 multidrug-resistant *M. tuberculosis* was uncommon, but by the mid-1990's increasing outbreaks were seen in hospitals and prisons, with death rates ranging from 50 to 80 percent. In a 2008 global report on anti-TB drug resistance, the World Health Organization noted that in 2006 the number of multidrug-resistant TB cases (those resistant to isoniazid and rifampicin, two of the first-line drugs used to treat the disease) had reached a record high of 489,139. Cases of the disease that were also resistant to second-line drugs were reported in 45 countries.

Likewise, multiple drug resistance in *Streptococcus pyogenes*, the so-called flesh-eating streptococci, was once rare, but erythromycin- and clindimycin-resistant strains have developed. In Italy, for example, antibiotic resistance in *S. pyogenes* increased from about 5 percent to more than 40 percent in some areas between 1993 and 1995.

Many old pathogens have become major clinical problems because of increased antibiotic resistance. *Salmonella* serotypes, for example, including those causing typhoid fever, have been discovered with resistance to at least five antibiotics, including ampicillin, chloramphenicol, streptomycin, sulfanilamide, and tetracycline. The number of these resistant isolates in England rose from 1.5 percent in 1989 to more than 34 percent in 1995. A Danish study found that persons suffering from resistant *Salmonella* infections had a higher mortality rate than those with drug-susceptible forms.

Gonorrhea, caused by *Neisseria gonorrhoeae*, is one of the most common sexually transmitted diseases. Physicians began using a class of broad-spectrum antibiotics called fluoroquinolones when *N. gonorrhoeae* developed resistance to penicillin, tetracycline, and streptomycin. By the late 1990's, however, resistance to the fluoroquinolone ciprofloxacin had developed in Hawaii and on the West Coast of the United States. By 2004 elevated resistance to ciprofloxacin was seen among men who had sex with men. Two years later, resistance was present in all regions of the country, including the heterosexual population. The U.S. Centers for Disease Control stopped recommending fluoroquinolones for the treatment of gonorrhea in 2007 and began monitoring gonorrhea patients being treated with the cephalosporin class of drugs for signs of treatment failure due to resistance.

Some antibiotic resistance appears to be linked to antimicrobial use in farm animals. Shortly after antibiotics appeared, it was discovered that subtherapeutic levels could promote growth in animals and treat acute infections in such settings as aquaculture. One antimicrobial drug, avoparcin, is a glycopeptide (a compound containing sugars and proteins) that is used as a feed additive. Vancomycin-resistant enterococci such as *Enterococcus faecium* were first isolated in 1988 and appeared to be linked to drug use in animals. Antibiotic resistance in enterococci was found to be more prevalent in farm animals exposed to antimicrobial drugs, and prolonged exposure to oral glycoproteins in tests led to vancomycin-resistant enterococci in 64 percent of the subjects. In 1997 the European Union banned the use of avoparcin, which is similar to vancomycin, as an animal feed additive. In June, 2010, the U.S. Food and Drug Administration issued draft guidelines to promote more prudent use of medically important antimicrobials in food-producing animals.

How Antibiotic Resistance Occurs

Antibiotic resistance occurs because the antibiotics exert a selective pressure on the bacterial pathogens. The selective pressure eliminates all but a few bacteria that can persist through evasion or mutation. One reason that the duration of antibiotic treatments is often several weeks is to ensure that bacteria that have evaded the initial exposure are killed. The rare bacteria that are resistant can persist and grow regardless of continued antibiotic exposure. Terminating antibiotic treatment early, once symptoms disappear, has the unfortunate effect of stimulating antibiotic resistance without completely eliminating the original cause of infection.

Mutations that promote resistance occur with different frequencies. For example, spontaneous resistance of *M. tuberculosis* to cycloserine and viomycin may occur in 1 in 1,000 cells, resistance to kanamycin may occur in only 1 in 1 million cells, and resistance to rifampicin may occur in only 1 in 100 million cells. Consequently, 1 billion bacterial cells will contain several individuals resistant to at least one antibiotic. Using multiple antibiotics further reduces the likelihood that an individual cell will be resistant to all antibiotics. However, it can cause multiple antibiotic resistance to develop in bacteria that already have resistance to some of the antibiotics.

Bacterial pathogens may not need to mutate spontaneously to acquire antibiotic resistance. There are several mechanisms by which bacteria can acquire the genes for antibiotic resistance from microorganisms that are already antibiotic-resistant. These mechanisms include conjugation (the exchange of genetic information through direct cell-to-cell contact), transduction (the exchange of genetic information from one cell to another by means of a virus), transformation (the acquisition of genetic information through the taking up of deoxyribonucleic acid, or DNA, directly from the environment), and transfer of plasmids, small circular pieces of DNA that frequently carry genes for antibiotic resistance. Exchange of genes for antibiotic resistance on plasmids is one of the most common means of developing or acquiring resistance in hospitals because of the heavy antibiotic use in these settings.

The genes for antibiotic resistance take many forms. They may make the bacteria impermeable to the antibiotic. They may subtly alter the target of the antibiotic within the cell so that it is no longer affected. The genes may code for production of an enzyme in the bacteria that specifically destroys the antibiotic. For example, fluoroquinolone antibiotics inhibit DNA replication in pathogens by binding to the enzyme required for replication. Resistant bacteria have mutations in the amino acid sequences of this enzyme that prevent the antibiotic from binding to this region. Some resistant pathogens produce an enzyme called penicillinase, which degrades the antibiotic penicillin before it can prevent cell wall formation.

New Strategies

The heavy use of antibiotics has led to increases in morbidity, mortality caused by previously controlled infectious diseases, and health costs. Some of the recommended ways of dealing with this public health problem include changing antibiotic prescription patterns, changing both doctor and patient attitudes about the necessity for antibiotics, increasing the worldwide surveillance of drug-resistant bacteria, improving techniques for susceptibility testing, and investing in the research and development of new antimicrobial agents.

Gene therapy, which remains an experimental technique, is regarded as one promising solution to antibiotic resistance. In gene therapy, the genes expressing part of a pathogen's cell are injected into the patient to stimulate a heightened immune response.

Some old technologies are also being revisited. There is increasing interest in using serum treatments, in which antibodies raised against a pathogen are injected into a patient to cause an immediate immune response. Previous serum treatment techniques have yielded mixed results, but with the advent of monoclonal antibodies and the techniques for producing them, serum treatments in future can be made much more specific and the antibodies delivered in much higher concentrations.

Another experimental treatment technique involves manipulating viruses genetically to make them target specific pathogens. Once injected into a patient, the reprogrammed viruses begin a focused attack on the pathogen. Viruses attack all living organisms, including bacteria, but are extremely specific. For this therapeutic technique, certain genes are removed to prevent the virus from causing disease or degrading anything other than the target pathogenic organism.

Mark Coyne
Updated by Karen N. Kähler

FURTHER READING

Biddle, Wayne. *A Field Guide to Germs.* 3d ed. New York: Anchor Books, 2010.

Fong, Ignatius W., and Karl Drlica, eds. *Antimicrobial Resistance and Implications for the Twenty-first Century.* New York: Springer, 2008.

Lashley, Felissa R., and Jerry D. Durham, eds. *Emerging Infectious Diseases: Trends and Issues.* 2d ed. New York: Springer, 2007.

Levy, Stuart B. *The Antibiotic Paradox: How the Misuse of Antibiotics Destroys Their Curative Power.* 2d ed. Cambridge, Mass.: Perseus, 2002.

Shnayerson, Michael, and Mark J. Plotkin. *The Killers Within: The Deadly Rise of Drug-Resistant Bacteria.* Boston: Little, Brown, 2002.

SEE ALSO: Antibiotics; Biotechnology and genetic engineering; Chloramphenicol; Horizontal gene transfer; Runoff, agricultural; Sewage treatment and disposal.

Balance of nature

CATEGORY: Ecology and ecosystems
DEFINITION: Ecological concept that nature, in its undisturbed state, achieves constant equilibrium
SIGNIFICANCE: The concept of the balance of nature has never been legitimated in science as either a hypothesis or a theory, but it persists as a designation for a healthy environment.

Greek natural philosophers in the fifth and sixth centuries B.C.E. attempted to explain naturalistically how nature works rather than depending upon myths. The atomistic theory of Leucippus and Democritus stated that matter can be transformed but is never created or destroyed. The Pythagoreans heard musical harmony in the universe. Hippocratic medicine taught that the balance of humors within the body produces health and an imbalance produces disease, and Greek physicians believed in the healing power of nature. Within this worldview, ecological balance would have been a compelling expectation.

Herodotus, the father of history, wrote about not only the human histories of Greece, Persia, and Egypt but also their geographies and natural histories. He was influenced by ideas in natural philosophy, and he was concerned with concrete examples that might illustrate generalities. In organizing his information on the lions, snakes, and hares of Arabia, he asked why predatory species do not eat up all of their prey. His answer was that a superintending "Providence" had created the different species with different capacities for reproduction. Predatory species, such as lions and snakes, produce fewer offspring than species that they eat, such as hares. From Egypt, Herodotus obtained a report about a mutually beneficial relationship between Nile crocodiles and plovers: Crocodiles allow plovers to sit on their teeth and eat the leeches that infest their mouths.

Plato lived after the natural philosophers and Herodotus, but he lacked their trust in sensory data, and therefore he explained nature with naturalistic myths. In a dialogue called *Prōtagoras* (c. 390 B.C.E.; English translation, 1804), Plato asked Herodotus's question about why some species do not eat up the others, but he asked it more abstractly and not for its own sake. The point of this creation myth was to explain why humans do not have specialized traits such as wings or claws. The gods assigned to Epimetheus the task of creating each species with traits that would enable it to survive. He had given out all the specialized traits before he got around to creating humans, and his brother Prometheus had to save humanity by giving it reason and fire. The secondary point of the myth supplements what Herodotus had concluded about differences in reproduction with a conclusion about differential traits that ensure survival.

In the writings of Aristotle and his colleagues in Athens, full-fledged science emerged. These scholars realized the importance of both collecting data and ordering it so as to allow the drawing of conclusions. These scholars, however, focused on physiology and anatomy and neglected to look for what modern-day persons call ecological explanations of how nature works. For example, they explained that the greater number of offspring in hares than in lions is because of their size. Since hares are smaller, it is easier for more of them to grow within a female than for multiple larger lion cubs to grow in their mother. The balance-of-nature concept was not distinct enough to require either defending or refuting.

SCIENTIFIC REVOLUTION

The ancient Romans excelled in engineering, not science. They were mostly content with abbreviated Latin versions of Greek science. Roman writings are still worth mentioning, however, because of their influence on later European naturalists. Aelianus compiled a popular natural history book that, among other things, explained that jackdaws are friends to the farmer because jackdaws eat the eggs and young of locusts that would eat the farmer's produce. The philosopher Cicero wrote an influential book, *De natura deorum* (44 B.C.E.; *On the Nature of the Gods*, 1683), in which he saw the work of Providence in endowing plants with the capacity to feed humans and animals and still be able to have seeds left over to ensure their own reproduction. Another philosopher, Plotinus, pondered the evil of suffering when predators kill animals for food. He decided that the existence of predation allows a greater diversity of life to exist than would be possible if all animals ate plants. These miscellaneous observations were insufficient for a theory, but they kept alive the notion of a balance of nature.

During the scientific revolution, fresh observations and conclusions appeared. Most significant, John Graunt, a merchant, analyzed London's baptismal and death records in 1662 and discovered the balance in the sex ratio and the regularity of most causes of

death (excluding epidemics). England's chief justice, Sir Matthew Hale, was interested in Graunt's discoveries, but he nevertheless decided that the human population, in contrast to animal populations, must have steadily increased throughout history. He surveyed the known causes of animal mortality and in 1677 published the earliest explicit account of the balance of nature.

English scientist Robert Hooke studied fossils and in 1665 concluded that they represented the remains of plants and animals, some of which were probably extinct. John Ray, a clergyman and naturalist, argued in response that the extinction of species would contradict the wisdom of the ages, by which he seems to have meant the balance of nature. Ray also studied the hydrologic cycle, which is a kind of environmental balance of water. In the late seventeenth century, Antoni van Leeuwenhoek, one of the first investigators to make biological studies with a microscope, discovered that parasites are more prevalent than anyone had suspected and that they are often detrimental or even fatal to their hosts. Before that, it was commonly assumed that the relationship between host and parasite was mutually beneficial.

Richard Bradley, a botanist and popularizer of natural history, pointed out in 1718 that each species of plant has its own kind of insect and that there are even different insects that eat the leaves and bark of a tree. His book *A Philosophical Account of the Works of Nature* (1721) explored aspects of the balance of nature more thoroughly than had been done before. Ray's and Bradley's books may have inspired the comment in Alexander Pope's *An Essay on Man* (1733-1734) that all the species are so closely interdependent that the extinction of one would lead to the destruction of all living nature.

TOWARD A SCIENCE OF ECOLOGY

Swedish naturalist Carolus Linnaeus was an important protoecologist. In his essay *Oeconomia Naturae* (1749; *The Economy of Nature*, 1749) he attempted to organize the aspect of natural history dealing with the balance of nature, but he realized that one must study not only ways that plants and animals interact but also their habitats. He knew that while balance had to exist, there occurred over time a succession of plants, beginning with a bare field and ending with a forest. In *Politia Naturae* (1760; *Governing Nature*, 1760) he discussed the checks on populations that prevent some species from becoming so numerous that they eliminate others. He noticed the competition among different species of plants in a meadow and concluded that feeding insects kept them in check. French naturalist Comte de Buffon developed a dynamical perspective on the balance of nature from his studies of rodents and their predators. Rodents can increase in numbers to plague proportions, but then predators and climate reduce their numbers. Buffon also suspected that humans had exterminated some large mammals, such as mammoths and mastodons.

However, a later Frenchman, Jean-Baptiste Lamarck, published a book on evolution called *Philosophie zoologique* (1809; *Zoological Philosophy*, 1914), which cast doubt on extinction by arguing that fossils represent only early forms of living species: Mammoths and mastodons evolved into African and Indian elephants. In developing this idea, he minimized the importance of competition in nature. An English opponent, the geologist Charles Lyell, argued in 1833 that species do become extinct, primarily because of competition among species. Charles Darwin was inspired by his own investigations during a long voyage around the world and by his reading of the works of Linnaeus and Lyell. In his revolutionary book *On the Origin of Species by Means of Natural Selection: Or, The Preservation of Favoured Races in the Struggle for Life* (1859), Darwin argued an intermediate position between those of Lamarck and Lyell: Species do evolve into different species, but in the process, some species do indeed become extinct.

Darwin's theory of evolution might have brought an end to the balance-of-nature concept, but it did not. Instead, American zoologist Stephen A. Forbes developed an evolutionary concept of the balance of nature in his essay "The Lake as a Microcosm" (1887), in which he observes that although the reproductive rate of aquatic species is enormous and the struggle for existence among them is severe, "the little community secluded here is as prosperous as if its state were one of profound and perpetual peace." Forbes emphasized the stabilizing effects of natural selection.

ECOLOGY

The science of ecology became formally organized in the period from the 1890's through the 1910's. One of its important organizing concepts was that of "biotic communities." Frederic E. Clements, an American plant ecologist, wrote a large monograph titled *Plant Succession* (1916), in which he drew a morpho-logical and developmental analogy between organisms and plant communities. Both the individual and the community have a life history during which each changes its anatomy and physiology. This superorganismic concept was an extreme version of the balance of nature that seemed plausible as long as one believed that a biotic community was a real entity rather than a convenient approximation of what one sees in a pond, a meadow, or a forest. However, the studies of Henry A. Gleason in 1917 and later indicated that plant species merely compete with one another in similar environments; he concluded that Clements's superorganism was poetry, not science.

While the balance-of-nature concept was giving way to ecological hypotheses and theories, Rachel Carson decided that she could not argue her case in *Silent Spring* (1962) without it. She admitted, "The balance of nature is not a *status quo*; it is fluid, ever shifting, in a constant state of adjustment." Nevertheless, for Carson the concept represented a healthy environment, which humans could upset. Her usage of the phrase has persisted within the environmental movement.

In 1972 English medical chemist James Lovelock developed a new balance-of-nature idea, which he called Gaia, named for a Greek earth goddess. His reasoning owed virtually nothing to previous balance-of-nature notions, which focused on the interactions of plants and animals. His concept emphasized the chemical cycles that flow from the earth to the waters, atmosphere, and living organisms. Lovelock soon had the assistance of a zoologist named Lynn Margulis, and the studies they conducted together convinced them that biogeochemical cycles are not random; rather, they exhibit homeostasis, just as some animals exhibit homeostasis in body heat and blood concentrations of various substances. Lovelock and Margulis argued that living beings, rather than inanimate forces, mainly control the earth's environment. In 1988 three scientific organizations sponsored a conference to evaluate their ideas; the conference was attended by 150 scientists from all over the world. Although science more or less understands how homeostasis works when a brain within an animal controls it, no one has succeeded in satisfactorily explaining how homeostasis can work in a world "system" that lacks a brain. The Gaia hypothesis is as untestable as were earlier balance-of-nature concepts.

Frank N. Egerton

FURTHER READING

Egerton, Frank N. "Changing Concepts of the Balance of Nature." *Quarterly Review of Biology* 48 (June, 1973): 322-350.

_____. "The History and Present Entanglements of Some General Ecological Perspectives." In *Humans as Components of Ecosystems*, edited by Mark J. McDonnell and S. T. A. Pickett. New York: Springer, 1993.

Judd, Richard W. *The Untilled Garden: Natural History and the Spirit of Conservation in America, 1740-1840.* New York: Cambridge University Press, 2009.

Kirchner, James W. "The Gaia Hypotheses: Are They Testable? Are They Useful?" In *Scientists on Gaia*, edited by Stephen H. Schneider and Penelope J. Boston. Cambridge, Mass.: MIT Press, 1991.

Kricher, John C. *The Balance of Nature: Ecology's Enduring Myth.* Princeton, N.J.: Princeton University Press, 2009.

Milne, Lorus J., and Margery Milne. *The Balance of Nature.* 1960. Reprint. New York: Alfred A. Knopf, 1970.

SEE ALSO: Carson, Rachel; Darwin, Charles; Ecology as a concept; Ecosystems; Food chains; Gaia hypothesis; Lovelock, James.

Basel Convention on the Control of Transboundary Movements of Hazardous Wastes

CATEGORIES: Treaties, laws, and court cases; waste and waste management

THE CONVENTION: International agreement designed to limit the movement of hazardous wastes across national boundaries

DATE: Opened for signature on March 22, 1989

SIGNIFICANCE: The Basel Convention has been effective in raising public awareness of the transporting of hazardous wastes, and it has somewhat slowed the movement of such wastes from country to country. The convention is not always strictly enforced, however, and some nations continue to transport hazardous wastes across national boundaries illegally.

Spurred by interest generated by several incidents during the 1980's in which ships carrying waste materials sailed from port to port in unsuccessful

Preamble to the Basel Convention

The Basel Convention's preamble, excerpted below, sets out the context and international legal framework within which the convention was conceived.

The Parties to this Convention . . .

Affirming that States are responsible for the fulfilment of their international obligations concerning the protection of human health and protection and preservation of the environment, and are liable in accordance with international law,

Recognizing that in the case of a material breach of the provisions of this Convention or any protocol thereto the relevant international law of treaties shall apply,

Aware of the need to continue the development and implementation of environmentally sound low-waste technologies, recycling options, good housekeeping and management systems with a view to reducing to a minimum the generation of hazardous wastes and other wastes,

Aware also of the growing international concern about the need for stringent control of transboundary movement of hazardous wastes and other wastes, and of the need as far as possible to reduce such movement to a minimum,

Concerned about the problem of illegal transboundary traffic in hazardous wastes and other wastes,

Taking into account also the limited capabilities of the developing countries to manage hazardous wastes and other wastes . . .

Convinced also that the transboundary movement of hazardous wastes and other wastes should be permitted only when the transport and the ultimate disposal of such wastes is environmentally sound, and

Determined to protect, by strict control, human health and the environment against the adverse effects which may result from the generation and management of hazardous wastes and other wastes,

HAVE AGREED AS FOLLOWS. . . .

efforts to be allowed to unload their cargoes, the international community, under the auspices of the United Nations Environment Programme, negotiated an agreement designed to limit the export of hazardous waste from one country to another. The Basel Convention on the Control of Transboundary Movements of Hazardous Wastes defines hazardous waste as used oils, biomedical wastes, persistent organic pollutants such as pesticides, polychlorinated biphenyls, and various other chemicals. The terms of the convention require that prior informed consent be obtained from a country before such wastes can be exported to that country. The goal of the Basel Convention is to minimize the movement of such wastes across national boundaries.

Since it entered into force in May, 1992, the Basel Convention has enabled countries to deal with attempts to move hazardous wastes across their borders. It was soon recognized that the industrialized nations are the major exporters of hazardous waste and that the less industrialized countries of Africa, Latin America, and Asia had been the primary recipients of such materials. In 1995, at the second meeting of the convention's signatory member states, agreement was reached to prohibit the movement of hazardous wastes from industrialized countries to less industrialized countries; this amendment is commonly referred to as the Basel Ban. Although many industrialized countries abide by the Basel Ban, it has not yet been ratified. Several countries, including the United States, that ratified the original agreement have yet to ratify the Basel Ban.

The Basel Convention and the Basel Ban have been effective in slowing the migration of hazardous wastes (which some estimate to be more than 8.5 million tons of waste per year) from industrialized to developing countries. Nonetheless, some critics of the convention argue for the use of developing countries as dumping grounds for industrialized countries, noting that such exchanges have economic benefits for both parties. This attitude ignores the economic costs associated with dealing with hazardous waste in developing countries, such as the health care costs that arise when illnesses are caused by contact with hazardous materials.

Even with the adoption of the Basel Convention, some parties still seek to dispose of hazardous wastes in cheap fashion. Several suspicious shipwrecks have occurred in the Mediterranean involving ships later found to have been carrying cargoes—such as radioactive materials—that would have been subject to the convention. In some cases these wrecks have led to fishing bans, as occurred along the coast of Italy in 2007.

Although the Basel Convention has been effective in raising public awareness of the transporting of hazardous wastes, it has not always been stringently enforced. The movement of hazardous wastes has been somewhat limited by the convention, but until tougher measures are adopted to reduce or somehow safely dispose of hazardous wastes at their points of origin, such wastes will continue to be exported, often illegally, from industrialized countries. Some of these wastes will arrive at developing countries, imposing costs on them; other wastes will be disposed of at sea, imposing costs on everyone.

John M. Theilmann

FURTHER READING

Mukerjee, Madhusree. "Poisoned Shipments." *Scientific American*, February, 2010, 14-15.

Pellow, David Naguib. *Resisting Global Toxics.* Cambridge, Mass.: MIT Press, 2007.

United Nations Environment Programme. *Minimizing Hazardous Wastes: A Simplified Guide to the Basel Convention.* Geneva: Author, 2002.

SEE ALSO: Environmental justice and environmental racism; Environmental law, international; Hazardous waste; *Khian Sea* incident; Nuclear and radioactive waste; Ocean dumping.

Beaches

CATEGORY: Preservation and wilderness issues

DEFINITION: Sandy or rocky areas on the shores of bodies of water

SIGNIFICANCE: Beaches have significant value for recreational purposes, and important industries, such as those related to swimming and surfing, depend on them. Beaches worldwide are threatened by accelerating erosion, as well as by a projected rise in sea level.

Beaches are accumulations of loose sedimentary material found along the shorelines of bodies of water. They extend landward from the low-tide level to a place where there is a marked change in appearance or topography of the land surface, such as a cliff, a row of sand dunes, or human-made structures such as seawalls. The two major subdivisions of a beach are the backshore, which extends seaward from the foot of a cliff, dunes, or seawall; and the foreshore, which usually has a steeper slope that continues down to water level. The boundary between the backshore and the foreshore is generally marked by a change in slope or by a small scarp excavated by wave activity. The seaweed and other debris that accumulate at this point are known as the wrack.

The backshore is considered to be the inactive part of a beach because there is little evidence for wave activity here. It is the part of a beach most used by humans. Human-built additions such as volleyball nets, lifeguard stands, and campfire pits are found on the backshore; grasses or trailing vines may grow down from cliff, dunes, or seawall to extend partway across the backshore.

The foreshore is the active part of the beach. It is covered daily by the rise and fall of the tides; no grasses or trailing vines are found here because the waves and tides would sweep them away immediately. All traces of human activity, such as footprints and lost toys, are removed by the waves as well. Upon going out, the tide leaves the surface of this part of the beach perfectly smooth, broken only at its lower end by a slight dip known as the low-tide step. This is a gentle trough excavated by the breakers at low tide. Waves grind up shells and pebbles here in a so-called wave mill, and the resulting fragments become new materials for the beach. They grow finer-grained as the wind and waves carry them upward toward the dunes.

The widths of beaches vary greatly around the world. Beaches at the foot of cliffs or seawalls may be just a few feet wide, whereas those on low-relief coasts can extend seaward for many miles.

ACCELERATING BEACH EROSION

Oceanographers calculate that 70 percent of the world's beaches are being cut back by erosion, 10 percent are growing forward, and the rest remain unchanged. Erosion of a beach occurs when it loses more sediment than it gains from the sources that feed it. Losses may take place when beach materials are carried inland during storm surges, when they are carried seaward as big waves cut back the coast during severe storms, or when they are carried along the shoreline by the longshore current. This current is present along coasts whenever waves strike the shoreline at an acute angle rather than coming in head-on.

During severe storms the loss of beach materials can be rapid; evidence for this is seen in collapsed seawalls and cliffed dunes on the backshore or in salt marsh deposits and tree stumps uncovered on the foreshore as the beach moves inland over a previously vegetated area. Aerial photographs are generally used to determine how much material has been removed from beaches, with recent photographs compared to photographs taken previously.

Beach erosion accelerated in the early twenty-first

century because of several factors: the active removal of materials by waves, wind, and currents; rising sea level, which resulted in the migration of shorelines inland; and increasing human modification of the coasts. Humans' changes to coastlines that affect beach erosion include both the construction of seawalls, groins, offshore breakwaters, and jetties and the bringing in of new sand in a procedure known as beach nourishment. Seawalls are built at the backs of beaches to prevent erosion, but they may reflect the storm waves, causing them to carry sediment seaward. Groins, which are rock ribs extending seaward along the foreshore to trap sand on their up-drift sides, cause erosion on their down-drift sides, resulting in alternating scallops of erosion and deposition along the coast. Breakwaters, which are offshore structures built parallel to the coast, and jetties, which are walls ex-

Currents That Affect Beaches

Longshore currents are those parallel to a seashore and extending from the shoreline through the breakers. They occur because waves approach coasts at a small angle and bend in the shallows. The speed of a longshore current is related to wave size and angle of approach to a shore. During quiet weather, longshore currents move at under half a mile per hour, but during storms they can move ten to twelve times that speed. The combined actions of waves and longshore currents transport a lot of sediment along shallows bordering a shore. Longshore currents move in either direction along a beach, depending on the direction of wave approach, itself a result of wind direction. Thus waves suspend sediment, and longshore currents transport it along beaches.

Tides, the regular rise and fall of sea level due to the gravitational fields of the moon and the sun, cause daily changes in ocean levels of 1 to 50 feet (0.30 to 15 meters). Tidal currents transport large amounts of sediment and erode rock, and tidal rise and fall distribute wave energy across shores by changing water depth. In estuaries, tides create the speeds needed to move sand. On open coasts—that is, on beaches—tides do not move the water fast enough for sediment transfer. However, the rise and fall of tides along open coasts indirectly affect sediment movement, because their landward movement or retreat causes shorelines to move. This changes the region where waves and longshore currents operate. Beach slope is also crucial, with gently sloped beaches having the largest shoreline changes during tide cycles.

tending seaward to stabilize inlets, both trap migrating sand, resulting in erosion on their down-drift sides. Even beach nourishment, a highly popular practice, has its negative effects. Because the sand that is added to beaches is dredged from offshore or trucked in from the land, it rarely has the same texture and composition of the original beach sand. The result is that this new sand usually erodes faster than the original sand that it replaces.

Donald W. Lovejoy

FURTHER READING

Bird, Eric. *Coastal Geomorphology: An Introduction.* 2d ed. Hoboken, N.J.: John Wiley & Sons, 2008.

Davis, Richard A., Jr., and Duncan M. FitzGerald. *Beaches and Coasts.* Malden, Mass.: Blackwell, 2004.

Neal, William J., Orrin H. Pilkey, and Joseph T. Kelley. *Atlantic Coastal Beaches: A Guide to the Ripples, Dunes, and Other Natural Features of the Seashore.* Missoula, Mont.: Mountain Press, 2007.

SEE ALSO: Climate change and oceans; Dredging; Erosion and erosion control; Sea-level changes; Wetlands.

Bees and other pollinators

CATEGORIES: Agriculture and food; animals and endangered species

DEFINITION: Flying insects and other biological agents that facilitate plant reproduction by moving pollen among seed plants

SIGNIFICANCE: Pollinators such as bees are extremely important to the health of the planet because without them most plant species would disappear. Pollinating species are in decline around the world, a trend that threatens biodiversity and food supplies for many animals, including humans.

Bees are the best-known group of pollinators; however, other insects such as butterflies, wasps, beetles, and ants are pollinators, too, along with some species of bats and birds. Plants and their pollinators coevolved, forming symbiotic relationships in which plants provided food for the pollinators, which in turn aided the plants' reproduction. Many other species evolved over time to take advantage of this fundamental relationship between plant and pollinator, in-

cluding humans and other animals that survive by eating the results of the fertilized plants, such as nuts, seeds, and berries. Pollinators also contribute to the web of life by ensuring the continued survival of plants through the process of pollination; the loss of pollinators can affect the biodiversity and well-being of a whole ecosystem. By the early twenty-first century, declines in pollinator populations around the world led environmental groups and government agencies to include pollinators in their conservation efforts.

POLLINATOR DECLINE

The decline of pollinating species has many causes, all for the most part related to human activity. One cause is the use (and misuse) of pesticides meant to kill other insects, such as mosquitoes; such pesticides also indiscriminately kill off other insect populations, many of them pollinator species. Another cause of pollinator decline is the large-scale transport of pollinators around the world; this practice also results in the large-scale transport of parasites and diseases that affect them, as well as the introduction of invasive species that can compete with or destroy native pollinator populations.

Human expansion into previously wild or relatively untouched spaces contributes to pollinator decline through habitat loss and degradation, as development brings with it pollution, hive destruction, and fragmentation of traditional "nectar corridors." Wild pollinators have fewer places to nest, mate, and roost safely, and their historic ranges have been destroyed or partitioned in such a way that it is far more difficult for them to find the nectar they need to survive. Many pollinators are already on endangered or threatened species lists, including more than one hundred avian species and dozens of mammals.

THE IMPORTANCE OF BEES

The thousands of species of bees in the world are widely known for pollination and for their making of honey and beeswax. Bees are the primary pollinators of flowers around the world and are also responsible for the pollination of approximately one-third of the human food supply. Many agricultural crops, ranging from watermelons to cashews, rely on pollination by bees. European honeybees, in particular, are extremely important to agriculture because of humans' ability to manage the species.

Pollination management, also called contract pollination, entails transporting honeybees in their hives to crops that are in need of pollination. Modern monoculture farming has created a situation in which the normal pollinators of some kinds of crops cannot pollinate them, either because of the huge size of the fields or because the pollinators have declined or died off. Honeybees, however, can be put in the fields in such large numbers that they make up for the fact that they are often less efficient pollinators than the native species that previously would have pollinated a particular crop. The economic viability of monoculture enterprises often depends entirely on pollination management, and most professional beekeepers in the twenty-first century focus on contract pollination rather than honey as their primary source of revenue.

The same factors that have caused declines in other pollinator populations since the late twentieth century have affected honeybees as well. In addition, across North America and Western Europe since at least 2006, honeybees have disappeared in what is called colony collapse disorder (CCD), the cause of which remains a mystery. CCD is characterized by the worker bees flying off and never returning to the hive or colony, leaving behind a healthy queen, an unhatched brood, and food supplies. Given the importance of honeybees to agriculture around the world, scientists have investigated many possible causes for CCD; possibilities include a parasitic mite, insect diseases, and the use or misuse of insecticides. Some researchers have reported evidence that it may be viral. These possibilities are aggravated by other factors negatively affecting honeybee populations, such as poor diet related to monoculture, which can lead to malnutrition or impaired immune system, and migratory beekeeping, which spreads parasites and diseases among populations.

Many pollinators are very small, but their importance to terrestrial life is great. Pollinators are a major part of the web of life, and what happens to them can affect every other creature in an ecosystem. The plants, whether they are cacti in the desert of Mexico or blueberries in Scotland, need their pollinators to survive—without pollinators a plant cannot reproduce and its species will go extinct. If this occurs, the

A beekeeper checks one of the hives he has brought to pollinate apple and pear trees at Quarry Hill Orchards in Berlin Heights, Ohio. (AP/Wide World Photos)

pollinators adapted to that plant will no longer have a food supply and will either starve to death or switch to another source of nectar, inevitably intruding on another animal's food supply. As the plants disappear for lack of pollinators, animals such as birds, small mammals, and lizards that eat the fruit, nuts, and seeds produced by pollinated plants will also begin to starve. This effect will ripple through the food web until it reaches larger animals, including human beings.

Megan E. Watson

FURTHER READING

Buchmann, Stephen L., and Gary Paul Nabhan. *The Forgotten Pollinators.* Washington, D.C.: Island Press, 1996.

James, Rosalind R., and Theresa L. Pitts-Singer, eds. *Bee Pollination in Agricultural Ecosystems.* New York: Oxford University Press, 2008.

Waser, Nickolas M., and Jeff Ollerton, eds. *Plant-Pollinator Interactions: From Specialization to Generalization.* Chicago: University of Chicago Press, 2006.

SEE ALSO: Biodiversity; Conservation; Ecosystems; Food chains; Habitat destruction; Killer bees; Monoculture; Open spaces; Pesticides and herbicides.

Benefit-cost analysis

CATEGORIES: Resources and resource management; land and land use

DEFINITION: Method of determining the benefit of undertaking a program or project based on its costs

SIGNIFICANCE: Benefit-cost analysis is used to determine the value of many types of projects and their impacts, including environmental management projects and the impacts of such projects as airport construction.

Benefit-cost analysis (also called cost-benefit analysis) is an idea that can be traced back to an article by nineteenth century French economist Jules Dupuit. Early use of the concept in the United States is seen in the water development projects of the U.S. Army Corps of Engineers, where it was used to prioritize projects concerning the nation's waterways. The term arose in the context of the environment and natural resources in the Flood Control Act of 1936, which states that the "federal government should improve or participate in the improvement of navigable waters or their tributaries, including watersheds thereof, for flood control purposes if the benefits to whomever they may accrue are in excess of the estimated costs." Hence the term "benefit-cost analysis" was established; it was later embraced by economists during the 1950's. Two important concepts in economics, the Pareto principle and Kaldor-Hicks efficiency, are related to benefit-cost analysis.

In benefit-cost analysis, the monetary measure of all expected benefits of proposed projects or activities is compared with the costs of undertaking those projects or activities. This typically involves using the "time value of money" formula, which converts the future expected streams of costs and benefits to a present value amount. This tool helps planners and policy makers to determine whether a project is worth undertaking by weighing the advantages and the disadvantages in monetary terms.

The kinds of projects for which benefit-cost analyses may be conducted include highways, training programs, and health care systems. Benefit-cost analyses can provide answers to questions about whether environmental management projects—such as constructing a new airport terminal, installing wind turbines on the Scottish coast, or building a new dam on a particular river—will provide benefits that outweigh the respective costs to the environment.

In conducting benefit-cost analysis, the analyst finds, quantifies, and adds up the values of the benefits of a course of action. The analyst then identifies, quantifies, and subtracts all the costs associated with the activities or project. The difference between the two indicates whether the planned action or project should or should not be undertaken. To conduct such an analysis well, the analyst must not only make sure all the costs and benefits are included and properly quantified but also avoid double counting.

One key problem in undertaking an environmental benefit-cost analysis lies in trying to assign monetary values to intangibles associated with the project. For example, a functioning ecosystem is of economic value to society, but its aesthetic functions, which may be lost because of a project and its future effects, fall outside the market system; such functions are often excluded from decision-making tools such as benefit-cost analysis. The construction of a dam or reclamation of land has benefits of water, energy, and fish supply. These benefits can be traded on the market, and a market price can be used to represent their economic value. Where benefits exist that cannot be traded on the market, shadow pricing may be used to place economic values on the benefits; this process is subjective, however.

In spite of its shortcomings for environment-related applications, benefit-cost analysis provides planners and policy makers with a method for putting all relevant benefits and costs of proposed projects on a common temporal footing so that they can make informed decisions. It also provides all parties concerned with greater understanding of the economic costs of their decisions and thus allows for arguments to be made both for and against a change based on benefit-cost considerations.

Josephus J. Brimah

Further Reading

Boardman, Anthony E., et al. *Cost-Benefit Analysis.* 4th ed. Upper Saddle River, N.J.: Prentice Hall, 2010.

Brent, Robert J. *Cost-Benefit Analysis for Developing Countries.* Northampton, Mass.: Edward Elgar, 2000.

Pearce, David. "The Limits of Cost-Benefit Analysis as a Guide to Environmental Policy." In *Economics and Environment: Essays on Ecological Economics and Sustainable Development.* Northampton, Mass.: Edward Elgar, 1998.

Puttaswamaiah, K., ed. *Cost-Benefit Analysis: Environmental and Ecological Perspectives.* New Brunswick, N.J.: Transaction, 2002.

See also: Accounting for nature; Best available technologies; Contingent valuation; Ecological economics; Environmental economics; Risk assessment.

Berry, Wendell

Categories: Activism and advocacy; agriculture and food

Identification: American author of books on conservation and agrarianism

Born: August 5, 1934; Henry County, Kentucky

Significance: Berry's integrated professions of farmer, writer, and critic of industrial development have placed him among the major figures of the twentieth century in both conservation and literature.

Born to a tobacco farm family during the Great Depression, Wendell Berry grew up in a simple environment of small farms that practiced crop diversification, organic fertilizing, and use of draft animals. Berry's family had deep roots in the community, as did their neighbors. Farmers in Henry County, Kentucky, were largely self-sufficient, depending little on resources beyond their region.

Berry's rural upbringing affected every facet of his adult life. After receiving bachelor's and master's degrees in English from the University of Kentucky at Lexington, he was awarded a prestigious Wallace Stegner Fellowship in creative writing at Stanford University in 1958. This opportunity moved Berry into a circle of scholars and writers, notably Stegner, who at that time was a prominent novelist and conser-

vationist. In 1960 Berry returned to Henry County. He spent a brief time in France and Italy on a Guggenheim Fellowship and later held teaching positions at New York University, Georgetown College, the University of Cincinnati, Bucknell University in Pennsylvania, and Stanford University. He has lived most of his life, however, with his wife Tanya Amyx Berry (they married in 1957) in his native north central Kentucky, where the couple farms. He was an English professor at the University of Kentucky from 1964 to 1977 and from 1987 to 1993.

Berry is an award-winning writer of more than forty books of poetry, fiction, and essays. He is considered by many to be one of the most important nature poets of his generation. He gained widespread recognition for his early volumes *The Broken Ground* (1964) and *Openings* (1968), both of which center on rural themes that he would continue to explore in later works such as *Given* (2005), *The Mad Farmer Poems* (2008), and *Leavings* (2009). Berry's writing on environmental issues focuses mainly on agriculture and

Author Wendell Berry. (©Dan Carraco/Courtesy, North Point Press)

the simple life. In his most successful novel, *The Memory of Old Jack* (1974), a ninety-two-year-old farmer relives his simple, agrarian life in flashbacks. Old Jack reflects the longing of Berry to return to his pre-World War II life of rural self-sufficiency. Among Berry's best-known nonfiction works are *The Long-Legged House* (1969), *The Hidden Wound* (1970), *A Continuous Harmony* (1972), *The Unsettling of America: Culture and Agriculture* (1977), and *The Gift of Good Land* (1981). In later essays such as those found in *Citizenship Papers* (2003), he examines the environmental, sociological, and political consequences of global industrialism.

Berry's writings emphasize the connectedness of human beings with the rest of nature. He is critical of the destruction of the land by mechanized monoculture farming, use of pesticides and fertilizers, clear-cutting of forests, and strip mining. He decries the movement away from the family farm to corporate farming, asserting that the corporate sector has killed rural America.

Berry views farming as an art and prioritizes ecology over economics, which values efficiency and specialization as means of maximizing income in the short term. In his view, technology has dehumanized agriculture by replacing the self-fulfilling labor of farmers and their families. In addition, technology has driven people from their land into the cities, generating social and environmental problems in urban areas.

Berry believes that a return to sustainable agriculture is an ecological imperative for maintaining a high quality of life. In his writings he praises the Amish for their stewardship of the land and their farming on a scale appropriate to the needs of their communities. He likewise admonishes Christians to heed the biblical message that "the earth is the Lord's and the fullness thereof."

Ruth Bamberger
Updated by Karen N. Kähler

FURTHER READING

Berry, Wendell. *The Art of the Commonplace: The Agrarian Essays of Wendell Berry*. Edited by Norman Wirzba. Washington, D.C.: Counterpoint, 2002.

Bonzo, J. Matthew, and Michael R. Stevens. *Wendell Berry and the Cultivation of Life: A Reader's Guide*. Grand Rapids, Mich.: Brazos Press, 2008.

Peters, Jason. *Wendell Berry: Life and Work*. Lexington: University Press of Kentucky, 2007.

Smith, Kimberly K. *Wendell Berry and the Agrarian Tradition: A Common Grace*. Lawrence: University Press of Kansas, 2003.

SEE ALSO: Back-to-the-land movement; Environmental ethics; Monoculture; Organic gardening and farming; Social ecology; Strip farming; Sustainable agriculture.

Best available technologies

CATEGORIES: Atmosphere and air pollution; water and water pollution

DEFINITION: The most efficient control and treatment techniques that are economically achievable by various industries to limit adverse environmental impacts

SIGNIFICANCE: As concerns regarding all forms of pollution have increased around the world, many governments—including those of the United States and many European nations—have responded with environmental laws requiring industries to employ the best available techniques to reduce polluting emissions and discharges.

Major U.S. industries receiving national permits, usually from the Environmental Protection Agency (EPA), for new or modified emissions and point source discharges regulated by the Clean Air Act and the Clean Water Act are required to use the best available technologies to reduce environmental pollution. Best available technologies, also known as best available control technologies and best available techniques, include state-of-the-art controls on emissions and discharges into environmental media such as air, water, and soil. The term "available" means that the technology exists or is capable of being developed. A best available technology, however, must be affordable based on a cost-benefit analysis.

In addition to best available technology standards, the environmental laws in the United States include requirements for best conventional technologies. The best available technology mandates generally apply to pollutants that are considered to be toxic or are labeled as nonconventional because no determination has yet been made concerning their toxicity. However, once the EPA determines that a pollutant is toxic, an industry has three years to meet the best available technology standards that apply to that pollutant. Conventional pollutants, such as agricultural chemicals that have the ability to accumulate in body fat but have not been deemed toxic, are governed by the best conventional technology standards.

The permitting agency reviews each major industrial application for a stationary source permit to allow emissions into the air or discharges into the water on a case-by-case basis. Environmental, energy, and economic impacts are all considered in determining which processes and techniques an industry can afford while still protecting the environment. A variance from a best available technology standard may be granted as part of the permitting process to an industry that has agreed to use the best technology that is financially achievable, even if the technology is not the latest available. The technology used, however, must not create further pollution. For example, an industry will not receive a waiver to the best available technology requirements under the Clean Water Act unless it can show that the discharge controls and techniques it will utilize will not interfere with the quality of public water supplies, recreational activities, or wildlife, fish, and shellfish populations.

Under the Clean Air Act some of the best available technology standards include production and treatment techniques and fuel cleaning and combustion systems for various regulated air pollutants. Under the Clean Water Act effluent guidelines are based on best available technologies to achieve the goal of eliminating the discharge of all pollutants into the waters of the United States. Some of the factors considered in the cost-benefit analysis to determine if the effluent technologies are economically feasible include assessment of existing equipment, facilities, and techniques used in effluence reduction; the cost to achieve further effluent reduction; and other environmental impacts that are not related to water quality but that may occur if new technologies are employed, such as increased energy requirements. In addition to effluent guidelines, the Clean Water Act requires the use of best available technologies to minimize environmental impacts for cooling water intake structures that might be used in industries that produce heated water, such as nuclear power plants and food and beverage manufacturing facilities employing pasteurization.

Carol A. Rolf

FURTHER READING

Malone, Linda A. *Emanuel Law Outlines: Environmental Law.* 2d ed. Frederick, Md.: Aspen, 2007.

Morag-Levine, Noga. *Chasing the Wind: Regulating Air Pollution in the Common Law State.* Princeton, N.J.: Princeton University Press, 2005.

Revesz, Richard L. *Environmental Law and Policy: Statutory and Regulatory Supplement.* 2009-2010 ed. New York: Foundation Press, 2009.

SEE ALSO: Air-pollution policy; Appropriate technology; Automobile emissions; Benefit-cost analysis; Carbon dioxide air capture; Clean Air Act and amendments; Clean Water Act and amendments; Pollution permit trading.

Bhopal disaster

CATEGORIES: Disasters; pollutants and toxins

THE EVENT: Release of a highly toxic gas from a pesticide production plant that killed thousands of people and impaired the health of hundreds of thousands

DATES: December 2-3, 1984

SIGNIFICANCE: The escape of methyl isocyanate from a pesticide production plant in Bhopal, India, resulted in the world's worst chemical disaster. The abandoned plant site, which was not properly remediated, continues to pose an environmental threat to nearby residents by contaminating their drinking-water source.

Union Carbide India, which was jointly owned by Union Carbide and the Indian public, ran a pesticide production plant in Bhopal, a city in central India with a population of approximately one million people. A heavily populated shantytown surrounded the plant. Methyl isocyanate (MIC), a very reactive and toxic chemical used in the production of pesticides, was stored at the plant.

As a cost-cutting measure the facility was understaffed, and the plant workers were poorly trained. The plant's signage and manuals were in English, but many of the workers were not English speakers. In 1982, U.S. engineers who conducted a safety audit of the plant noted sixty-one hazards in the unkempt facility. Thirty were deemed critical hazards; of these, eleven were found in the MIC/phosgene units. The audit warned that a major toxic release could result.

On the evening of December 2, 1984, the plant was closed for inventory reduction and routine maintenance. Around 11:30 P.M., workers realized when

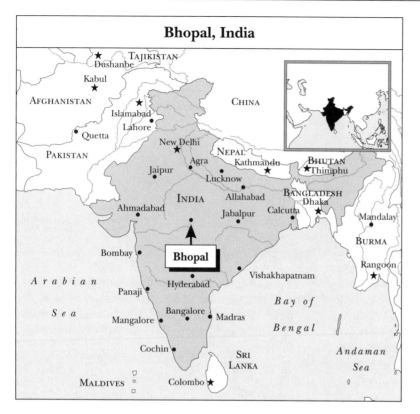

Bhopal, India

gas, while others awoke gasping for breath with their eyes and throats burning. The gas cloud quickly spread over an area of 65 square kilometers (25 square miles), engulfing panicked residents who were trying to flee. In a matter of minutes people began to collapse and die. Pregnant women caught in the cloud spontaneously miscarried.

According to the Indian Council of Medical Research, roughly three thousand people were initially killed, and another fifteen thousand later died from chemical-related illnesses. Approximately fifty thousand would suffer permanent disability. All told, the release affected the health of some half a million survivors. Their children would represent a second generation of victims, as many would be born with birth defects.

As a result of the Bhopal disaster, India significantly strengthened its regulation of hazardous industries. One of the most important actions the government took was the passage of its Environmental Protection Act of 1986.

their eyes started to tear and burn that an MIC leak had occurred. They reported the leak to their supervisor and located a section of open piping that they believed to be the source. They took action to contain this presumed source, and the supervisor went on break. By 12:15 A.M., the pressure and temperature in the MIC storage tank had risen to dangerous levels.

When MIC comes into contact with water, a spontaneous reaction results, releasing heat. In the presence of a variety of catalysts, including iron ions, three molecules of MIC will join together to form a trimer. This reaction also releases heat. It is thought that both of these reactions occurred in the MIC storage tank. The heat released during the chemical reactions raised the temperature of the MIC, which increased the rates of the MIC reactions, releasing even more heat. MIC has a low boiling point, and this heat of reaction caused the MIC to vaporize into a gas. The gas expanded, increasing the pressure inside the storage tank until it burst a rupture disk on the line leading to the pressure release valve. When that safety valve was forced open, the gases from the tank began to escape and formed a lethal cloud that moved across Bhopal.

Some people were killed in their sleep by the toxic

A Series of Failed Safeguards

Culpability for the toxic release was found to extend from high levels of management at Union Carbide to workers at the plant. Four months after the disaster, Union Carbide released a report detailing failings that led to the MIC escape. Warren Anderson, the chairman of Union Carbide, acknowledged that conditions at the plant were so poor that it should not have been in operation. The report recounted that sometime before midnight on December 2, 1984, a considerable quantity of water entered the tank and started the heat-releasing reactions. Union Carbide claimed that a disgruntled employee had intentionally introduced the water. Plant workers said it occurred accidentally when water being used to clean pipes leaked into the tank. The tank was equipped with a refrigeration system designed to keep the contents at a low temperature, but it had been shut down months before to cut costs. At the higher storage temperature, the MIC reacted more rapidly with the water.

An alarm that should have sounded when the tank temperature started to rise failed to go off because it had not been reset for the higher storage temperature resulting from the lack of refrigeration. Union Carbide officials hypothesized that the water and high temperature in the tank caused rapid corrosion of its stainless-steel walls. This led to iron contamination that would have catalyzed the heat-releasing trimerization reaction. The temperature in the tank probably rose to at least 200 degrees Celsius (392 degrees Fahrenheit). A control-room operator manually started a sodium hydroxide vent scrubber designed to neutralize leaking MIC, but the sodium hydroxide failed to circulate. The gas rushed to the flare tower, where any MIC escaping the scrubber should have been burned off. This was the last safety control, but it was also out of service. A basic design fault of the plant was that the safety systems were designed to deal with only minor leaks, so even if they had been operative, the volume of gas released would have overwhelmed them.

Although the release started about 12:30 A.M., the plant's public alarm was not sounded until 2:00 A.M.

So many small leaks had triggered it in previous months that the plant had shut it off to avoid disrupting the neighborhood. The local police were informed of the leak only after it had been stopped and after the alarm had been sounded.

An Ongoing Tragedy

The Indian government sued Union Carbide for $3.3 billion but accepted a settlement in February, 1989, that awarded $470 million to the victims. The next of kin of those killed were each awarded $2,000. Each injured survivor received roughly $500, an amount quickly consumed by medical costs. The tens of thousands who could not navigate the process of filing settlement claims or who were too ill to stand in line for hours to register their claims got nothing. The settlement did not address environmental damages.

A lack of clear, continuing corporate liability, combined with negligence on the part of the Indian government, exacerbated the plight of Bhopal residents. Union Carbide and another company that took over the facility in 1994 performed some cleanup but did

Child victims of the Bhopal toxic gas leak receive treatment. (©Kapoor Baldev/CORBIS)

not fully remediate the 4.5-hectare (11-acre) site. When the state government took over the property in 1998, hundreds of tons of abandoned toxics remained on-site. In addition to toxic wastes dumped at and around the plant, hundreds of tons of pesticides and other chemicals had been left behind in the abandoned facility. The derelict buildings provided insufficient protection against the elements, allowing rainwater to mix with the chemicals and wash them into the soil. Eventually, the contaminants migrated into the groundwater and thus into the water supplies on which residents relied for drinking and washing needs.

In 2001 Dow Chemical acquired Union Carbide. Because Dow had not operated the plant or caused the accident, it would not acknowledge any responsibility to the affected population. Bhopal residents insisted that Dow had purchased not only Union Carbide's assets but also its liabilities, and they remained unmoved by the contention that Union Carbide's liability ended with the 1989 settlement.

Bhopal citizens have continued to demand that the American-based Dow remediate the site and address the problems that the abandoned facility has created: poisoned cattle, damaged crops, and an array of human health problems including cancers, neurological disorders, mental illness, and birth defects. The victims have also continued to demand action from their government, which they believe has placed greater priority on India's economic advancement and positive relations with multinational corporations than on justice. In 2009 Bhopal victims' groups commemorated the twenty-fifth anniversary of the lethal gas release with vigils and demonstrations.

Francis P. Mac Kay
Updated by Karen N. Kähler

FURTHER READING

Amnesty International. *Clouds of Injustice: Bhopal Disaster 20 Years On.* London: Author, 2004.

D'Silva, Themistocles. *The Black Box of Bhopal: A Closer Look at the World's Deadliest Industrial Disaster.* Victoria, B.C.: Trafford, 2006.

Fortun, Kim. *Advocacy After Bhopal: Environmentalism, Disaster, New Global Orders.* Chicago: University of Chicago Press, 2001.

Hanna, Bridget, Ward Morehouse, and Satinath Sarangi, eds. *The Bhopal Reader: Remembering Twenty Years of the World's Worst Industrial Disaster.* New York: Apex Press, 2005.

SEE ALSO: Environmental illnesses; Environmental justice and environmental racism; Environmental law, international; Hazardous and toxic substance regulation; Italian dioxin release; Pesticides and herbicides; Shantytowns.

Bicycles

CATEGORY: Atmosphere and air pollution

DEFINITION: Pedal-driven vehicles consisting of light frames mounted on two wheels, one in front of the rider and the other behind, steered by means of handlebars

SIGNIFICANCE: The bicycle is the most efficient form of human-powered transportation and is virtually nonpolluting. Many environmentalists advocate the increased use of bicycles for daily transportation in the United States and other developed nations where bicycle use tends to be confined to recreation.

German inventor Karl Drais produced the precursor of the modern bicycle in 1817. The device, known as the *Laufmaschine* (running machine), had a frame that the rider straddled between two wheels of the same size, similar in appearance to today's bicycle. The *Laufmaschine* had no pedals—it was propelled by the rider's feet pushing against the ground. In 1865, the velocipede appeared; this improved on the earlier device with pedals attached to the front wheel. Shortly thereafter, in 1870, the high-wheel bicycle appeared. Also known as the penny-farthing, it had a large front wheel and a much smaller rear wheel (the name penny-farthing was inspired by the relative sizes of the old British penny and farthing coins). The large front wheel allowed faster cycling speeds than were possible on the velocipede, but the high front wheel made it difficult to mount and ride.

By the end of the nineteenth century, the bicycle had evolved to a form that was similar in appearance to the modern bicycle. This vehicle, dubbed the safety bicycle, had two equal-sized wheels with pedals mounted midframe that were connected by a chain to the rear wheel. The safety bicycle sported pneumatic (inflatable) tires, which replaced the hard-riding, solid-rubber tires used on the high-wheel bicycle.

MODERN BICYCLES

The varieties of bicycles available in the twenty-first century range from basic cruisers to lightweight racing models. An example of a basic cruiser is the single-speed Flying Pigeon, which is widely used as transportation in China. Basic cruisers may have three or more gears. Beyond the basic cruiser are the road bike, designed for use on paved roads, and the mountain bike, intended to be used off-road on hilly terrain. Both types have one, two, or three front chain rings connected to the pedals. Power from the front chain ring is transmitted through a chain to the rear-wheel derailleur, which may have up to ten different gears (cogs); thus bicycles may have up to thirty different gear ratios. Low ratios are used for hill climbing, and high ratios allow for pedaling at high speeds. Both road and mountain bikes come in tandem models, which accommodate two riders. A variation on traditional models is the recumbent bicycle; to use a recumbent bike, the rider sits low to the ground on a seat mounted on the frame between the two wheels rather than straddling a saddle mounted on the frame.

Bicycle riders, or cyclists, must overcome two primary forces: gravity and air resistance. Cyclists have an advantage over runners and walkers in overcoming gravity—gearing aids them in climbing hills, and the downhills give them a chance to recharge. Weight is an important factor in hill climbing—a relatively small rider on a lightweight bicycle has a definite advantage. In regard to air resistance, a stiff headwind can make forward progress difficult for cyclists even on a level surface. In the absence of wind, air resistance increases exponentially with speed. Crouching on a bicycle decreases wind resistance. Recumbent bikes have less wind resistance because of their low profile. These bikes are often equipped with aerodynamic windshields, which further reduce wind resistance. When cyclists ride as a group, they commonly engage in a practice known as drafting, in which they trail each other closely to take advantage of the reduced wind resistance behind other riders. After a turn at the front of the pack, the lead riders drop to the rear, and less fatigued riders take the lead.

BENEFITS OF CYCLING

Many environmentalists promote the nonpolluting benefits of bicycle use, as well as the fact that bicycles do not consume fossil fuels. Some experts have estimated that if everyone in the United States who lives 16 kilometers (10 miles) or less from work were to travel to work every day on a bicycle rather than in a motor vehicle, the nation could become independent of foreign oil. Cycling is also a pleasant form of exercise that can improve physical fitness, and cycling clubs around the world enhance the enjoyment of this activity for many enthusiasts.

Cyclists ride along a rode in Quezon City, north of the capital city of Manila, in the Philippines on November 7, 2010, as part of a demonstration by about one thousand cyclists aimed at drawing attention to the need for more bicycle lanes and pedestrian walkways to make Manila more environmentally friendly. (AP/Wide World Photos)

Beyond recreational riding, many cyclists engage in highly competitive activities. Many bicycle races are conducted in many different nations; some take place on paved roads, some on off-road routes, and some on oval tracks called velodromes. One well-known road race is the annual Tour de France, in which professional teams of cyclists race throughout France, including routes that take them up and down the Alps, covering some 3,200 kilometers (2,000 miles) in approximately three weeks.

Robin L. Wulffson

FURTHER READING

Armstrong, Lance. *Comeback 2.0: Up Close and Personal.* New York: Simon & Schuster, 2009.

Hurst, Robert. *The Art of Cycling: A Guide to Bicycling in Twenty-first-Century America.* Guilford, Conn.: Falcon, 2004.

Peveler, Willard. *The Complete Book of Road Cycling and Racing.* Camden, Maine: Ragged Mountain Press, 2008.

Sovndal, Shannon. *Cycling Anatomy.* Champaign, Ill.: Human Kinetics, 2009.

SEE ALSO: Air pollution; Alternative energy sources; Carbon footprint; Carpooling; Clean Air Act and amendments; Ecological footprint.

Bikini Atoll bombing

CATEGORIES: Disasters; nuclear power and radiation

THE EVENT: Hydrogen bomb test conducted by the United States at the Bikini Atoll in the Marshall Islands

DATE: March 1, 1954

SIGNIFICANCE: The hydrogen bomb test at the Bikini Atoll produced an explosion that was more powerful than had been anticipated, and an unexpected shift in the wind resulted in radioactive fallout landing in populated regions of the Marshall Islands, on ships participating in the test, and on a Japanese fishing boat.

The hydrogen bomb that the United States tested at the Bikini Atoll was code-named *Bravo*. It had an explosive force of approximately fifteen megatons. However, because of its new design, it used only about the same amount of the rare isotope uranium 235 as was used in the bomb dropped in 1945 on Hiroshima, Japan, which had an explosive yield one thousand times weaker. Although the residents of Bikini had been relocated before the test, the radiation spread to two inhabited islands, Rongelap and Utrick, located about 160 kilometers (100 miles) and 480 kilometers

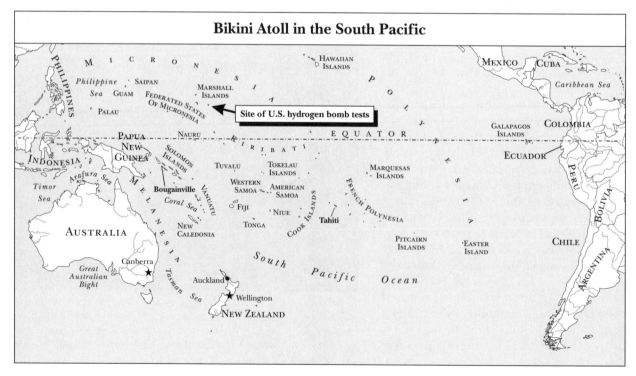

Bikini Atoll in the South Pacific

(300 miles) from Bikini, respectively. U.S. military personnel on Rongerik Atoll were also exposed to radiation, as was the crew of a U.S. Navy gasoline tanker en route from the Marshall Islands to Hawaii.

The crew of a Japanese tuna fishing boat, the *Lucky Dragon*, received the most serious radiation exposure. At the time of the explosion, the boat was located about 160 kilometers (100 miles) from Bikini, apparently outside the "danger zone" established by U.S. authorities. For about three hours following the blast, the crew observed white, sandy dust falling onto the deck of their boat. Most of the twenty-three members of the *Lucky Dragon* crew later indicated they had suffered from skin inflammation and nausea over the next few days. On March 14, when the boat returned to port in Japan, the crew members were hospitalized and treated for radiation exposure. Medical tests indicated that they had suffered bone marrow damage, resulting in anemia, as well as temporary sterility. The exposed areas of their skin were also damaged.

The radiotelegraph operator of the *Lucky Dragon* died in September, 1954. The cause of his death has been disputed. There were claims that he died as a direct result of radiation exposure, but U.S. officials said his death was caused by hepatitis contracted from the blood transfusions he received to treat the radiation exposure. The other crew members of the *Lucky Dragon* all recovered.

By the time the crew's radiation exposure was discovered, tuna from boats that had been fishing in the region of the bomb test had been shipped to the markets. Tests indicated that some of this tuna was radioactive. By the time officials realized this, some radioactive tuna had already been sold. About one hundred people in Japan consumed this contaminated fish before it was removed from the marketplace. No adverse health effects were reported, but when word of the contamination spread, many consumers simply stopped buying fish. Tuna canneries on the West Coast of the United States were alerted to the possibility of contamination, and radiation monitoring was instituted for a short time.

In January, 1955, the U.S. government donated two million dollars to the government of Japan to compensate for the injuries to the crew of the *Lucky Dragon* and for damage to the Japanese fishing industry.

George J. Flynn

FURTHER READING

International Atomic Energy Agency. *Radiological Conditions at Bikini Atoll: Prospects for Resettlement.* Vienna: Author, 1998.

Niedenthal, Jack. *For the Good of Mankind: A History of the People of Bikini and Their Islands.* 2d ed. Majuro, Marshall Islands: Bravo, 2002.

SEE ALSO: Hiroshima and Nagasaki bombings; Nuclear testing; Nuclear weapons production; Radioactive pollution and fallout.

Bioassays

CATEGORY: Ecology and ecosystems

DEFINITION: Tests that use biological organisms to detect the presence of given chemical substances or determine the biological activity of known substances in particular environments

SIGNIFICANCE: Bioassays enable environmental scientists to evaluate the effects of the chemicals used in pesticides as well as the resistance of plants to particular pests.

In many instances, a scientist may suspect that a certain chemical is present in a given environment but may not have access to a specific piece of equipment designed to measure the presence of the chemical. In some cases, an experimental protocol for the detection of the chemical may not exist. In either of these cases, the scientist may be able to detect the presence of the chemical by using a biological organism that responds in a specific manner when exposed to that particular chemical agent. At other times, a scientist may know that a certain chemical is present but not know how a particular organism will respond when exposed to the agent. In this case, the scientist will expose the test organism to the chemical and measure a particular physiological response.

Bioassays are utilized in many different areas of the biological sciences, including environmental studies. Some bioassay methods work better than others. A good bioassay meets two basic criteria. First, it is specific for a given physiological response. For example, if a given chemical is responsible for inhibiting the feeding response of a particular insect, then the bioassay for that chemical should measure only the inhibition of feeding of that insect and not some other physiological response to the chemical. Second, a good bioassay measures the same response in the laboratory that is observed in the field. Again, if a particular chemical inhibits the feeding response in the field, then the laboratory bioassay for that chemical should also inhibit feeding. An ongoing need exists for the development of accurate bioassay methods as well as the improvement of existing techniques.

Many different bioassays are used in environmental studies. One of the most common is the measure of the median lethal dose (LD_{50})—the concentration or dose of a chemical that will result in the deaths of one-half of a population of organisms—of a new pesticide on species of pest and nonpest organisms. To conduct this bioassay, the test species is exposed to a wide range of different concentrations of the chemical. The concentration of the pesticide that kills one-half of the test organisms represents the LD_{50}.

Another common environmental bioassay is the measure of resistance of plants to a particular insect pest. In order to reduce the dependence on chemical insecticides, plant breeders are continually trying to develop insect-resistant plants, either through traditional breeding programs or by using biotechnology to transfer resistance genes to susceptible crop strains. Bioassays are used to measure the degree of success of these attempts. In these bioassays, the same numbers of susceptible and resistant plants are subjected to infestation by equal numbers of the insect pest for which the breeder is trying to develop resistance. The two groups of plants are observed, and the degree of resistance, if any, is recorded.

D. R. Gossett

FURTHER READING

Ohkawa, H., H. Miyagawa, and P. W. Lee, eds. *Pesticide Chemistry: Crop Protection, Public Health, Environmental Safety.* New York: Wiley-VCH, 2007.

Rand, Gary M., ed. *Fundamentals of Aquatic Toxicology: Effects, Environmental Fate, and Risk Assessment.* 3d ed. Boca Raton, Fla.: CRC Press, 2008.

SEE ALSO: Agricultural chemicals; Biopesticides; Biotechnology and genetic engineering; Pesticides and herbicides; Sustainable agriculture.

Biocentrism

CATEGORY: Philosophy and ethics

DEFINITION: Life-centered stance that rejects the view that only human beings and their interests matter, while recognizing the moral standing of all living creatures

SIGNIFICANCE: Biocentrism has played a key role in the development of environmental ethics since that discipline was founded during the 1970's and had already been influential among important earlier thinkers, including Albert Schweitzer and Mahatma Gandhi.

Biocentrists hold that all living creatures have goods of their own and that their flourishing, or attaining their goods, is intrinsically valuable. This value should thus be taken into consideration whenever decisions affecting the flourishing of any creatures are being made. During the 1970's, Norwegian philosopher Arne Naess wanted to include living systems (such as habitats and ecosystems) within the scope of biocentrism as intrinsically valuable entities, but this view is now classified as "ecocentrism," and subsequent biocentrists have restricted moral standing to individual living creatures.

The most widespread kind of biocentrism was presented during the early 1980's by Paul Taylor, who advocated a life-centered ethic of respect for nature. In this ethic, the realization of the good of every living creature is intrinsically valuable and to be pursued for its own sake. In Taylor's version of biocentrism, not only is human superiority denied but also all living things are held to be equally worthy of respect, irrespective of differences of capacities, and to have the same moral importance. Taylor recommends defensible policies of conservation and social justice, but he has difficulty deriving them from these principles.

Another kind of biocentrism, not committed to equal respect for all living creatures, was proposed in 1981 by Robin Attfield and later by Gary E. Varner, who both defend the intrinsic value of the good of every creature, whether sentient or nonsentient. Unlike Taylor, however, Attfield upholds Peter Singer's principle of equal consideration for equal interests, applying it to the entire realm of life; thus the satisfaction of greater interests takes priority over that of lesser ones when they conflict. (The capacities of living creatures for health and for being harmed are held to distinguish them from artifacts, which have no goods of their own.) Attfield integrates this account of intrinsic value into a consequentialist recognition of beneficial practices, general compliance with which makes actions right through generating or preserving greater value than would otherwise prevail.

More recently, James Sterba has defended a Taylor-like commitment to species equality through species-neutral principles authorizing resort to self-defense in certain circumstances (whether greater capacities are at stake or not). These principles are held to be defensible not by enhancing value but as part of a deontological ethical system. Simultaneously Sterba rejects the extension of biocentrism to ecosystems, holding that biotic communities have no clear goods of their own; individual living creatures constitute the limits of ethical egalitarianism.

Biocentrism in any form supports more radical policies to preserve habitats and curtail global warming than do anthropocentric stances. The biocentric viewpoint remains influential in debates about both normative principles and practical decision making.
Robin Attfield

FURTHER READING

Attfield, Robin. *Environmental Ethics*. Malden, Mass.: Blackwell, 2003.
_____, ed. *The Ethics of the Environment*. Burlington, Vt.: Ashgate, 2008.
Sterba, James. "A Biocentrist Strikes Back." *Environmental Ethics* 20, no. 4 (Winter, 1998): 361-376.
Taylor, Paul. *Respect for Nature: A Theory of Environmental Ethics*. Princeton, N.J.: Princeton University Press, 1986.
Varner, Gary E. "Biocentric Individualism." In *Environmental Ethics: What Really Matters, What Really Works*, edited by David Schmidtz and Elizabeth Willott. New York: Oxford University Press, 2002.

SEE ALSO: Anthropocentrism; Ecocentrism; Naess, Arne; Nature preservation policy; Speciesism.

Biodiversity

CATEGORY: Ecology and ecosystems
DEFINITION: Biological variety in a given environment
SIGNIFICANCE: Biodiversity—including diversity among species, among ecosystems, and among individuals within species—is generally assumed to enhance ecological stability.

In 1993 the Wildlife Society defined biodiversity as "the richness, abundance, and variability of plant and animal species and communities and the ecological processes that link them with one another and with soil, air, and water." Included in this concept is the recognition that life on earth exists in great variety and at various levels of organization. Many kinds of specialists—including organismic biologists, population and evolutionary biologists, geneticists, and ecologists—investigate biological processes that are encompassed by the concept of biodiversity. Conservation biologists are concerned with the totality of biodiversity, including the process of speciation that forms new species, the measurement of biodiversity, and factors involved in the extinction process. The primary thrust of their efforts, however, is the development of strategies to preserve biodiversity.

The biodiversity paradigm connects classical taxonomic and morphological studies of organisms with modern techniques employed by those working at the molecular level. It is generally accepted that biodiversity can be approached at three levels of organization, commonly identified as species diversity, ecosystem diversity, and genetic diversity. Some also recognize biological phenomena diversity.

SPECIES DIVERSITY

No one knows how many species inhabit the earth. Estimates range from five million to several times that number. Each species consists of individuals that are somewhat similar and capable of interbreeding with other members of their species but are not usually able to interbreed with individuals of other species. The species that occupy a particular ecosystem are a subset of the species as a whole. Ecosystems are generally considered to be local units of nature; ponds, forests, and prairies are common examples.

Conservation biologists measure the species diversity of a given ecosystem by first conducting a careful, quantitative inventory. From such data, scientists may determine the "richness" of the ecosystem, which is simply a reflection of the number of species present. An island with three hundred species would thus be considered to be 50 percent richer than another with only two hundred species. Some ecosystems, especially tropical rain forests and coral reefs, are much richer than others. Among the least rich are tundra regions and deserts.

A second aspect of species diversity is "evenness," defined as the degree to which all of the various elements are present in similar percentages of the total species. As an example, consider two forests, each of which has a total of twenty species of trees. Suppose that the first forest has a few tree species represented by rather high percentages and the remainder by low percentages. A second forest with species more evenly distributed would rate higher on a scale of evenness.

Why Preserve Biodiversity?

Biodiversity is defined as the total number of species within an ecosystem, and also as the resulting complexity of interactions among them. It measures the "richness" of an ecological community. Among an estimated 8 to 10 million unique and irreplaceable species existing on Earth, fewer than 1.6 million have been named. A tiny fraction of this number has been studied. Over thousands of years, organisms in a community have been molded by forces of natural selection exerted by other living species as well as by the nonliving environment that surrounds them. The result is a highly complex web of interdependent species whose interactions sustain one another and provide the basis for the very existence of human life as well.

Loss of biodiversity poses a serious challenge to the sustenance of many communities and ecosystems. For example, the destruction of tropical rain forests by clear-cut logging produces high rates of extinction of many species. Most of these species have never been named, and many never even discovered. As species are eliminated, the communities of which they were a part may change and become unstable and more vulnerable to damage by diseases or adverse environmental conditions. Aside from the disruption of natural food webs, potential sources of medicine, food, and raw materials for industry are also lost. As Harvard professor Edward O. Wilson once said, "The loss of species is the folly our descendants are least likely to forgive us."

Species diversity, therefore, is a value that combines both species richness and species evenness measures. Values obtained from a diversity index are used in comparing species diversity among ecosystems of both the same and different types. They also have implications for the preservation of ecosystems; other things being equal, it would be preferable to preserve ecosystems with high diversity indexes, thus protecting a larger number of species.

Considerable effort has been expended to predict species diversity as determined by the nature of the area involved. For example, island biogeography theory suggests that islands that are larger, are nearer to other islands or continents, and have more heterogeneous landscapes would be expected to have higher species diversity than those possessing alternate traits. Such predictions apply not only to literal islands but also to other discontinuous ecosystems; examples would be alpine tundra of isolated mountaintops and ponds several miles apart.

The application of island biogeography theory to designing nature preserves was proposed by Jared Diamond in 1975. His suggestion began the "single larger or several smaller," or SLOSS, area controversy. Although island biogeography theory would, in many instances, suggest selecting one large area for a nature preserve, it is often the case that several smaller areas, if carefully selected, could preserve more species.

The species diversity of a particular ecosystem is subject to change over time. Pollution, deforestation, and other types of habitat degradation invariably reduce diversity. Conversely, during the extended process of ecological succession that follows disturbances, species diversity typically increases until a permanent, climax ecosystem with a large index of diversity results. Ecologists generally assume that more diverse ecosystems are more stable than those with less diversity. Certainly, the more species present, the greater the opportunity for various interactions, both with other species and with the environment. Examples of interspecific reactions include mutualism, predation, and parasitism. Such interactions apparently help to integrate a community into a whole, thus increasing its stability.

ECOSYSTEM DIVERSITY

Ecology can be defined as the study of ecosystems. From a conservation standpoint, ecosystems are important because they sustain their particular assemblages of living species. Conservation biologists also consider ecosystems to have intrinsic value beyond the species they harbor; therefore, it would be ideal if representative global ecosystems could be preserved. This is far from realization, however. Just deciding where to draw the line between interfacing ecosystems can be a problem. For example, the water level of a stream running through a forest is subject to seasonal fluctuation, causing a transitional zone characterized by the biota (that is, the flora and fauna) from both adjoining ecosystems. Such ubiquitous zones negate the view that ecosystems are discrete units with easily recognized boundaries.

The protection of diverse ecosystems is of utmost importance to the maintenance of biodiversity, but ecosystems throughout the world are threatened by global warming, air and water pollution, acid deposition, ozone depletion, and other destructive forces. At the local level, deforestation, thermal pollution, urbanization, and poor agricultural practices are among the problems affecting ecosystems and therefore reducing biodiversity. Both global and local environmental problems are amplified by rapidly increasing world population pressures.

In the process of determining which ecosystems are most in need of protection, many scientists have recognized that the creation of a system for naming and classifying ecosystems is highly desirable, if not imperative. Efforts are being made to establish a system similar to the hierarchical system applied to species that was developed by Swedish botanist Carolus Linnaeus during the eighteenth century. A comprehensive classification system for ecosystems is far from complete, but freshwater, marine, and terrestrial ecosystems are recognized as main categories, with each further divided into particular types. Though tentative, the system that has been established thus far has made possible the identification and preservation of a wide range of representative threatened ecosystems.

In 1995 conservation biologist Reed F. Noss of Oregon State University and his colleagues identified more than 126 types of ecosystems in the United States that are threatened or critically endangered. The following list illustrates their diversity: southern Appalachian spruce-fir forests; eastern grasslands, savannas, and barrens; California native grasslands; Hawaiian dry forests; caves and karst systems; old-growth forests of the Pacific Northwest; and southern forested wetlands.

Not all ecosystems can be saved. Establishing priorities involves many considerations, some of which are economic and political. Ideally, choices would be made on merit, taking into consideration the rarity, size, and number of endangered species that ecosystems include as well as other objective, scientific criteria.

GENETIC AND BIOLOGICAL PHENOMENA DIVERSITY

Most of the variations among individuals of the same species are caused by the different genotypes (combinations of genes) the individuals possess. Such genetic diversity is readily apparent in cultivated or domesticated species such as cats, dogs, and corn, but it also exists, though usually to a lesser degree, in wild species. Genetic diversity can be measured only through the use of exacting molecular laboratory procedures. Such tests detect the amount of variation in the deoxyribonucleic acid (DNA) or isoenzymes (chemically distinct enzymes) possessed by various individuals of the species in question.

A significant degree of genetic diversity within a population or species confers a great advantage. This diversity is the raw material that allows evolutionary processes to occur. When a local population becomes too small, it is subject to a serious decline in vigor from increased inbreeding. This leads, in turn, to a downward, self-perpetuating spiral in genetic diversity and further reduction in population size. Extinction may be imminent. In the grand scheme of nature, this is a catastrophic event; never again will that particular genome (set of genes) exist anywhere on the earth. Extinction is the process by which global biodiversity is reduced.

The term "biological phenomena diversity" refers to the numerous unique biological events that occur in natural areas throughout the world. Examples include the congregation of thousands of monarch butterflies on tree limbs at Point Pelee in Ontario, Canada, as they await favorable conditions before continuing their migration, and the return of hundreds of loggerhead sea turtles each April to Padre Island in the Gulf of Mexico in order to lay their eggs.

Although biologists have been concerned with protecting plant and animal species for decades, only recently has conservation biology emerged as an identifiable discipline. Conceived in a perceived crisis of biological extinctions, conservation biology differs from related disciplines, such as ecology, in that it is advocative in nature, insisting that the maintenance of biodiversity is intrinsically good. Conservation biology is a value-laden science, and some critics consider it akin to a religion with an accepted dogma.

The prospect of preserving global ecosystems and the life processes they make possible, all necessary for maintaining global diversity, is not promising. Western culture does not give environmental concerns a high priority. Among those who do, greater concern is often expressed about the immediate health effects of environmental degradation than about the loss of biodiversity. The impetus necessary to save ecosystems and all their inhabitants—including humans—will develop only when education in basic biology and ecology at all levels is extended to include an awareness of the importance of biodiversity.

Thomas E. Hemmerly

FURTHER READING

Chivian, Eric, and Aaron Bernstein, eds. *Sustaining Life: How Human Health Depends on Biodiversity.* New York: Oxford University Press, 2008.

Ehrlich, Paul R., and Anne H. Ehrlich. *Extinction: The Causes and Consequences of the Disappearance of Species.* New York: Ballantine Books, 1981.

Hunter, Malcolm L., Jr. *Fundamentals of Conservation Biology.* 3d ed. Hoboken, N.J.: Wiley-Blackwell, 2006.

Novacek, Michael J., ed. *The Biodiversity Crisis: Losing What Counts.* New York: New Press, 2001.

Primack, Richard B. *Essentials of Conservation Biology.* 5th ed. Sunderland, Mass.: Sinauer Associates, 2010.

Ray, Justina C., et al., eds. *Large Carnivores and the Conservation of Biodiversity.* Washington, D.C.: Island Press, 2005.

Wilson, Edward O. *Biophilia.* 1984. Reprint. Cambridge, Mass.: Harvard University Press, 2003.

_____. *The Diversity of Life.* New ed. New York: W. W. Norton, 1999.

Zeigler, David. *Understanding Biodiversity.* Westport, Conn.: Praeger, 2007.

SEE ALSO: Biodiversity action plans; Convention on Biological Diversity; Ecology as a concept; Ecosystems; Extinctions and species loss; Global Biodiversity Assessment; Rain forests.

Biodiversity action plans

CATEGORIES: Treaties, laws, and court cases; ecology and ecosystems

DEFINITION: Government-devised plans for protecting and restoring threatened ecological systems and biological species

SIGNIFICANCE: Biodiversity action plans are important tools that governments use in protecting and restoring threatened ecosystems. These plans seek to implement the 1992 United Nations Convention on Biological Diversity, which demonstrates worldwide recognition that biodiversity is intrinsically valuable and deserves to be protected or restored through careful planning.

The schemes for protecting and restoring threatened species of plants and animals and their ecosystems known as biodiversity action plans (BAPs; also known as national biodiversity strategies and action plans, or NBSAPs) gained worldwide recognition with the signing of the United Nations Convention on Biological Diversity (CBD) at the Earth Summit in Rio de Janeiro, Brazil, in 1992. BAPs represent the steps that nations are taking to implement the provisions of the convention. However, although most of the world's nations have ratified the CBD, only a handful have actually developed substantive BAPs. Among those that have are the United States, Australia, the United Kingdom, Tanzania, and Uzbekistan.

A well-conceived BAP includes several components: plans for the carrying out of inventories and documentation of selected species and specific habitats, with particular emphasis placed on population distribution and conservation status within certain ecosystems; realistic targets or indicators for conservation and restoration; plans covering funding and time lines for achievement of specific goals; and plans for the establishment of partnerships among private and public institutions and agencies that will work together to achieve the goals set.

OBSTACLES AND CRITICISMS

To implement a BAP effectively, a nation must overcome a number of obstacles; the difficulty of the process may explain in part why so few countries have attempted to develop such plans. In some parts of the world, for example, undertaking complete inventories of plant and animal species is not realistic. Scientists have estimated that only about 10 percent of the world's species have been characterized and documented; most of those still unknown include plants and lower animals such as insects. An ideal BAP includes the assessment of species population estimates over time so that the variability and degree of vulnerability of species can be determined; it also includes descriptions of species' ranges, habitats, behaviors, breeding practices, and interactions with other species. The collection of such fundamental information can be a daunting task. Another factor preventing some nations from developing BAPs is the cost involved. Depending on the size of the country, the cost of preparing a solid BAP can easily come to the equivalent of millions of U.S. dollars, with about 10 percent of the initial cost factored in for annual maintenance. It is therefore not surprising that the call for BAPs has been criticized by some developing countries.

In addition to the difficulty and expense involved in the implementation of BAPs, many developing countries are unwilling to create such plans because, they argue, the plans obviously favor the consideration of wildlife and plant protection over food production and industrial growth; in some cases, BAPs may even represent impediments to population growth. Most of the Middle Eastern countries and many African nations have shown little interest in participating in a substantive way in such plans. Others have simply opted to create pro forma plans that expend little on research and even less on the management of natural resources. In contrast, the European Union has chosen to divert the purpose of BAPs, instead implementing the CBD through a set of economic development policies while paying special attention to the protection of certain ecosystems.

It has become increasingly clear that what is at stake is the very definition of "biodiversity" itself. According to the CBD, biodiversity is the variation of life-forms within a given ecosystem; it is a combination of ecosystem structure and function as well as components (species, habitat, and genetic resources). The CBD states:

> In addressing the boundless complexity of biological diversity, it has become conventional to think in hierarchical terms, from the genetic material within individual cells, building through individual organisms, populations, species, and communities of species, to biosphere overall. . . . At the same time, in seeking to make management interventions as efficient as possible, it is essential to take a holistic view of biodiversity and address the interactions that species have with

each other and their nonliving environment, i.e., to work from an ecological perspective.

At the World Summit on Sustainable Development in 2002, delegates adopted the objectives of the CBD and designated 2010 the Year of Biodiversity.

Nader N. Chokr

FURTHER READING

Chivian, Eric, and Aaron Bernstein, eds. *Sustaining Life: How Human Health Depends on Biodiversity.* New York: Oxford University Press, 2008.

O'Riordan, Tim, and Susanne Stoll-Kleemann, eds. *Biodiversity, Sustainability, and Human Communities: Protecting Beyond the Protected.* New York: Cambridge University Press, 2002.

Ray, Justina C., et al., eds. *Large Carnivores and the Conservation of Biodiversity.* Washington, D.C.: Island Press, 2005.

Wilson, Edward O. *The Diversity of Life.* New ed. New York: W. W. Norton, 1999.

Zeigler, David. *Understanding Biodiversity.* Westport, Conn.: Praeger, 2007.

SEE ALSO: Biodiversity; Convention on Biological Diversity; Earth Summit; Ecosystems; Extinctions and species loss; Global Biodiversity Assessment; Habitat destruction; World Summit on Sustainable Development.

Biofertilizers

CATEGORIES: Biotechnology and genetic engineering; agriculture and food

DEFINITION: Fertilizers consisting of either naturally occurring or genetically modified microorganisms

SIGNIFICANCE: Biofertilizers provide a means by which biological systems can be utilized to supply plant nutrients such as nitrogen to agricultural crops. The use of biofertilizers could reduce the dependence on chemical fertilizers, which are often detrimental to the environment.

Plants require adequate supplies of the thirteen mineral nutrients necessary for normal growth and reproduction. These nutrients, which must be supplied by the soil, include both macronutrients (those nutrients required in large quantities) and micronutrients (those nutrients required in smaller quantities). As plants grow and develop, they remove these essential mineral nutrients from the soil. Since normal crop production usually requires the removal of plants or plant parts, the nutrients are continuously removed from the soil. Therefore, the long-term agricultural utilization of any soil requires periodic fertilization to replace lost nutrients.

Nitrogen is the plant nutrient that is most often depleted in agricultural soils, and most crops respond to the addition of nitrogen fertilizer by increasing their growth and yield; therefore, more nitrogen fertilizer is applied to cropland than any other fertilizer. In the past, nitrogen fertilizers have been limited to either manures, which have low levels of nitrogen, or chemical fertilizers, which usually have high levels of nitrogen. The excess nitrogen in chemical fertilizers often runs off into nearby waterways, causing a variety of environmental problems.

Biofertilizers offer a potential alternative: They supply sufficient amounts of nitrogen for maximum yields yet have a positive impact on the environment. Biofertilizers generally consist of either naturally occurring or genetically modified microorganisms that improve the physical condition of soil, aid plant growth, or increase crop yield. Such fertilizers provide an environmentally friendly way to increase plant health and yields with reduced input costs, new products and additional revenues for the agricultural biotechnology industry, and cheaper products for consumers.

While biofertilizers could potentially be used to supply a number of different nutrients, most of the interest thus far has focused on enhancing nitrogen fertilization. The relatively small amounts of nitrogen found in soil come from a variety of sources. Some nitrogen is present in all organic matter in soil; as this organic matter is degraded by microorganisms, it can be used by plants. A second source of nitrogen is nitro-

Nitrogen-fixing Rhizobia *bacteria nodules on soybean roots supply the plants with sufficient amounts of nitrogen for maximum yields and do not have the negative impacts on the environment that chemical fertilizers have.* (©Visuals Unlimited/CORBIS)

gen fixation, the chemical or biological process of taking nitrogen from the atmosphere and converting it to a form that can be utilized by plants. Bacteria such as *Rhizobia* can live symbiotically with certain plants, such as legumes, which house nitrogen-fixing bacteria in their roots. The *Rhizobia* and plant root tissue form root nodules that house the nitrogen-fixing bacteria; once inside the nodules, the bacteria use energy supplied by the plant to convert atmospheric nitrogen to ammonia, which nourishes the plant. Natural nitrogen can also be supplied by free-living microorganisms, which can fix nitrogen without forming a symbiotic relationship with plants. The primary objective of biofertilizers is to enhance any one or all of these processes.

One of the major goals in the genetic engineering of biofertilizers is to transfer the ability to form nodules and establish effective symbiosis to nonlegume plants. The formation of nodules in which the *Rhizobia* live requires plant cells to synthesize many new proteins, and many of the genes required for the expression of these proteins are not found in the root cells of plants outside the legume family. Many research programs have been devoted to efforts to transfer such genes to nonlegume plants so that they can interact symbiotically with nitrogen-fixing bacteria. If this is accomplished, *Rhizobia* could be used as a biofertilizer for a variety of plants.

There is also much interest in using the free-living, soil-borne organisms that fix atmospheric nitrogen as biofertilizers. These organisms live in the rhizosphere (the region of soil in immediate contact with plant roots) or thrive on the surface of the soil. Since the exudates from these microorganisms contain nitrogen that can be utilized by plants, increasing their abundance in the soil could reduce dependence on chemical fertilizers. Numerous research efforts have been designed to identify and enhance the abundance of nitrogen-fixing bacteria in the rhizosphere. Soil microorganisms primarily depend on soluble root exudates and decomposed organic matter to supply the energy necessary for fixing nitrogen; hence there is also an interest in enhancing the biodegradation of organic matter in the soil. This research has largely centered on inoculating the soil with cellulose-degrading fungi and nitrogen-fixing bacteria or applying organic matter to the soil, such as straw that has been treated with a combination of the fungi and bacteria.

D. R. Gossett

Further Reading

Akinyemi, Okoro M. *Agricultural Production: Organic and Conventional Systems.* Enfield, N.H.: Science Publishers, 2007.

Black, C. A. *Soil-Plant Relationships.* New York: John Wiley & Sons, 1988.

Crispeels, M. J., and D. E. Sadava. *Plants, Genes, and Agriculture.* Sudbury, Mass.: Jones and Bartlett, 1994.

Lynch, J. M. *Soil Biotechnology: Microbiological Factors in Crop Production.* Malden, Mass.: Blackwell, 1983.

Salisbury, F. B., and C. W. Ross. *Plant Physiology.* Pacific Grove, Calif.: Brooks/Cole, 1985.

Yadav, A. K., S. Ray Chaudhuri, and M. R. Motsara, eds. *Recent Advances in Biofertilizer Technology.* New Delhi: Society for Promotion and Utilisation of Resources and Technology, 2001.

See also: Agricultural chemicals; Biopesticides; Biotechnology and genetic engineering; Genetically altered bacteria; Sustainable agriculture.

Biogeochemical cycles

Categories: Weather and climate; atmosphere and air pollution

Definition: Movements of chemical elements among parts of the earth, including the atmosphere, the earth's crust, oceans, and living things

Significance: The movement and location of chemical elements among the various systems that make up the earth affect the planet's climate, the diversity and range of species, the impact and intensity of geological events, and a host of other matters related to the earth as a whole and subsystems within it.

Modern geoscientists consider the earth to comprise a set of interacting open systems: the atmosphere (the layers of air that envelop the earth), the biosphere (the areas that support and are filled with living things), the lithosphere (the earth's crust), and the hydrosphere (the water found on the earth). These various spheres contain within them important chemical elements. The six most abundant elements are carbon, nitrogen, oxygen, hydrogen, phosphorus, and sulfur; these elements, alone or in combination with others, serve as critical macronutrients to living

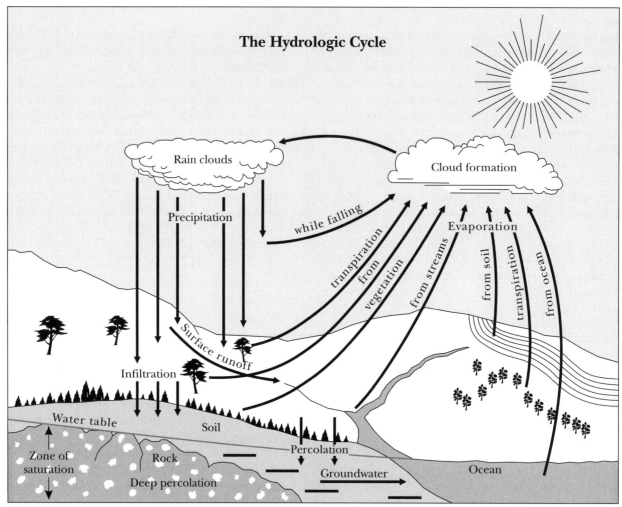

The Hydrologic Cycle

Source: U.S. Department of Agriculture, *Yearbook of Agriculture* (Washington, D.C.: Government Printing Office, 1955).

things. Other important macronutrients are potassium, calcium, iron, and magnesium. So-called micronutrients are present in smaller amounts and play critical roles in sustaining life on earth: boron (used by green plants), copper (critical to the functioning of some enzymes), and molybdenum (vital for the functioning of nitrogen-fixing bacteria). Other chemical elements, such as ammonia, also are involved in biogeochemical cycles.

The atmosphere and the oceans serve as important reservoirs where the key gas elements (carbon, nitrogen, oxygen, and hydrogen) are mainly stored. Soils and sedimentary rocks serve as principal reservoirs for the storage of phosphorus and sulfur. Throughout the earth these various chemicals are moved about by different physical, chemical, and human-induced

processes that alter the respective amounts of elements in various places and within both living and nonliving things.

An example of a biogeochemical cycle is the hydrologic (water) cycle that moves oxygen and hydrogen, along with other chemicals, from reservoir to reservoir, principally at or near the surface of the earth. This includes the proportionally large amounts of water residing within living things as well as the water found in the oceans, the atmosphere, surface water, groundwater, glaciers, and soils. During various periods of the planet's history the relative percentages of oxygen, hydrogen, and other chemical elements found in the hydrosphere have changed dramatically. These changes have in turn affected the overall temperature and climate of the earth and the types of environments available for living things; they have also sometimes caused physical changes to the other spheres (atmosphere, biosphere, and lithosphere).

Another biogeochemical cycle is the carbon cycle. Most of the original carbon in the earth's systems came out of the earth's mantle in the form of carbon dioxide gas released by volcanoes. The carbon dioxide in the atmosphere is taken up by various means, including dissolution in seawater in the form of bicarbonate ions and absorption by photosynthetic organisms such as algae and plants that convert it into sugar and other organic chemicals. As this carbon enters the food chain it ultimately finds its way into the tissues of animals. It is estimated that some 63 billion tons of carbon move every year worldwide from the atmosphere into life-forms. In this sense the carbon that has been so transformed has moved from the reservoir of the atmosphere into the reservoir of the biosphere.

Human beings' burning of fuels that are rich in carbon has resulted in increases in the amounts of carbon dioxide and other greenhouse gases in the earth's atmosphere, beyond the rate that would have occurred from natural processes. Most climatologists and atmospheric scientists believe that these greenhouse gases contribute to increasing the earth's overall atmospheric temperature and that the resulting net global warming has the potential to produce catastrophic effects over time.

Dennis W. Cheek

FURTHER READING

Bashkin, Vladimir N., with Robert W. Howarth. *Modern Biogeochemistry*. Norwell, Mass.: Kluwer Academic, 2003.

Libes, Susan M. *Introduction to Marine Biogeochemistry*. 2d ed. New York: Academic Press, 2009.

SEE ALSO: Carbon cycle; Climate change and oceans; Greenhouse effect; Greenhouse gases.

Biomagnification

CATEGORIES: Ecology and ecosystems; animals and endangered species

DEFINITION: Accumulation of toxic contaminants in organisms as they move up through the food chain

SIGNIFICANCE: Recognition of the detrimental environmental effects of biomagnification led to the adoption of several procedures to prevent the accumulation of toxic materials in higher organisms along the food chain. Some pesticides were banned outright, and others were modified to prevent their accumulation in the environment.

As members of each level of the food chain are progressively eaten by those organisms found in higher levels of the chain, the potential grows for increased concentrations of toxic chemicals to accumulate within the tissues of the higher organisms. Not all chemicals, potentially toxic or not, are equally likely to undergo this accumulation, known as biomagnification. However, molecules susceptible to biomagnification have certain characteristics in common. They are resistant to natural microbial degradation and therefore persist in the environment. They are also lipophilic—that is, they tend to accumulate in the fatty tissue of organisms. In addition, chemicals must be biologically active in order to have effects on the organisms in which they are found. Such compounds are likely to be absorbed from food or water in the environment and stored within the membranes or fatty tissues.

The process of biomagnification usually begins with the spraying of pesticides for the purpose of controlling insect populations. Industrial contamination, including the release of heavy metals, can be an additional cause of such pollution. Biomagnification results when these chemicals contaminate the water supply and are absorbed into the lipid membranes of microbial organisms. This process, often referred to as bioaccumulation, results in the initial concentration of the chemical in the organisms in a form that is not naturally excreted with normal waste material. The levels of the chemical in the organisms may reach anywhere from one to three times the level found in the surrounding environment. Since the nature of the chemical is such that it is neither degraded nor excreted, it remains within the organisms.

As the organisms on the bottom of the food chain

are eaten and digested by members of the next level in the chain, the concentration of the accumulated material significantly increases; at each subsequent level, the concentration may reach one order of magnitude (a tenfold increase) higher. Consequently, the concentrations of the pollutant at the top of the environmental food chain, such as in fish or carnivorous birds, and potentially even humans, may be as much as one million times as high as the original, presumably safe, levels in the environment. For example, studies of dichloro-diphenyl-trichloroethane (DDT) levels in the 1960's found that zooplankton at the bottom of the food chain had accumulated nearly one thousand times the level of the pollutant in the surrounding water. Ingestion of the plankton by fish resulted in concentration by another factor of several hundred. By the time the fish were eaten by predatory birds, the level of DDT was concentrated by a factor of more than two hundred thousand.

POLLUTANTS SUBJECT TO BIOMAGNIFICATION

DDT is characteristic of most pollutants subject to potential biomagnification. It is relatively stable in the environment, persisting for decades. It is soluble in lipids and readily incorporated into the membranes of organisms. While DDT represents the classic example of biomagnification of a toxic chemical, it is by no means the only representative of potential environmental pollutants. Other pesticides with similar characteristics include aldrin, chlordane, parathion, and toxaphene. In addition, cyanide, polychlorinated biphenyls (PCBs), and heavy metals—such as selenium, mercury, copper, lead, and zinc—have been found to concentrate within the food chain.

Some heavy metals are inherently toxic or may undergo microbial modification that increases their toxic potential. For example, mercury does not naturally accumulate in membranes and was therefore not originally viewed as a significant danger to the environment. However, some microorganisms are capable of adding a methyl group to the metal, producing methylmercury, a highly toxic material that does accumulate in fatty tissue and membranes.

Since pesticides are, by their nature, biologically active compounds, which reflects their ability to control insects, they are of particular concern if subject to biomagnification. DDT remains the classic example of how bioaccumulation and biomagnification may have effects on the environment. Initially introduced as a pesticide for control of insects and insect-borne disease, DDT was not thought to be particularly toxic. However, biomagnification of the chemical was found to result in the deaths of birds and other wildlife. In addition, DDT contamination was found to result in the formation of thin eggshells in birds, which greatly reduced the numbers of birds successfully hatched. Before the use of DDT was banned in the 1960's, the population levels of predatory birds such as eagles and falcons had fallen to a fraction of the levels found prior to use of the insecticide. Though it was unclear whether there was any direct effect on the human population in the United States, the discovery of elevated levels of DDT in human tissue contributed to the decision to ban the use of the chemical.

Several procedures have been adopted since the 1960's to prevent the biomagnification of toxic materials. In addition to outright bans, pesticides are often modified to prevent their accumulation in the environment. The chemical structures of most synthetic pesticides are easily degraded by microorganisms found in the environment. Ideally, a pesticide should survive no longer than a single growing season before being rendered harmless by the environmental flora. Often such chemical changes require only simple modification of a compound's basic structure.

Richard Adler

FURTHER READING

Atlas, Ronald, and Richard Bartha. *Microbial Ecology.* 4th ed. San Francisco: Benjamin Cummings, 1998.
Carson, Rachel. *Silent Spring.* 40th anniversary ed. Boston: Houghton Mifflin, 2002.
Colborn, Theo, Dianne Dumanoski, and John P. Myers. *Our Stolen Future: Are We Threatening Our Fertility, Intelligence, and Survival?* New York: Plume, 1996.
Karasov, William H., and Carlos Martínez del Rio. *Physiological Ecology: How Animals Process Energy, Nutrients, and Toxins.* Princeton, N.J.: Princeton University Press, 2007.
Primack, Richard B. *Essentials of Conservation Biology.* 5th ed. Sunderland, Mass.: Sinauer Associates, 2010.
Willmer, Pat, Graham Stone, and Ian Johnston. *Environmental Physiology of Animals.* Malden, Mass.: Blackwell Science, 2000.

SEE ALSO: Agricultural chemicals; Dichloro-diphenyl-trichloroethane; Heavy metals; Mercury; Pesticides and herbicides; Polychlorinated biphenyls.

Biomass conversion

CATEGORY: Energy and energy use

DEFINITION: Process of converting biological organic material, such as plant material and animal waste, into fuels

SIGNIFICANCE: The conversion of biomass into fuels provides alternatives to the use of fossil fuels and thus can contribute to the reduction of air pollution. Biomass conversion also serves to reduce the amount of wastes that must be disposed of in other ways.

Plants and algae use solar energy and transform carbon dioxide (CO_2) from the atmosphere into their biomass (also called primary biomass). Human beings have converted biomass into energy for centuries. For example, burning biomass in the form of wood is the oldest form of such conversion. Biomass can also be converted into other energy sources or fuels—for example, through fermentation to alcohols (ethanol or butanol) or biogas and through gasification to a substitute for natural gas. In addition, biomass such as plant oil can be transformed by the chemical reaction of transesterification into biodiesel, a diesel fuel substitute. The processing of primary biomass by organisms creates secondary biomass sources, such as animal manure and other wastes. Several countries around the world use incinerators to convert this kind of biomass into electricity. Biomass is produced naturally (for example, in forests) and agriculturally (for example, agricultural residues and dung).

The processes used to convert biomass into fuels can have both positive and negative environmental impacts, but the positive influences on the environment outweigh the negative ones. It is widely recognized that the use of fossil fuels is the leading cause of global climate change due to carbon dioxide release. Biomass conversion does not result in net CO_2 emissions because it releases only the amount of CO_2 absorbed in the biomass during plant growth. Biomass conversion into energy is thus a favorable option for reducing CO_2 emissions.

Biomass conversion also offers a means of reducing wastes, such as agricultural residues and human wastes. A great number of wastes result from the cultivation of crops such as corn, and these wastes can be turn into ethanol, which is used as a transportation fuel. Conversion of biomass from landfills produces biogas, which is two-thirds methane. Methane is a very powerful greenhouse gas, and thus a contributor to global warming, but it is also a very good fuel. It can be burned in electrical generators to produce electricity, and it can be used as a fuel for vehicles. To prevent the methane produced by landfills from being released into the atmosphere, some municipalities have installed "gas wells" in landfills to tap the methane for use as fuel.

Biogas can also be generated from wastewater and from animal waste. In 2006 the city of San Francisco contemplated a plan to extend its recycling program to include conversion of dog feces into methane to produce electricity and to heat homes; given the city's dog population of 120,000, this initiative was seen to have the potential to generate significant amounts of fuel while reducing waste. The plan was never initiated, however, as the city opted to focus on other recycling programs.

Sergei A. Markov

FURTHER READING

Bourne, Joel K. "Green Dreams." *National Geographic*, October, 2007, 38-59.

Hall, David O., and Joanna I. House. "Biomass: A Modern and Environmentally Acceptable Fuel." *Solar Energy Materials and Solar Cells* 38 (1995): 521-542.

Wright, Richard T. *Environmental Science: Toward a Sustainable Future*. 10th ed. Upper Saddle River, N.J.: Prentice Hall, 2008.

SEE ALSO: Alternative fuels; Ethanol; Landfills; Methane; Renewable energy; Synthetic fuels; Waste management.

Biomes

CATEGORY: Ecology and ecosystems

DEFINITION: Major ecological zones shaped primarily by climate and characterized by the types of plant and animal life they support

SIGNIFICANCE: The complexity of biomes exposes them to varying and sometimes cascading changes from both natural and human-initiated events. The consequences for the larger environment can range from minimal to catastrophic.

Each of the earth's biomes has a typical ecological pattern, with its flora and fauna sharing similar characteristics, such as leaf forms and survival strate-

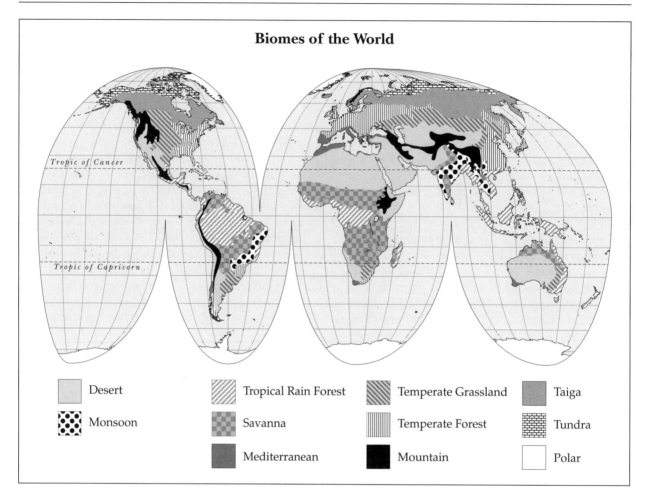

Biomes of the World

Legend:
- Desert
- Monsoon
- Tropical Rain Forest
- Savanna
- Mediterranean
- Temperate Grassland
- Temperate Forest
- Mountain
- Taiga
- Tundra
- Polar

gies. A study of any biome, on any continent, will show multiple ways its species carry out the essential tasks of life. Biomes are large and complex, with species forming intricate networks of life within subsets of the biomes.

Land-based major biomes include the following: tropical rain forest, tropical dry forest, savanna (tropical grassland), desert, temperate grassland, Mediterranean scrub or chaparral, temperate deciduous forest, temperate mixed and coniferous forest, boreal forest (taiga), and Arctic tundra. Marine biomes are basically characterized as oceanic or freshwater, although the lack of vegetation-climate links and the gradients between coastal and deep-sea habitats make the simplicity of this scheme problematic. Finally, human activities have so transformed large portions of the earth that some biologists treat anthropogenic (human-dominated) biomes as a separate concept. These biomes range from dense settlement patterns in urban landscapes to cropland, rangeland, and forest.

INFLUENCES ON BIOMES

Biomes are sometimes viewed as giant ecosystems. Because of biomes' size and complexity, however, both scientists and lay observers find it more useful to view biomes as made up of many different ecosystems and habitats.

Land-based biomes are shaped largely by their climates. The most important climatic factors are latitude and humidity (total precipitation, seasonal rainfall patterns, ambient moisture); sometimes elevation is important also. Cross-cutting these climatic factors—though influenced by them—are such features as terrain, soil type and nutrient status, and prevailing winds. All these affect the kinds of vegetation that can flourish in a region, thus determining the types and amounts of animal life the plant cover will support.

Generally biomes closer to the equator support more biodiversity, in the form of different species. An estimated 50 percent of all life-forms on earth are native to tropical forests. Tropical rain forests also have many layers of life systems, from the canopy ecosystems with their monkeys, marmosets, and tropical birds, such as macaws, down through several canopy-gap layers that shelter opossums and woodpeckers, to the forest floor, the base station for jaguars and frogs, among other jungle dwellers. What the tropical forests do not have is a rich nutrient base. They have poor soil quality, and most of the available nutrients are held within the trees themselves rather than fixed in the soil. ·

In contrast, temperate deciduous forests lack the great variety of species, and of levels, found in tropical rain forests. The life processes in temperate deciduous forests are highly regulated by seasonal temperature changes. As leaves fall to the ground, they carry calcium and other minerals to the forest floor; this litter turns into humus and becomes the nutrient-rich soil on which broad-leaved trees thrive.

Another mix of resources and life strategies is found in desert biomes. Most deserts do get some water and support some life, but anything growing there must adapt to little moisture and hot, unshaded land that is poor in nutrients. Desert plants store nutrients in underground organs such as their root systems, and cacti have expandable stems for storing water from sudden but infrequent downpours. Desert dwellers such as the kangaroo rat get their moisture from eating seeds. Their underground burrows retain three times as much humidity as exists aboveground.

ENVIRONMENTAL ISSUES AND BIOMES

For the nonhuman organisms living in any given biome, the environmental features of that biome are essential to life. Human beings, in contrast, interact with biomes by finding ways to use the biomes' natural resources to sustain them. Human technologies and large numbers of humans, however, can upset a delicately balanced biome far beyond its ability to recover. Even in the ancient world, large-scale grazing and firewood collection apparently turned once-wooded coastal areas into the Mediterranean scrub that bears that region's name.

Throughout history, humans have cleared forests or modified grasslands for use in agriculture. Sometimes the environmental damage done by such processes is limited. For example, the prairie grasslands of the American Great Plains were originally roamed by bison, and the cattle herds that replaced the bison filled the same ecological niche. Other human-caused changes have brought near disaster to entire biomes.

Human activities have altered every one of the earth's biomes. Among the most dire changes has been the destruction of tropical forests caused by timber harvesting and by the clearing of trees and other plant life to make way for ranching or industry. The damage caused when a biome is altered is not confined to the loss of diverse natural resources within that biome; rather, such change threatens the entire biosphere because it interferes with the role the biome plays in balancing the earth's systems. Loss of rain forest, for example, damages the atmosphere because of the accompanying loss of oxygen renewal that was provided by the trees.

The ocean biome is central to regulating heat and rainfall around the world, and, through its plankton, is the source of most photosynthesis on the earth. The crumbling of coral reefs and the huge trash vortex reported in the North Pacific in the late 1990's are early warning signals about the environmental threats to this biome's future. Freshwater bodies are susceptible to damage from industrial and fertilizer runoff. Other natural biomes face their own unique threats of rapid environmental degradation.

Anthropogenic biomes are by definition shaped by human activity, but humans can make conscious decisions that can reduce the negative impacts of such activity on the environment. Within cities, for example, humans can counter the effects of vast, heat-retaining stretches of pavement by creating rooftop gardens, rain gardens, and other green spaces.

Emily Alward

FURTHER READING

Butz, Stephen D. *Science of Earth Systems.* 2d ed. Clifton Park, N.Y.: Thomson Delmar Learning, 2008.

Whitfield, Philip, Peter D. Moore, and Barry Cox. *Biomes and Habitats.* New York: Macmillan, 2002.

Woodward, Susan L. *Biomes of Earth: Terrestrial, Aquatic, and Human-Dominated.* Westport, Conn.: Greenwood Press, 2003.

_____. *Introduction to Biomes.* Westport, Conn.: Greenwood Press, 2009.

SEE ALSO: Biodiversity; Biosphere; Biosphere 2; Boreal forests; Climax communities; Coniferous forests; Deciduous forests; Grazing and grasslands; Habitat destruction; Rain forests; Resource depletion; Savannas.

Biopesticides

CATEGORY: Biotechnology and genetic engineering

DEFINITION: Biological agents that are used to control insect and weed pests

SIGNIFICANCE: Biopesticides have significant advantages over commercial pesticides in that they appear to be environmentally safer, given that they do not accumulate in the food chain and they have only slight effects on ecological balances.

Pests are any unwanted animals, plants, or microorganisms. When the environment has no natural resistance to a pest and when no natural antagonists are present, pests can run rampant. For example, the fungus *Endothia parasitica*, which entered New York State in 1904, caused the nearly complete destruction of the American chestnut tree because no natural control was present.

Biopesticides represent the biological, rather than the chemical, control of pests. Many plants and animals are protected from pests by passive means. For example, plant rotation is a traditional method of insect and disease protection in which the host plant is removed for a period long enough to reduce pathogen and pest populations.

Biopesticides have several significant advantages over commercial pesticides. They appear to be ecologically safer than commercial pesticides because they do not accumulate in the food chain. Some biopesticides also provide persistent control, because pests require more than a single mutation to adapt to them and because they can become an integral part of a pest's life cycle. In addition, biopesticides have only slight effects on ecological balances because they do not affect nontarget species. Finally, biopesticides are compatible with other control agents. The major drawbacks to using biopesticides are that, in comparison with chemical pesticides, biopesticides work less efficiently and take more time to kill their targets.

Viruses, bacteria, fungi, protozoa, mites, and flowers have all been used as biopesticides. Viruses have been developed against insect pests such as *Lepidoptera*, *Hymenoptera*, and *Dipterans*. These viruses cause hyperparasitism. Gypsy moths and tent caterpillars, for example, periodically suffer from epidemic virus infestations.

Many saprophytic microorganisms that occur on plant roots and leaves can protect plants against microbial pests. *Bacillus cereus* has been used as an inoculum on soybean seeds to prevent infection by the fungal pathogen *Cercospora*. Some microorganisms used as biopesticides produce antibiotics, but the major mechanism for protection is probably competitive exclusion of a pest from sites on which the pest must grow. For example, *Agrobacterium radiobacter* antagonizes *Agrobacterium tumefaciens*, which causes crown gall disease. Two bacteria—*Bacillus* and *Streptomyces*—added as biopesticides to soil help control the damping off disease of cucumbers, peas, and lettuces caused by *Rhizoctonia solani*. *Bacillus subtilis* added to plant tissue also controls stem rot and wilt rot caused by the fungus *Fusarium*. *Mycobacteria* produce cellulose-degrading enzymes, and their addition to young seedlings helps control fungal infection by *Pythium*, *Rhizoctonia*, and *Fusarium*. *Bacillus* and *Pseudomonas* are bacteria that produce enzymes that dissolve fungal cell walls.

The best examples of microbial insecticides are *Bacillus thuringiensis* (*B.t.*) toxins, which were first used in 1901. They have had widespread commercial production and use since the 1960's and have been successfully tested on 140 insect species, including mosquitoes. *B.t.* produces insecticidal endotoxins during sporulation and also produces exotoxins contained in crystalline parasporal protein bodies. These protein crystals are insoluble in water but readily dissolve in an insect's gut. Once dissolved, the proteolytic en-

Comparison of the Properties of *Bacillus Thuringiensis* and *Bacillus Popilliae* as Microbial Biocontrol Agents

	BACILLUS THURINGIENSIS	BACILLUS POPILLIAE
Pest controlled	Lepidoptera (many)	Coleoptera (few)
Pathogenicity	low	high
Response time	immediate	slow
Formulation	spores and toxin crystals	spores
Production	in vitro	in vivo
Persistence	low	high
Resistance in pests	developing	reported

Source: Data adapted from J. W. Deacon, *Microbial Control of Plant Pests and Diseases* (1983).

zymes paralyze the gut. *Bacillus* spores that have also been consumed germinate and kill the insect. *Bacillus popilliae* is a related bacterium that produces an insecticidal spore that has been used to control Japanese beetles, a pest of corn.

Saprophytic fungi can compete with pathogenic fungi. Among the fungi used as biopesticides are *Gliocladium virens, Trichoderma hamatum, Trichoderma harzianum, Trichoderma viride,* and *Talaromyces flavus.* For example, *Trichoderma* competes with the pathogens *Verticillium* and *Fusarium. Peniophora gigantea* antagonizes the pine pathogen *Heterobasidion annosum* through three mechanisms: It prevents the pathogen from colonizing stumps and traveling down into the root zone, it prevents the pathogen from traveling between infected and uninfected trees along interconnected roots, and it prevents the pathogen from growing up to stump surfaces and sporulating.

Nematodes are pests that interfere with commercial button mushroom (*Agaricus bisporus*) production. Several types of nematode-trapping fungi can be used as biopesticides to trap, kill, and digest the nematode pests. The fungi produce structures such as constricting and nonconstricting rings, sticky appendages, and spores, which attach to the nematodes. The most common nematode-trapping fungi are *Arthrobotrys oligospora, Arthrobotrys conoides, Dactylaria candida,* and *Meria coniospora.*

Protozoa have occasionally been used as biopesticide agents, but their use has suffered because of such difficulties as slow growth and complex culture conditions associated with their commercial production. Predaceous mites are used as a biopesticide to protect cotton from other insect pests such as the boll weevil.

Dalmatian and Persian insect powders contain pyrethrins, which are toxic insecticidal compounds produced in *Chrysanthemum* flowers. Synthetic versions of these naturally occurring compounds are found in products used to control head lice. Molecular genetics has also been used to insert the gene for the *B.t.* toxin into cotton and corn. *B.t.* cotton and *B.t.* corn both express the gene in their roots, which provides them with protection from root worms. Ecologists and environmentalists have expressed concern that constantly exposing pests to the toxin will cause insect resistance to develop rapidly and thus reduce the effectiveness of traditionally applied *B.t.*

Mark Coyne

FURTHER READING

Churchill, B. W. *Biological Control of Weeds with Plant Pathogens.* Edited by R. Charudattan and H. Walker. New York: John Wiley & Sons, 1982.

Deacon, J. W. *Microbial Control of Plant Pests and Diseases.* Research Triangle Park, N.C.: Instrumentation Systems & Automation, 1983.

Metz, Matthew, ed. *Bacillus Thuringiensis: A Cornerstone of Modern Agriculture.* Binghamton, N.Y.: Haworth Press, 2003.

Ohkawa, H., H. Miyagawa, and P. W. Lee, eds. *Pesticide Chemistry: Crop Protection, Public Health, Environmental Safety.* New York: Wiley-VCH, 2007.

SEE ALSO: Agricultural chemicals; *Bacillus thuringiensis*; Biotechnology and genetic engineering; Genetically altered bacteria; Integrated pest management; Organic gardening and farming; Pesticides and herbicides; Sustainable agriculture.

Biophilia

CATEGORY: Philosophy and ethics

DEFINITION: A philosophical hypothesis that addresses the relationship between human beings and other organisms

SIGNIFICANCE: The biophilia hypothesis may be viewed as a bridge between the sciences and the humanities. It has been employed in the sciences, philosophy, and literature regarding investigations of biodiversity, habitat conservation, and ecosystem studies.

The word "biophilia," first coined by the German psychologist Erich Fromm, is a neologism combining two Greek terms (*bio,* meaning "life" and *philias,* meaning "friendly love"); it refers to the natural human inclination to focus on life and lifelike operations. Biophilia became a key concept in environmental studies in 1984 when Harvard entomologist Edward O. Wilson published a book titled *Biophilia.* In this volume, Wilson argues that human sympathy toward living things can be explained in terms of evolutionary developments in human physiology and psychology. Taking this natural affinity for life into account, Wilson asserts that conservation of species and their habitat is valuable to human beings not merely in the instrumental sense of resource management and profit but in an intrinsic sense as well, insofar as the diversity of life is essential to the human psyche.

Because of its focus on life and lifelike processes, biophilia is most often associated with biocentric views in environmental ethics, since it holds that land is valuable insofar as diversity in ecosystems is a requirement for diversity of life. However, the results of biophilia—particularly when viewed in light of coevolutionary data (to which Wilson appeals)—reveal that ecosystem health depends on equilibrium in biodiversity. In this way, biophilia may be said to fall in line with more ecocentric views, such as the deep ecology and land ethic perspectives.

At the core of the biophilia hypothesis is the belief that the value human beings place on life, both human and nonhuman, will increase in proportion to the amount of understanding humans glean from studying other organisms. Without such understanding, it is argued, phobias regarding nature will endure. With this deeper understanding, phobias will be replaced by philias. Taking up the metaphor of historian Leo Marx, Wilson attributes this tension between biophobia and biophilia to a dilemma known as "the machine in the garden," wherein human beings cannot survive in the wild without technologies that have the potential to destroy wilderness. Thus he and other conservationists have embraced biophilia in calling for a more intelligent, limited use of technologies that have the potential to harm habitats.

Because it brings together biology and psychology in these ways, biophilia may be viewed philosophically as a bridge between the sciences and the humanities. While the sciences tend to aim at discovery of fact and the humanities at the transmission of value, both may be needed for the deep understanding of nature that biophilia pursues. In this way, the implications of biophilia may go beyond a mere division of labor between the sciences and the humanities toward a view of *Homo sapiens* as the "poetic species." Accordingly, the love for living things that brings together those human capacities for art and for science is found in human beings' biology and in their relationship to other organisms.

Christopher C. Kirby

FURTHER READING

Kellert, Stephen R., and Edward O. Wilson, eds. *The Biophilia Hypothesis*. Washington, D.C.: Island Press, 1993.

Wilson, Edward O. *Biophilia*. 1984. Reprint. Cambridge, Mass.: Harvard University Press, 2003.

_____. *The Diversity of Life*. New ed. New York: W. W. Norton, 1999.

SEE ALSO: Biocentrism; Biodiversity; Conservation biology; Deep ecology; Ecocentrism; Ecosystems; Environmental ethics.

Biopiracy and bioprospecting

CATEGORY: Resources and resource management

DEFINITION: Extraction of biological resources from areas of biodiversity

SIGNIFICANCE: The practice of extracting biological resources from regions of the world known for their great biological diversity, often carried out by scientists working for corporations or educational institutions, is the subject of ongoing debate. Many environmentalists and indigenous peoples see such resource extraction as a form of exploitation.

With the signing of the United Nations Convention on Biological Diversity at the Earth Summit in Rio de Janeiro, Brazil, in 1992, participatory nations agreed to no longer consider biological resources the "common heritage of mankind" but conceded the rights to distribute such resources to the individual nations that housed them. Around the same time, the terms "biopiracy" and "bioprospecting" began to be used to describe the acquisition of these newly protected resources. The two terms refer to essentially the same thing, the extraction of biological resources from areas of biodiversity, but they have decidedly different tones, the former having been coined by opponents of such activity and the latter being preferred by the practitioners of this type of resource extraction.

Biological resources include whole organisms such as crops or livestock, chemical compounds that can be purified from specific organisms that produce them, and even the genetic material taken from organisms that can then be used to produce desired proteins, usually in conjunction with some form of genetic engineering. These resources hold value in that they can be used to improve agricultural yields, perform certain industrial processes, or serve various pharmaceutical applications. The debate over the appropriate acquisition of these resources led to the split in the terms used to describe the same activity. "Biopiracy" brings to mind a swashbuckler who pillages resources without regard to the victims; "bioprospecting," in contrast, conveys the image of a gold miner staking out a claim and then working it, with no guarantee of ultimate success.

Bioprospecting

The image of the gold-rush prospector is perhaps most appropriate for one particular type of resource collection: the biodiversity-driven, or random-collection, approach. Scientists taking this approach sample large amounts of organisms for a desired chemical activity or genetic attribute without prior knowledge of precisely where to look. The screened organisms are typically plants, microorganisms, insects, or marine invertebrates. This is called the biodiversity-driven approach because mass sampling is best done in areas with wide ranges of different organisms living in close proximity.

Just as modern mining methods include scientific means for discovering deposits of minerals, however, bioprospecting often makes use of prior knowledge to narrow the pool of organisms being tested. This knowledge falls into three main categories: chemotaxonomic, ecological, and ethnobotanical/ethnopharmacological. The use of chemotaxonomic knowledge involves the sampling of organisms that belong to the same taxonomic class as an organism that is already known to have a desired property. An example would be screening a number of bacteria from the class Actinobacteria, the taxonomic group known to be responsible for the production of streptomycin, for antibiotic properties. Ecological knowledge is knowledge that can be gained from field observations of the interactions between particular organisms. Certain plants and animals, for example, produce chemical compounds called secondary metabolites that they use to defend themselves against predator attack. A scientist taking an ecological approach to bioprospecting may detect such interactions and choose species for further testing based on these observations.

The use of ethnobotanical/ethnopharmacological knowledge is the most controversial approach, as it

Brazilian environmental official confiscates an alligator that was illegally hunted in an alleged act of biopiracy. (AP/Wide World Photos)

seeks to capitalize on the medical practices of indigenous peoples who inhabit the areas of interest. Ethnobotanical knowledge focuses on plants that have traditionally been used for healing purposes by indigenous peoples, whereas ethnopharmacological knowledge is broader, encompassing all traditional drugs as well as their biological activities. Using such knowledge, scientists can screen specific organisms for desired properties with a much higher degree of success than is seen with randomly sampled collections.

Biopiracy

Much of the world's biodiversity lies in the tropical regions, often in developing countries that have historically experienced oppression by wealthier nations. It is not surprising, therefore, that indigenous peoples in these regions tend to be wary of the academic institutions and multinational corporations that engage in what these entities may view as simple bioprospecting. Often, indigenous peoples have concerns regarding the entire practice of treating biodiversity as a biological resource, including the patenting of living organisms and profiting from biological materials that for many years previously were exchanged freely among those who reaped the benefits. Even if these concerns are allayed, questions often remain about who should be compensated for traditional knowledge that leads to a "discovery," as well as what would constitute a fair level of compensation.

Although no entity has been prosecuted officially for biopiracy under the Convention on Biological Diversity, many allegations of biopiracy have been made, and a number of planned bioprospecting projects have been abandoned after information about them became public and protests ensued. It may be partially because of such controversies that bioprospecting activities actually decreased in the decades following the convention's signing, as many companies turned away from using natural resources and instead developed synthetic processes, such as combinatorial chemistry to produce lead compounds that could be screened for a desired activity.

James S. Godde

FURTHER READING

Godde, James S. "Genetic Resources." In *Encyclopedia of Global Resources*, edited by Craig W. Allin. Pasadena, Calif.: Salem Press, 2010.

Hamilton, Chris. "Biodiversity, Biopiracy, and Benefits: What Allegations of Biopiracy Tell Us About Intellectual Property." *Developing World Bioethics* 6 (2006): 158-173.

Soejarto, D. D., et al. "Ethnobotany/Ethnopharmacology and Mass Bioprospecting: Issues on Intellectual Property and Benefit-Sharing." *Journal of Ethnopharmacology* 100, nos. 1-2 (August, 2005): 15-22.

Tan, G., C. Gyllenhaal, and D. D. Soejarto. "Biodiversity as a Source of Anticancer Drugs." *Current Drug Targets* 7, no. 3 (March, 2006): 265-277.

Tedlock, Barbara. "Indigenous Heritage and Biopiracy in the Age of Intellectual Property Rights." *Explore: The Journal of Science and Healing* 2, no. 3 (May, 2006): 256-259.

SEE ALSO: Biodiversity; Convention on Biological Diversity; Development gap; Gene patents; Genetically engineered pharmaceuticals; Paclitaxel.

Bioregionalism

CATEGORY: Philosophy and ethics

DEFINITION: Environmental movement that holds that local populations should be self-sustaining based on the resources of their surrounding bioregions and largely self-governing

SIGNIFICANCE: Bioregionalists are concerned with reversing the alienation from the land that is evident in the modern global economy, with protecting the environment from unsustainable human exploitation, and with ensuring that natural resources remain abundant and diverse for future generations.

Bioregionalism rests on two basic existential principles: that humanity is but one component of the "web of life" (that is, the ecosystem) and that humans best pursue their own welfare by living in balance with the local environment. Bioregionalism began as a philosophical offshoot of the environmental movement during the 1970's. Among its early proponents were environmental activist Peter Berg, ecologist Raymond F. Dasmann, and poet Gary Snyder. They particularly worried that technology, consumerism, postmodern culture, and global economics were producing a rootless population estranged from a sense of home and community. Bioregionalism is one of the "relocalization" movements seeking to reverse this trend.

A bioregion is an area defined by natural rather than political boundaries. It most commonly comprises a particular watershed, the plants, animals, climate, hydrology, and ecology of which give it a distinctive character. In 1978 Berg and Dasmann argued that people must concentrate on "living-in-place," which entails satisfying the necessities of life and enjoying life's pleasures as they are available in a particular area, as well as ensuring their long-term availability. The approach is pragmatic: It assumes that people who live long in one place come to know it thoroughly, come to care about it, and want to take care of it.

Bioregionalism is an eclectic movement. It fosters an awareness of local economic and cultural assets by applying lessons from physical geography, ecology, ecosystem management, sustainable agriculture, economics, literature, and political theory. For example, the bioregionalist approach to architecture involves the use of local materials and labor, as well as designs that reflect both regional traditions and the surrounding landscape while satisfying the requirements of present-day life. Bioregionalist political theory stresses participatory democracy and the resolution of social problems through the efforts of voluntary, nonprofit groups. Bioregionalist economics is centered on locally produced goods and services in place of imports and encourages recycling.

Bioregionalism is neither essentially hostile to technology nor divorced from global civilization. It acknowledges that bioregions are parts of larger economic, political, and cultural contexts but emphasizes local resources, both physical and intellectual. Moreover, bioregionalism is not doctrinaire. A variety of philosophical approaches and different kinds of activism—not all of them congruent—participate in or derive from bioregionalism. Among the organizations and movements it has fostered are grassroots environmental efforts to preserve natural features (such as Oregon's Friends of Trees), watershed conservancy, the "locavore" effort (a movement devoted to the promotion of eating only locally produced food), coordinated resource management plans, farmers' markets, Green political parties, community-based alternative energy projects, and educational and research institutions, such as the Bioregional Congress.

Critics contend that bioregionalism is utopian and impractical, at least for large populations. Nonetheless, the movement attracted increasing interest during the late twentieth and early twenty-first centuries among persons interested in establishing self-sufficient economies, in preserving wildlife, and in safeguarding air and water against pollution.

Roger Smith

FURTHER READING

McGinnis, Michael Vincent. *Bioregionalism.* New York: Routledge, 1999.

Sale, Kirkpatrick. *Dwellers in the Land: The Bioregional Vision.* Athens: University of Georgia Press, 2000.

Thayer, Robert L., Jr. *LifePlace: Bioregional Thought and Practice.* Berkeley: University of California Press, 2003.

SEE ALSO: Biocentrism; Biomes; Biophilia; Deep ecology; Eat local movement; Ecocentrism; Environmental ethics; Green movement and Green parties; Land-use planning; Land-use policy; Snyder, Gary.

Bioremediation

CATEGORIES: Ecology and ecosystems; waste and waste management

DEFINITION: Waste management technology that employs naturally occurring plants, microorganisms, and enzymes or genetically engineered organisms to clean contaminated environments

SIGNIFICANCE: The environmentally beneficial and inexpensive waste management strategy of bioremediation enables the degradation of toxic organic and inorganic compounds into environmentally harmless products.

Bioremediation uses biological agents to degrade or decompose toxic environmental compounds into less toxic forms. It is a beneficial and inexpensive waste management strategy that is environmentally friendly in comparison with other remediation technologies. The products of waste decomposition are usually simple inorganic nutrients or gases.

Bioremediation works because, as a general rule, all naturally occurring compounds in the environment are ultimately degraded by biological activity. Toxic and industrial wastes, and even some chemically synthesized compounds that do not naturally occur, can also be decomposed because parts of their structures resemble naturally occurring compounds that are sources of carbon and energy for biological systems. Wastes are either metabolized, in which case they are used as a source of carbon and energy, or cometabolized, in which case they are simply modified so that they lose their toxicity or are bound to organic material in the environment and rendered unavailable.

Bioremediation can occur in situ (at the contaminated site) or ex situ, in which case contaminated soil or water is removed to a treatment facility where bioremediation takes place under controlled environmental conditions. Bioremediation can use organisms that naturally occur at a site, or it can be stimulated through the addition of organisms, sometimes genetically engineered organisms, to the contaminated site in a process known as seeding. The first organism ever patented was a genetically engineered bacterium that had been designed to degrade the components of oil.

TECHNIQUES

Numerous approaches to bioremediation have been developed. One of the simplest is to fertilize a contaminated site to optimal nutrient levels and allow naturally occurring biodegrading populations to increase and become active. Organic contaminants have been mixed with decomposed and partially decomposed organic material and composted as a bioremediation process. In a method analogous to the activated sludge process in wastewater treatment, contaminants are mixed in slurries and aerated to promote their decomposition. It is possible to obtain biosolids that are specially adapted for slurry systems because they have previously been exposed to similar organic wastes.

In situ restoration of contaminated groundwater is often accomplished through the injection of nutrients and oxygen into the aquifers to promote the population and activity of indigenous microorganisms. Trichloroethylene (TCE), for example, is cometabolized by methane-oxidizing bacteria and can be bioremediated through the injection of oxygen and methane into contaminated aquifers to stimulate the activity of these bacteria. Nitrate-contaminated aquifers have been successfully treated through the pumping of readily available carbonlike methanol or ethanol into the aquifers to stimulate denitrifying bacteria, which subsequently convert the nitrate to harmless nitrogen gas.

Bioreactors have been used in which the contaminant is mixed with a solid carrier, or the organisms are immobilized to a solid surface and continuously exposed to the contaminant. This has been used with both bacteria and fungi. For example, *Phanerochaete chysosporium*, which produces an extracellular peroxidase and hydrogen peroxide (H_2O_2), has been used to cleave various organic contaminants such as dichloro-diphenyl-trichloroethane (DDT) in bioreactors.

Highly chlorinated organic contaminants such as TCE and polychlorinated biphenyls (PCBs) resist degradation aerobically, but the contaminants can be dechlorinated by anaerobic bacteria, which decreases their toxicity and makes them easier to decompose. High concentrations of PCBs in the Hudson River in New York have been dechlorinated to less toxic forms by anaerobic bacteria. Methanogens—anaerobic bacteria that produce methane—have been observed to dechlorinate TCE in anaerobic bioreactors.

One of the problems with some wastes is that they are mixed with radioactive materials that are highly toxic to living organisms. One solution to this problem has been the genetic engineering of radiation-resistant bacteria so that they also have the ability to bioremediate. For example, *Deinococcus radiodurans*, a bacterium that can survive in nuclear reactors, has been genetically engineered to contain genes for the metabolism of toluene, which will enable it to be used in the bioremediation of radiation- and organic waste-contaminated sites.

PHYTOREMEDIATION

Phytoremediation is a special type of bioremediation in which plants—grasses, shrubs, trees, and algae—are used to biodegrade or immobilize environmental contaminants, usually metals. Types of phytoremediation include phytoextraction, in which the contaminant is extracted from soil by plant roots; phytostabilization, in which the contaminant is immobilized in the vicinity of plant roots; phytostimulation, in which the plant root exudates stimulate rhizosphere microorganisms that bioremediate the contaminant; phytovolatilization, in which the plant helps to volatilize the contaminant; and phytotransformation, in which the plant root and its enzymes actively transform the contaminant. For example, horseradish peroxidase is a plant enzyme that is used to oxidize and polymerize organic contaminants. The polymerized contaminants become insoluble and relatively unavailable.

Plants such as Indian mustard (*Brassica juncea*) and loco weed (*Astragalus*) are heavy metal accumulators and remove selenium and lead from soil. The aboveground plant parts are harvested to dispose of the metals. Algae are used to accumulate dissolved selenium in some treatments. Poplar trees have even been genetically engineered to contain a bacterial methyl reductase that lets them methylate and volatilize arsenic, mercury, and selenium absorbed by their roots.

EXAMPLES

A 1992 U.S. Environmental Protection Agency (EPA) survey indicated that of 132 well-documented bioremediation studies, 75 involved petroleum or related compounds, 13 involved wood preservatives such as creosote, 7 involved agricultural chemicals, 5 examined tars, 4 treated munitions such as trinitrotoluene (TNT), and the rest involved miscellaneous compounds. As this list suggests, bioremediation of oil spills has been the single best example of successful bioremediation in practice.

In March, 1989, the *Exxon Valdez* oil tanker spilled millions of gallons of crude oil in Prince William Sound, Alaska. On many beaches, the EPA authorized the use of simple bioremediation techniques, such as stimulating the growth of indigenous oil-degrading bacteria by adding common inorganic fertilizers. Beaches cleaned by this method did as well as beaches cleaned by mechanical methods. In another instance of successful bioremediation, selenium-contaminated soil in the Kesterson National Wildlife Refuge in California was partially decontaminated through the method of supplying indigenous fungi with organic substrates such as casein and waste orange peels. This promoted as much as 60 percent selenium volatilization in less than two months.

Mark Coyne

FURTHER READING

Alexander, Martin. *Biodegradation and Bioremediation.* 2d ed. San Diego, Calif.: Academic Press, 1999.

Atlas, Ronald M., and Jim Philp, eds. *Bioremediation: Applied Microbial Solutions for Real-World Environmental Cleanup.* Washington, D.C.: ASM Press, 2005.

Fingas, Merv. *The Basics of Oil Spill Cleanup.* 2d ed. Boca Raton, Fla.: CRC Press, 2001.

Frankenberger, William, and Sally Benson, eds. *Selenium in the Environment.* Boca Raton, Fla.: CRC Press, 1994.

Singh, V. P., and R. D. Stapleton, Jr., eds. *Biotransformations: Bioremediation Technology for Health and Environmental Protection.* New York: Elsevier Science, 2002.

Skipper, H. D., and R. F. Turco, eds. *Bioremediation: Science and Applications.* Madison, Wis.: American Society of Agronomy, 1995.

SEE ALSO: Biotechnology and genetic engineering; Detoxification; *Diamond v. Chakrabarty*; Genetically altered bacteria; Genetically modified organisms; Waste management.

Biosphere

CATEGORY: Ecology and ecosystems

DEFINITION: The zone within which all life on earth exists

SIGNIFICANCE: The conceptualization of all life on earth as part of one large, integrated system, the biosphere, helps to encourage understanding of the interrelatedness of all ecosystems and biotic communities.

The concept of the biosphere was introduced in the nineteenth century by Austrian geologist Eduard Suess. The biosphere is the zone, approximately 20 kilometers (12 miles) thick, that extends from the floors of the earth's oceans to the tops of the mountains, within which all life on the planet exists. It is thought to be more than 3.5 billion years old, and it supports nearly one dozen biomes, regions of similar climatic conditions within which distinct biotic communities reside.

Compounds of hydrogen, oxygen, carbon, nitrogen, potassium, and sulfur are cycled among the four major spheres—biosphere, lithosphere, hydrosphere, and atmosphere—to make the materials that are essential to the existence of life. The most critical of these compounds is water, and its movement between the spheres is called the water cycle. Dissolved water in the atmosphere condenses to form clouds, rain, and snow. The annual precipitation for a region is one of the major controlling factors in determining the terrestrial biome that can exist. The water cycle follows the precipitation through various paths leading to the formation of lakes and rivers. These flowing waters interact with the lithosphere to dissolve chemicals as they flow to the oceans, where about one-half of the biomes on earth occur. Evaporation of water from the oceans resupplies the vast majority of the moisture existing in the atmosphere. This cycle supplies water continuously for the needs of both the terrestrial and the oceanic biomes.

The biosphere is also dependent on the energy that is transferred from the various spheres. The incoming solar energy is the basis for all life. Light enters the life cycle as an essential ingredient in the photosynthesis reaction. Plants take in carbon dioxide, water, and light energy, which is converted into chemical energy in the form of sugar, with oxygen generated as a by-product. Most animal life reverses this process during the respiration reaction, where chemical energy is released to do work by the oxidation of sugar to produce carbon dioxide and water.

The incoming solar energy also has a dramatic interaction with the water cycle and the worldwide distribution of biomes. Because of the earth's curvature, the equatorial regions receive a greater amount of solar heat than do the polar regions. Convective movements in the atmosphere (such as winds, high- and low-pressure systems, and weather fronts) and the hydrosphere (such as water currents) are generated during the redistribution of this heat. The weather patterns and general climates of earth are responses to these energy shifts. The seven types of climates are defined by mean annual temperature and mean annual precipitation, and there is a strong correspondence between the climate at a given location and the biome that will flourish.

DESERT BIOME

The major deserts of the world are located between 20 to 30 degrees latitude north and south of the equator. The annual precipitation in a desert biome is less than 25 centimeters (10 inches) per year. Deserts are located in northern and southwestern Africa, parts of the Middle East and Asia, Australia, the southwestern United States, and northern Mexico.

Deserts are characterized by life that is unique in its ability to capture and conserve water. Deserts show the greatest extreme in temperature fluctuations of all biomes: Daytime temperatures can exceed 49 degrees Celsius (120 degrees Fahrenheit), and night temperatures can drop to 0 degrees Celsius (32 degrees Fahrenheit). Most of the animals that live in desert biomes are active at night and retreat to underground burrows or crevices during the day to escape the heat. The water cycle in deserts rarely provides surface water, so plant life usually finds water through a wide distribution of shallow roots to capture the near-surface infiltration or a deep taproot system that finds

groundwater located below the surface of dry stream-beds. The plant life is characterized by scattered thorny bushes, shrubs, and occasional cacti. Animal life consists of an abundance of reptiles (mostly lizards and snakes), rodents, birds (many predatory types such as owls and hawks), and a wide variety of insects.

Deserts and semideserts cover approximately one-third of the land surface on earth. They continue to grow in size because of human influences such as deforestation and overgrazing.

GRASSLAND BIOME

Grasslands are found in a wide belt of latitudes higher than those in which desert biomes exist. Large grassland regions occur in central North America, central Russia and Siberia, subequatorial Africa and South America, northern India, and Australia. This biome flourishes in moderately dry conditions, having an annual rainfall between 25 and 150 centimeters (10 and 60 inches). Precipitation and solar heating are unevenly divided throughout the year, providing a wet, warm growing season and a cool, dry dormant season.

The animal life in grassland regions is characterized by large grazing mammals, such as wild horses, bison, antelopes, giraffes, zebras, and rhinoceroses, as well as smaller herbivores, such as rabbits, prairie dogs, mongooses, kangaroos, and warthogs. This abundance of herbivores allows for a large development of secondary and tertiary consumers in the food chain, such as lions, leopards, cheetahs, wolves, and coyotes. Grasslands have rich soils that provide the fertile growing conditions for a wide variety of tall and short grasses. Within a single square meter of this healthy soil, several hundred thousand living organisms can be found, from microbes to insects, beetles, and worms. The profusion of these smaller life-forms fosters an abundance of small birds.

Grasslands have been environmentally stressed as humans have converted them to farmland because of their rich soils and to rangeland because of the grass supply. It is estimated that only 25 percent of the world's original grasslands remain undisturbed by human development. Worldwide, overgrazing and mismanagement of rangeland have caused large tracts of fertile grassland to become desert or semidesert.

TROPICAL RAIN-FOREST BIOME

Rain forests receive heavy rainfall almost daily, with an annual average of more than 240 centimeters (95

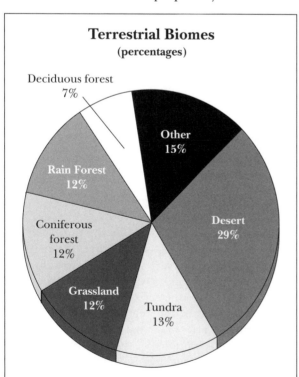

Terrestrial Biomes
(percentages)

Deciduous forest 7%
Other 15%
Rain Forest 12%
Coniferous forest 12%
Desert 29%
Grassland 12%
Tundra 13%

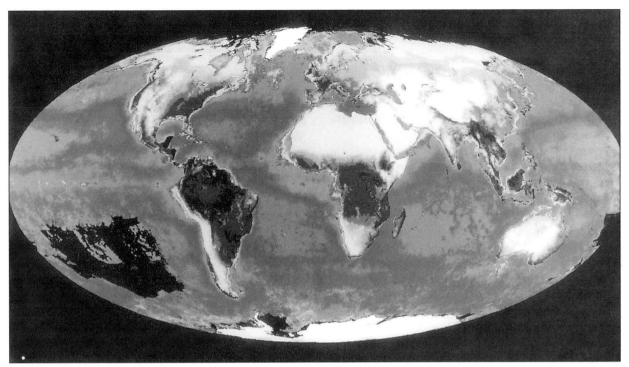

Composite image of the earth's biosphere shows the planet's heaviest vegetative biomass in the dark sections, known to be rain forests. (NASA)

inches). Temperatures in these areas are fairly constant from day to day and season to season, with an annual mean value of about 28 degrees Celsius (82 degrees Fahrenheit). The combination of plentiful rain and high temperatures causes high humidity, allowing some plants to utilize the atmosphere for their water supply via "air roots." This biome is also unique because the chemical nutrients needed to sustain life within it are almost entirely contained in the lush vegetation of the biosphere itself and not in the upper layers of soil of the lithosphere. The soils are thin and poor in nutrients.

The tropical rain forests contain a wider diversity of plant and animal species per unit area than any other biome. It is estimated that nearly two-thirds of all the plants and insects found on earth are contained in tropical rain forests. This enormous biodiversity is accommodated in part because each form of plant or animal occupies a specialized niche based on its ability to thrive with a particular level of sunlight that corresponds to a given height above the forest floor within the forest canopy. Numerous exotic insects, amphibians, reptiles, birds, and small mammals can coexist within a single canopy level. The plants growing in tropical rain forests currently provide ingredients found in 25 percent of the world's prescription

and nonprescription drugs. It is estimated that at least three thousand tropical plants contain cancer-fighting chemicals.

Tropical rain forests are being destroyed at a fast rate for farmland, timber operations, mining, and grazing. It has been estimated that most of the animal and plant diversity of the rain forests will be lost by the year 2050. Further, if the high rate of destruction continues, the release of carbon dioxide into the atmosphere from the burning of biomass and fossil fuels may no longer be offset by plant consumption, and the balance of the carbon cycle will shift toward a higher concentration of carbon dioxide in the atmosphere.

TEMPERATE FOREST BIOME

Temperate forests exist in areas where temperatures change dramatically during the four distinct seasons. Temperatures fall below freezing during the winter, with warmer, more humid conditions during the summer. Rainfall averages between 75 and 200 centimeters (30 and 80 inches) per year. This biome is often divided into forests with broad-leaved deciduous (leaf-shedding) trees and those with coniferous (cone-bearing) trees. Deciduous forests develop in regions with higher precipitation values, whereas the needle-like evergreen leaves of conifers have scales and thick, waxy coatings that allow them to flourish at the lower end of the precipitation range.

Deciduous forests develop more solid canopies with widely branching trees such as elm, oak, maple, ash, beech, and other hardwood varieties. The forest floor often contains an abundance of ferns, shrubs, and mosses. Coniferous forests are usually dominated by pine, spruce, fir, cedar, and hemlock trees. Some deciduous trees, such as aspen and birch, often occur with the conifers. The coniferous forest floor is so acidic from decomposing evergreen needles that often only lichens and mosses can grow.

Temperate forests are home to diverse animal life. Common mammals of deciduous forests include squirrels, chipmunks, porcupines, raccoons, opossums, deer, foxes, black bears, and mice. Snakes, toads, frogs, and salamanders exist alongside a large bird population of thrushes, warblers, woodpeckers, owls, and hawks. The larger animals of the coniferous forests include moose, elk, wolves, lynx, grouse, jays, and migratory birds.

Forests have traditionally been viewed as a limitless timber resource by the lumber industry. The vast forests of Europe were cleared one thousand years ago. This liquidation of temperate forest continues in Siberia, where 25 percent of the world's timber reserves exist.

TUNDRA BIOME

Tundras occur in areas near the Arctic ice cap and extend southward across the far northern parts of North America, Europe, and Asia. During the majority of the year, these largely treeless plains are covered with ice and snow and are battered by bitterly cold winds. Tundras are covered with thick mats of mosses, lichens, and sedges (grasslike plants). Because the winters are long and dark, tundra vegetation grows during the three months of summer, when there is almost constant sunlight.

Bogs, marshes, and ponds are common on the summer landscape because permafrost, a thick layer of ice that remains beneath the soil all year long, prevents drainage of melted waters. These wet areas provide perfect breeding grounds for mosquitoes, deerflies, and blackflies during the brief summer. These insects in turn serve as a source of food for migrating birds. Larger mammals, such as caribou, reindeer, musk ox, and mountain sheep, migrate in and out of the tundra. Some animals, such as lemmings,

arctic hares, grizzly bears, and snowy owls, can be found in the tundra during all times of the year.

The tundra is the earth's most fragile terrestrial biome. Vegetation disturbed by human activity can take decades to replenish itself. Roads and pipelines must be constructed on bedrock or layers of added gravel; otherwise, they will melt the upper layers of the permafrost.

OCEANIC BIOMES

Oceans cover 70 percent of the earth's surface and contain 97.6 percent of the water of the hydrosphere. They play a primary role in regulating the earth's distribution of heat, and they are central to the water cycle. Oceans are instrumental in the survival of all life on earth. In addition, oceans house more than 250,000 species of marine plants and animals that occur as six common biomes.

Coral reefs are coastal biomes that develop on continental shelves in regions of clear, tropical waters. Collectively, coastal biomes make up only 10 percent of the world's ocean area but contain 90 percent of all ocean species. These are the regions where most commercial fishing is done. The vast open oceans contain only 10 percent of all oceanic species. Vegetation is mostly limited to free-floating plankton. Exotic bottom fauna exist on deep hydrothermal vents. Animal life includes whales, dolphins, tuna, sharks, flying fish, and squids.

A number of transitional biomes also exist at the ocean-land interface. The intertidal biome can be composed of sandy beaches or more rocky zones that are covered by water only during periods of high tide. A variety of crustaceans and mollusks are found on wet sandy beaches, whereas rock tidal pools contain kelp, Irish moss, and rockweed, all of which compete for space with snails, barnacles, sea urchins, and starfish.

Toby Stewart and Dion Stewart

FURTHER READING

Butz, Stephen D. *Science of Earth Systems.* 2d ed. Clifton Park, N.Y.: Thomson Delmar Learning, 2008.

Ehrlich, Paul R., and Anne H. Ehrlich. *Extinction: The Causes and Consequences of the Disappearance of Species.* New York: Ballantine Books, 1981.

Krebs, Robert E. "Biosphere: Envelope of Life." In *The Basics of Earth Science.* Westport, Conn.: Greenwood Press, 2003.

McNeely, Jeffrey A., et al. *Conserving the World's Biological Diversity.* Washington, D.C.: Island Press, 1990.

Smil, Vaclav. *The Earth's Biosphere: Evolution, Dynamics, and Change.* Cambridge, Mass.: MIT Press, 2002.

Weiner, Jonathan. *The Next One Hundred Years.* New York: Bantam Books, 1990.

Wilson, E. O., ed. *Biodiversity.* Washington, D.C.: National Academy Press, 1990.

Woodward, Susan L. *Introduction to Biomes.* Westport, Conn.: Greenwood Press, 2009.

SEE ALSO: Balance of nature; Biodiversity; Biosphere reserves; Ecosystems; Gaia hypothesis; Global Biodiversity Assessment.

Biosphere reserves

CATEGORY: Preservation and wilderness issues

DEFINITION: UNESCO-designated sites where the preservation of natural resources is integrated with research and the sustainable management of those resources

SIGNIFICANCE: The worldwide network of biosphere reserves is the only international network of protected areas that also emphasizes sustainable development and wise use of natural resources; thus these sites enable the examination and testing of the objective of integrating conservation and development.

Through discussions that started in 1970, the United Nations Educational, Scientific, and Cultural Organization (UNESCO) initiated the Man and the Biosphere Programme to establish sites where the preservation of natural resources would be integrated with research and the sustainable management of those resources. The first biosphere reserve was designated in 1976, and by May, 2009, 553 such sites existed across 107 countries. In 1995 UNESCO convened a conference in Spain and developed the Seville Strategy for Biosphere Reserves, which was designed to strengthen the international network and encourage the use of the sites for research, monitoring, education, and training. In the early years of the program, preservation was stressed. The adoption of the Seville Strategy by UNESCO emphasized the role of people in the use of their natural resources.

Pelicans take off from Lake Razim in the Danube Delta in Romania. The delta has been designated a UNESCO biosphere reserve. (AP/Wide World Photos)

Each biosphere reserve contains a legally protected core area where there has been minimal disturbance by people. Only uses that are compatible with the preservation of biological diversity are permitted in the protected core. Surrounding the core is a managed-use or buffer zone; research and environmental education are examples of activities suitable for the buffer zone. Surrounding the buffer zone is a zone of cooperation or transition zone. The boundaries of the transition zone are loosely defined and often include local towns and communities. Economic activities such as farming, logging, mining, and recreation occur within the transition zone and are not restricted by the biosphere reserve.

Only the boundaries of the core area are legally defined. The designation of a biosphere reserve does not alter the legal ownership of the land or water that is included within its zones. UNESCO does not have jurisdiction over any nation's biosphere reserves. In many cases the areas within reserves reflect a mosaic of landownership, including federal, state, local, and private ownership. Even the core area may be privately owned, as long as it is managed for its preservation.

In the United States, the core areas of some biosphere reserves are within national parks, such as Glacier and Yellowstone, whereas other biosphere reserves are composed of clusters of core areas, such as the ten units within the California Coastal Range Biosphere Reserve. The management and administration of a biosphere reserve often involves a number of interested citizens, government agencies, and owners.

The worldwide network of biosphere reserves represents the only international network of protected areas that also emphasizes sustainable development and wise use of natural resources. Hence they are sites where the objective of integrating conservation and development can be examined, demonstrated, and tested. Research at these sites serves to solve practical problems in resource management.

William R. Teska

FURTHER READING

Hanna, Kevin S., Douglas A. Clark, and D. Scott Slocombe, eds. *Transforming Parks and Protected Areas: Policy and Governance in a Changing World.* New York: Routledge, 2008.

Sourd, Christine. *Explaining Biosphere Reserves.* Paris: UNESCO, 2004.

SEE ALSO: Biodiversity; Biosphere; Conservation; Nature preservation policy; Preservation; United Nations Educational, Scientific, and Cultural Organization.

Biosphere 2

CATEGORY: Ecology and ecosystems

IDENTIFICATION: An environmental research facility in Arizona originally constructed as a self-sustaining, closed ecological system

SIGNIFICANCE: Two Biosphere 2 missions during the 1990's explored the viability of maintaining and inhabiting a closed, human-made ecological system for prolonged periods. In the period since such colonization experiments ended, the facility has been a useful site for university-run environmental research.

Located on a 16-hectare (40-acre) campus near Oracle, Arizona, Biosphere 2 is a 1.27-hectare (3.14-acre) glass-and-steel structure that resembles a series of connected greenhouses. A basement "technosphere" of electrical, mechanical, and plumbing systems maintains Biosphere 2's temperature, humidity, and airborne particulate levels. Space Biospheres and Ventures, a private, for-profit firm founded by Biosphere 2 inventor John Allen and bankrolled by philanthropist Edward Bass, began constructing the facility in 1986 as a prototype for a self-sustaining space colony. Biosphere 2 was designed to mimic the earth—Biosphere 1—by being closed to exchanges of material or organisms from the outside environment but open to energy and information exchanges.

On September 26, 1991, Biosphere 2 was sealed with a team of four men and four women inside; the mission ended exactly two years later. A second, seven-person mission was conducted between March 6 and September 6, 1994. One month into the second mission, Bass dissolved his business partnership with Allen, citing project mismanagement. Federal marshals served a restraining order that ousted on-site management, and Bass's Decisions Investments Corporation took over the property.

Biosphere 2 was intended as a century-long experiment to determine whether it is possible to maintain a self-sustaining human colony in a hostile environment. Crew members had to grow their own food, recycle the water they drank, and produce the oxygen they breathed. They were responsible for the health of the plant and animal species inside the sealed structure, and they had to maintain the complex apparatus that kept Biosphere 2 functioning.

The biomes of earth were represented within Biosphere 2 by a tropical rain forest, a desert, a savanna grassland, mangrove wetlands, and an ocean with a coral reef, plus an area of intensive agriculture. It proved to be impossible to keep the biomes separate: The desert, for example, was too wet. Also, the structure shut out 40 percent of the sun's light, and the ceiling was sometimes so hot that the treetops burned. Nineteen out of twenty-five vertebrate species inside the structure died, while others (such as ants and cockroaches) swarmed out of control.

Among the difficulties encountered during the first two-year experiment was a lack of oxygen: The integrity of the sealed environment had to be broken when it became apparent that the organically rich soil had consumed more oxygen than predicted. The biospherians produced about 80 percent of the food they needed but suffered crop failures because of unanticipated depredation by mites and other insects. One crew member had to be taken out of Biosphere 2 for emergency surgery on a wounded finger. Upon her return, she brought some supplies with her. Such incidents drew criticism that the giant terrarium was not as self-sufficient as intended, but the project directors maintained that they had never expected perfection in this first model and that finding flaws in the design was precisely the purpose of the experiment.

Inevitably, crew members also experienced interpersonal friction. While the professionalism of the biospherians saw them through their differences, conflicts nevertheless had the potential to cause serious disruption of the mission. By the end of their two-year stay, three of the biospherians were taking part in therapy sessions by telephone.

Biosphere 2 attracted nationwide attention. The structure's thousands of windows allowed spectators to view almost everything going on inside. This public exposure made Biosphere 2 an excellent tool for generating interest in ecological concerns as well as in the human potential in space.

Columbia University assumed management of Biosphere 2 in 1996 and operated it as a research and education facility, conducting ecosystems studies focused on the effects of elevated carbon dioxide concentrations on plants, fish, and coral reefs. Co-

lumbia ended its association with the facility in 2003. In 2007 the Biosphere 2 campus and surrounding property were sold to a development company, which leased the core complex to the University of Arizona. The university continues use of Biosphere 2 as a center for environmental research and education. Within the facility, scientists can manipulate environmental variables and obtain high-resolution measurements, making Biosphere 2 a unique laboratory for large-scale experiments in earth systems and environmental change.

Robert B. Bechtel
Updated by Karen N. Kähler

FURTHER READING

Marino, B. D. V., and Howard T. Odum, eds. *Biosphere 2: Research Past and Present.* Amsterdam: Elsevier Science, 1999.
Poynter, Jane. *The Human Experiment: Two Years and Twenty Minutes Inside Biosphere 2.* New York: Thunder's Mouth Press, 2006.
Reider, Rebecca. *Dreaming the Biosphere: The Theater of All Possibilities.* Albuquerque: University of New Mexico Press, 2009.

SEE ALSO: Biomes; Biosphere; Ecosystems.

Biotechnology and genetic engineering

CATEGORY: Biotechnology and genetic engineering

DEFINITIONS: Biotechnology is the use of living organisms, or substances obtained from such organisms, to produce products or processes of value to humankind; genetic engineering is the manipulation of deoxyribonucleic acid (DNA) and the transfer of genes or gene components from one species to another

SIGNIFICANCE: Biotechnology has made tremendous advances possible in human and veterinary medicine, agriculture, food production, and other fields. However, debates continue regarding the potential of biotechnology, in particular genetic engineering, to produce organisms that may disrupt ecosystems, negatively affect human health, or be used in ethically inappropriate ways.

The term "biotechnology" is relatively new, but the practice of biotechnology is as old as civilization. Civilization did not evolve until humans learned to produce food crops and domestic livestock through the controlled breeding of selected plants and animals. The pace of modifying organisms accelerated during the twentieth century. Through carefully controlled breeding programs, plant architecture and fruit characteristics of crops were modified to facilitate mechanical harvesting. Plants were developed to produce specific drugs or spices, and microorganisms were selected to produce antibiotics such as penicillin and other useful medicinal and food products.

The ability to utilize artificial media to propagate plants led to the development of a technology called tissue culture. In some plant tissue culture, the tissue is treated with the proper plant hormones to produce masses of undifferentiated cells called callus tissue, which can also be separated into single cells to establish a cell suspension culture. Specific drugs or other chemicals can be produced with callus tissue and cell suspensions, or this tissue can be used to regenerate entire plants. Tissue culture technology is used as a propagation tool in commercial-scale plant production.

Numerous advances have also occurred in animal biotechnology. Artificial insemination, the process in which semen is collected from the male animal and deposited into the female reproductive tract through artificial techniques rather than natural mating, emerged as a practical procedure roughly a century ago, although as early as 1784 Italian biologist Lazzaro Spallanzani successfully inseminated a dog. Males in species such as cattle can sire hundreds of thousands of offspring through artificial insemination, whereas they could sire only fifty or fewer through natural means.

Embryo transfer is a technique used in humans to facilitate conception after in vitro fertilization, a procedure in which eggs are surgically removed from the ovaries and manually combined with sperm in a laboratory. Once fertilization and cell division are confirmed, the embryos are placed in the uterus. The eggs may be supplied by a woman who is unable to conceive naturally but who can carry a child to term. They may also be provided by an egg donor to a woman who cannot otherwise get pregnant; or a woman who cannot carry a child to term may supply eggs to be fertilized in vitro and implanted in a surrogate mother. Superovulation is the process in which females that are to provide eggs are injected with hormones to stimulate increased egg production. Embryo splitting is the mechanical division of an embryo into identical twins, quadruplets, sextuplets, and so on. Both superovulation and embryo splitting have made routine embryo transfers possible. In livestock, embryo transfer technology is used to combine the sperm from a superior male animal and several eggs, each of which can then be split into several offspring, from a superior female. The resulting embryos can then be transferred to the reproductive tracts of inferior surrogate females.

RECOMBINANT DNA TECHNOLOGY

Biotechnological advances have enabled scientists to tap into the world gene pool. This technology has great potential, and its full magnitude is far from being fully realized. Theoretically, it is possible to transfer one or more genes or gene segments from any organism in the world into any other organism. Because genes ultimately control how an organism functions, gene transfer can have a dramatic impact on agricultural resources and human health.

Research has provided the means by which genes can be identified and manipulated at the molecular and cellular levels. This identification and manipulation depend primarily on recombinant DNA technology. In concept, recombinant DNA methodology is fairly easy to comprehend, but in practice it is rather complex. The genes in all living cells are very similar in that they are all composed of the same chemical, deoxyribonucleic acid, or DNA. The DNA of all cells, whether from bacteria, plants, lower animals, or humans, is very similar, and when DNA from a foreign species is transferred into a different cell, it functions exactly as the native DNA functions; that is, it codes for protein.

The simplest protocol for this transfer involves the use of a vector, usually a piece of circular DNA called a plasmid, which is removed from a microorganism such as a bacterium and cut open by an enzyme called a restriction endonuclease or restriction enzyme. A section of DNA from the donor cell that contains a previously identified gene of interest is cut out from the donor cell DNA by the same restriction endonuclease. The section of donor cell DNA with the gene of interest is then combined with the open plasmid DNA, and the plasmid closes with the new gene as part of its structure. The recombinant plasmid (DNA from two sources) is placed back into the bacterium, where it will replicate and code for protein just as it did in the donor cell. The bacterium can be cultured and the gene product (protein) harvested, or the bacterium can be used as a vector to transfer the gene to another species, where it will also be expressed. This transfer of genes, and therefore of inherited traits, between different species has revolutionized biotechnology and provides the potential for genetic changes in plants and animals that have not yet been envisioned.

BIOTECHNOLOGY AND AGRICULTURE

Biotechnology has had a tremendous impact on agriculture. Traditional breeding programs may be too slow to keep pace with the needs of a rapidly expanding human population. Biotechnology provides a means of developing higher-yielding crops in one-third of the time it takes to develop them though traditional plant breeding programs because the genes for desired characteristics can be inserted directly into a plant without having to go through several generations to establish the trait. Also, there is often a need or desire to diversify agricultural production in a given area, but soil or climate conditions may severely limit the amount of diversification that can take place. Biotechnology can provide the tools to help solve this problem: Crops with high cash value can be developed to grow in areas that would not support unmodified versions of such crops. In addition, biotechnology can be used to increase the cash value of crops, as plants can be developed that can produce new and novel products such as antibiotics, hormones, and other pharmaceuticals.

As public pressure has grown for crop production to be more friendly to the environment, biotechnology has been touted as an important tool for the development of a long-term, sustainable, environmentally friendly agricultural system. Biotechnology is already being used to develop crops with improved resistance to pests. For example, a gene from the bacterium *Bacillus thuringiensis* (*B.t.*) codes for an insecticidal protein that kills insects but is harmless to other organisms. When this gene is transferred from the bacterium to a plant, insect larvae are killed if they eat from the leaves or roots of the plant. A number of *B.t.* plants have been developed, including cotton and potatoes. Crop varieties engineered for improved pest resistance have the potential to reduce reliance on pesticides; however, insect pests have developed resistance to some of these crops.

Biotechnology also plays an important role in the livestock industry. Bovine somatotropin, a hormone that stimulates growth in cattle, is harvested from recombinant bacteria and injected in dairy cattle to enhance milk production. However, questions have arisen as to whether overstimulating milk production is humane or healthy for cows, and fears regarding the health implications for humans consuming milk that contains bovine hormone residues have made many people seek organic dairy products free from artificial hormones. Some countries do not allow the use of these hormones in milk intended for human consumption.

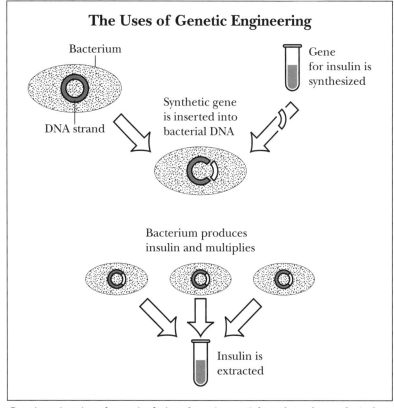

The Uses of Genetic Engineering

Bacterium

Gene for insulin is synthesized

DNA strand

Synthetic gene is inserted into bacterial DNA

Bacterium produces insulin and multiplies

Insulin is extracted

Genetic engineering, the manipulation of genetic material, can be used to synthesize large quantities of drugs or hormones, such as insulin.

Researchers are exploring the possibilities of genetically engineering animals that can resist disease or produce novel and interesting products such as pharmaceuticals. The cloning of Dolly the sheep in Scotland in 1996 opened a whole new avenue in the use of biotechnology for livestock production. The use of cloning technology in conjunction with surrogate mothers provides the means to produce a whole herd of genetically superior animals in a short period of time. However, reproductive cloning is expensive, its success rate is low, and many cloned animals have been found to be unhealthy and short-lived.

BIOTECHNOLOGY AND MEDICINE

DNA technology also has a direct impact on human health and is used to manufacture a variety of gene products that are utilized in the clinical treatment of diseases. Several human hormones produced by this methodology are already in use. The hormone insulin, for instance, which is used to treat insulin-dependent diabetics, was the first major success in using a product of recombinant technology. Recombinant DNA-produced insulin has been used to treat diabetic patients since 1982. Genetic engineering has also been used to synthesize protropin, a human growth hormone (HGH) employed in the treatment of growth failure conditions such as hyposomatotropism. Without treatment with HGH, people suffering from these conditions do not produce enough growth hormone to achieve a typical adult height.

Somatostatin, another pituitary hormone, has also been produced through recombinant DNA techniques. This hormone controls the release of insulin and HGH. Small proteins called interferons normally produced by cells to combat viral infections have been produced using recombinant DNA methodology, as have some vaccines against viral diseases. Recombivax HB, the first of these vaccines, is used in vaccinating against hepatitis B, an incurable and sometimes fatal liver disease.

The potential for the future application of gene therapy has also been enhanced by advances in biotechnology. Among the forms of gene therapy currently being considered are gene surgery, in which a mutant gene that may or may not be replaced by its normal counterpart is excised from the DNA; gene repair, in which defective DNA is repaired within the cell to restore the genetic code; and gene insertion, in which a normal gene complement is inserted in cells that carry a defective gene.

Gene surgery and gene repair techniques are extremely complex and remain in the experimental stages. Gene insertion can potentially be done in germ-line cells such as the egg or sperm, the fertilized ovum or zygote, the fetus, or the somatic cells (nonreproductive cells) of children or adults. Although zygote therapy holds the most promise, as this technique could eliminate genetic disease, gene insertion in zygotes also represents a means by which traits such as strength or intelligence might be enhanced and the genetic traits of future generations artificially selected, a possibility that raises a host of ethical questions. Germ-line genetic modification has been performed successfully in laboratory animals, but unwanted mutations with serious or lethal consequences have also sometimes resulted.

Gene insertion into somatic cells does not make changes that are passed on to subsequent generations, so it does not present the ethical dilemma that germ-line manipulation does. In this technique, a gene or gene segment is inserted into specific organs or tissues as a treatment for an existing condition. In human clinical trials, somatic gene therapy has shown success in treating advanced melanoma, myeloid disorders, inherited childhood blindness, and severe combined immunodeficiency. However, the carrier molecules used to deliver the therapeutic gene to the target cells have the potential to provoke a serious or fatal immune response in the patient.

ENVIRONMENTAL ISSUES

The potential benefits of biotechnology for human health, agriculture, and the environment are accompanied by potential drawbacks. Since the first recombinant DNA experiments in 1973, numerous social, ethical, and scientific questions have been raised about the possible detrimental effects of genetically engineered organisms on public health and the environment. The major environmental concerns are related to containment, or how to prevent genetically engineered organisms from escaping into the environment.

In the mid-1970's U.S. scientists invoked a self-imposed moratorium on genetic engineering experiments until the government could establish committees to develop safety guidelines that would apply to all recombinant DNA experimentation in the United States. This resulted in the formulation of guidelines specifying the degree of containment required for various types of genetic engineering experiments.

Two types of containment, biological and physical, are addressed by the guidelines. "Physical containment" refers to the methods required to prevent an engineered organism from escaping from the laboratory; "biological containment" refers to the techniques used to ensure that an engineered organism cannot survive outside the laboratory. The guidelines associated with containment, particularly physical containment, are sometimes difficult to monitor and enforce.

Some observers have noted that despite the rigors of the containment guidelines, the possibility remains that an engineered organism will eventually escape into the environment. Should this occur, the organism could cause environmental damage as great as or greater than that caused in the past by the introduction of foreign species to new habitats. For example, the introduction of rabbits to Australia dramatically upset the ecological balance on that continent. Hence field experiments with genetically engineered organisms must be strenuously controlled and monitored.

Although numerous safe field trials have been conducted with genetically engineered organisms, such as *B.t.* plants, widespread opposition to such practices remains. There appear to be few risks that cannot be ascertained within the laboratory associated with the release of genetically engineered higher plants, but opponents have expressed the fear that engineered genes could possibly be transferred by cross-pollination to other species of plants. Such a transfer could, for example, produce a highly vigorous species of weed. In addition, such gene transfers could potentially result in a plant that produces a toxin that would be detrimental to other plants, animals, or humans.

Because viruses and bacteria are major components of numerous natural biochemical cycles and readily exchange genetic information in a variety of ways, it is even more difficult to envision all the ramifications associated with releasing these genetically altered organisms into the environment. Field testing of genetically engineered organisms will always involve some element of risk, and assessment of the risks of such testing is easier for some species, such as higher plants, than for other species, such as bacteria.

A clear need exists for rigid controls, and minimizing the risks also requires integral cooperation among industry, governments, and regulatory organizations. Under the Cartagena Protocol on Biosafety, which entered into force in 2003, before an importing nation may release living modified organisms (LMOs) into the environment, the country into which the LMO is to be imported must first give its informed consent. The importer must clearly identify the LMO, detail its traits and characteristics, and explain its proper handling, storage, transport, and use.

With advances in the cloning of plants and animals, environmentalists and others have expressed concerns about losses in genetic variability. In nature, species survival is dependent on the genetic variability, or diversity, of the population. Genetic variability obtained through normal sexual reproduction provides a species with the ability to adapt to changes in the environment; because the environment is continually changing, loss of genetic variability usually leads to extinction of the species. Because cloning results in genetically identical individuals, the cloning of large numbers of animals or plants of particular species at the expense of those produced through sexual reproduction can lead to the loss of genetic variability and thus to eventual extinction of those species.

D. R. Gossett
Updated by Karen N. Kähler

FURTHER READING

Chrispeels, Maarten J., and David E. Sadava, eds. *Plants, Genes, and Crop Biotechnology.* 2d ed. Sudbury, Mass.: Jones and Bartlett, 2003.

Drlica, Karl. *Understanding DNA and Gene Cloning: A Guide for the Curious.* 4th ed. Hoboken, N.J.: John Wiley & Sons, 2004.

Field, Thomas G., and Robert E. Taylor. *Scientific Farm Animal Production: An Introduction to Animal Science.* 10th ed. Upper Saddle River, N.J.: Prentice Hall, 2011.

Grace, Eric S. *Biotechnology Unzipped: Promises and Realities.* Rev. 2d ed. Washington, D.C.: National Academies Press, 2006.

Groves, M. J., ed. *Pharmaceutical Biotechnology.* 2d ed. Boca Raton, Fla.: Taylor & Francis, 2006.

Hill, Walter E. *Genetic Engineering: A Primer.* London: Taylor & Francis, 2002.

SEE ALSO: *Bacillus thuringiensis*; Cloning; Dolly the sheep; Genetically altered bacteria; Genetically engineered pharmaceuticals; Genetically modified foods; Genetically modified organisms.

Birth defects, environmental

CATEGORY: Human health and the environment

DEFINITION: Problems in fetal development related to environmental pollutants to which parents are exposed before conception or to which mothers are exposed during pregnancy

SIGNIFICANCE: Embryotoxic agents encountered through environmental exposure can cause a variety of birth defects, ranging from growth retardation to malformations and death. The severity of any particular birth defect results from a combination of factors, including the gestational age at which the fetus is exposed to the agent and the effective dose of the toxic substance.

According to a 2006 report issued by the March of Dimes Foundation, roughly eight million of the world's children are born with birth defects every year. The U.S. Centers for Disease Control and Prevention reports that birth defects affect about one in every thirty-three babies born in the United States annually, and that these babies account for more than 20 percent of the country's infant deaths. Some birth defects are caused by environmental exposure to teratogens (substances that cause developmental malformations). According to the U.S. National Research Council, environmental exposures cause 3 percent of all birth defects and developmental disabilities; an interaction between genes and the environment may account for at least another 25 percent.

In the United States, the National Birth Defect Registry, assembled through responses to a questionnaire designed by Birth Defect Research for Children (formerly the Association of Birth Defect Children), details maternal and paternal exposure to environmental agents, including chemicals, radiation, pesticides, lead, and mercury. In March, 1998, the U.S. Congress passed the Birth Defects Prevention Act, which continued the National Birth Defects Prevention Study, a research project begun in 1996 to gather data on birth defects, including possible environmental causes.

TERATOGENIC AGENTS

Chemical agents known to cause environmental birth defects in animals and humans include compounds of heavy metals (especially mercury, lead, and thallium), urethane, dioxin-like chemicals, various steroids, sex hormones (including xenoestrogens and antiandrogenic pesticides), and trypan blue, an agent once used to treat mange. Chemical agents disturb intracellular chemistry; embryonic changes usually precede placental changes.

Testing of some agents has shown increases in the frequency of common malformations. For example, dioxin and similar chemicals have been observed to increase cleft palate and hydronephrosis (a swelling of the kidneys caused by obstructed urine flow) in mice. A minimum of two species of animals, studied at a minimum of three dose levels, is essential for teratogenic testing. Strong teratogenic agents usually produce similar malformations in different species. Agents that produce demonstrable birth defects in animals, especially in higher primates such as baboons, are presumed capable of producing birth defects in a human fetus.

Prenatal exposure to ionizing radiation is another environmental factor known to cause birth defects. For this reason, pregnant women are asked to inform radiologic technicians of their condition before X-ray sessions so that the technicians can minimize risk to the fetuses. Exposure to large, fluctuating electromagnetic fields (EMFs) has also been a suspected cause of environmental birth defects. Speculation that high-voltage electric power transmission lines might cause environmental birth defects or childhood cancer has led to many epidemiological studies using small mammals. Animal studies conducted over a number of years have not established any link between high-voltage lines and birth defects. Studies into the genotoxic potential (ability to alter genetic material) of EMFs have been conducted with bacteria, fruit flies, and mice. Although some cell research has suggested that EMF exposure may negatively affect the ability of cells to repair normal damage to genetic material, most evidence has suggested that EMF exposure is not genotoxic.

EXAMPLES

Clusters of environmental birth defects within human populations sometimes arise as a result of exposure of multiple persons to a release of a teratogenic agent, especially when the food chain is contaminated. When consumed by pregnant women, organic mercury is a potent teratogen. The best-known widespread occurrence of organic mercury poisoning took place in the 1950's and early 1960's in Japan's Minamata Bay, where industrial effluents containing high levels of mercury were pumped into the ocean for many years. This mercury was assimilated into seafood consumed by the local population, resulting in many cases of congenital Minamata disease.

In contrast to the environmental health problems at Minamata Bay, which took years to emerge, damage to human health was immediately apparent in Bhopal, India, after the catastrophic December, 1984, industrial accident that released a lethal cloud of methyl isocyanate gas over a crowded shantytown. Some pregnant women spontaneously aborted upon exposure to the gas. A 1987 study found that, of a sample of 865 surviving pregnant women who lived within a kilometer (0.6 mile) of the accident site when the incident occurred, only 57 percent gave birth to live children. Of those babies, 14 percent died within a month of delivery, an infant mortality rate about five times greater than what was common in the community before the accident. Many were monstrously deformed. In the decades after the release, birth defects were common in children born to gas victims. These defects included cleft palate, joined or extra fingers, retarded growth, and mental problems. Soil and groundwater contamination from chemicals abandoned at the industrial site continues to poison area residents, giving rise to still more birth defects.

Use of chemical agents during wartime is also associated with widespread birth defects. Birth Defect Research for Children has collected data since 1990 concerning an elevated incidence of birth defects and developmental disabilities in the children of U.S. veterans of the Vietnam War. Heavy use of the dioxin-tainted defoliant Agent Orange in Vietnam is associated with a suite of deformities and ailments among the children of veterans. These include spina bifida; oral clefts; learning, attention, and behavioral disorders; and increased susceptibility to chronic childhood infections and to cancers later in life. Vietnamese researchers have found that, in comparison with background levels in their country, rates of birth defects are four times higher in areas that were sprayed with Agent Orange during the war.

Anita Baker-Blocker
Updated by Karen N. Kähler

FURTHER READING

Hansen, Deborah Kay, and Barbara D. Abbott, eds. *Developmental Toxicology.* 3d ed. New York: Informa Healthcare, 2009.

Moore, Keith L., and T. V. N. Persaud. *Before We Are Born: Essentials of Embryology and Birth Defects.* 7th ed. Philadelphia: Saunders/Elsevier, 2008.

Schardein, James L. *Chemically Induced Birth Defects.* 3d ed. New York: Marcel Dekker, 2000.

Shenoy, Rathika, and Nutan Kamath. "Teratogenicity of Environmental Pollutants: An Overview." In *Focus on Birth Defects Research,* edited by Janet V. Engels. New York: Nova Science, 2006.

Yu, Ming-Ho. "Environmental Change and Health." In *Environmental Toxicology: Impacts of Environmental Toxicants on Living Systems.* Boca Raton, Fla.: CRC Press, 2001.

SEE ALSO: Agent Orange; Bhopal disaster; Chelyabinsk nuclear waste explosion; Chernobyl nuclear accident; Environmental illnesses; Love Canal disaster; Minamata Bay mercury poisoning; Superfund; Sydney Tar Ponds.

Black lung

CATEGORY: Human health and the environment
DEFINITION: Chronic respiratory disease caused by long-term inhalation of coal and mineral dusts in closed coal-mining environments
SIGNIFICANCE: Improvements of conditions for workers in coal mines have helped to reduce the incidence of coal workers' pneumoconiosis, commonly known as black lung, but the disease has not been eradicated.

Since the Industrial Revolution, coal has provided energy for industrialized societies. With escalating demand, coal mining developed into a prevalent industry in many regions, such as Wales in Great Britain and the Appalachian Mountains of the United States. By the nineteenth century, the economies of such regions were dependent on coal.

The burning of coal produces gaseous and particulate (soot) pollution, and cities in the nineteenth century were characterized by black, soot-coated buildings and coal residue on other surfaces. Explosions caused by coal dust became common in coal mines and storage facilities. Slower to appear from the use of coal for energy were the greenhouse effect, acid rain, and coal workers' pneumoconiosis, also called black lung, caused by chronic inhalation of coal dust. This particle-induced fibrosis-emphysema produces lesions in respiratory bronchioles that interfere with the absorption and transport of oxygen in the lungs.

Its symptoms are consistent with similar lung diseases: weakness and poor health, shortness of breath and oxygen starvation, heart disease, immune system irregularities, and lung cancers. It resembles diseases caused by fine airborne, respirable particles of asbestos (asbestosis); cotton, wood, and other plant-based dusts (farmers' lung); and silicas such as quartz, glass, and sand (silicosis).

The greatest incidence of black lung occurs in underground mining. Drilling, pulverizing, loading, and transporting coal generate large dust concentrations, which are breathed into the workers' lungs. Numerous safeguards have been instituted in coal min-

United Mine Workers member Rich Fuller listens to speakers at a 2003 rally held by miners in Charleston, West Virginia, to oppose proposed changes in mine safety rules that would affect the monitoring of coal dust levels in underground mines. (AP/Wide World Photos)

ing to minimize the risks of workers' developing black lung. Mine passages are ventilated by fans and baffles, and sophisticated routing produces multiregion flows that keep the air at face level fresh. Mines also utilize wetting and scrubber systems to keep coal wet and water mists to remove dust from the air.

More recent developments include ventilation and filtering systems worn on the body—for example, helmets that provide continuous streams of filtered air across workers' faces. Research is ongoing in the development of drill bits and other components that can minimize the dust produced. The replacement of steel bits with bits coated with tungsten carbide, polycrystalline diamond, or other hard ceramic or metallic films has received attention, as has optimization of thread geometry and bit speed.

Since the Federal Coal Mine Health and Safety Act of 1969 and the Mine Act amendments of 1977, federal and state agencies (for example, the Mine Safety and Health Administration, or MSHA) have mandated that mine operators undertake efforts aimed at preventing black lung. Operators are required to sample mine air for deviations from permissible exposure limits (PELs) of coal and silica dusts, methane, carbon monoxide, sulfur dioxide, and hydrogen sulfide; to report these deviations; and to take appropriate measures to correct them.

Operators are also required to provide medical screenings for employees over time. Periodic chest X rays, measurements of lung capacity, cardiovascular checkups, and general blood, urine, and endocrine analyses warn of early signs of black lung. Worker health data are gathered through periodic questionnaires, providing insight into the overall risks in a given operation. Emphasis has been placed on education regarding black lung and the regulatory infrastructure that now addresses it. Numerous government agencies offer information and services related to the disease, its prevention, its control, and compensation or support for those affected.

The MSHA's PEL for coal dust is 2 milligrams per cubic meter (mg/m^3) for unaffected workers and 1 mg/m^3 for workers with any signs of black lung. The National Institute for Occupational Safety and Health (NIOSH) recommends an exposure level no greater than 1 mg/m^3 for all workers. The Occupational Safety and Health Administration (OSHA) has set a slightly different PEL of 2.4 mg/m^3 for coal dust with less than 5 percent silica. In 1968-1969 the average dust concentration in underground coal mines was 6 mg/m^3, but since the Coal Act and the Mine Act, these averages have dropped to below 2 mg/m^3.

The silica content of coal dust affects the epidemiology of black lung as well as permissible exposure limits. Less than 1 mg/m^3 of silica can cause silicosis, and the current PEL for silicon is 0.1 mg/m^3, with reduction to 0.05 mg/m^3 being considered. Silica and mineral dusts are generated during initial drilling through rock to reach the coal in coal deposits that are interspersed with bedrock.

Significant progress was made over the last three to four decades of the twentieth century in fighting black lung and associated problems in coal mines. Government intervention, technological developments, and education should further improve the health, safety, and environmental status of the coal mining industry.

Robert D. Engelken

FURTHER READING

Goodell, Jeff. *Big Coal: The Dirty Secret Behind America's Energy Future.* New York: Houghton Mifflin, 2006.

Levine, Linda. *Coal Mine Safety and Health.* Washington, D.C.: Congressional Research Service, 2008.

Meyers, Robert A., ed. *Coal Handbook.* New York: Marcel Dekker, 1981.

Ripley, Earle, Robert Redmann, and Adele Crowder. *Environmental Effects of Mining.* Boca Raton, Fla.: CRC Press, 1996.

Witschi, Hanspeter, and Paul Nettesheim, eds. *Mechanisms in Respiratory Toxicology.* London: Chapman & Hall, 1982.

SEE ALSO: Coal; Environmental health; Environmental illnesses.

Black Sea

CATEGORIES: Places; water and water pollution; ecology and ecosystems

IDENTIFICATION: Large inland sea bounded by the Eurasian nations of Turkey, Bulgaria, Romania, Ukraine, Russia, and Georgia

SIGNIFICANCE: Because of a lack of oxygen, the lower levels of the Black Sea are virtually lifeless, and the abundant life of the sea's upper levels has been gravely endangered by pollution, overharvesting, and the introduction of destructive nonnative species. Progress has been made in addressing these problems, however, as the result of dedicated international efforts.

It is believed that the Black Sea was formed when the rising waters of the Mediterranean Sea swamped a freshwater lake about 7,500 years ago. The sea has a surface area of approximately 448,000 square kilometers (173,000 square miles) and reaches a maximum depth of approximately 2,200 meters (7,200 feet). Five important rivers—the Kuban, Don, Dnieper, Dniester, and Danube—flow into it, and in turn its brackish surface water flows out into the Mediterranean through the Bosporus strait, the Sea of Marmara, and the Dardanelles strait. Colder, saltier water flows in beneath this current from the Mediterranean to fill the Black Sea's depths. Within the Black Sea itself, circulation between the two layers is poor.

Because the Black Sea's lower levels do not contain enough dissolved oxygen to enable bacteria to decompose the organic matter carried into the sea by the rivers, the water has grown increasingly eutrophied (saturated with nutrients) and anoxic (oxygen-depleted), creating the largest "dead zone" of virtually lifeless water on the planet. Bacteria have evolved under these conditions, to react with naturally occurring sulfate ions in the seawater to produce the world's largest concentration of deadly hydrogen sulfide gas. Significant deposits of ammonia and methane are also present.

Fish and shellfish such as anchovy, mullet, tuna, mussels, and oysters were once found in abundance in the upper levels of the Black Sea, but in the latter decades of the twentieth century their numbers fell dramatically. This decline was caused in part by overfishing, but pollution from agriculture, industry, and shipping (particularly of crude oil) played a more important role. Cultural (human-caused) eutrophication of the sea's shallower waters contributed to a precipitous decline in fields of *Cystoseira* and *Phyllophora* algae, species essential to the survival of the food chain. In 1982 the alien carnivorous sea jelly *Mnemiopsis leidyi* was found for the first time in the Black Sea, and within a few years its rapidly growing numbers had destroyed a large proportion of the native species' eggs and larvae. During the 1990's radioactive substances resulting from nuclear power generation and the Chernobyl nuclear accident of 1986 began to enter the Black Sea as well.

By the end of the twentieth century, however, the Black Sea's ecology began to show a partial recovery owing to the region's economic slowdown and a resulting decrease in pollution. The sea had been recognized as a major environmental disaster area, and the six nations on its shores had drafted the Convention for the Protection of the Black Sea Against Pollution in 1992. This convention was succeeded by a number of other international efforts focusing on specific environmental problems, including the Black Sea Ecosystem Recovery Project of 2004-2007, which dealt primarily with eutrophication and hazardous waste.

In the future the Black Sea's deposits of ammonia and methane may be utilized to produce fertilizer and to generate electricity, and its hydrogen sulfide may be tapped as a source of hydrogen gas for fuel. The extraction technologies involved are expected to have the added advantage of returning purified water to the sea.

Grove Koger

FURTHER READING

Ascherson, Neal. *Black Sea*. New York: Hill & Wang, 1995.
Kostianoy, Andrey G., and Aleksey N. Kosarev, eds. *The Black Sea Environment*. Berlin: Springer, 2008.
Land, Thomas. "The Black Sea: Economic Developments and Environmental Dangers." *Contemporary Review* 278 (March, 2001): 144-151.

SEE ALSO: Asia; Cultural eutrophication; Danube River; Dead zones; Europe; Eutrophication; Introduced and exotic species; Nuclear and radioactive waste; Ocean pollution; Oil spills.

Black Wednesday

CATEGORIES: Disasters; atmosphere and air pollution

THE EVENT: Severe episode of photochemical smog in Los Angeles

DATE: September 8, 1943

SIGNIFICANCE: After the Black Wednesday incident helped to demonstrate the extent of the health dangers posed by smog, government actions aimed at addressing the problem were undertaken.

The Los Angeles, California, basin is prototypical of a smog-producing area, with mountains rising on the east and north and persistent high-pressure weather systems during summer and early fall. Subsidence in upper levels of the high-pressure atmosphere results in air being compressed and heated. Westerly winds blowing over a cold ocean current carry cool, moist air in beneath the subsiding air, forming a temperature inversion at an altitude lower than the mountain peaks. With cool, moist air beneath hot, dry air, rising currents can ascend only to the inversion level. They then spread laterally and descend when they run into the natural barrier of the mountains. This situation places a lid over rising air, which causes pollutants to remain at low elevations for several days.

When Los Angeles experienced a severe episode of photochemical smog on September 8, 1943, the *Los Angeles Times* dubbed the day Black Wednesday. The human-made origins of Black Wednesday included increased industrialization and population growth in the Los Angeles area as a result of World War II. These factors led to a rise in the number of automobiles, with an accompanying increase in automotive emissions, as well as a higher yield of effluent from industrial smokestacks. In addition, a shortage of natural rubber led to the manufacture of synthetic rubber. A plant in the Los Angeles neighborhood of Boyle Heights produced a synthetic called butadiene, which appeared to be a significant source of air pollution. Other minor sources contributing to the photochemical smog were backyard incinerators used to burn refuse and the smudge pots, or orchard heaters, used by citrus growers.

The consequences for humans of exposure to photochemical smog can vary. Carbon monoxide (CO) from automotive exhaust can result in a change in physiology: Exposure to low levels of CO can produce impaired functions, whereas exposure to high concentrations may end in death. There seems to be a positive correlation between polyaromatic hydrocarbons and lung, skin, and scrotal cancers. The main complaints of those exposed to smog, however, are burning eyes and irritated throats. Peroxyacetyl nitrates (PAN) are the primary eye irritants, although acrolein and formaldehyde also contribute to the problem. Finally, reduced visibility is a product of aerosols contained in smog.

Vegetation is distressed by several components of photochemical smog. Ozone assaults leaf palisades, resulting in destruction of chlorophyll, lowered photosynthesis and respiration rates, and development of dark spots on leaf surfaces. Plant exposure to sulfur dioxide causes gas to combine with water to form sulfite ions, which cause leaves to darken, grow flaccid, and become dry. This condition eventually induces death of tissue. The effects of photochemical smog on animals were not studied or recorded during the Black Wednesday event, but animals are believed to experience impacts of exposure to smog that are similar to those experienced by humans.

Photochemical smog also alters inanimate objects. Rapid cracking and eventual deterioration of stretched rubber can result from exposure to smog. Likewise, breakdowns of natural and synthetic fabrics and fading of dyes can be traced to smog elements.

In the years following the Black Wednesday incident, government and civic organizations began to take steps aimed at curbing air pollution in Southern California. The use of backyard incinerators was banned, and increasingly strict laws were passed concerning automobile emissions.

Ralph D. Cross

FURTHER READING

Carle, David. *Introduction to Air in California.* Berkeley: University of California Press, 2006.

Jacobs, Chip, and William J. Kelly. *Smogtown: The Lung-Burning History of Pollution in Los Angeles.* Woodstock, N.Y.: Overlook Press, 2008.

Vallero, Daniel. *Fundamentals of Air Pollution.* 4th ed. Boston: Elsevier, 2008.

SEE ALSO: Air pollution; Automobile emissions; London smog disaster; Smog.

Bonn Convention on the Conservation of Migratory Species of Wild Animals

CATEGORIES: Treaties, laws, and court cases; animals and endangered species

THE CONVENTION: International agreement protecting animal species that migrate across national borders

DATE: Opened for signature on June 23, 1979

SIGNIFICANCE: The Convention on the Conservation of Migratory Species of Wild Animals encourages the protection of migratory species by nations around the world. Most of the convention's success has been in fostering regional daughter agreements that protect particular migratory species of concern.

The Convention on the Conservation of Migratory Species of Wild Animals (also known as CMS or the Bonn Convention) arose as the result of recommendations made at the 1972 United Nations Conference on the Human Environment held in Stockholm, Sweden. Opened for signing in Bonn, Germany, in 1979, the treaty entered into force in 1983. By 2010 the convention had been signed by 113 nations, and many other nonsignatory nations were participating in CMS-sponsored agreements. The United Nations Environment Programme provides the CMS Secretariat.

The convention recognizes that the conservation status of a migratory species is vulnerable to threats in any states that species occupies or passes through. Appendix I to the convention lists endangered species for which concerted conservation action is required to ensure their survival. Parties to the convention (that is, signatory nations) are encouraged to forbid the taking of these species, to take steps to enhance their welfare, to remove obstacles to their migration, and to manage uses of the species by indigenous cultures. Appendix II lists migratory species that are not necessarily endangered but would benefit from conservation action.

CMS encourages states whose territories include the biogeographic ranges of migratory species to enter into one of four types of daughter agreements. In order of formality from least to most, these are designated as Action Plans, Memoranda of Understanding (MOUs), agreements (all lowercase in the CMS text), and AGREEMENTS (all capitals in the convention text).

Action Plans require the least of signatories and generally affirm that species of concern would benefit from research and conservation action and that parties will endeavor to engage in such action. Action Plans are also negotiated as components of the other kinds of agreements. Of the few stand-alone Action Plans that have been developed, one protects birds that use the migration routes within the Central Asian Flyway and another protects several species of African antelope.

MOUs are typically nonenforceable and do not require obligatory actions by parties, but they can help draw official and public attention to species of concern and result in conservation efforts on the species' behalf. The MOUs that have been negotiated under the auspices of CMS cover a range of species, from West African elephants to flamingoes in the Andes to marine turtles in the Indian Ocean and off Southeast Asia.

Lowercase agreements tend to require action by their signatories that may or may not be obligatory, depending on the agreement. Agreements have been negotiated to protect gorillas in Central Africa, birds that migrate between Africa and Eurasia, European bats, and whales in European waters. Uppercase AGREEMENTS necessitate adequate a priori knowledge of migratory patterns and population status and require signatories to follow strict guidelines outlined in the CMS text. By the early years of the twenty-first century no AGREEMENTS had been negotiated.

The main criticism of CMS has been that its area of responsibility is too large for a single treaty, with most progress made by agreements negotiated under the convention but not by the main convention itself. It has also been criticized for fostering weak, nonenforceable agreements, but supporters of the convention note that the range of formalities with which agreements can be made encourages participation by states that perhaps would otherwise be hesitant to engage in conservation. The effectiveness of CMS is limited by the absence of several large states important to migratory species, including the United States, Canada, China, Russia, and Brazil, although several of these do participate in daughter agreements.

Adam B. Smith

FURTHER READING

Cioc, Mark. *The Game of Conservation: International Treaties to Protect the World Migratory Animals.* Athens: Ohio University Press, 2009.

DeSombre, Elizabeth R. *Global Environmental Institutions.* New York: Routledge, 2006.

SEE ALSO: Conservation biology; Convention on Biological Diversity; Convention on International Trade in Endangered Species; Elephants, African; Environmental law, international; Extinctions and species loss; Habitat destruction; Hunting; International Union for Conservation of Nature; Migratory Bird Act; Mountain gorillas; Ramsar Convention on Wetlands of International Importance.

Bookchin, Murray

CATEGORIES: Activism and advocacy; urban environments

IDENTIFICATION: American ecological activist, author, and anarchist thinker

BORN: January 14, 1921; New York, New York

DIED: July 30, 2006; Burlington, Vermont

SIGNIFICANCE: Bookchin, the creator of the concept of social ecology, suggested in the 1960's that the prosperity of the post-World War II United States had been bought at the price of serious harm to the environment.

Like German socialist philosopher Karl Marx, Murray Bookchin argued that the human race cannot survive in a civilization based on life in the modern city, bureaucratic decision-making structures, and industrialized labor. Bookchin, however, built on Marx's insights to argue further that both socialism and capitalism are heedless of modern industry's impacts on the environment, and so neither socialism nor capitalism can be the basis for a sustainable society. He posited a close interconnection between human domination over nature and human beings' domination over one another, arguing that since the propensities to dominate nature and to dominate other humans sprang up together, they must be eliminated together.

Bookchin published his first two books, *Our Synthetic Environment* (1962) and *Crisis in Our Cities* (1965), under the pseudonym Lewis Herber. In these works he argued that it would not be an uprising of the proletariat but rather an uprising of an antiauthoritarian younger generation that would resolve the environmental crisis by dissolving all social hierarchies. This has not come to pass, of course, and some have criticized Bookchin's faith in the younger generation as naïve. His insight concerning the connection between social problems and the earth's environment, however—that the existence of hierarchy, in addition to the misuse of technology, has brought humankind to the brink of disaster—is seen as a valuable contribution to environmental thought.

Throughout his career, Bookchin developed and refined his argument that a link exists between the "destructive logic behind a hierarchical social structure" and environmental crisis. He strongly criticized environmentalists who are satisfied to save endangered species or ban harmful chemicals yet support underlying social structures that produce new toxins and similar problems. He called for a return to life as it was before the Industrial Revolution, in particular citing what he saw as the organic and harmonious way of life of past societies such as the Plains Indians in North America. Some critics also deemed this notion naïve.

In addition to his writing, Bookchin served as professor of social ecology at Ramapo College in New Jersey and as director of the Institute for Social Ecology at Goddard College in Rochester, Vermont. Among his books are *Toward an Ecological Society* (1980), *The Ecology of Freedom: The Emergence and Dissolution of Hierarchy* (1982), *Remaking Society* (1989), *The Philosophy of Social Ecology* (1990), *The Rise of Urbanization and the Decline of Citizenship* (1987), and *Social Ecology and Communalism* (2007).

Anne Statham

FURTHER READING

Barry, John. "Murray Bookchin, 1921-   ." In *Fifty Key Thinkers on the Environment*, edited by Joy A. Palmer. New York: Routledge, 2001.

Bookchin, Murray. "What Is Social Ecology?" In *Earth Ethics: Introductory Readings on Animal Rights and Environmental Ethics*, edited by James P. Sterba. 2d ed. Upper Saddle River, N.J.: Prentice Hall, 2000.

White, Damian. *Bookchin: A Critical Appraisal*. London: Pluto Press, 2008.

SEE ALSO: Carson, Rachel; Ecological economics; Environmental ethics; Social ecology.

Boreal forests

CATEGORIES: Forests and plants; ecology and ecosystems

DEFINITION: Coniferous forests within the Subarctic land biome, existing in a nearly continuous band throughout the northern regions of North America, Europe, and Asia

SIGNIFICANCE: Boreal forests play an important role in the cultural identities of people living in many parts of the world, particularly Canada and Siberia. These forests, which store large amounts of carbon and thus are a crucial element in the global carbon cycle, are threatened by resource development and climate change.

Boreal forests, or taiga, make up the world's largest land biome, covering vast parts of the Northern Hemisphere near 50 degrees of latitude and forming an ecologically sensitive habitat between the Arctic tundra and the temperate forest. Boreal forests extend through most of inland Canada, parts of Scandinavia, Russia (particularly Siberia), northern Mongolia and parts of central Asia, and northern Japan. They extend southward to parts of the Scottish highlands and the northern continental United States. Boreal forests are characterized by the dominance of conifers, such as firs, pines, spruces, hemlocks, and larches, and by a mean annual temperature between −5 and 5 degrees Celsius (23 and 41 degrees Fahrenheit).

Boreal forests typically have short, cool summers and long, cold winters. Precipitation can vary from 20 to 200 centimeters (approximately 8 to 79 inches) per year and falls mostly as snow. Because there is little evaporation, the ground remains moist during the summer growing season. Owing to the high latitude, summer days are very long and winter days are very short.

In addition to conifers, mosses and lichens form an important part of the boreal forest ecosystem. Some broad-leaved deciduous trees, such as birch, aspen, and willow, are also present in boreal forests.

While low in biodiversity when compared with other biomes, boreal forests are home to a variety of animals as well as plants, many of them endangered. Animals found in boreal forests include reindeer or caribou; carnivores such as bear, wolverine, lynx, fox, weasel, and wolf; and more than three hundred species of birds.

Boreal forests, or taiga, are located in the northern portion of the Northern Hemisphere. (©Irina Bekulova/Dreamstime.com)

Boreal forests rely on natural cycles of fire and insect damage followed by new tree growth; in North American boreal forests, these cycles are typically seventy to one hundred years long. Boreal forests store large amounts of carbon, perhaps more than the world's tropical and temperate forests combined.

The primary threats to boreal forests are logging, development of oil and natural gas reserves, and climate change. Large areas of Siberian and Canadian boreal forests have been logged, often by clearcutting. Substantial reserves of oil and natural gas exist under boreal forests in Alaska, Canada, and Russia. As world oil demand and prices rise and technologies for working in very cold climates improve, these reserves become possible candidates for development. It is unclear whether slow-growing boreal forests would be able to recover from the environmental impacts that accompany oil and natural gas extraction.

Although boreal forests require fire and insect infestations for their natural cycle of death and renewal, the increased frequency of both that is associated with climate change may have negative effects. Also, as temperatures rise, deciduous broad-leaved trees can outcompete conifers in southern regions.

Several countries have enacted measures to protect boreal forests through improved management of resource development and conservation. For example, Canadian logging companies are generally required to replant or encourage natural forest renewal, and in 2010 the Canadian federal government enlarged the areas of boreal forests to be protected from development.

Melissa A. Barton

FURTHER READING

Elliot-Fisk, Deborah L. "The Taiga and Boreal Forest." In *North American Terrestrial Vegetation*, edited by Michael G. Barbour and William Dwight Billings. New York: Cambridge University Press, 2000.

Hari, Pertti, and Liisa Kulmala, eds. *Boreal Forest and Climate Change.* Berlin: Springer, 2008.

Henry, J. David. *Canada's Boreal Forest.* Washington, D.C.: Smithsonian Institution Press, 2001.

SEE ALSO: Arctic National Wildlife Refuge; Asia; Biodiversity; Biomes; Coniferous forests; Deciduous forests; Ecosystems; Europe; Logging and clear-cutting; National forests; North America; Old-growth forests.

Borlaug, Norman

CATEGORIES: Activism and advocacy; agriculture and food

IDENTIFICATION: American plant pathologist and environmental activist

BORN: March 25, 1914; Cresco, Iowa

DIED: September 12, 2009; Dallas, Texas

SIGNIFICANCE: Borlaug, who became known as the father of the Green Revolution, pioneered efforts to develop high-yield crops to increase food production throughout the world.

Norman Borlaug credited his childhood experiences on his family's farm with providing him with a practical approach to agriculture. In the 1930's Borlaug studied forest management and plant pathology at the University of Minnesota, earning a doctorate by 1942. In 1943 the Rockefeller Foundation, which U.S. secretary of agriculture Henry Wallace had convinced to fund agricultural aid for Mexican farmers, hired Borlaug to breed disease-immune crops that could be grown in varied climates. He perfected a strain of high-yield dwarf spring wheat.

While Borlaug was experimenting with plant breeding, the post-World War II global population rapidly increased. Some environmentalists, such as Paul R. Ehrlich, predicted that not enough food could be produced and mass starvation would occur. Borlaug believed that his wheat could prevent such a disaster. In 1963 the Rockefeller Foundation and the Mexican government established the International Maize and Wheat Improvement Center (known as CIMMYT, for its name in Spanish, Centro Internacional de Mejoramiento de Maíz y Trigo) and named Borlaug director of the Wheat Improvement Program. He traveled to India and other developing nations to share his high-yield agricultural techniques. Critics thought that Borlaug should plant indigenous crops rather than Western grains, but he emphasized that wheat provides necessary calories and nutrients.

The Indian and Pakistani governments resisted Borlaug's efforts until famine in their countries became extreme. Delayed seed shipments and the outbreak of war between India and Pakistan also hindered Borlaug's work. He persisted, however, and yields increased approximately 70 percent during the first season. By 1968 Pakistan was agriculturally self-sufficient and increased its yields from 3.4 million tons of wheat in 1963 to 18 million by 1997. India boosted its yields from 11 million tons to 60 million tons, even briefly exporting wheat.

The expansion of food production spearheaded by Borlaug, which saved hundreds of millions of people from starvation, was called the Green Revolution. Borlaug was awarded the Nobel Peace Prize in 1970 for his humanitarian efforts to secure the basic human right of freedom from hunger. This honor did not ensure continued support of his work, however. When he expressed interest in using his techniques to assist African agriculture, Borlaug's CIMMYT patrons

Plant pathologist and environmental activist Norman Borlaug in 1996, examining some sorghum test plants. (AP/Wide World Photos)

stopped his funding because of a backlash among environmentalists that protested high-yield agriculture, claiming that its use of inorganic fertilizers and irrigation damages the environment. Borlaug countered with the argument that high-yield agriculture preserves habitats from slash-and-burn techniques used to create farmland. He criticized theorists who, in his opinion, do not comprehend the reality of what is economically and politically possible in developing countries. With colleagues Haldore Hanson and R. Glenn Anderson, he wrote *Wheat in the Third World* (1982).

Borlaug's projects in African nations were funded by Sasakawa-Global 2000, which was financed by former U.S. president Jimmy Carter and Japanese industrialist Ryoichi Sasakawa. Continuing the development of high-yield crop strains at CIMMYT in the 1990's, Borlaug wrote articles, lectured, and testified before the U.S. Congress about the opposition to his work posed by environmental lobbyists. After receiving government support in Ethiopia and Poland, he planned similar agricultural programs in the former Soviet Union and Latin America. Warning that arable land is finite, Borlaug stressed that uncontrolled population growth could result in starvation in the twenty-first century and demanded the storage of food reserves.

Elizabeth D. Schafer

FURTHER READING

Borlaug, Norman. "Are We Going Mad?" In *The Ethics of Food: A Reader for the Twenty-first Century*, edited by Gregory E. Pence. Lanham, Md.: Rowman & Littlefield, 2002.
Hesser, Leon. *The Man Who Fed the World: Nobel Peace Prize Laureate Norman Borlaug and His Battle to End World Hunger.* Dallas: Durban House, 2006.
Paarlberg, Robert. *Starved for Science: How Biotechnology Is Being Kept out of Africa.* Cambridge, Mass.: Harvard University Press, 2008.

SEE ALSO: Agricultural revolution; Alternative grains; Green Revolution; High-yield wheat; Population growth; Slash-and-burn agriculture.

Boulder Dam. *See* Hoover Dam

Bovine growth hormone

CATEGORIES: Agriculture and food; biotechnology and genetic engineering
DEFINITION: Protein secreted by the anterior pituitary gland in cattle that controls the rate of skeletal and visceral growth
SIGNIFICANCE: The administration of growth hormone to dairy cattle to increase milk production is controversial because of unsubstantiated allegations about the safety of the milk produced. It is also argued, however, that increasing the efficiency of milk production would benefit the environment by reducing the amount of feed needed and the waste produced per unit of production.

Growth hormone is produced by all vertebrate animals to control their rate of growth. Prior to the development of biotechnology, acquiring growth hormone was a laborious process, involving extraction from the pituitary glands of cadavers or slaughterhouse animals. All early studies of the hormone were dependent on these sources. Advances in biotechnology, however, now permit the efficient production of synthetic growth hormone, referred to as recombinant growth hormone. Recombinant human growth hormone is used to treat growth retardation conditions in children.

Recombinant bovine growth hormone, also known as recombinant bovine somatotropin, or rBST, is administered to beef cattle to increase their growth rate and administered to dairy cattle to increase their milk production. The Monsanto Company developed and tested the first such product, which it called Posilac. In 1993, the U.S. Food and Drug Administration (FDA) approved the use of Posilac, which increases milk production by at least 10 percent and has not been found to be harmful to cattle or human consumers of the milk so produced. Over the intervening years, the FDA has continued to affirm the safety of the product, despite continuing controversy about it and bans on its use by European, Canadian, and other governments.

To be effective, rBST must be injected because, as a protein, it would be degraded by the digestive system if administered orally. The milk of injected cows is indistinguishable from that of noninjected cows, with the exception of slightly elevated insulin-like growth factor (IGF-1), although the latter remains within the physiological range of human breast milk. In any event, as a protein, IGF-1 is degraded in the human digestive system. Nevertheless, concerns have been raised about the health risks to consumers of milk from treated cows. None of these have been substantiated by any reputable scientific review in the United States or elsewhere. It has been argued that the bans by the European Union and Canada are based on political and economic considerations, with the governments protecting quota-based milk-production systems that would be disrupted by this product. From an environmental perspective, the ability to have fewer cows (using less land, consuming less feed, and generating less waste, including the greenhouse gas methane) producing the same amount of milk that would be produced by larger numbers of untreated cows is a definitive benefit.

The cattle injected with rBST are metabolically equivalent to genetically improved cattle and must be fed and treated as such. Like any high-producing cows, they have a slightly increased incidence of mastitis, an infection of the mammary gland, although it remains lower per gallon of milk produced. Mastitis is treated with antibiotics, and any milk with antibiotic residue must be withheld from the market. The assertion that rBST-treated cows are increasing the antibiotic content of the milk supply is unsupported by any evidence. Nevertheless, a number of American grocery chains and food companies have indicated that they will not sell dairy products from cows injected with rBST. However, given that the milk of rBST-treated cows is virtually indistinguishable from that of untreated cows, it is difficult, if not impossible, to verify the compliance of their suppliers.

James L. Robinson

FURTHER READING

Grace, Eric S. *Biotechnology Unzipped: Promises and Realities*. Rev. 2d ed. Washington, D.C.: Joseph Henry Press, 2006.

Hammond, Bruce G. "The Food Safety Assessment of Bovine Somatotropin (bST)." In *Food Safety of Proteins in Agricultural Biotechnology*, edited by Bruce G. Hammond. Boca Raton, Fla.: CRC Press, 2008.

Schacter, Bernice. *Issues and Dilemmas of Biotechnology: A Reference Guide*. Westport, Conn.: Greenwood Press, 1999.

SEE ALSO: Biotechnology and genetic engineering; Cattle; Genetically engineered pharmaceuticals; Green marketing; Greenhouse gases; Methane.

Bovine spongiform encephalopathy. *See* **Mad cow disease**

BP *Deepwater Horizon* oil spill

CATEGORIES: Disasters; water and water pollution

THE EVENT: Explosion and fire on the *Deepwater Horizon* drilling platform in the Gulf of Mexico that claimed eleven lives and triggered the largest accidental marine oil spill in history and the largest oil spill of any kind in the United States

DATES: April 20-July 15, 2010

SIGNIFICANCE: The oil spill that began with an explosion on British Petroleum's *Deepwater Horizon* drilling platform raised questions about the safety of deepwater drilling, the adequacy of the corporate response to the disaster and of governmental regulation of offshore oil drilling, and the possibility of long-term damage to the Gulf of Mexico's ecosystem.

The *Deepwater Horizon* was a semisubmersible drilling platform owned by Transocean and under lease to British Petroleum (BP). In April, 2010, it was located in the Gulf of Mexico approximately 84 kilometers (52 miles) southeast of Venice, Louisiana, where it was completing work on the exploratory Macondo 252 well. Oil had been found 5.5 kilometers (18,000 feet) below the seafloor and 7 kilometers (23,000 feet) below the drilling platform. The drill hole had been—or was being—cemented to seal the well so that the drill pipe could be removed and the *Deepwater Horizon* could be moved to a new location. The cement failed, allowing gas and oil under high pressure to escape the reservoir and rise through the drill pipe casing and up the riser pipe to the drilling platform. A blowout preventer located on top of the wellhead was designed to cut through the drill casing and seal the wellhead in cased of an emergency. The blowout preventer also failed, and about 11:00 P.M. central daylight time on April 20 the escaping gas reached the surface and exploded, setting the *Deepwater Horizon* on fire. Most of the workers on the platform were evacuated without serious injury, but eleven who had been in close proximity to the explosion died.

Without any mechanism to stop the flow of oil to the platform, fireboats were unable to extinguish the

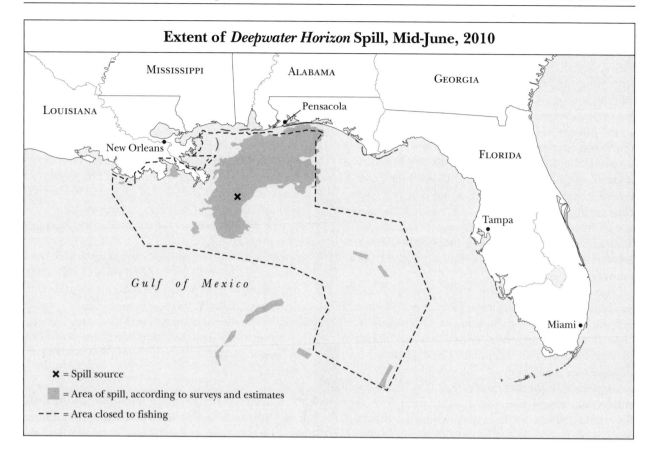

Extent of *Deepwater Horizon* Spill, Mid-June, 2010

flames. The drilling platform burned for about thirty-six hours and then sank, twisting and breaking the riser pipe that had connected the platform to the well-head some 1.5 kilometers (5,000 feet) beneath the sea surface.

MAGNITUDE OF THE SPILL

On April 24, unmanned submarines working for BP detected oil flow from the wellhead and the collapsed riser pipe. The magnitude of the flow was estimated to be about 1,000 barrels per day (BPD), a figure repeated by both company and governmental officials. (One barrel of oil equals 42 gallons, or approximately 159 liters.) On April 26 a scientist with the National Oceanic and Atmospheric Administration (NOAA) estimated the flow at roughly 5,000 BPD based on satellite imagery of the oil slick. Nongovernmental scientists using similar methodologies provided estimates as high as 26,500 BPD.

After BP released video images of the underwater leak on May 12, independent experts reported estimates of up to 50,000 BPD. Despite the existence of potentially better methodologies, government offi-

cials adhered to their estimate of 5,000 BPD until May 27, when a government task force, the Flow Rate Group (FRG), released its first estimate: 12,000 to 25,000 BPD. FRG estimates were increased to 20,000 to 40,000 BPD on June 12 and to 35,000 to 60,000 BPD on June 15. By that time oil was being washed up on the shores of the coasts of Louisiana, Mississippi, Alabama, and Florida.

In the aftermath of the spill, both government and independent scientists appeared to agree that the actual flow rate was approximately 60,000 BPD and that the total release of oil into the Gulf was approximately 5 million barrels, roughly twenty times the volume of the 1989 *Exxon Valdez* oil spill in Prince William Sound. This was the largest oil spill in the history of the Gulf of Mexico, exceeding the 3 million gallons discharged into the Gulf after the 1979 explosion and fire on a PEMEX semisubmersible drilling platform off the coast of Mexico.

EMERGENCY RESPONSE

Within one week of the initial explosion, oil had reached the Mississippi Delta, and the nation became aware that the "Gulf oil spill" threatened serious economic and environmental damage. The event was better described as a discharge, a blowout, a flow, or a leak, but the term "spill" was almost universally adopted. In the space of a few days, President Barack Obama announced that the United States would use all available resources to contain the spill. Fishing was prohibited in the affected areas, and a moratorium was declared on further deepwater drilling in the Gulf of Mexico pending an investigation. President Obama announced that British Petroleum was responsible for the spill and that it would be held responsible for the cleanup. The U.S. Coast Guard commandant, Admiral Thad Allen, was named incident commander for the federal response, and it was announced that a national commission would be formed to study the disaster and make recommendations. BP chief executive officer Tony Hayward declared that BP would take full responsibility; he pledged to stop the leak, repair the damage to the Gulf, and pay all legitimate claims for damages.

From the beginning experts agreed that the permanent solution would be a relief well that would intersect the drill hole below the lowest level of drill pipe casing and above the top of the petroleum reservoir. Cement pumped through the relief well would seal the reservoir permanently. BP began drilling such a relief well on May 2. It was anticipated that the relief well would be completed in August, but waiting that long to stop the flow of oil was not a viable option.

BP pursued multiple strategies to stanch the flow. Several attempted quick fixes were relatively unsuccessful. Remotely controlled underwater vehicles working for the company failed to close valves on the blowout preventer. Surface oil slicks were burned on several occasions, but the volume of oil consumed was relatively small. U.S. Air Force planes were enlisted to spray chemical dispersants on surface slicks while BP injected dispersants underwater in an effort to break up the oil flow at the source. More than 1.8 million gallons of dispersants were used—

almost 800,000 gallons near the wellhead. Local fishing boats were hired to skim floating oil, and miles of booms were deployed in an effort to prevent slicks from contaminating ecologically sensitive coastlines. In many areas wind and waves rendered these strategies ineffective. Efforts were also undertaken to construct artificial barrier islands to protect fragile coastlines. Every strategy was controversial, and some worked at cross-purposes. Chemical dispersants, for example, made the use of booms and skimming less effective but also arguably less necessary. At the height of the crisis thousands of people and hundreds of vessels were employed in efforts to mitigate the environmental damage of the spill.

By early May it was apparent that at least three significant leaks were coming from a section of broken riser pipe that lay crumpled on the ocean floor still attached to the failed blowout preventer on the wellhead. On May 4 remotely controlled underwater vehicles successfully sawed off the free end of the leaking riser pipe and installed a shutoff valve, reducing the number of leaks to two, but without significantly diminishing the flow of oil. On May 7 and 8 one of three custom-built coffer dams was lowered over the largest leak on the ocean floor. This concrete and metal box,

A brown pelican is cleaned at the Fort Jackson Wildlife Rehabilitation Center at Buras, Louisiana. The bird was rescued after it was soaked by oil spilled into the Gulf of Mexico as a result of the explosion on the BP Deepwater Horizon *oil platform three weeks previously.* (AP/Wide World Photos)

12 meters (40 feet) high, was designed to capture the plume of escaping oil so that it could be pumped to the surface. The coffer dam failed, however, when a frozen mixture of gas and water clogged the system. A smaller version, dubbed "top hat," was lowered on May 11 but never deployed. Instead, BP chose to insert a 15.2-centimeter (6-inch) pipe directly into the leaking 53.3-centimeter (21-inch) riser pipe.

On May 16 BP announced that it was capturing most of the leaking oil, but the following day the estimate of oil captured was reduced to 1,000 BPD, approximately one-fifth of BP's estimated leak rate. Plans were announced for a "junk shot" to plug the leak by injecting the well with a high-pressure mixture of cement and solids such as shredded tires. It was never executed, however; instead BP chose the "top kill," which was designed to stop the flow by pumping drilling mud into the blowout preventer. This procedure failed to stop the flow, even after "junk shot" solids were added to the mixture.

By June 1 BP was working to saw off the broken riser pipe just above the blowout preventer and attach a cap connected to a new riser pipe. This strategy entailed significant risk because cutting off the bent riser pipe would increase the flow of oil into the Gulf. The cap was connected, but the fit was loose. Over time the fraction of escaping oil that was recovered slowly increased, approaching 50 percent. A significant fraction of the captured oil was burned at the surface. After about a month, while efforts continued to drill relief wells, BP removed the cap and replaced it with what amounted to a blowout preventer on top of the previous blowout preventer. After eighty-seven days, the flow of oil was stopped on July 15.

In August, as work on the relief wells continued, BP announced a successful "static kill." Tons of drilling mud followed by cement were pumped into the wellhead, providing increased assurance that the flow would not resume. The first relief well intersected Macondo 252 on September 16, and crews cemented the blown-out well from the bottom. The federal incident commander declared Macondo 252 officially sealed on September 19.

CONSEQUENCES

The economic damage associated with the BP oil spill is difficult to quantify. One study estimated the short-term damage to the Gulf fishing industry at $115 to $172 million. Severe economic impacts were also associated with the deepwater drilling morato-

rium and the spill's damage to the "Louisiana brand." Perhaps the clearest economic indicator of damage done was reduced investor confidence in BP. Between April 21 and June 25, 2010, the value of BP stock declined by 55 percent, representing a reduction of $67 billion in market capitalization. By the time the well was sealed, BP had reportedly spent more than $11 billion on the capping and cleanup operations, and it had created a $20 billion escrow account for payment of damages. In the autumn of 2010 BP announced that it had taken a pretax charge of $32.2 billion and had plans to sell up to $30 billion in assets. It had canceled its stock dividend. It is likely that BP, and possibly other companies involved, will eventually face penalties that could amount to billions of dollars under the Clean Water Act and other federal statutes.

BP's long-term liability will depend in part on the environmental damage caused by the spill, which is even more difficult to measure than economic damage. Studies are expected to continue for years if not decades. The fate of the spilled oil remains the subject of scientific controversy. No one knows with any degree of certainty what fraction evaporated, sank to the bottom of the Gulf, or remained suspended in the water column. The environmental consequences of the unprecedented intensive use of chemical dispersants remain unclear, but preliminary analyses by the Environmental Protection Agency indicated that the environmental benefits of dispersant use outweighed the environmental cost.

The totality of environmental damage to the Gulf of Mexico from the *Deepwater Horizon* spill will certainly be significant, but the ecological significance of a major oil spill may depend significantly on the location. The ecological damage done by the *Exxon Valdez* grounding in Prince William Sound was disproportionate to the size of the spill. By contrast, the vastly larger PEMEX spill in the Gulf of Mexico is generally regarded as having caused relatively little environmental damage. Despite the attention given to the acute BP spill, the chronic damage from agricultural runoff throughout the Mississippi River basin probably remains the most significant ecological threat to the Gulf.

The BP oil spill raised important questions about the safety of deepwater drilling, the industry's preparedness for spills, and the government's supervision of industry behavior. Early reports by the national commission studying the spill indicated that the industry and the government were both poorly prepared for a spill of such magnitude. BP and its contractors had bypassed safety measures, and government agencies had routinely approved work that did not meet legal standards.

Craig W. Allin

FURTHER READING

Barstow, David, et al. "Between Blast and Spill, One Last, Flawed Hope." *New York Times,* June 1, 2010, A1.

Fausset, Richard. "Oil Spill's Legacy." *Los Angeles Times,* September 19, 2010, A1.

Jernelöv, Arne. "The Threats from Oil Spills: Now, Then, and in the Future." *AMBIO: A Journal of the Human Environment* 39, no. 6 (2010): 353-366.

Jonsson, Patrik. "Gulf Oil Spill: Where Has the Oil Gone?" *Christian Science Monitor,* July 27, 2010.

Ornitz, Barbara E., and Michael A. Champ. *Oil Spills First Principles: Prevention and Best Response.* New York: Elsevier, 2002.

SEE ALSO: *Exxon Valdez* oil spill; Monongahela River tank collapse; Oil spills; PEMEX oil well leak; Santa Barbara oil spill.

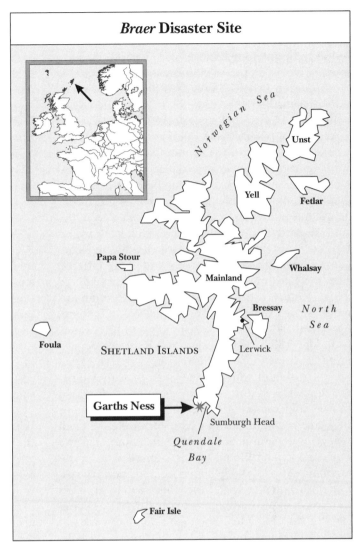

Braer Disaster Site

Braer oil spill

CATEGORIES: Disasters; water and water pollution

THE EVENT: Grounding of the oil tanker *Braer* at the Shetland Islands, resulting in the spilling of its cargo of crude oil into the sea

DATE: January 5, 1993

SIGNIFICANCE: The oil spilled when the *Braer* ran aground killed or sickened thousands of shellfish and led to severe economic setbacks for the local fishing industry.

When the oil tanker *Braer* ran aground at the southern tip of the Shetland Islands, located about 210 kilometers (130 miles) north of the Scottish mainland, more than one million barrels of crude oil were spilled into the Atlantic Ocean. Under the severe wind and wave conditions that prevailed around the Shetland Islands at the time, the spilled oil thoroughly mixed into the turbulent seawater and rapidly

dispersed. Ten days after the spill, the concentrations in the vicinity of the wreck had fallen from several hundred parts per million (ppm) to 4 ppm, still about two thousand times the normal level.

Being less dense than water, most of the oil floated. The lightest, most volatile hydrocarbons started to evaporate, decreasing the volume of the spill but polluting the air. Subsequently, a slow decomposition process occurred, caused by sunlight and bacterial action. After several months, the oil mass was reduced by approximately 80 percent.

Shortly after the *Braer* oil spill, all fishing activities were prohibited in the surrounding areas. For most shellfish species, this ban remained in effect for more than two years, until the spring of 1995, when they were judged to be free of any significant levels of oil

194 • Brazilian radioactive powder release   ENCYCLOPEDIA OF ENVIRONMENTAL ISSUES

contamination. Shortly after fishing resumed, however, the catches of lobsters and queen scallops were found to be very poor, and the proportion of young lobsters and scallops was abnormally low.

Supported by the Shetland Fishermen's Association, the North Atlantic Fisheries College carried out a number of laboratory trials in 1996 and 1997 to determine what effects crude oil may have on lobsters and scallops and to investigate whether the *Braer* oil spill could have had adverse effects on these stocks. During the experiments, researchers used Norwegian Gullfaks crude oil, the same type as the *Braer*'s cargo, to simulate the spill conditions. Results of the study showed that in the short term, adult scallops and lobsters could survive exposure to relatively high concentrations of oil, but lobster eggs and larvae suffered high mortality rates. Although the oil did not kill the adult lobsters, it did cause major behavioral abnormalities, including significant reductions in feeding, movement, responsiveness to stimuli, and aggression.

In May, 1997, the Scottish Office Fisheries Department released a document indicating that the *Braer* oil spill polluted a much wider area than previously thought. The report showed that levels of oil in prawns and mussels from a 1,036-square-kilometer (400-square-mile) zone of excluded fishing were still rising in 1996, and there were indications that the tides had spread the oil underwater all around the Shetland Islands' 1,450-kilometer (900-mile) coastline. Fishermen reported that their fishing grounds were ruined, with nothing replacing what was being caught. Into the late 1990's, considerable work was still being undertaken to rectify the damage done by the *Braer* oil spill, particularly by the large quantities of oil that had become incorporated into subtidal sediments.

Alvin K. Benson

FURTHER READING

Burger, Joanna. *Oil Spills*. New Brunswick, N.J.: Rutgers University Press, 1997.

Clark, R. B. *Marine Pollution*. 5th ed. New York: Oxford University Press, 2001.

National Research Council. *Oil in the Sea III: Inputs, Fates, and Effects*. Washington, D.C.: National Academies Press, 2003.

SEE ALSO: *Amoco Cadiz* oil spill; *Argo Merchant* oil spill; *Exxon Valdez* oil spill; Oil spills; *Sea Empress* oil spill; Tobago oil spill; *Torrey Canyon* oil spill.

Brazilian radioactive powder release

CATEGORIES: Disasters; nuclear power and radiation

THE EVENT: Accidental release of radioactive powder in Goiania, Brazil

DATES: September-October, 1987

SIGNIFICANCE: When radioactive powder was released accidentally in Goiania, hundreds of people were contaminated, four of whom died. The incident raised the issue of responsibility for the disposal of radioactive waste.

On September 13, 1987, two men found an old radiation-therapy machine in an abandoned medical clinic in Goiania, a Brazilian state capital with a population of one million people. When they took the machine apart, they found a cylindrical lead container inside, which they sold to a local junk dealer. Unaware that the canister contained radioactive cesium 137, an isotope of cesium used in the treatment of cancer, the junk dealer opened the canister to investigate the curious blue light visible inside. He shared his discovery with friends, who were fascinated by the blue powder and the shimmer it made on their skin. The powder was rapidly distributed throughout the community as people gave samples to friends as gifts. Children played with it as if it were a new toy. Symptoms of radiation poisoning began days later.

Radiation destroys the reproductive mechanisms of cells, affecting to the greatest extent those cells that divide the most rapidly. These include bone marrow, gastrointestinal tract, skin, and hair cells. Effects include burns, hemorrhaging, and, because bone marrow cells are involved in the immune response of the body, decreased white blood cell count, with consequent susceptibility to infections.

A team of doctors was dispatched to the area, and helicopters equipped with radiation detection equipment determined the "hot spots" in the city. Tons of materials were found to have traces of the powder, including furniture, buses, money, and animals. The doctors found that 244 people had been contaminated with doses up to 600 rads (1 rad is roughly equivalent to seven chest X rays) and immediately hospitalized 54 people. The patients were thoroughly washed to remove any excess cesium that may still have been on their skin, and they were fed Prussian blue, a compound known to complex with cesium and block its further absorption into the body. The

doctors treated infections with antibiotics and hemorrhaging with blood-clotting factors. The Brazilian government enlisted the aid of the International Atomic
Energy Agency (IAEA) and the Radiation Emergency
Center Training Sites (REACTS), a World Health Organization unit that responds to radiation incidents in
the Western Hemisphere.

Some controversy arose over the use of an experimental treatment involving granulocyte-macrophage
colony-stimulating factor (GM-CSF) on severely ill patients. GM-CSF is one of five hormones that increase
the production of white blood cells in bone marrow.
The treatment is used in cancer patients because it
offsets the effects of radiation and chemotherapy, allowing the use of larger doses. The six patients treated
with GM-CSF in Brazil represented the worst cases of
contamination. Four of them died, but the other two
seemed to respond well to the treatment. By the end
of the year, twenty-eight people remained hospitalized with radiation sickness, but there were no further
fatalities.

The Brazilian Nuclear Energy Commission came
under strong criticism for the accident, which caused
widespread panic among the citizens of Goiania and
other cities in Brazil. There was confusion regarding
which government agency had the responsibility to license, monitor, and ensure proper disposal of radioactive waste in the country. The Brazilian Ministry of
Health and the Ministry of Labor were also implicated
as negligent of their responsibilities with regard to the
monitoring of radioactive materials.

Suzanne Jones and Massimo D. Bezoari

FURTHER READING

Friis, Robert H. "Ionizing and Nonionizing Radiation." In *Essentials of Environmental Health.* Sudbury,
Mass.: Jones and Bartlett, 2007.
Gusev, Igor A., Angelina K. Guskova, and Fred A.
Mettler. *Medical Management of Radiation Accidents.*
2d ed. Boca Raton, Fla.: CRC Press, 2001.
Yard, Charles Richard. "Lost Radiation Sources: Raising Public Awareness About the Hazards Associated with Industrial and Medical Radiography
Sources." *Journal of Environmental Health* 58, no. 10
(1996).

SEE ALSO: Hazardous waste; Nuclear accidents; Nuclear and radioactive waste; Radioactive pollution and
fallout.

Breeder reactors

CATEGORIES: Energy and energy use; nuclear power
and radiation

DEFINITION: Nuclear reactors designed to create
more fuel than they use in the production of
energy

SIGNIFICANCE: Properly managed and maintained
breeder reactors produce useful energy, generate
less waste than conventional light-water fission reactors, and reduce greenhouse gas emissions by reducing the use of fossil fuels.

As opposed to normal nuclear fission reactors that
use uranium 235 as their energy source, breeder
reactors can make use of the much more abundant
uranium 238 or thorium 232. Whereas a typical fission
reactor uses only about 1 percent of the natural uranium 235 that starts its fuel cycle, a breeder reactor
consumes a much larger percentage of the initial fissionable material. In addition, if the price of uranium
is more than two hundred dollars per kilogram, it is
cost-efficient to reprocess the fuel so that almost all of
the original fissionable material produces useful energy. Breeder reactors are designed to produce from
1 percent to more than 20 percent more fuel than
they consume. The time required for a breeder reactor to generate enough material to fuel a second nuclear reactor is referred to as the doubling time; the
typical doubling time targeted in power plant design
is ten years.

Scientists have proposed two main types of breeder
reactors: fast-breeder reactors and thermal breeder
reactors. The fast-breeder reactor uses fast neutrons
given off by fission reactions to breed more fuel from
nonfissionable isotopes. The most common fast-
breeding reaction produces fissionable plutonium
239 from nonfissionable uranium 238. The liquid
metal fast-breeder reactor (LMFBR) breeds plutonium 239 and uses liquid metal, typically sodium, for
cooling and for heat transfer to water to generate
steam that turns a turbine to produce electricity. The
thermal breeder reactor uses thorium 232 to produce
fissile uranium 233 after neutron capture and beta
decay.

Unlike fossil-fuel-powered plants, breeder reactors
do not generate carbon dioxide or other greenhouse
gas pollutants. Environmental concerns related to
any type of breeder reactor include nuclear accidents
that could emit radiation into the atmosphere and the

difficulty of safely disposing of radioactive waste by-products. In addition, for plutonium-based breeders a major concern is the possibility of the diversion of bred plutonium for nuclear weapon production. This concern can be addressed through the intermixing of actinide impurities with the plutonium; such impurities make little difference to reactor operation, but they make it extremely difficult for anyone to use the bred plutonium to manufacture a nuclear weapon.

France has been the most prominent nation in the implementation of breeder reactors. The Superphénix breeder reactor built on the Caspian Sea was used for power generation and desalination of seawater from 1985 to 1996. India has plans to build a large fleet of breeder reactors, including a large prototype LMFBR that uses a plutonium-uranium oxide mixture in the fuel rods. Russia, China, and Japan are also developing breeders. The Russian BN-600 fast-breeder reactor has experienced several sodium leaks and fires. If the sodium coolant in the central part of an LMFBR core were to overheat and bubble, core melting could accelerate in the event of an accident and release radioactive material into the environment. A strong reactor containment building and a reactor core designed so that the fuel rod bundles are interspersed within a depleted uranium blanket that surrounds the core could greatly decrease this effect.

Alvin K. Benson

FURTHER READING

Bodansky, David. *Nuclear Energy: Principles, Practices, and Prospects.* 2d ed. New York: Springer, 2004.

Mosey, David. *Reactor Accidents: Institutional Failure in the Nuclear Industry.* Sidcup, England: Nuclear Engineering International, 2006.

Muller, Richard A. *Physics for Future Presidents: The Science Behind the Headlines.* New York: W. W. Norton, 2008.

SEE ALSO: Alternative energy sources; Nuclear accidents; Nuclear and radioactive waste; Nuclear power; Nuclear weapons production; Power plants; Radioactive pollution and fallout; Superphénix.

Brent Spar occupation

CATEGORIES: Activism and advocacy; water and water pollution

THE EVENT: Occupation of an abandoned oil storage platform by Greenpeace protesters

DATES: April 30-May 23, 1995

SIGNIFICANCE: As the result of their occupation of the *Brent Spar*, which Royal Dutch Shell had planned to sink, Greenpeace protesters were able to influence the company to adopt alternative disposal methods.

On April 30, 1995, fourteen Greenpeace volunteers boarded the abandoned oil storage platform *Brent Spar*, which was anchored off the western coast of Norway in the North Sea. The *Brent Spar*, a floating cylinder moored to the sea bottom, had been built in 1976, but it had been taken out of commission in 1991. Its owner, Royal Dutch Shell, intended to lower the platform into the Atlantic Ocean and then sink it under more than six thousand feet of water.

Greenpeace had engaged in months of negotiations with Shell prior to the occupation. Representatives of the environmental activist group had urged Shell officials to develop a plan for dismantling the platform on land, citing studies that had concluded that on-land disposal and recycling of oil rigs is environmentally preferable to sea dumping. Shell replied with studies of its own asserting that deep-sea disposal is the more environmentally sound option.

When it became clear that Shell planned to proceed with the ocean dumping, Greenpeace sent fourteen volunteers to the platform. It also stepped up its publicity campaign, drawing support from the European Union (EU) commissioner for the environment, Denmark's minister for environment and energy, and other important sources. In May, the European Parliament passed a resolution opposing the dumping of the *Brent Spar.*

On May 22, Shell dispatched a team to the *Brent Spar* to remove the protesters, but bad weather delayed the evacuation. The volunteers ended their occupation the next day but vowed to continue the fight in the courts. However, an English judge refused to hear the Greenpeace case, stating that the English courts had no jurisdiction over the North Sea matter.

In June, German chancellor Helmut Kohl raised the question of the *Brent Spar* with British prime minister John Major. By then, Germany had joined the list

Water cannons from a supply boat bombard the Brent Spar *oil platform in an attempt to prevent a helicopter, from which this photo was taken, from dropping Greenpeace activists on the platform.* (AP/Wide World Photos)

of nations opposing the dumping of oil platforms in the sea. Shell nevertheless proceeded with its plan. In late June, therefore, Greenpeace activists again boarded the platform by helicopter. Meanwhile, demonstrations against Shell took place in Great Britain, Denmark, the Netherlands, Germany, and Switzerland. In Germany, a Shell gas station was firebombed, and the company's sales fell more than 15 percent.

On June 20, Shell announced that it would not dump the *Brent Spar* in the sea. Nine days later, the member nations of the Oslo Paris Commission (OSPAR), in a vote of eleven to two (Norway and the United Kingdom opposing), decided to impose a moratorium on sea disposal of oil platforms. The following month, Shell received permission from the Norwegian government to store the *Brent Spar* in an inlet on Norway's west coast while working out plans

for its dismantling. The platform was finally dismantled in 1999, with parts of it repurposed as the foundation for a new ferry terminal at Mekjarvik, Norway.

Christopher Kent

FURTHER READING

Chasek, Pamela S., et al. *Global Environmental Politics.* 4th ed. Boulder, Colo.: Westview Press, 2006.

Entine, Jon. "Shell, Greenpeace, and *Brent Spar*: The Politics of Dialogue." *Case Histories in Business Ethics*, edited by Chris Megone and Simon J. Robinson. New York: Routledge, 2002.

Jordan, Grant. *Shell, Greenpeace, and the Brent Spar.* New York: Palgrave, 2001.

SEE ALSO: Greenpeace; Oil spills; Seabed disposal; Water pollution.

Brockovich, Erin

CATEGORIES: Activism and advocacy; human health and the environment

IDENTIFICATION: Legal clerk and environmental activist

BORN: June 22, 1960; Lawrence, Kansas

SIGNIFICANCE: Brockovich helped construct a legal case against the Pacific Gas and Electric Company for its role in polluting the drinking water of Hinkley, California, with chromium 6. The clients in the case received the largest settlement ever made in the United States in a direct-action lawsuit.

Erin Brockovich was born Erin Pattee; her mother, Betty Jo O'Neal-Pattee, was a journalist, and her father, Frank Pattee, was an industrial engineer. She graduated from Lawrence High School in 1978 and earned an associate in applied arts degree from Wade Business College in Dallas in 1980. She then became a management trainee for the Kmart retail chain and moved to California. After a period in which she married and divorced twice (Steve Brockovich was her second husband), had three children, and held a variety of jobs, she became a secretary for the law firm of Masry & Vititoe of Northridge, California, in 1991.

Brockovich was assigned to work on a pro bono case for a home owner in Hinkley, California, and quickly developed a rapport with the client, who was bringing a lawsuit that alleged contamination of the town's drinking water with chromium 6. Residents of the area, it was alleged, were experiencing above-average numbers of miscarriages and cancers. Brockovich discovered that from 1952 to 1966 the Pacific Gas and Electric Company (PG&E) used chromium 6 to fight corrosion in the cooling tower of its natural gas pumping station in Hinkley. Wastewater containing chromium 6 was then pumped into unlined ponds, from which it leached into the aquifer that supplied Hinkley's drinking water. Brockovich pursued the case tenaciously and signed up more than six hundred additional persons to participate in the lawsuit. In 1996, after winning an initial round in court, the clients agreed to PG&E's offer to settle all their claims against the company for $333 million—a record high settlement for a direct-action lawsuit in the United States.

Thomas R. Feller

SEE ALSO: Aquifers; Drinking water; Groundwater pollution; Wastewater management; Water pollution; Water quality.

Brower, David

CATEGORIES: Activism and advocacy; preservation and wilderness issues

IDENTIFICATION: American environmental activist and writer

BORN: July 1, 1912; Berkeley, California

DIED: November 5, 2000; Berkeley, California

SIGNIFICANCE: Brower, who was vigorously involved in battles concerning environmental issues for more than fifty years, was one of the twentieth century's most influential and controversial environmental activists and writers.

Environmental activist and author David Brower. (AP/Wide World Photos)

David Ross Brower served as the first executive director of the Sierra Club from 1952 to 1969. He is credited by many with helping the San Francisco-based organization grow from two thousand to seventy-seven thousand members and developing it into a powerful national organization. He led the club in aggressive campaigns against U.S. government projects to develop wild areas, most notably fights that successfully stopped the construction of the Echo Park Dam, which would have flooded part of Dinosaur National Monument in Utah in the 1950's, and two different dams across the Colorado River in the Grand Canyon in the 1960's and 1970's. His enterprising tactics included full-page advertisements in *The New York Times* and the *San Francisco Chronicle*, which resulted in the Internal Revenue Service reclassifying the nonprofit Sierra Club as a lobbying organization and removing its tax-deductible status.

For more than twenty-five years, Brower focused much of his passion and energy on the Glen Canyon Dam in northeastern Arizona. "Glen Canyon died, and I was partly responsible for its needless death," Brower wrote in Eliot Porter's *The Place No One Knew: Glen Canyon on the Colorado* (1963). In the mid-1950's the U.S. Bureau of Reclamation was planning to construct dams across the Colorado River in the Grand Canyon and Glen Canyon, Arizona, and across the Green and Yampa rivers in Utah. Following the directives of the Sierra Club board of directors, Brower agreed to drop the club's opposition to the Colorado River dams if the Bureau of Reclamation would discontinue plans to build the two dams in Utah. The bureau agreed to the deal and moved forward to build the Glen Canyon Dam.

Before the dam construction was completed, Brower and the Sierra Club decided the compromise had been a mistake. They blamed their decision on a lack of familiarity with the spectacular beauty of Glen Canyon. (In 1996 the Sierra Club directors unanimously passed a motion by Brower to support draining Lake Powell, the reservoir behind the Glen Canyon Dam, and return the Colorado River flow to the most natural state possible. Brower did not advocate dismantling the dam; rather, he stated, it should be left "as a tourist attraction, like the Pyramids, with passers-by wondering how humanity ever built it, and why.")

In the mid-1960's Brower and the Sierra Club successfully led an effort to prevent construction of the Bureau of Reclamation's proposed Marble Canyon Dam in the Grand Canyon and helped cripple the bureau's effort to build Bridge Canyon Dam farther downstream in the Grand Canyon. By 1969, however, the majority of the Sierra Club's board of directors found Brower's tactics too reckless, both financially and politically, and they removed him as executive director. He then formed the preservation-oriented Friends of the Earth and the League of Conservation Voters, both of which flourished under his leadership. He also facilitated the establishment of independent Friends of the Earth organizations in other countries. In 1982, after conflicts with members of the Friends of the Earth's professional staff and its directors, Brower moved on to form another group, Earth Island Institute, the stated mission of which was to globalize the environmental movement. He returned to the Sierra Club as a director in 1983 and was reelected in 1986 and 1995. In the fall of 1994 Brower helped develop the Ecological Council of Americas to improve cooperation among organizations in the Western Hemisphere that were attempting to integrate environmental and economic needs.

In the 1990's Brower called on the federal government to replace the U.S. Bureau of Land Management with a new agency called the National Land Service. Its mission would be to protect and restore private and public land in the United States. He also strongly advocated the creation of a national biosphere reserve system.

Throughout his life, Brower pushed the edges of environmental thought of the day. He pioneered ideas and methods to preserve the environment and create a global approach to issues. For many years Brower advocated the establishment of international natural reserves in areas of rich biodiversity and ecosystems. The United Nations Educational, Scientific, and Cultural Organization (UNESCO) has established such a system of World Heritage Sites.

Brower also advocated a method known as CPR to guard against the destruction of natural areas and biodiversity: "C" is for *conservation*, or the rational use of resources: "P" represents the *preservation* of threatened, endangered, and yet undiscovered species; and "R" stands for *restoration* of lands already damaged by human activities. Many of his tactics and ideas seemed radical when he introduced them but later became standard practice among mainstream environmentalists. Russell Train, chairman of the Council on Environmental Quality during President Richard Nixon's administration, once said, "Thank

God for Dave Brower; he makes it so easy for the rest of us to be reasonable."

Louise D. Hose

FURTHER READING

Brower, David. *Let the Mountains Talk, Let the Rivers Run: A Call to Save the Earth.* New ed. San Francisco: Sierra Club Books, 2007.

_____. "The Sermon." In *Speaking of Earth: Environmental Speeches That Moved the World*, edited by Alon Tal. New Brunswick, N.J.: Rutgers University Press, 2006.

McPhee, John. *Encounters with the Archdruid.* 1971. Reprint. New York: Farrar, Straus and Giroux, 2000.

Porter, Eliot. *The Place No One Knew: Glen Canyon on the Colorado.* Edited by David Brower. Commemorative ed. Layton, Utah: Gibbs Smith, 2000.

Stoll, Steven. *U.S. Environmentalism Since 1945: A Brief History with Documents.* New York: Palgrave Macmillan, 2007.

SEE ALSO: Biosphere reserves; Echo Park Dam opposition; Friends of the Earth International; Glen Canyon Dam; League of Conservation Voters; Preservation; Sierra Club.

Brown, Lester

CATEGORIES: Activism and advocacy; agriculture and food

IDENTIFICATION: American agricultural scientist and author

BORN: March 28, 1934; Bridgeton, New Jersey

SIGNIFICANCE: Brown founded the Worldwatch Institute, an environmental think tank the mission of which is to analyze the state of the earth and to act as "a global early warning system."

Lester Brown was raised on a small tomato farm in Bridgeton, New Jersey. He joined the 4-H Club and the Future Farmers of America at his local school. When he was fourteen, he and his brother purchased a used tractor and a small plot of land to grow tomatoes. Within a brief time, they became two of the most successful tomato farmers on the East Coast. Brown graduated from Rutgers University in 1955 with a degree in agricultural science and immediately put his education to practical use. He worked for six months in a small farming community in India, becoming inti-mately acquainted with hunger problems created by population growth and unsustainable agricultural practices.

In 1959 Brown earned a master's degree in agricultural economics and soon after joined the U.S. Department of Agriculture as an international agricultural analyst. After leaving that post in 1969, he helped organize the Overseas Development Council, a private group devoted to analyzing issues relevant to relations between the United States and developing countries.

Brown's educational and career background prepared him well for his work and leadership in the Worldwatch Institute. While living and working in developing nations, he became acutely aware of such problems as the extensive poverty caused by economic systems dependent on cash crops for export to wealthy industrial countries and the use of agricultural practices that cause deforestation and desertification. He realized that food security could replace military security as the major concern of governments in the twenty-first century.

Shortly after Brown established the Worldwatch Institute in 1974, he and other staff members initiated the Worldwatch Papers, which focus on population growth and the resulting stress on natural resources, transportation trends, and the human and environmental impact of urbanization. In 1984 Brown established the institute's annual report, *State of the World*, a comprehensive overview of specific global environmental issues. This publication, now available in more than twenty-five languages, is used by political leaders, educators, and citizens as a resource for information on environmental problems and ways to address them. In 1992 Brown inaugurated the publication of *Vital Signs: The Trends That Are Shaping Our Future.* This annual handbook features environmental, economic, and social statistical indicators on trends likely to influence the world's future.

In Brown's view, global environmental problems should be addressed through international efforts funded by taxes on currency exchanges, taxes for pollution emissions, and a greater involvement by the United Nations. He has advocated a shift away from national spending on unsustainable economic growth and toward investing in research and development that enhances environmental quality and protection of natural resources.

Brown has published numerous books, some in partnership with the Worldwatch Institute and some published by the Earth Policy Institute, which Brown founded in 2001 to raise awareness of environmental issues. His books include *In the Human Interest: A Strategy to Stabilize World Population* (1974), *Building a Sustainable Society* (1981), *Who Will Feed China? Wake-Up Call for a Small Planet* (1995), and *Plan B: Rescuing a Planet Under Stress and a Civilization in Trouble* (2003). In these and other writings, Brown warns about the ecological dangers of overexploiting the earth's resources and urges those in the world's developed nations to change their lifestyles. He also advises governments and scientists to cooperate in finding solutions to environmental problems.

Ruth Bamberger

FURTHER READING

Brown, Lester. "Worldwatch." In *Life Stories: World-Renowned Scientists Reflect on Their Lives and on the Future of Life on Earth,* edited by Heather Newbold. Berkeley: University of California Press, 2000.

Nelson, David E. "In Praise of Lester Brown." *Futurist* 42, no. 6 (2008).

Wallis, Victor. "Lester Brown, the Worldwatch Institute, and the Dilemmas of Technocratic Revolution." *Organization and Environment* 10, no. 2 (1997): 109-125.

SEE ALSO: Population growth; Sustainable agriculture; Sustainable development; Urban sprawl; Worldwatch Institute.

Brownfields

CATEGORY: Land and land use

DEFINITION: Abandoned lots or other properties that were once used for commercial or industrial purposes and have the potential to contain harmful contaminants or pollutants

SIGNIFICANCE: Governments have increasingly encouraged the redevelopment of brownfields, requiring that such redevelopment be preceded by the cleanup of any potentially hazardous substances in the soil and water on these abandoned industrial sites. In addition to the benefits of such cleanup for the environment, brownfield redevelopment reduces urban sprawl and improves property values.

Brownfields are found in cities and towns throughout the world, usually located in industrial areas. The potential hazards and liability issues associated with such sites' past uses become important when these areas are targeted for redevelopment. In order for municipalities to redevelop brownfields, they must determine liability for any problems on the sites and take any cleanup measures required by law. In the United States, the Comprehensive Environmental Response, Compensation, and Liability Act, widely known as Superfund, mandates liability laws in regard to the cleanup and redevelopment of brownfields and also sets cleanup and development standards. The redevelopment of brownfields can be beneficial to cities, promoting urban and community revitalization through sustainable reuse of land, reducing sprawl, increasing local property taxes, promoting jobs and economic development, and protecting the environment from harmful substances and pollutants.

BARRIERS TO REDEVELOPMENT

In the United States, when redevelopment of a brownfield is being considered, the U.S. Environmental Protection Agency (EPA), a state agency, or both must determine the kind and extent of cleanup required and help to determine the liability for the cleanup. Cleanup efforts for brownfield redevelopment are often costly, and it can be difficult for developers to obtain loans to develop potentially hazardous land. Because this is the case, federal, state, and municipal governments may offer some financial contributions under the EPA's Brownfields Initiative, the U.S. Department of Housing and Urban Development's (HUD's) Brownfields Economic Development Initiative, or state and municipal funding programs.

The potential presence of harmful substances and pollutants can increase the difficulty of assessing and cleaning up a brownfield site. Depending on their previous uses, brownfields may contain many hazardous substances, such as petroleum, lead, asbestos, hydrocarbon, pesticides, tributyltins, solvents, and diesel fuels; cleanup is thus imperative before such sites are developed. Sites that are determined to be harmful enough to be hazardous to individuals living and working around them are not considered brownfields; rather, these are designated as Superfund sites and are placed on either the Superfund National Priority List or a state priority list for cleanup.

LAWS AND REGULATIONS

The term "brownfields" was first introduced in 1992 during a congressional hearing. The first legislation concerning brownfields was passed in 1995 when the EPA created its Brownfields Program. Through this program, the EPA seeks to motivate communities, states, and organizations to work together to improve local brownfield sites and use them as resources for redevelopment. Since the 1990's U.S. states and municipalities, along with the federal government, have created programs to provide tax incentives and grants for developers, organizations, and governments to redevelop brownfield sites.

The Small Business Liability Relief and Brownfields Revitalization Act, enacted in 2002, provides support for small businesses to redevelop brownfields through financial assistance and state and local program promotion. HUD's Brownfields Economic Development Initiative provides grants to cities to promote economic and community development; this program is directed toward development in low- and moderate-income neighborhoods, with the aim of promoting economic stability, job creation or retention, business improvement, and increased property taxes.

BENEFITS OF BROWNFIELD REDEVELOPMENT

As land pressures and urban sprawl have become increasingly problematic, brownfield redevelopment has grown in importance. The redevelopment of brownfields allows developers to use existing properties within cities rather than create developments that sprawl outward from urban centers, possibly disrupting functioning ecosystems. Additionally, brownfield redevelopment projects often increase the property values in their surrounding areas because of the improvements made to the previously abandoned sites. Brownfield redevelopment also reduces the potential for harm to humans and other forms of life from environmental hazards and pollution, as cleanup of all hazardous substances is a requirement of redevelopment of any brownfield site.

Courtney A. Smith

FURTHER READING

Davis, Todd S. *Brownfields: A Comprehensive Guide to Redeveloping Contaminated Property*. 2d ed. Chicago: American Bar Association, 2002.

De Sousa, Christopher. *Brownfields Redevelopment and the Quest for Sustainability*. Boston: Elsevier, 2008.

Dixon, Tim, et al., eds. *Sustainable Brownfield Regeneration: Liveable Places from Problem Spaces*. Malden, Mass.: Blackwell, 2007.

Witkin, James B. *Environmental Aspects of Real Estate and Commercial Transactions: From Brownfields to Green Buildings*. 3d ed. Chicago: American Bar Association, 2005.

SEE ALSO: Community gardens; Environmental impact assessments and statements; Land clearance; Land pollution; Land-use planning; Land-use policy; Restoration ecology; Superfund; Sustainable development.

Brucellosis

CATEGORY: Human health and the environment

DEFINITION: Infectious disease transmitted from animals to humans by gram-negative bacteria belonging to the genus *Brucella*

SIGNIFICANCE: Brucellosis is uncommon in the United States and other industrialized countries, but it is found frequently in many geographic regions of the world, including the Mediterranean basin, Eastern Europe, the Middle East, Asia, Africa, the Caribbean, South America, and Central America.

Bacteria belonging to the genus *Brucella* are non-motile, exist as non-spore-forming short rods, and live inside their host cells. Different species tend to associate and infect specific hosts: *Brucella abortus* in cattle, bison, and buffalo; *B. canis* in dogs; *B. melitensis* in sheep, goats, and camels; *B. suis* in pigs, with its other bacterial strains found in European hares, reindeer, caribou, and rodents. Four species are known to cause infections in humans: *B. abortus*, *B. canis*, *B. suis*, and *B. melitensis* (the most common infectious agent in humans).

Animal infection is transmitted through contact with infected animal parts and body fluids, including venereal transmission. Human infection occurs through direct inoculation on skin wounds; inhalation of aerosolized bacteria (classified as a class B bioterrorism agent); occupational exposure in laboratories; exposure to fomites (contaminated objects); ingestion of raw meat, unpasteurized milk, and other dairy products; and possibly venereal transmission.

Brucellosis causes premature births in all animal species. In humans it also causes acute, nonspecific symptoms that include undulating fever, chills, weakness, headaches, depression, weight loss, and joint and muscle pains. Chronic complications include sacroiliitis, meningitis, endocarditis, and liver abscess; in men, epididymo-orchitis may occur.

Brucellosis is diagnosed through isolation and characterization in specialized growth cultures and through various serologic tests. The usual treatment regimen for human infection involves administration of the antibiotics doxycycline and rifampin for six weeks to prevent relapse. No treatment has been developed for infected cattle or pigs. Antibiotic treatment has been used with some success in infected dogs and rams.

Prevention of human brucellosis can be achieved through the control of the infection in animals, pasteurization of milk and related products, and avoidance of raw meat ingestion. Animal brucellosis can be prevented through disinfection of contaminated farms and implementation of effective animal health and sanitation programs, which should include steps to ensure that any animals introduced into existing herds, kennels, or other animal groupings are brucellosis-free.

Miriam E. Schwartz

SEE ALSO: Antibiotics; Mad cow disease.

Brundtland, Gro Harlem

CATEGORIES: Activism and advocacy; ecology and
    ecosystems
IDENTIFICATION: Norwegian politician, physician,
    and environmental advocate
BORN: April 20, 1939; Oslo, Norway
SIGNIFICANCE: Brundtland has been called the
    "Green Goddess" because of the innovative environmental programs she initiated during her career as prime minister of Norway.

Gro Harlem Brundtland grew up in a politically active family. She graduated from Oslo University Medical School in 1963 and completed a master's degree in public health at Harvard University two years later. She served as a medical officer in Norwegian public health offices and participated in party politics.

In 1974, Brundtland was appointed Norway's minister of the environment. One year later she was named deputy leader of the Norwegian Labor Party. Elected to the Storting, Norway's parliament, in 1977,

Brundtland became prime minister and leader of the Norwegian Labor Party from February to October, 1981. She was the youngest person and the first woman to achieve that office. Brundtland was also the first environmental minister to become a country's leader, contemplating political and environmental problems together. A popular prime minister, Brundtland was reelected three times, serving from 1986 to 1989, 1990 to 1993, and 1993 to 1996.

In 1983 Brundtland was chosen to chair the World Commission on Environment and Development sponsored by the United Nations. The twenty-three-member commission worked for three years devising a global agenda for environmental protection, enhancement, and management. Brundtland traveled around the world to assess environmental conditions. The 1987 report *Our Common Future*, also known as the Brundtland Report, explained that environmental deterioration is caused by poverty. It supported the idea of sustainable development—that is, development that does not destroy resources to attain economic growth. Brundtland emphasized that citizens of the world's nations should be active in decision making and accountable for environmental quality. The report raised public awareness of the environment but was criticized by some for being overly optimistic and for offering simplistic solutions.

Brundtland spoke at diplomatic gatherings to convince the world community to accept the report's proposals. She criticized economic development that depletes nonrenewable resources crucial for future generations and urged countries to stop such practices. In an address to the United Nations in 1988, Brundtland stressed the need for international environmental cooperation by citing the example of industrial pollution, which crosses national borders and causes acid rain in other countries. "The environment is where we all live," she said, "and development is what we all do in attempting to improve our lot within that abode. The two are inseparable." Brundtland urged coordinated political support of environmental issues because of her belief that the world shares one economy and environment and thus politicians should adopt global solutions to environmental problems, such as formulating environmentally sound energy policies.

Considered a champion of human rights and environmental quality in Norway and elsewhere, Brundtland has been the recipient of such awards as the Third World Prize. She has been praised for providing

Gro Harlem Brundtland with United Nations secretary-general Javier Pérez de Cuéllar at a press conference after Brundtland presented the report of the World Commission on Environment and Development at the headquarters of the United Nations Food and Agriculture Organization in Rome. (AP/Wide World Photos)

opportunities for women in her cabinet and other government positions during her time as prime minister. Despite her environmental rhetoric, Brundtland was criticized for her controversial support of Norwegian whaling in 1993. She resigned as prime minister in October, 1996. In July, 1998, she was named director general of the World Health Organization, a position in which she served until 2003, focusing on the worldwide prevention of disease, especially the improvement of children's health care. In 2007, she was appointed by Secretary-General Ban Ki-moon to serve as one of three United Nations special envoys on climate change.

Elizabeth D. Schafer

FURTHER READING

Brundtland, Gro Harlem. *Madam Prime Minister: A Life in Power and Politics.* New York: Farrar, Straus and Giroux, 2002.

Palmer, Joy A. "Gro Harlem Brundtland, 1939-   ." In *Fifty Key Thinkers on the Environment,* edited by Joy A. Palmer. New York: Routledge, 2001.

SEE ALSO: Environmental economics; International Whaling Commission; *Our Common Future;* Sustainable development; United Nations Commission on Sustainable Development; United Nations Conference on the Human Environment.

Brundtland Commission

CATEGORIES: Organizations and agencies; resources and resource management
IDENTIFICATION: Body formed by the United Nations to propose long-term environmental strategies for achieving sustainable development through international cooperation
DATE: Established in 1983
SIGNIFICANCE: The work of the Brundtland Commission raised global awareness of the concept of sustainable development and thus led to the Earth Summit in 1992, which produced the Rio Declaration on Environment and Development as well as conventions on biological diversity and on climate change.

The World Commission on Environment and Development—commonly known as the Brundtland Commission, for its chair, Gro Harlem Brundtland—was established because of growing worldwide recognition of the impacts of human development on the environment. Publication of the commission's 1987 report, titled *Our Common Future* (often referred to as the Brundtland Report), propelled issues of sustainable development to the forefront of international policy debates and decision making.

Our Common Future defines "sustainable development" as "development that meets the needs of the present without compromising the ability of future generations to meet their own needs." The report emphasizes that such development is sustainable socially and environmentally as well as economically, and it urges support for efforts to inaugurate sustainable systems of agriculture, industry, and energy generation that satisfy these requirements. A central theme of the report is the importance of maintaining sustainable levels of population. The report also explicitly states that living creatures have intrinsic value as well as instrumental value (a biocentric stance); the international community, however, replaced this biocentric approach with an anthropocentric one in the Rio Declaration on Environment and Development, which was produced by the 1992 Earth Summit.

Robin Attfield

SEE ALSO: Agenda 21; Anthropocentrism; Biocentrism; Brundtland, Gro Harlem; Earth Summit; International Institute for Sustainable Development; *Our Common Future*; Rio Declaration on Environment and Development; Sustainable development; Sustainable forestry; United Nations Commission on Sustainable Development.

The Brundtland Commission's Report

The report of the Brundtland Commission, published as Our Common Future, *includes the following summary statement of the status of the relationship between nature and humanity near the end of the twentieth century.*

Over the course of this century, the relationship between the human world and the planet that sustains it has undergone a profound change.

When the century began, neither human numbers nor technology had the power radically to alter planetary systems. As the century closes, not only do vastly increased human numbers and their activities have that power, but major, unintended changes are occurring in the atmosphere, in soils, in water, among plants and animals, and in the relationships among all of these. The rate of change is outstripping the ability of scientific disciplines and our current capabilities to assess and advise.

Bureau of Land Management, U.S.

CATEGORIES: Organizations and agencies; land and land use
IDENTIFICATION: Federal agency responsible for managing the public lands of the United States
DATE: Established in 1946
SIGNIFICANCE: The Bureau of Land Management is charged with managing public lands in ways that are consistent with both multiple-use concepts and sustained-yield principles.

The U.S. federal government's original policy concerning public lands was to encourage their disposal. The most widely known method for this disposal was through homesteading as a result of the Homestead Act of 1862, which was overseen by the General Land Office, a forerunner of the Bureau of Land Management (BLM) created in 1812. The land that remained after settlement and the designation of national parks, national forests, and wildlife refuges was available for public use. Abuses of these public lands became widespread, however, and by the 1930's there was need for correction. Extensive overgrazing of livestock constituted one of the most serious misuses of public lands. As a result of these abuses, the Grazing Service was created as part of the U.S. Department of the Interior to manage some 32.4 million hectares (80 million acres) under the provisions of the Taylor Grazing Act in 1934.

In 1946 the Grazing Service became the BLM. Part of the BLM's continuing responsibilities were based on the need to evaluate damage, classify public lands for grazing purposes, and assess fees for grazing. Concerns about environmental quality grew during the 1960's and 1970's, and these increased concerns extended to the public lands. As a result, the BLM was granted more authority under the provisions of the Federal Land Policy and Management Act of 1976, which encouraged the BLM to manage public lands in ways that are consistent with both multiple-use concepts and sustained-yield principles.

The multiple-use approach to land-use planning has a lengthy history in resource management in the United States, particularly in the forestry area. Because of the potential for land to be used for a wide variety of purposes, such as timber production, grazing, and recreation, legislators recognized that careful planning is needed and that there are strong advantages to managing public lands in a way that ensures that resources are sustained and the environment is protected. These principles were articulated in the Multiple Use-Sustained Yield Act of 1960, and they have, in turn, become a part of BLM policy.

In the twenty-first century, most lands managed by the BLM are in Alaska and the other states west of the Mississippi River. However, the management of onshore oil drilling, gas production, and mineral development on federal lands is also part of the BLM's responsibilities. As a result, the bureau maintains an office to deal with oil and mineral policies on public land east of the Mississippi River.

Jerry E. Green

FURTHER READING

Allen, Leslie. *Wildlands of the West: The Story of the Bureau of Land Management.* Washington, D.C.: National Geographic Society, 2002.

Skillen, James. *The Nation's Largest Landlord: The Bureau of Land Management in the American West.* Lawrence: University Press of Kansas, 2009.

SEE ALSO: Federal Land Policy and Management Act; Forest and range policy; Grazing and grasslands; Land-use policy; Multiple-use management; Sagebrush Rebellion; Wise-use movement.

Burroughs, John

CATEGORIES: Activism and advocacy; preservation and wilderness issues

IDENTIFICATION: American nature writer

BORN: April 3, 1837; near Roxbury, New York

DIED: March 29, 1921; en route from California to New York

SIGNIFICANCE: Through his best-selling books, Burroughs raised Americans' awareness of the beauty of nature and the importance of preserving it.

John Burroughs grew up on a farm in the Catskill Mountains in New York, spending as much time as he could outdoors. Around the age of twenty, he decided that he would try to earn his living as a writer. After a brief teaching career, he spent ten years as a clerk in the U.S. Treasury Department in Washington, D.C. As a sideline, he published magazine essays about natural history and philosophy, always working to sharpen his writing skills. During these years he published his first book, *Notes on Walt Whitman as Poet and Person* (1867), the first biography of the great poet, who had also been a government clerk and who was Burroughs's personal friend.

Burroughs's first volume of essays, *Wake-Robin* (1871), was representative of the twenty-two collections that would follow: It featured close observations of natural history and commentary about simple country life, was made up mostly of essays (including such titles as "Birds' Nests" and "In the Hemlocks") that had been previously published in magazines, and won immediate acclaim. To Burroughs's first readers, the genre of nature writing was new and captivating, and Burroughs soon became its most popular practitioner. Sure now that he could live by his pen, he left his government job and moved back to New York, establishing a small fruit farm on the banks of the Hudson River in 1873. He continued to publish essays in some of the most popular magazines of his day and collected them into new books approximately every two years. His titles reveal something of the simple wonder that informs these books: *Fresh Fields* (1885), *Bird and Bough* (1906), and *Under the Apple-Trees* (1916).

Burroughs's work remained popular throughout his lifetime—a rare achievement for a writer. He formed lasting friendships with many of his admirers, including John Muir, Thomas Edison, and Henry

John Burroughs, right, poses in Alaska in 1899 with naturalist and preservationist John Muir, one of the many admirers of Burroughs's nature writings. (Library of Congress)

Ford. Two more friendships led to books: *John James Audubon* (1902), an appreciation and biography, and *Camping with President Roosevelt* (1906).

After Burroughs's death in 1921, the John Burroughs Association was founded and the John Burroughs Sanctuary was established to preserve his property and many of his books in West Park, New York. Burroughs continues to be acknowledged as a pioneer and a master of the genre of nature writing, and many of his books remain in print. The John Burroughs Association presents two annual awards for nature writing published in the preceding year: one to the author of an outstanding natural history essay and one to the author of a distinguished book of natural history. In 1997 Burroughs was named a charter member of the Ecology Hall of Fame in Santa Cruz, California.

Cynthia A. Bily

FURTHER READING

Walker, Charlotte Zoë, ed. *Sharp Eyes: John Burroughs and American Nature Writing.* Syracuse, N.Y.: Syracuse University Press, 2000.

Warren, James Perrin. *John Burroughs and the Place of Nature.* Athens: University of Georgia Press, 2006.

SEE ALSO: Audubon, John James; Muir, John; Roosevelt, Theodore; Thoreau, Henry David.

C

CAFE standards. *See* **Corporate average fuel economy standards**

Cancer clusters

CATEGORY: Human health and the environment

DEFINITION: Occurrences of larger-than-expected numbers of cases of cancer within groups of people in particular geographic areas over periods of time

SIGNIFICANCE: Environmental factors such as chemicals released into air or water can cause various types of cancer. If clusters of certain types of cancers are found, scientists may be able to trace the cancers to environmental causes, which can then be removed.

The concept of cancer clusters became widely understood in 2000, when the Hollywood film *Erin Brockovich* was a success with audiences. The real Brockovich, whose story inspired the motion picture, was a file clerk in a law office who discovered a cancer cluster in Southern California related to hexavalent chromium (chromium 6) released by the Pacific Gas and Electric Company. Many other cancer clusters have been noted, including lung cancers among miners, vaginal cancers in young women whose mothers took the synthetic estrogen diethylstilbestrol (DES) while pregnant, and radiation-related cancers in "downwinders," people in southern Utah and Nevada who lived downwind of nuclear bomb testing in the Nevada desert in the 1950's and 1960's.

Epidemiologists may investigate the possibility of a cancer cluster when the same type of cancer is diagnosed in a number of individuals who have something in common in their environment. Perhaps they live together or near each other (such as family members or neighbors) or spend time together in the same place (such as coworkers). One kind of indicator of a possible cancer cluster is a rare type of cancer appearing in high numbers of people; another is a type of cancer appearing where it is usually rare (for example, a cancer that usually appears in adults appearing in children). Epidemiologists evaluate the cancer's frequency and distribution in the population at large and determine whether a higher incidence of this type of cancer appears in the identified population.

Public health departments have formed cancer registries that collect data about the different types of cancers appearing in given areas. Public health officials then examine these registries to determine whether unusual types or striking amounts of cancers are being reported, perhaps in particular places or among people who work in the same industry. Constant surveillance of the data collected by cancer registries helps health officials to determine whether and where cancer clusters exist.

Actual scientific confirmation of a cancer cluster is a rare event. Often, not enough cases of cancer are present to make possible a meaningful statistical comparison to the rest of the population. Other problems in confirming a cluster include inability to assess accurately a person's exposure to the agent that is thought to cause the cancer, given that many people move in and out of neighborhoods and jobs frequently. Also, cancer is often caused by a combination of circumstances, including lifestyle choices, rather than by one particular event or agent, and cancers may take a long time to develop. The tendency of cancers to spread in the body makes it difficult to determine in which organ any given cancer originated and thus to assess whether the cancer in question was truly caused by a suspected environmental agent. In attempts to confirm cancer clusters, public health officials might collect information from medical records; take water, air, and soil samples; and test patients' bodily fluids and compare the results with those of others in the environment.

Marianne M. Madsen

FURTHER READING

Aldrich, Tim E., and Thomas H. Sinks. "Things to Know and Do About Cancer Clusters." *Cancer Investigations* 20, nos. 5-6 (2002): 810-816.

Brown, Phil. *Toxic Exposures: Contested Illnesses and the Environmental Health Movement.* New York: Columbia University Press, 2007.

Centers for Disease Control and Prevention. "Guidelines for Investigating Clusters of Health Events." *Morbidity and Mortality Weekly Report,* July 27, 1990, 1-23.

Nash, Linda. *Inescapable Ecologies: A History of Environment, Disease, and Knowledge.* Berkeley: University of California Press, 2006.

Thun, Michael J., and Thomas H. Sinks. "Understanding Cancer Clusters." *CA: A Cancer Journal for Clinicians* 54 (2004): 273-280.

SEE ALSO: Alar; Asbestosis; Brockovich, Erin; Carcinogens; Centers for Disease Control and Prevention; Environmental illnesses; Hazardous and toxic substance regulation; Love Canal disaster.

Captive breeding

CATEGORY: Animals and endangered species

DEFINITION: Selective breeding of an endangered or threatened species in captivity

SIGNIFICANCE: Captive breeding is sometimes the only way to save a species from extinction. Captive-bred animals are often returned to their native habitats once their populations have sufficiently recovered and the animals have been properly conditioned to survive in the wild.

As conditions in zoos steadily improved throughout the twentieth century and captive animals began breeding, scientists realized that the breeding of threatened and endangered species in captivity could save some species that would otherwise become extinct. Many zoos began to shift their priorities from entertainment to wildlife conservation in the late 1970's.

Initially, zoo animals were allowed to breed without consideration of their genetic or health status—the only guiding principle involved was "more is better." Zoo populations, however, are far too small to sustain healthy breeding; depending on the species, a healthy breeding population might range from seventy-five to four hundred animals. The inbreeding that resulted from early zoo policies led to fewer offspring, increased rates of birth defects, and susceptibility to disease. In response, the Association of Zoos and Aquariums established the Species Survival Plan (SSP) in 1981 to increase the number of animals in the breeding pool. A computerized mating system, the SSP maintains genetic records of all captive animals of particular species, allowing zoos to ex1change animals without fear of inbreeding. By 2009 there were 116 SSP programs covering 172 species.

In the 1990's captive breeding began to lose some of its appeal as the costs began to outweigh the bene-

A wild Arabian oryx in Israel's Negev region. This species was saved from extinction by captive-breeding programs. (©Dreamstime.com)

fits. In vitro fertilization of gorillas can cost up to $75,000, and the cost of the California Condor Recovery Plan was an estimated $20 million between 1974 and the late 1990's. Critics of captive-breeding programs—and zoos in general—argue that such large amounts of money could be better spent preserving animals' natural habitats. Opponents of the U.S. Endangered Species Act (1973) have also used the existence of captive-breeding programs as an argument against habitat protection laws.

Another drawback of SSPs and captive breeding in general is the problem of surplus animals. Breeding endangered species may be a good public relations tactic, but the culling of surplus animals is definitely not. One large issue faced by captive-breeding programs is what is to be done with animals that are genetically inferior or past breeding age, or those that have been hybridized in ways that would contaminate the gene pool. Contraceptive implants and separation are two common solutions, but some surplus animals end up in game parks and hunting preserves.

Despite its drawbacks, captive breeding remains the only option for some species, the most famous of which is the California condor. In an extraordinary measure of intervention, the last 9 known condors in the wild were captured in 1985, bringing the total number in captivity to 27. By 1992 captive breeding had brought the population up to 134, and scientists gradually began returning birds to their natural habitat. By 2010, 348 condors were known to be living, 187 of them in the wild. Other species that have been saved by captive breeding include the black-footed ferret, the Arabian oryx, Przewalski's horse, and Père David's deer; the first three of these have been reintroduced, in limited numbers, into their original habitats.

P. S. Ramsey

FURTHER READING

Loftin, Robert W. "Captive Breeding of Endangered Species." In *Preserving Wildlife: An International Perspective*, edited by Mark A. Michael. Amherst, N.Y.: Humanity Books, 2000.

Moir, John. *Return of the Condor: The Race to Save Our Largest Bird from Extinction*. Guilford, Conn.: Lyons Press, 2006.

SEE ALSO: Condors; Endangered species and species protection policy; Extinctions and species loss; Whooping cranes; Zoos.

Carbon capture and storage. *See* Carbon dioxide air capture

Carbon cycle

CATEGORIES: Atmosphere and air pollution; weather and climate
DEFINITION: Pathways by which carbon moves through the environment
SIGNIFICANCE: The balance of the carbon cycle determines the atmospheric concentration of carbon dioxide, which, through its role as a greenhouse gas, modulates the earth's temperature. Human activities that affect the carbon cycle thus have an effect on global climate.

Carbon is naturally exchanged between the atmosphere and the oceans, the terrestrial biosphere and soils, and the solid earth. The preindustrial balance of these carbon exchanges led to an atmospheric carbon dioxide concentration of 280 parts per million. Human industrial activities have added carbon dioxide to the atmosphere, increasing the atmospheric concentration to 380 parts per million.

Carbon is naturally removed from the atmosphere by photosynthesis on land and by dissolution of atmospheric carbon dioxide in the oceans. Carbon is introduced into the atmosphere by respiration and combustion of terrestrial organic matter, by outgassing of carbon dioxide from the oceans, and as a by-product of human industrial activities. Terrestrial plants remove about 15 percent of the atmosphere's carbon dioxide each year through photosynthesis. About half of this carbon is respired by these same plants as they release energy for their internal metabolic processes. The other half of the carbon removed from the atmosphere by terrestrial photosynthesis is primarily returned to the atmosphere through respiration of organic matter by decomposers. Most of this carbon is returned to the atmosphere as carbon dioxide through aerobic respiration, but in low-oxygen wetland environments carbon can be respired anaerobically and released to the atmosphere as methane, a much more potent greenhouse gas. This methane is then broken down to carbon dioxide in the atmosphere over a timescale of about eight years.

Approximately 12 percent of the atmosphere's car-

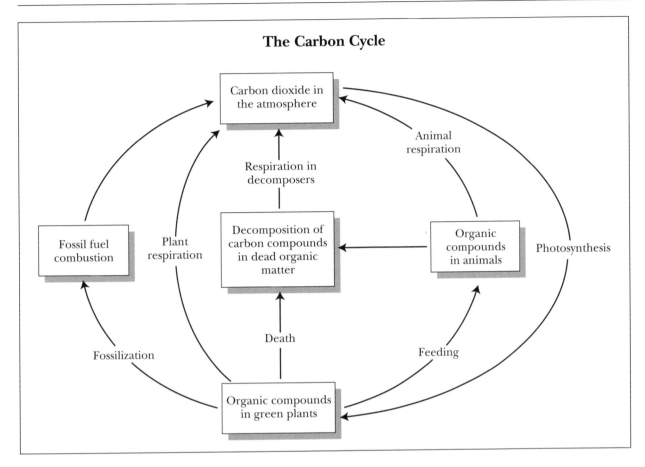

bon dioxide is exchanged with the oceans each year. Carbon is more soluble in cold water than in hot water, and so the oceans take up carbon at high northern latitudes and emit carbon in the Tropics. Photosynthesis in the surface ocean binds dissolved carbon into organic matter. Much of this carbon is returned to the surface ocean by breakdown of this organic matter, but some of this material is packaged into large enough clumps of organic matter that it sinks under its own weight and is redissolved in the deep ocean. This process, known as the biological pump, acts to move carbon (and nutrients) from the surface ocean to depth. In ocean upwelling zones, carbon-rich water from depth is brought up to the surface, producing a source of carbon to the atmosphere. Exchanges of carbon between the solid earth and the other reservoirs are important only on geologic timescales.

Human industrial activity introduces an amount of carbon to the atmosphere each year equivalent to about 1 percent of the total atmospheric carbon content. The rate of atmospheric carbon dioxide increase is, however, only half the rate at which industrial activity introduces carbon dioxide to the atmosphere. The other half is removed from the atmosphere by dissolution in the oceans and uptake by the land surfaces.

Carbon-cycle models predict that global warming will trigger processes that will alter the natural carbon cycle in a way that will decrease the ability of natural processes to remove industrial carbon from the atmosphere and accelerate the rate of increase of atmospheric greenhouse gases. Warming of the oceans decreases the solubility of carbon dioxide in water. Decomposition of organic material in soils to produce carbon dioxide proceeds more rapidly at higher temperatures. Permanently frozen soils at high latitudes (permafrost) globally hold more carbon than the atmosphere, and this carbon becomes exposed to decomposition when the permafrost thaws.

Alexander R. Stine

FURTHER READING

Field, Christopher B., and Michael R. Raupach, eds. *The Global Carbon Cycle: Integrating Humans, Climate, and the Natural World.* Washington, D.C.: Island Press, 2004.

Houghton, R. A. "The Contemporary Carbon Cycle." *Treatise on Geochemistry* 8, no. 10 (2003): 473-513.

Wigley, T. M. L., and D. S. Schimel, eds. *The Carbon Cycle.* New York: Cambridge University Press, 2000.

SEE ALSO: Biogeochemical cycles; Carbon dioxide; Carbon footprint; Deforestation; Ecosystem services; Fossil fuels; Global warming; Greenhouse effect; Greenhouse gases; Methane; Positive feedback and tipping points.

Carbon dioxide

CATEGORIES: Pollutants and toxins; atmosphere and air pollution

DEFINITION: Chemical compound in which molecules are composed of one carbon atom and two oxygen atoms

SIGNIFICANCE: The increase of carbon dioxide in the earth's atmosphere that has been occurring since the Industrial Revolution, if not earlier, has been linked to global warming. International efforts undertaken to limit emissions of carbon dioxide have met with varying degrees of success.

Carbon dioxide (CO_2), a gas that is essential to life on earth, is generated by the burning of fossil fuels such as wood, oil, and coal and by the decomposition of organic matter. Approximately 57 percent of greenhouse gases, of which CO_2 is one, come from the burning of fossil fuels. Land clearing also contributes to CO_2 in the atmosphere. A large amount of CO_2 is found in deposits in the floors of the earth's oceans. When this gas is released into the atmosphere, it remains and helps to trap solar radiation in a process that has been shown to contribute to climate change.

The amount of carbon dioxide in the atmosphere has varied over the life of the planet; the amount has been both much lower and much higher than that present during the early twenty-first century. The CO_2 level in the atmosphere was relatively stable, however, from the end of the last ice age until the nineteenth century, averaging 280 parts per million (ppm). From the start of the Industrial Revolution to the middle of the twentieth century, CO_2 levels grew at a steady rate,

and the rate of growth increased in the latter half of the twentieth century. When measurements were first undertaken at the Mauna Loa Observatory in Hawaii in the 1950's, the measured concentration of CO_2 was 315 ppm; by 2008 the concentration stood at 385 ppm. Some scientists estimate that the amount of CO_2 in the earth's atmosphere could be as high as 1,000 ppm by the year 2100 if emissions continue to increase unchecked. Such a level could produce global temperatures that are 5 degrees Celsius (9 degrees Fahrenheit) warmer than in the early twenty-first century, a level not seen for several million years. Even a moderate increase to 440 ppm is expected to produce a temperature increase of 3 degrees Celsius (5.4 degrees Fahrenheit).

Industrialized countries such as the United States and Germany are major producers of all greenhouse gases, especially CO_2, owing to their high levels of use of fossil fuels. Some industrialized nations have turned increasingly to energy sources that produce less CO_2, such as natural gas. However, industrializing nations such as India and China are also major producers of CO_2, largely because they burn a great deal of coal, which produces more CO_2 than other fossil fuels.

Somewhat surprisingly, the industrializing nation of Brazil is a significant producer of CO_2 as the result of the clearance of massive areas of land in the Amazon basin for agriculture; this process produces CO_2 as the biomass decomposes or is burned. Other, less industrialized nations in Asia, Africa, and South America also produce CO_2 as they harvest large amounts of timber and clear land for agriculture or mining.

The Kyoto Protocol of 1997, an international agreement aimed at limiting and reducing CO_2 emissions, applied only to industrialized countries and countries such as China and India that have exhibited the fastest rate of growth in CO_2 emissions. Although some countries have tried to reduce CO_2 emissions, the United States initially refused to ratify the agreement and, during the presidential administration of George W. Bush, was slow to try to limit emissions, arguing that industrial production was more important than potential climate change. The delegates to the 2009 United Nations Climate Change Conference, held in Copenhagen, Denmark, attempted to address the issue of regulating CO_2 and other greenhouse gas emissions further, but little was accomplished.

John M. Theilmann

Further Reading

Houghton, John. *Global Warming*. 3d ed. New York: Cambridge University Press, 2004.

Smail, Vaclav. *Cycles of Life*. New York: Scientific American, 1997.

Volk, James. *CO_2 Rising*. Cambridge, Mass.: MIT Press, 2008.

See also: Air pollution; Air-pollution policy; Carbon cycle; Carbon footprint; Climate change and oceans; Coal-fired power plants; Deforestation; Fossil fuels; Kyoto Protocol; United Nations Framework Convention on Climate Change.

Carbon dioxide air capture

Category: Atmosphere and air pollution

Definition: The trapping or elimination of carbon dioxide emitted from industrial or commercial sources before it can enter the atmosphere

Significance: Carbon dioxide capture and sequestration or storage is anticipated to play a bridging role between carbon dependence and a sustainable low-carbon energy future by serving as the critical enabling technology that will lead to a significant reduction in CO_2 release into the air while allowing industrial processes such as the burning of coal for power generation to continue to meet global energy needs.

The term "carbon dioxide air capture" is used to describe a set of technologies aimed at preventing the carbon dioxide (CO_2) emitted from industrial or commercial sources from entering the atmosphere. The processes involved are also commonly referred to as carbon dioxide capture and sequestration or carbon capture and storage (CCS). The geoengineering technique of scrubbing (absorbing) CO_2 from ambient air is also sometimes referred to as CSS, as are biological techniques that employ organisms (such as plankton) and organic matter to capture CO_2 from the air.

In CCS, the process of capturing CO_2 from a large emissions source is often coupled with the subsequent compression of the gas, storage of the gas (for example, by injecting it into deep underground geological formations or into deep ocean masses called saline aquifers, or by converting it into the form of mineral carbonate), and recycling of the gas to enhance industrial processes, such as that seen in CO_2-assisted enhanced oil recovery (EOR), in which CO_2 gas is injected into an oil-bearing stratum under high pressure to cause oil to be displaced upward.

The CCS process begins with the capture of CO_2 generated by a power station or other industrial facility, such as a cement factory, steelworks, or oil refinery. The CO_2 can be captured before, during, or after the source's combustion (burning) of fossil fuels. Precombustion capture involves the separation of fossil fuels into hydrogen and CO_2 before they are burned. For example, in the instance of coal, the process involves the conversion of coal into a synthetic gas (syngas) made up of carbon monoxide and hydrogen. The syngas is reacted further with steam to produce a CO_2-hydrogen mix. Further processing produces a mix with a high concentration of CO_2, which is separated out. The remaining hydrogen is then utilized as a CO_2-free energy source that produces only heat and water vapor when combusted. Precombustion capture technology has been widely implemented in the fertilizer industry as well as in natural gas forming.

In oxy-fuel combustion, burning of the fossil fuel in oxygen instead of air results in an exhaust gas that is CO_2-free. This technology is commonly used in the glass furnace industry. In postcombustion capture, the CO_2 is separated from flue (exhaust) gases after the combustion of fossil fuels. The CO_2 content is usually much lower than in the gas that is separated during oxy-fuel combustion or precombustion capture, with the volume of CO_2 in the range of 3 percent to 15 percent by volume.

Challenges to Implementation

One challenge facing CCS is the demonstration of its efficacy and safety on an industrial scale at competitive cost. While CCS is known to be safe and is well understood in terms of the fundamental science and technical requirements, no evidence has been gathered regarding the process's long-term impacts on the environment (for example, the safety of storing CO_2 in geological formations) or possible danger to humans (for instance, if CO_2 leaks from storage).

CCS applied to a modern power plant could reduce CO_2 emissions by up to 90 percent compared to an equivalent plant with no CCS devices, but the implementation of CO_2 capture is significantly more expensive than the use of traditional systems of emissions control. For example, capturing and compressing CO_2 increases the fuel requirement of a coal-fired plant by as much as 25-40 percent. It is estimated that the cost of energy produced by a new power plant with CCS is from 21 percent to 91 percent higher than that produced by a non-CCS power plant.

Aside from the cost and technical challenges of CCS, a regulatory framework needs to be established to support CCS and to clarify at regional, national, and international levels the long-term rights, liabilities, and technical requirements associated with the use of CCS technologies. Moreover, before investors, scientists, politicians, and industries can be persuaded that CCS is a worthwhile investment, agreements need to be reached on a price on carbon emissions and on whether a carbon tax, cap-and-trade regime, or other carbon-trading/taxation framework will be implemented.

Rena Christina Tabata

FURTHER READING

Hanjalić, K., R. van de Krol, and A. Lekić, eds. *Sustainable Energy Technologies: Options and Prospects.* London: Springer, 2008.

Kutz, Myer. *Environmentally Conscious Fossil Energy Production.* Hoboken, N.J.: John Wiley & Sons, 2009.

Rackley, Steve. *Carbon Capture and Storage.* Boston: Butterworth-Heinemann, 2009.

Rojey, Alexandre. *Energy and Climate: How to Achieve a Successful Energy Transition.* Chichester, England: John Wiley & Sons, 2009.

Shiosani, Fereidoon P. *Generating Electricity in a Carbon-Constrained World.* Burlington, Mass.: Academic Press, 2009.

SEE ALSO: Air pollution; Carbon dioxide; Chicago Climate Exchange; Coal; Coal-fired power plants; Global warming; Greenhouse gases.

Carbon footprint

CATEGORIES: Resources and resource management; energy and energy use

DEFINITION: Total amount of greenhouse gases produced to directly and indirectly support a certain human activity or a representative lifestyle, usually expressed in equivalent tons of carbon dioxide

SIGNIFICANCE: The use of carbon footprint estimates allows individuals, organizations, and nations to understand their relative contribution to emissions of greenhouse gases, which have been linked to global warming. In turn, these estimates are used in the purchase of carbon offsets and in gauging progress toward goals set under international agreements such as the Kyoto Protocol.

The concept of the carbon footprint provides a useful way for individuals and organizations to understand their impact in contributing to global warming. The calculation of a carbon footprint has been extended to include assessment of the greenhouse gas contributions made by the production and shipment of individual products. The general concept is that once the magnitude of a carbon footprint is known, a strategy can be devised to reduce that footprint. Examples of reduction methods include technological developments, carbon capture, and improved process and product management. Additionally, carbon offsetting through the development of alternative projects, such as solar or wind power, is a common way to reduce carbon footprints.

The concept and name of the carbon footprint originate from the concept of the ecological footprint. The carbon footprint is related to both the ecological footprint and the overarching life-cycle assessment (LCA) approach. A carbon footprint is composed of two elements: the primary, or direct, footprint and the secondary, or indirect, footprint. The primary footprint is the amount of direct carbon dioxide (CO_2) emissions stemming from the burning of fuels—that is, for domestic energy consumption and transportation. The secondary footprint is the amount of indirect CO_2 emissions caused over the whole life cycle of products, including the manufacture and eventual breakdown of the products.

Calculation of a carbon footprint generally begins with CO_2 emissions based on fuel consumption.

Average Carbon Footprints Around the World

NATION	AVERAGE CARBON FOOTPRINT PER CAPITA, 2004 (METRIC TONS)
Kuwait	38.00
United States	20.40
Australia	16.30
Russia	10.50
Germany	9.79
United Kingdom	9.79
Japan	9.84
Italy	7.69
China	3.84
India	1.21

Source: Data from U.S. Department of Energy.

Greenhouse gases other than CO_2 are typically translated to CO_2 equivalents through conversion rates, based on mass. Common greenhouse gases include methane, nitrous oxide, and chlorofluorocarbons (CFCs). To allow comparisons among the varied effects of different gases on the environment, scientists have developed methods of determining the greenhouse potency (global warming potential) of gases relative to that of carbon dioxide.

The calculation of carbon footprints for products is becoming mainstream in people's understanding of products and lifestyle adjustments to reduce greenhouse gas emissions. These carbon footprint analyses are also useful in guiding regulation. For example, the U.S. Environmental Protection Agency has calculated carbon footprints for paper, plastic, glass, cans, computers, and tires, among other materials. Researchers in Australia, Korea, and the United States have addressed the carbon footprints of paved roads. The United Kingdom Carbon Trust has worked with manufacturers on assessing foods, shirts, and detergents.

The ready availability of carbon footprint calculators on the Internet has led many individuals to analyze their own carbon footprints. These calculators ask questions concerning lifestyle (such as how many times a year an individual takes airline flights) to arrive at an approximate carbon footprint.

On an international scale, the 1997 Kyoto Protocol defines targets and timetables for reductions in greenhouse gas emissions. An understanding of carbon footprints is necessary to the implementation of the flexible mechanisms defined under this international agreement (certified emission reduction, joint implementation, and emissions trading), which require nations to determine necessary levels of reduction and to maintain benchmarking activities.

Jennifer F. Helgeson

FURTHER READING

Brown, Marilyn A., Frank Southworth, and Andrea Sarzynski. *Shrinking the Carbon Footprint of Metropolitan America.* Washington, D.C.: Brookings Institution Press, 2008.

Sim, Stuart. *The Carbon Footprint Wars: What Might Happen If We Retreat from Globalization?* Edinburgh: Edinburgh University Press, 2009.

Weber, Christopher L., and H. Scott Matthews. "Quantifying the Global and Distributional Aspects of American Household Carbon Footprint." *Ecological Economics* 66 (June, 2008): 379-391.

Wiedmann, Thomas, and Jan Minx. "A Definition of 'Carbon Footprint.'" In *Ecological Economics Research Trends*, edited by Carolyn C. Pertsova. Hauppauge, N.Y.: Nova Science, 2007.

SEE ALSO: Carbon dioxide; Ecological footprint; Fossil fuels; Global warming; Life-cycle assessment.

Carbon monoxide

CATEGORIES: Pollutants and toxins; atmosphere and air pollution

DEFINITION: Odorless and colorless toxic gas in which molecules consist of one carbon atom and one oxygen atom

SIGNIFICANCE: The air pollutant carbon monoxide is a component of automobile exhaust emissions, but the gas is perhaps most dangerous indoors, where it can cause accidental death by asphyxiation.

Carbon monoxide (CO), an indoor and outdoor air pollutant, is formed by incomplete combustion of fossil fuels when insufficient oxygen is present to convert carbon compounds to nontoxic carbon dioxide. Indoors, CO may come from unvented kerosene and gas space heaters; leaking or improperly vented chimneys, furnaces, gas water heaters, wood stoves, gas stoves, and fireplaces; generators and other gasoline-powered equipment; automobile exhaust from attached garages; and tobacco smoke. Outdoors, CO is present in small amounts naturally in the atmosphere, chiefly as a product of volcanic activity, but also from naturally occurring fires.

Motor vehicle exhaust contributes about 56 percent of all CO emissions in the United States. The highest levels of CO occur in areas with heavy traffic congestion, and in cities, 85-95 percent of all CO emissions may come from motor vehicle exhaust. Other nonroad engines and vehicles (such as construction equipment and boats) contribute about 22 percent of CO emissions nationwide. Additional sources of CO emissions include industrial processes (metals processing and chemical manufacturing) and residential wood burning. The highest levels of outdoor CO typically occur during the winter months, for three primary reasons: Motor vehicles need more fuel to start at cold temperatures, some emissions-control devices (oxygen sensors and catalytic converters) operate less efficiently in the cold, and temperature inversion conditions, which trap CO near the ground beneath a layer of warm air, are more frequent.

CO enters the bloodstream through the lungs and binds to the blood protein hemoglobin, the main function of which is to transport oxygen to body tissues, including vital organs such as the heart and brain. The CO and hemoglobin combine to form the compound carboxyhemoglobin, and the hemoglobin is no longer available for transporting oxygen. In closed indoor environments, the level of carboxyhemoglobin in the body can rise to toxic levels and eventually result in asphyxiation.

People with heart disease are especially sensitive to CO poisoning because of its reduction of oxygen transport through the blood; they may experience chest pain if they breathe the gas. Infants, elderly persons, and individuals with respiratory diseases are also particularly sensitive to CO. CO can affect healthy individuals by impairing exercise capacity, visual perception, manual dexterity, learning ability, and the ability to perform complex tasks.

During the early 1970's, the U.S. Environmental Protection Agency (EPA) set two national air-quality standards for CO in motor vehicle emissions: a one-hour standard of 35 parts per million (the highest allowable level of CO measured in ambient air over a one-hour period) and an eight-hour standard of 9 parts per million. The EPA also mandated reductions in emissions from large industrial facilities. Advances in vehicle technologies and the development of cleaner fuels have resulted in reductions in CO emissions. Across the United States, air-quality stations measure the levels of CO (and other pollutants) in the air on a regular basis. The measurements are compared with the EPA standards, and areas that have CO levels that exceed the standards are required to develop and carry out plans to reduce their CO emissions. As a result of these steps, CO emissions from automobiles, motorcycles, and light- and heavy-duty trucks have been reduced by more than 40 percent since 1970. The greatest reduction has been the nearly 60 percent decrease in automobile emissions of CO.

Bernard Jacobson

FURTHER READING

Kleinman, Michael T. "Carbon Monoxide." In *Environmental Toxicants: Human Exposures and Their Health Effects*, edited by Morton Lippmann. 3d ed. Hoboken, N.J.: John Wiley & Sons, 2009.

Penney, David G., ed. *Carbon Monoxide Poisoning*. Boca Raton, Fla.: CRC Press, 2008.

SEE ALSO: Air pollution; Automobile emissions; Catalytic converters; Indoor air pollution; Smog.

Carcinogens

CATEGORIES: Human health and the environment; pollutants and toxins

DEFINITION: Substances or physical agents that cause or worsen cancer

SIGNIFICANCE: The effects of human exposure to carcinogens in the environment may include the development of different types of illness, deaths, and economic obligations on a national and global scale.

Cancer is a leading cause of death throughout the world. Environmentalists and others have raised concerns regarding the cancer-causing (carcinogenic) potential of exposure to a constantly growing number of both newly developed and long-existing chemicals in the environment. In addition, humans seem to be increasingly exposed to various sources of electromagnetic waves, such as microwaves, and some groups and individuals are concerned about the possible carcinogenicity of these physical phenomena.

Carcinogens can be categorized based on their origin as chemicals (naturally occurring or synthetic), physical agents, or infectious agents. Chemical carcinogens can be classed as compounds that occur naturally, such as aflatoxins, chromium 6 compounds, and arsenic compounds; and others that are largely synthetic in origin, such as benzene, formaldehyde, vinyl chloride, and dioxins. In addition there are carcinogenic minerals, such as asbestos. Other carcinogenic chemicals are elements or substances such as radon (a radioactive gas), beryllium, and cadmium. Some carcinogens—such as tobacco smoke and alcoholic beverages—are mixtures of compounds. Physical agents that are carcinogens include solar radiation (primarily ultraviolet radiation), gamma rays, and X rays.

Infectious disease agents that have been implicated as carcinogens include human papilloma virus (HPV), which can cause cervical cancer; *Helicobacter pylori* (*H. pylori*), a bacterium causally associated with stomach cancer; the hepatitis C and hepatitis B viruses, which can cause liver cancer; Epstein-Barr virus, which is associated with Burkitt's lymphoma; and human T-lymphotrophic virus type 1 (HTLV-1), which has been linked to leukemia in adults. The human immunodeficiency virus (HIV), which causes acquired immunodeficiency syndrome (AIDS), has been associated with Kaposi's sarcoma. Viruses have also been shown definitively to cause tumors in animals such as mice (mammary tumor virus), chickens (Rous sarcoma virus), and Tasmanian devils.

Exposure to carcinogens can be related to work environments, such as in the case of workers in the nuclear power and medical radioisotope industries. Sometimes carcinogenic agents happen to be concentrated in particular geographic regions; for example, widespread exposure to the carcinogen arsenic occurred in Bangladesh as the result of contaminated drinking water from a large number of wells that accessed groundwater in which arsenic was uncommonly abundant. Exposure to *H. pylori* is believed to occur through contaminated water supplies and thus is considered to be environmental in origin. Tobacco smoke, a major cause of lung cancer, is an environmental carcinogen to which people are widely exposed.

Carcinogens may lead to cancer by directly damaging deoxyribonucleic acid (DNA), as in the case of radiation; through conversion through metabolism; or through effects on metabolism. In the case of viruses, viral genetic material may be incorporated into host DNA at sites of oncogenes, which are genes whose altered function or disruption leads to cancer. Different carcinogens often lead to effects on different organs, thus *H. pylori* is associated mainly with stomach cancer, whereas tobacco smoking or use is associated with lung, oral, and laryngeal cancers (and also bladder, colon, and kidney cancers, among others), and asbestos is primarily linked to lung cancer. The time period and frequency of exposure to carcinogens as well as a person's genetic background can also influence the likelihood that a carcinogen causes cancer.

Oluseyi A. Vanderpuye

FURTHER READING

Hill, Marquita K. "Chemical Exposures and Risk Assessment." In *Understanding Environmental Pollution.* 3d ed. New York: Cambridge University Press, 2010.

McKinnell, Robert. G., et al. *The Biological Basis of Cancer.* 2d ed. New York: Cambridge University Press, 2006.

Ward, Elizabeth M. "Cancer." In *Occupational and Environmental Health: Recognizing and Preventing Disease and Injury,* edited by Barry S. Levy et al. 5th ed. Philadelphia: Lippincott Williams & Wilkins, 2006.

SEE ALSO: Arsenic; Asbestosis; Bioassays; Cancer clusters; Environmental health; Environmental illnesses; Hazardous and toxic substance regulation; Heavy metals; Radioactive pollution and fallout; Radon.

Carpooling

CATEGORY: Resources and resource management

DEFINITION: Shared use of vehicles for private journeys

SIGNIFICANCE: A popular method of reducing commuting costs for individuals, carpooling offers subtantial benefits to communities and to the environment by reducing traffic congestion, fuel consumption, and pollution of the atmosphere.

Carpooling is often an informal practice in which friends, neighbors, or colleagues ride together in their daily commutes. It can also involve extensive formal networks of drivers and riders. The forming of a car pool typically starts with the assembly of a list of people, such as employees of a given company or people living in the same area, interested in sharing rides to work or other destinations. The individuals on the list then contact one another and make arrangements accordingly. Fixed and recurring journeys, such as a shared commuting route, are the most common framework for organizing carpooling. Alternatively, in some cities certain roadside locations are popular sites where drivers riding alone in their vehicles can pick up passengers and thus take advantage of high-occupancy vehicle (HOV) highway lanes. The costs of carpooling journeys are usually shared among drivers and riders in various ways; participants may rotate the cars used, split rental or fuel charges, or pay the owner of the car.

Proponents of the wider adoption of carpooling schemes note that the practice of carpooling has several benefits. Among these are cost savings for carpoolers in terms of reduced fuel usage and car wear; carpooling is also generally cheaper than public transport because salaried drivers are not required. Most participants are relieved of the stresses of driving, and many carpoolers share driving duties, so all participants receive this benefit over time; the problem of finding parking spaces is also greatly reduced. From the community perspective, traffic congestion is reduced, as is the strain on parking services. Environmental benefits include reduced fuel consumption and an accompanying reduction in exhaust emissions. Other benefits connected with carpooling include increased mobility for nondrivers and for those in rural areas who lack access to public transport.

Local communities, including private companies and government authorities, support carpooling in various ways. The most visible of these is the provision of HOV lanes on busy routes; these specially marked lanes on freeways and other roads are reserved for public transportation vehicles and for private automobiles and trucks carrying multiple occupants. Preferential parking is also assigned to carpooling vehicles in some cities, and some employers offer financial incentives to encourage their employees to carpool. To support carpooling further, some employers provide guaranteed ride services—often in the form of prepaid taxi services—for occasions when carpooling arrangements break down and individual passengers are left without drivers.

An important development in carpooling is the transition from relatively static and prearranged carpooling to what has been called dynamic carpooling. Unlike habitual ride sharing, dynamic carpooling enables individuals to find drivers for one-time journeys at short notice. Such real-time and flexible matching of riders and drivers has become possible through technological innovations. Since the beginning of the twenty-first century, several pilot schemes in Europe and the United States have examined the feasibility of large-scale real-time matching. The aim is to establish dynamic carpooling as a flexible and convenient mode of transportation. An example is the Covoiturage carpooling scheme operating in France.

The success of carpooling schemes depends on the attainment of "critical mass"—a sufficiently large number of participants to ensure that both riders and drivers have confidence in the program. Carpooling schemes might struggle in this regard for several reasons. These include individual drivers' unwillingness to surrender the flexibility and freedom of a personal car, attachment to the personal space and privacy provided by a private car, fears about a possible lack of safety, and the perception of marginal utility—that is, once a car has been purchased and taxed, individual journeys seem to be inexpensive.

Andrew Lambert

FURTHER READING

Balbus, John, and Dushana Yoganathan Triola. "Transportation and Health." In *Environmental Health: From Global to Local*, edited by Howard Frumkin. 2d ed. Hoboken, N.J.: John Wiley & Sons, 2010.
Kemp, Roger L., ed. *Cities and Cars: A Handbook of Best Practices.* Jefferson, N.C.: McFarland, 2007.
Marzotto, Toni, Vicky Moshier Burnor, and Gordon Scott Bonham. *The Evolution of Public Policy: Cars and the Environment.* Boulder, Colo.: Lynne Rienner, 2000.
Moavenzadeh, F., and M. J. Markow. "Transportation Policy and Environmental Sustainability." In *Moving Millions: Transport Strategies for Sustainable Development in Megacities.* New York: Springer, 2007.

SEE ALSO: Automobile emissions; Gasoline and gasoline additives; Road systems and freeways; Urban sprawl.

Carrying capacity

CATEGORY: Population issues

DEFINITION: The number of individuals within a given population that an environment is able to sustain indefinitely

SIGNIFICANCE: The number of a species that can survive within a given area is limited by the area's biological carrying capacity. Although populations might exceed or dip below the carrying capacity temporarily, the numbers fluctuate around a stable line of equilibrium. Environmentalists have expressed concern that with continued human population growth, humankind may one day exceed the earth's carrying capacity.

In theory, all populations have the potential for exponential growth or the ability to increase indefinitely in number. In nature, however, there are not infinite numbers of every possible species. No population is able to grow exponentially for long because certain factors will cause the birthrate to decrease and the death rate to increase. Every population within a given area has a maximum number it can reach, and this population size is limited by the area's carrying capacity. Each additional individual introduced beyond the carrying capacity makes it more difficult for the existing population to survive. Therefore, at a certain point, population growth stops.

A number of limiting factors can cause population growth to stop, such as food and water availability. If there is only enough food to feed a set population, any individuals that exceed that capacity will not survive. Similarly, a limited amount of habitat is available for each species within an area. If an area does not have enough nesting sites to support a large population, individuals will migrate from that area or adapt until the numbers in the area are again within its carrying capacity. Another factor in the limiting of population size is interaction with predators. If the number of prey increases dramatically in an area, more predators are likely to enter the area, which in turn leads to a reduction in the number of prey. Conversely, if the number of predators exceeds the number of prey necessary to sustain them, the excess predators will die or be forced to migrate to new areas. Disease is another factor that can limit population size: In large populations diseases tend to affect more individuals, and this can result in large reductions in numbers.

As a species, human beings are biologically limited by environment just as are other species. For centuries the carrying capacity for humankind was naturally low. High rates of disease, high infant mortality rates, and short life expectancies kept populations in check. However, in time human beings learned cultivation and domestication. The need to find food became less of a limiting factor on population growth when humans learned to grow enough food to sustain themselves. Humans also discovered ways to use fossil fuels, learned how to treat diseases and injuries, and improved hygiene, all of which led to increases in population. With the inventions of new technologies it seemed as if the limits of biological carrying capacity no longer applied to humans. This is a controversial topic, however; many environmentalists and other observers assert that there will be a point at which the earth will no longer be able to sustain an increasing human population. They note that, in any case, continued growth in the human population will be accompanied by high costs, particularly in the numbers of other species the planet can sustain.

Kathryn A. Cochran

FURTHER READING

Brown, Lester R., and Hal Kane. *Full House: Reassessing the Earth's Population Carrying Capacity*. New York: W. W. Norton, 1994.

Jensen, Derrick. *Endgame*. 2 vols. New York: Seven Stories Press, 2006.

Manning, Robert E. *Parks and Carrying Capacity: Commons Without Tragedy*. Washington, D.C.: Island Press, 2007.

SEE ALSO: Balance of nature; Ecosystems; *Limits to Growth, The*; Malthus, Thomas Robert; Population-control and one-child policies; Population-control movement; Population growth; United Nations population conferences.

Carson, Rachel

CATEGORIES: Activism and advocacy; ecology and ecosystems

IDENTIFICATION: American author and environmentalist

BORN: May 27, 1907; Springdale, Pennsylvania

DIED: April 14, 1964; Silver Spring, Maryland

SIGNIFICANCE: As the author of *Silent Spring* (1962) and other best-selling books, Carson helped to spark the modern environmental movement.

Even before the publication of *Silent Spring* made her a household name, Rachel Carson had a notable career. She was born on May 27, 1907, in Springdale, Pennsylvania, approximately eighteen miles from Pittsburgh; her mother, the daughter of a Presbyterian minister, instilled a love of nature in her three children. At the age of eighteen, Carson entered the Pennsylvania College for Women (later Chatham College), where she contributed many works to the school newspaper. Midway through her college career, Carson changed her major from English to biology, and she was accepted to graduate school at The Johns Hopkins University.

Following her graduation from Pennsylvania College for Women in the spring of 1929, Carson studied under a scholarship at the Marine Biological Laboratory at Woods Hole on Cape Cod. She received her master's degree in marine zoology in 1932 and subsequently took a job with the Bureau of Fisheries, which later became the U.S. Fish and Wildlife Service. Carson continued her work for the bureau for sixteen years, rising to become editor in chief of the publications department.

In 1941, Carson's first book, *Under the Sea Wind*, a lyrical exploration of the sea and its life, was published to excellent reviews. Her 1951 *The Sea Around Us* proved even more successful, winning the National Book Award and the John Burroughs Medal; it remained on the best-seller lists for more than a year. Carson was awarded a Guggenheim Fellowship, but she returned the money after receiving substantial royalties from her second book. The resulting financial independence allowed her to resign from her government post and devote herself to her writing.

In 1955, she published *The Edge of the Sea*, another best seller. It was the 1962 publication of *Silent Spring*, however, that transformed Carson from a successful nature writer to a controversial public figure. *Silent Spring* and related articles by Carson alerted the public to the fact that pesticides, particularly dichloro-diphenyl-trichloroethane (DDT), were decimating the bird populations of North America. Although chemical companies and other vested interests attacked the work savagely, *Silent Spring* attracted a broad readership and had a profound effect on public policy.

Carson influenced environmental policy on two levels. First, *Silent Spring* and related works led to the virtual ban of the use of DDT. Public views of pesticides and toxic chemicals changed forever, and politicians responded with a generation of legislation regulating the use of pesticides and other chemicals. Despite the criticism it engendered, *Silent Spring* remains a defining critique of the indiscriminate use of

Author and environmentalist Rachel Carson. (Library of Congress)

chemicals, and the ecological dangers of modern technology have remained a scientific and public concern since its publication. At a broader level, Carson publicized the interdependence of humanity and nature. Her integration of scientific concepts into popular writings helped to educate the public on ecological principles and the beauty of natural systems. Carson's death in 1964 came at the height of her influence, but her vision of the need to appreciate, understand, and protect natural systems would remain forceful decades later.

Mark Henkels

FURTHER READING

Lytle, Mark Hamilton. *The Gentle Subversive: Rachel Carson, "Silent Spring," and the Rise of the Environmental Movement.* New York: Oxford University Press, 2007.

Murphy, Priscilla Coit. *What a Book Can Do: The Publication and Reception of "Silent Spring."* Amherst: University of Massachusetts Press, 2005.

Quaratiello, Arlene R. *Rachel Carson: A Biography.* Westport, Conn.: Greenwood Press, 2004.

SEE ALSO: Abbey, Edward; Agricultural chemicals; Balance of nature; Biomagnification; Bookchin, Murray; Brower, David; Dichloro-diphenyl-trichloroethane; Green movement and Green parties; Hazardous and toxic substance regulation; Pesticides and herbicides; *Silent Spring.*

Carter, Jimmy

CATEGORY: Preservation and wilderness issues

IDENTIFICATION: American politician who served as governor of Georgia and as president of the United States

BORN: October 1, 1924; Plains, Georgia

SIGNIFICANCE: During his political career, Carter made many decisions that demonstrated an environmentalist agenda.

Jimmy Carter was born into a family that owned a general store and farm in the small community of Plains, Georgia. He was appointed to the U.S. Naval Academy, graduating in 1946, and served in the Navy until his father's death in 1953, when he assumed his father's business responsibilities.

Carter expanded his family's businesses and successfully ran for local political office. He was elected to the Georgia Senate in 1962 and 1964. He was elected governor of Georgia in 1970 and served in the office from 1971 to 1975. Governor Carter reorganized the state government, consolidating many functions and putting all environmental agencies under the Department of Natural Resources. He rejected a U.S. Army Corps of Engineers plan to dam the Flint River, the last free-flowing large river in western Georgia. This was apparently the first such action ever taken by a governor of a U.S. state. He helped to arrange for the greater part of Cumberland Island to be given to the state of Georgia; it would later become Cumberland Island National Seashore. He also began acquiring land along the Chattahoochee River in Atlanta for state parks; these later became parts of the Chattahoochee River National Recreation Area.

Carter ran for the U.S. presidency in 1976 and won the election by a narrow margin over incumbent Gerald Ford. Early in his term, President Carter vetoed a public works bill that included nineteen water projects that he considered economically unjustified and environmentally unsound. Under intense political pressure he later signed a compromise bill that included nine of the projects. He also issued executive orders that directed federal agencies to protect or restore wetlands and floodplains wherever possible as a matter of government policy.

During the Carter administration a large number of important environmental acts were passed and signed into law by the president. These included the Alaska National Interest Lands Conservation Act (1980), the Fish and Wildlife Conservation Act (1980), and the Comprehensive Environmental Response, Compensation, and Liability Act, commonly referred to as Superfund (1980). It is remarkable that all of this was accomplished in the midst of a very difficult single term in office. President Carter's term began with a worldwide economic recession and instability in the international petroleum market, resulting in large national budget deficits, severe inflation, and high interest rates; it ended with the overthrow of the shah of Iran and the seizure of the U.S. embassy in Iran by the revolutionary Islamic government.

In response to the petroleum market problem, President Carter requested the creation of a cabinet-level Department of Energy. The mission of this department included research on reducing American

President Jimmy Carter, center right, tours the Three Mile Island nuclear plant on April 1, 1979, four days after the plant suffered an accident. Carter demonstrated concern for environmental issues throughout his term as president. (AP/Wide World Photos)

dependence on fossil fuels and the development of alternative energy resources. Over the decades since, this research has had very desirable environmental effects.

Robert E. Carver

FURTHER READING

Horowitz, Daniel. *Jimmy Carter and the Energy Crisis of the 1970s: A Brief History with Documents.* Boston: Bedford/St. Martin's Press, 2005.

Stine, Jeffrey K. "Environmental Policy During the Carter Presidency." In *The Carter Presidency: Policy Choices in the Post-New Deal Era*, edited by Gary M. Fink and Hugh Davis Graham. Lawrence: University Press of Kansas, 1998.

SEE ALSO: Alaska National Interest Lands Conservation Act; Department of Energy, U.S.; Environmental law, U.S.; Nuclear regulatory policy; Superfund.

Cash for Clunkers

CATEGORY: Atmosphere and air pollution
IDENTIFICATION: U.S. federal government program that briefly subsidized purchases of cleaner-running and more fuel-efficient vehicles
DATES: June 24-August 24, 2009
SIGNIFICANCE: The three goals of the Cash for Clunkers program were to help reduce American dependence on foreign oil, help reduce carbon emissions from passenger vehicles, and provide a stimulus to the American automobile industry. It at least partly achieved the first two goals, but claims about its success in achieving the third goal have been challenged.

Enacted on June 24, 2009, and officially known as the Car Allowance Rebate System (CARS), the popularly called Cash for Clunkers program made a total of $3 billion available to subsidize the private purchase of more fuel-efficient automobiles, trucks, sport utility vehicles (SUVs), and vans before it was terminated ahead of schedule on August 24 because its funding had been exhausted by popular demand. The subsidies were disbursed as credits toward the purchase of only new vehicles and, depending on the magnitude of the improvement in miles per gallon (mpg) offered by the new vehicle, the individual credits were either $4,500 (for major mpg gains) or $3,500 (for lesser gains). New purchases had to cost less than $45,000, trade-ins generally had to have combined city and highway mileage ratings of 18 mpg or less, and all trade-ins were to be scrapped, not resold.

A similar program had been pursued previously by the Slovak Republic in Europe, and other countries were considering adopting such progams when the U.S. Congress enacted Cash for Clunkers. The program's goals were to stimulate sales in the sagging American auto industry, to reduce American dependence on foreign energy by replacing "gas guzzlers" with more fuel-efficient vehicles, and to reduce carbon emissions emanating from older, fossil-fuel-burning vehicle engines.

The program unquestionably had some success in achieving the second and third of these goals. By definition, replacing low-mileage vehicles with higher-mileage vehicles, all else (such as driving distances) remaining even, reduced the U.S. demand for gasoline and resulted in lower carbon emissions per car on the road. To the extent that some of those cars were also hybrids, which run part of the time on their self-charging batteries, the benefit was enhanced. However, critics correctly noted that the program only subsidized at most slightly more than 690,000 dealer transactions involving more fuel-efficient vehicles—a proverbial drop in the ocean for a country with more than 250 million registered passenger vehicles. Morever, many American drivers were already being motivated to acquire more fuel-efficient vehicles by steep rises in gasoline prices in the months preceding initiation of the program.

Joseph R. Rudolph, Jr.

SEE ALSO: Automobile emissions; Corporate average fuel economy standards; Electric vehicles; Fossil fuels; Hybrid vehicles.

Vehicles turned in during the Cash for Clunkers program are marked to be scrapped. (AP/Wide World Photos)

Catalytic converters

CATEGORY: Atmosphere and air pollution

DEFINITION: Devices that convert carbon monoxide, hydrocarbons, and nitrogen oxides in automobile exhaust gases into less harmful substances

SIGNIFICANCE: The use of catalytic converters has significantly reduced air pollution caused by automobile exhaust.

Since their invention in the late nineteenth century, automobiles have revolutionized society. However, automobiles represented a significant new source of air pollution, particularly in urban areas. As early as 1943, the effects of automobile exhaust emissions on air quality were noted in the Los Angeles area. In 1952 A. J. Haagen-Smit showed that the interaction of nitrogen oxides and hydrocarbons from automobile exhaust with sunlight results in the formation of secondary pollutants, such as peroxyacetyl nitrates (PAN) and ozone. This new form of air pollution, termed photochemical or Los Angeles smog, was soon recognized as a major pollution problem in urban atmospheres.

In 1961 the state of California began regulating the release of pollutants from automobiles. Two years later, the first federal emission standards were imposed as part of the 1963 Clean Air Act. As further restrictions on automobile emissions were introduced in the 1960's and early 1970's, new devices were developed to aid in reducing emissions. Of these, the catalytic converter was the most important.

The first type of catalytic converter consisted of an inert ceramic support material coated with a thin layer of platinum or palladium metal. Carbon monoxide and hydrocarbons from the exhaust gases attached themselves to the metal surface, where they reacted with molecular oxygen and were converted into carbon dioxide and water. The use of a metal catalyst lowered the temperature and increased the rate at which reaction occurred, making possible the removal of almost all of these two pollutants from the exhaust gases.

Because lead, phosphorus, and other substances can coat the metal surfaces in catalytic converters and

prevent them from functioning, the Environmental Protection Agency (EPA) was granted the authority to regulate gasoline composition and fuel additives as part of the 1970 Clean Air Act amendments. A consequence of the introduction of catalytic converters in new automobiles in the mid-1970's was the gradual switch from leaded to unleaded gasoline. As a result, emission of lead from transportation, the major source of lead as an air pollutant, decreased by 97 percent between 1978 and 1987.

While the first-generation catalytic converters dramatically reduced emission of carbon monoxide and hydrocarbons from automobiles, they had no effect on nitrogen oxide emissions. Dual-bed catalytic converters, developed in the early 1980's, use a two-step process to reduce emission of nitrogen oxides, carbon monoxide, and hydrocarbons. Dual-bed catalytic converters were followed by three-way converters, which use platinum and rhodium as the metal catalyst. A feedback system with an oxygen sensor is used to adjust the mixture of air and gasoline sent to the automobile engine to ensure maximum removal of pollutants.

The introduction of catalytic converters significantly reduced the release of pollutants by automobiles. In the United States, emission of carbon monoxide and hydrocarbons from transportation sources decreased by more than 50 percent between 1970 and 1995, even though the number of cars and trucks more than doubled. Similar reductions in automobile pollution have occurred in other countries where catalytic converters have been introduced.

Jeffrey A. Joens

FURTHER READING

Halderman, James D., and Jim Linder. *Automotive Fuel and Emissions Control Systems.* Upper Saddle River, N.J.: Pearson/Prentice Hall, 2006.

Yamagata, Hiroshi. *The Science and Technology of Materials in Automotive Engines.* Boca Raton, Fla.: CRC Press, 2005.

SEE ALSO: Air pollution; Air-pollution policy; Automobile emissions; Black Wednesday; Clean Air Act and amendments; Smog.

Cattle

CATEGORY: Animals and endangered species
DEFINITION: Large domesticated mammals raised as livestock and as dairy animals
SIGNIFICANCE: The environmental impacts of the raising of cattle include the production of greenhouse gases that contribute to climate change, land degradation owing to increased erosion and overgrazing, and pollution of air and water caused by concentrated waste. These problems most often arise from the nature of the farming practices of cattle ranchers, and they can be alleviated somewhat by changes in management techniques.

The collective species *Bos primigenius* includes three subspecies: *Bos primigenius taurus*, the European cattle; *Bos primigenius indicus*, the zebu; and the extinct *Bos primigenius primigenius*, the ancestor to modern domestic cattle and also called the auroch. The term "cattle" is usually used to refer to domesticated *B. primigenius* that are bred for multiple uses, spanning the production of food to clothing to fuel, and also to serve as work animals. Cattle are chief sources of food (from meat and milk), labor, clothing (from leather), and fuel (from dung) in many cultures worldwide.

According to a 2006 report from the United Nations Food and Agriculture Organization (FAO), the world's livestock, most of which are cattle, are among the major causes of serious global environmental problems, including land degradation, air and water pollution, and loss of biodiversity. Further, livestock are responsible for 18 percent of total global greenhouse gas emissions, which have been linked to climate change. The potential for those who raise livestock to contribute to solving these environmental problems is thus very large, and experts argue that major improvements can be achieved at reasonable costs.

The grazing of livestock occupies almost 30 percent of the earth's terrestrial surface, and the agricultural production of feed for livestock takes up about one-third of all arable land. In addition to affecting land that has already been converted to agriculture, the raising of cattle is responsible for the direct acceleration of deforestation in some parts of the world where ranchers burn forests to expand grazing land. This is a pressing problem in Latin America, where nearly 70 percent of previously forested land is used as pasture and much of the rest is used for growing livestock feed crops such as soybeans and corn. Tropical soils are especially susceptible to degradation by overgrazing, compaction, and erosion because of high precipitation and low nutrient loads. Ranching is thus usually not a sustainable practice in the Tropics, but it continues because it promises quick economic returns.

Cattle fed in feedlots like this one produce great amounts of highly concentrated waste that creates environmental problems. (©Mauro Scarone Vezzoso/ iStockphoto.com)

Concentrated animal feeding operations (CAFOs), in which livestock are fed in small confined areas for maximum profit, produce great amounts of highly concentrated waste that ultimately ends up in the water and as gases in the air. FAO has estimated that livestock manure is the largest sectoral source of water pollutants. In addition to manure, chemicals from pesticides and fertilizers used for feed crops and antibiotics administered to livestock also end up in groundwater and surface water, where they contribute to high nutrient loading and algal blooms, medicinal pollution that affects aquatic biology, and high sediment loads that reduce water quality.

GREENHOUSE GAS EMISSIONS

Cattle produce methane through an anaerobic process in the gut called methanogenesis; the methane is released through belching and flatulence. Methane is an extremely effective greenhouse gas, having a warming effect 23 to 50 times greater than carbon dioxide; this causes concern as there are approximately 1.3 billion cattle worldwide, and the number only grows. According to FAO's 2006 report, the raising of cattle generates more greenhouse gases than do all forms of transportation. With increasing prosperity worldwide, people consume more meat and dairy products every year; the global production of both meat and dairy is expected to double by 2050.

A number of possible ways of reducing methane production in cattle have been proposed or are under study. These include the administration of bovine medicines similar to the antacid Alka-Seltzer, the use of new varieties of feed grasses, and targeted breeding that selects for less gassy cattle. Any successes in this area will certainly be helpful, but scientists note that a global reduction in numbers of domestic cattle must also accompany these techniques if cattle's large-scale production of methane is to be effectively reduced.

ENVIRONMENTAL SOLUTIONS

Humans certainly need not stop raising cattle entirely in order to ensure a healthy future for the environment, but some practices must change. Sustainable cattle ranching requires restoration of overgrazed and damaged land through soil conservation, the planting of trees, and protection of areas sensitive to erosion. Changing the ways cattle are fed to better reflect a natural diet can go a long way toward curbing greenhouse gas emissions. In addition, it has been argued that moving away from CAFOs to less intense feed operations

can have positive effects on the cows, their environment, and the quality of all products that come from the cows. Increased use of processed manure as fertilizer can reduce waste pollution and reclaims a resource that should be valued and used instead of dumped into the water supply.

Many cattle ranchers and dairy farmers have become dedicated to reducing the negative environmental impacts of raising cattle because for them the advantages of sustainable management systems outweigh the pressures for higher profit. Persons living in subsistence communities often have little choice in how they raise cattle, however—they clear land and overgraze because they will starve otherwise. It has been proposed that developed nations should provide incentives to poorer countries to ensure that practicing deforestation to create pastureland is not the only option the people have for income.

Another solution to the negative environmental impacts of cattle raising that is sometimes proposed, primarily by animal rights and environmental conservation groups, is vegetarianism. If humans were to limit their beef consumption, this would certainly reduce demand and therefore reduce the number of cattle worldwide that are overgrazing, belching methane, and compacting soil. As critics of this approach have noted, however, expecting large numbers of people to change their diets radically is unrealistic. This is especially true in the United States, where the culture of beef eating is quite strong. For this reason, many environmentalists, scientists, and others have increasingly suggested that people could help reduce the negative environmental impacts of cattle raising simply by reducing the amount of beef in their diets.

Jamie Michael Kass

FURTHER READING
Clutton-Brock, Juliet. *A Natural History of Domesticated Mammals.* 2d ed. New York: Cambridge University Press, 1999.
Soliva, Carla Riccarda, Junichi Takahashi, and Michael Kreuzer, eds. *Greenhouse Gases and Animal Agriculture.* Boston: Elsevier, 2006.
Steinfeld, Henning, et al. *Livestock's Long Shadow: Environmental Issues and Options.* Rome: United Nations Food and Agriculture Organization, 2006.

SEE ALSO: Deforestation; Desertification; Grazing and grasslands; Greenhouse gases; Intensive farming; Methane; Organic gardening and farming; Overgrazing of livestock.

Center for Health, Environment, and Justice

CATEGORIES: Organizations and agencies; human health and the environment
IDENTIFICATION: American nonprofit organization established to assist local communities with environmental issues
DATE: Founded in 1981
SIGNIFICANCE: When it was founded in 1981 as the Citizens Clearinghouse for Hazardous Waste, the organization that became the Center for Health, Environment, and Justice was part of an emerging grassroots movement among people concerned about protecting their local communities from the harmful consequences of environmental hazards.

During the late 1970's Lois Gibbs discovered that her child's school in the Love Canal neighborhood of Niagara Falls, New York, was built on thousands of tons of toxic chemicals. At the time no national organization existed to help communities with environmental issues. With no one to turn to for help, Gibbs organized the Love Canal Homeowners Association in 1978 to protest the situation in her neighborhood. This experience and the legal battles that followed led Gibbs to found the Citizens Clearinghouse for Hazardous Waste in 1981; the organization would later change its name to the Center for Health, Environment, and Justice (CHEJ). CHEJ subsequently grew into a national organization with Gibbs serving as executive director. Its main office is located in Falls Church, Virginia.

Since its establishment, CHEJ has remained a grassroots organization that focuses on helping communities coordinate their responses to environmental hazards. CHEJ seeks to assist local neighborhoods by providing necessary aid, information, resources, training, and strategic or technical assistance. In this way, the organization encourages individuals and communities to take social and political action.

CHEJ's stated mission is to "build healthy communities, with social justice, economic well-being, and democratic governance." This includes protecting consumers from hazardous or toxic products. CHEJ addresses its objectives through various campaigns and programs. Its BE SAFE campaign is a precautionary effort to prevent pollution, and its Child Proofing

Our Communities campaign helps to educate communities about strategies for protecting children from environmental hazards. CHEJ's Green Flag Schools Program for Environmental Leadership works with schools to educate students about how they can engage in environmental advocacy. PVC: The Poison Plastic is an example of a CHEJ campaign for safe and healthy consumer products. This ongoing national campaign is aimed at moving major corporations away from using PVC plastic, a substance that is harmful for both the environment and human health.

Jeff Cervantez

SEE ALSO: Air pollution; Environmental health; Environmental illnesses; Environmental justice and environmental racism; Gibbs, Lois; Land pollution; Love Canal disaster; Rainforest Action Network; Water pollution.

Centers for Disease Control and Prevention

CATEGORIES: Organizations and agencies; human health and the environment
IDENTIFICATION: U.S. federal agency that oversees all areas of public health
DATE: Founded on July 1, 1946, as the Communicable Disease Center
SIGNIFICANCE: The Centers for Disease Control and Prevention monitors, diagnoses, and works to control outbreaks of disease in the United States. The agency also collects and analyzes data concerning public health and works to educate industries as well as the public regarding possible environmental causes of illness.

The Centers for Disease Control and Prevention has undergone significant changes, both in its primary function and in its name. The agency from which it would eventually arise was founded in 1942 as the Office of National Defense Malaria Control Activities, which was established in Atlanta, Georgia, with the goal of controlling or eliminating malaria in the southern United States. That agency's name was altered several times during the World War II years, reflecting its growing functions: It was absorbed into the Office of Typhus Fever Control (1942-1943), which became the Office of Malaria Control in War Areas (1944). On July 1, 1946, it became the Communicable Disease Center (CDC), the name it carried for some twenty years. Between 1967 and 1973 the CDC was incorporated into the Public Health Service, with a concomitant evolution in its function as well as its title as the Public Health Service became an agency within the Department of Health, Education, and Welfare and then its successor, the Department of Health and Human Services. The agency took on the more familiar name of Center for Disease Control until 1980, and from 1980 to 1992 it was known as the Centers for Disease Control; it was renamed the Centers for Disease Control and Prevention in 1992.

The original functions of the agency, reflected in its early titles, were the control and eradication of infectious diseases such as malaria and typhus. Using the newly developed insecticide dichloro-diphenyl-trichloroethane, popularly known as DDT, the agency directed a spraying/eradication program between 1947 and 1949 for control of the mosquito vector throughout the South; millions of homes were treated, and the program proved successful. The number of reported cases of malaria was reduced by nearly 90 percent by 1949, and the disease was considered eradicated in the United States by 1951.

During the 1950's, the purview of the agency was broadened significantly, with a division devoted to venereal diseases incorporated in 1957 and a tuberculosis unit established in 1960. From the 1950's through the 1970's, the agency became increasingly involved in monitoring bacterial (cholera, salmonellosis) outbreaks as well as those associated with viral diseases (polio, hepatitis, mumps, rubella, influenza). The CDC sponsored frequent conferences and workshops with the aim of addressing problems of infectious disease as well as educating the public in these areas.

By the 1990's most major infectious diseases were under control in the United States, and the CDC moved to target other threats to the public health, such as chronic diseases and conditions associated at least in part with lifestyle: cancer, heart disease, workplace injuries, and environmental hazards. The agency also began to address the modern threat of bioterrorism. In order to carry out research on biological threats, the CDC maintains one of the few Biosafety Level 4 laboratories in the United States, used to study the most dangerous of biological agents. The CDC routinely interacts with equivalent agencies

in other nations, as well as with the World Health Organization, to address public health threats around the world.

Richard Adler

FURTHER READING

Meyerson, Beth E., Fred A. Martich, and Gerald P. Naeh. *Ready to Go: The History and Contributions of U.S. Public Health Advisors.* Research Triangle Park, N.C.: American Social Health Association, 2008.

Tulchinsky, Theodore H., and Elena A. Varavikova. *The New Public Health.* 2d ed. Burlington, Mass.: Elsevier, 2009.

SEE ALSO: Antibiotics; Asbestosis; Bacterial resistance; Birth defects, environmental; Brucellosis; Cancer clusters; Carcinogens; Chloramphenicol; Cloning; Environmental health; Environmental illnesses; Pandemics; Times Beach, Missouri, evacuation; World Health Organization.

Central America and the Caribbean

CATEGORIES: Places; ecology and ecosystems; resources and resource management

SIGNIFICANCE: Central America and the Caribbean are both so-called biodiversity hot spots, areas with extremely high biodiversity that is under threat from human activities such as tourism, deforestation, and unsustainable land use. Both regions experience tensions between environmental conservation and traditional modes of economic development.

Central America is located on the isthmus between North America and South America and encompasses the nations of Belize, Costa Rica, El Salvador, Guatemala, Honduras, Nicaragua, and Panama. Mexico is often considered to be part of Central America as well, and it and the other Central American nations are bordered by the Caribbean Sea. Central America contains diverse terrain, ranging from lowlands to volcanic highlands, with numerous forest types, mangrove swamps, savannas, fertile highlands, and mountain ranges.

The Caribbean region includes the Caribbean Sea and the islands of the West Indies, which are divided

Habitats and Selected Vertebrates of Central America

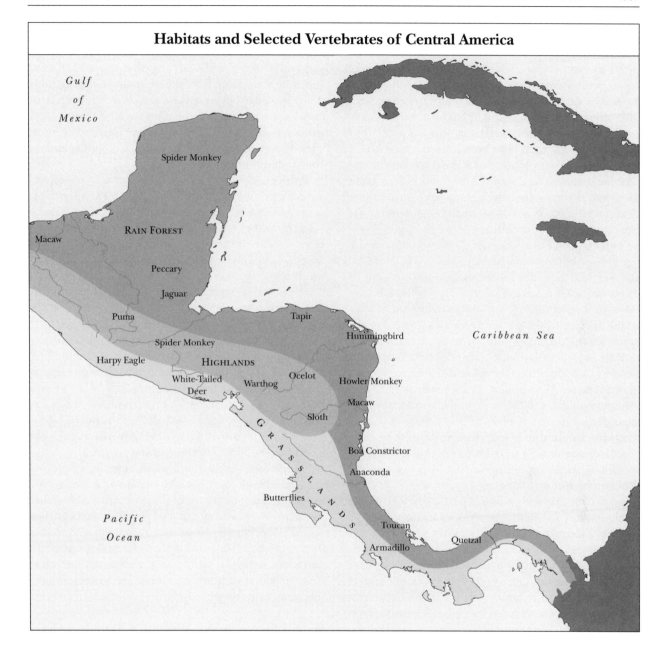

into the Greater Antilles (which contains most of the landmass of the West Indies), the Lesser Antilles, and the Bahamas and Turks and Caicos Islands, the last of which are located in the Atlantic Ocean. Central America and the Caribbean are often treated as a geographical region, and they face many similar challenges. There are twenty-seven territories in the Caribbean, which encompasses more than seven thousand islands, islets, reefs, and cays, many in volcanic island arcs.

High poverty rates throughout the region have created a strong economic impetus toward development, expansion, and promotion of tourism, all of which have had numerous negative impacts on the environment. Earthquakes, hurricanes, and volcanism also affect the region. Efforts to promote ecotourism and sustainable development, however, may in the long term help balance the economic needs of the region's growing population with the need for environmental protection.

Biodiversity

Both Central America and the Caribbean are located in Nearctic and Neotropical terrestrial ecozones, which means they contain plant and animal groups found in both North and South America. Central America has been estimated to be home to 7 percent of the world's biodiversity in only 0.5 percent of the world's land area. The area known as the Mesoamerican biodiversity hot spot is the third largest biodiversity hot spot in the world, with more than eight thousand endemic species of plants and animals (excluding insects). It also contains crucial habitat for Neotropical migratory birds.

The Mesoamerican Barrier Reef System, extending along the coasts of Mexico, Belize, Guatemala, and Honduras, is the second-largest barrier reef in the world, after Australia's Great Barrier Reef. It provides important habitat for four endangered or protected species of sea turtle (green sea turtle, loggerhead sea turtle, hawksbill turtle, and leatherback turtle), as well as manatees. The Bosawás Biosphere Reserve in Nicaragua is the second-largest rain forest in the Western Hemisphere at approximately 10,000 square kilometers (3,861 square miles), after the Brazilian Amazon.

Owing to the number of islands there, the Caribbean is home to at least thirteen thousand endemic species of plants and animals (excluding insects), native species found nowhere else. About half are endemic to single islands. Cuba in particular, with about 48 percent of the landmass in the Caribbean, is extremely biodiverse and contains thousands of endemic species. There are also numerous endemic marine organisms throughout the Caribbean, many with very limited ranges. The numbers of species of insects and spiders native to Central America and the Caribbean cannot be estimated reliably, but scientists place them at least in the thousands.

Terrestrial Environmental Issues

Numerous threats to the environment and biodiversity exist on land in Central America and the Caribbean, including the effects of deforestation, slash-and-burn agriculture, mining, urban development and pollution, habitat encroachment, and tourism. The Caribbean and Central America have some of the highest deforestation rates in the world. Only about 10 percent of the Caribbean's original vegetation remains; Cuba's forests (about 15 percent of original forest extent) are the largest remaining forests in the Caribbean. Approximately 13 percent of the land area in the Caribbean is protected, but much of this protected area has already experienced severe environmental impacts. The primary motivation behind deforestation in most of the Caribbean has been land clearance for sugarcane plantations, and sugarcane is still a primary export crop for many Caribbean countries. Forests are also cleared for cacao, coffee, and tobacco plantations.

Between 1980 and 1990, annual deforestation averaged 1.4 percent in Mesoamerica, with El Salvador affected the most. Deforestation in this area is motivated by timber, mineral, and oil extraction, as well as by subsistence agriculture. Because soils in this region quickly lose nutrients, subsistence farmers must constantly clear new land, often by slashing and burning. Slash-and-burn agriculture changes habitats and can cause soil erosion and watershed contamination when practiced on a large scale.

Expanding populations have led to rapid urban development, with negative effects on air and watershed quality, as well as depletion of water supplies. Mexico City, for example, has one of the worst levels of air pollution in the world, largely as the result of vehicle exhaust. Although aggressive pollution-control measures have resulted in some improvements, air pollution remains a serious issue.

Tourism has both positive and negative effects on the Caribbean and Central America. It brings in money that not only supports the population but also in some cases is used for conservation efforts that would not otherwise be possible. Tourism also results in increased waste production and pollution of air and water, however, and tourist resorts and the infrastructures that support them alter landscapes and often bring in nonnative species.

Marine Environmental Issues

The problems of the marine environment in the region are not disconnected from the terrestrial; problems on land affect the oceans and vice versa. For example, urban development often leads to water contamination in the marine environment from sewage and other by-products of urban areas. Deposits of soil that has eroded from the land because of deforestation can bury and kill coral reefs. Marine oil spills can contaminate beaches and rivers and kill coastal mangrove swamps, which provide habitat for fish as well as storm protection.

Coral reef loss is also a significant problem in the

Caribbean. In 2002 it was estimated that 30 percent of reefs in the Caribbean were threatened. Reefs are threatened directly by human activity, such as through marine pollution, but also by global climate change and disease. These threats often interact; for example, a reef already compromised by pollution is more susceptible to storm damage or disease.

One major threat to Caribbean coral reefs is coral bleaching, a stress response in which the coral expels its symbiotic zooxanthellae, which can eventually result in death. Major causes of coral bleaching include increased sediment runoff and sudden temperature increases. A worldwide coral bleaching event was correlated with an El Niño climate pattern.

In the Caribbean, tourism may pose an even greater threat to the marine environment than to the terrestrial environment. Pollution and trash from cruise ships, including fuel, treated and untreated sewage, and plastic, are frequently discharged into international waters. As the Caribbean has one of the highest densities of cruise ships in the world, this is a serious threat to marine habitats, particularly coral reefs—many of which are already damaged by coral bleaching, tropical storms, and soil erosion. Overfishing, as well as hunting of marine mammals, turtles, and seabirds, is also a threat to marine wildlife populations in the Caribbean and Central America.

CONSERVATION AND SUSTAINABLE DEVELOPMENT

As populations in the Caribbean and Central America increase, it becomes more important to find sustainable development strategies that will improve the standard of life for the human population while protecting the environment. Strategies that have been employed thus far include both the traditional conservation of areas and species by local governments and conservation efforts by international organizations such as the Nature Conservancy, the United Nations Educational, Scientific, and Cultural Organization (UNESCO), and the World Wide Fund for Nature (WWF). Some examples of the latter include the Bosawás Biosphere Reserve (UNESCO), the Belizean stretch of the Mesoamerican Reef (Nature Conservancy), and conservation work in Mexico's Chihuahuan Desert (WWF).

Other efforts have focused on pollution reduction,

such as Mexico City's aggressive program, which reduced air pollutant levels by 57-95 percent between 1991 and 2008, and on the development of renewable energy resources. For example, the Caribbean Renewable Energy Development Programme is a joint effort of thirteen Caribbean territories to remove barriers to the development of renewable energy resources.

Many countries in this region are exploring further development of sustainable ecotourism as an alternative to more environmentally destructive forms of economic development. However, not all tourism labeled as ecotourism is low impact, responsible, or sustainable, and some negatively affects local peoples. The development of economically and environmentally positive ecotourism opportunities under local control is a priority for many Caribbean and Central American environmental groups.

Melissa A. Barton

FURTHER READING

Baver, Sherrie L., and Barbara Deutsch Lynch, eds. *Beyond Sun and Sand: Caribbean Environmentalisms.* Piscataway, N.J.: Rutgers University Press, 2006.

Dallmeier, F., and J. A. Comiskey, eds. *Forest Biodiversity in North, Central, and South America, and the Caribbean: Research and Monitoring.* Washington, D.C.: Parthenon, 1998.

Hillstrom, Kevin, and Laurie Collier Hillstrom. *Latin America and the Caribbean: A Continental Overview of Environmental Issues.* Santa Barbara, Calif.: ABC-CLIO, 2004.

Pennington, R. Toby, and James A. Ratter. *Neotropical Savannas and Seasonally Dry Forests: Plant Diversity, Biogeography, and Conservation.* Boca Raton, Fla.: CRC Press, 2006.

Romero, Aldemaro, and Sarah E. West, eds. *Environmental Issues in Latin America and the Caribbean.* Dordrecht, Netherlands: Springer, 2005.

SEE ALSO: Air pollution; Amazon River basin; Biodiversity; Biosphere reserves; Cloud forests; Coral reefs; Ecotourism; Habitat destruction; Indigenous peoples and nature preservation; Rain forests; Savannas; Sea turtles; Slash-and-burn agriculture.

Ceres

CATEGORIES: Organizations and agencies; activism and advocacy

IDENTIFICATION: American nonprofit organization dedicated to injecting environmental considerations into investment decisions

DATE: Established in 1989

SIGNIFICANCE: Ceres promotes the well-being of the environment and human society by uniting investors and key stakeholders, environmental organizations, and public interest groups to work with corporations and capital markets to integrate sustainability and environmentally responsible practices into their day-to-day operations.

The 1989 *Exxon Valdez* oil spill in Prince William Sound was the catalyst for the formation of Ceres. Meeting in Chapel Hill, North Carolina, fifteen major environmental groups joined with investors and public pension fund managers to encourage greater corporate consideration of the environmental consequences of their actions. With participation by the New York and California state pension funds, the Social Investment Forum, and a coalition of more than two hundred Protestant and Roman Catholic groups, the initial members of that group represented more than $150 billion in invested capital. The group was originally known by the acronym CERES (for Ceres, the Roman goddess of agriculture), which stood for the Coalition for Environmentally Responsible Economies; it eventually dropped the longer name altogether and became simply Ceres.

Ceres was founded on the notion that government regulation alone is not sufficient to effect material improvements in the environment. Rather, progress requires fundamental changes in corporate behavior and more responsible attitudes toward the environment. According to Ceres, investors can influence this shift in beneficial ways if their investment decisions include consideration of corporations' environmental track records. Ceres's first project was the development of a corporate code of environmental ethics. Originally called the Valdez Principles as a reminder of the oil spill, the code later become known as the Ceres Principles. This ten-point code of conduct calls for corporations to protect the biosphere, use natural resources sustainably, reduce and responsibly dispose of wastes, conserve energy, reduce environmental

and health risks to employees and communities, offer safe products and services, restore the environment, inform the public of hazards, ensure management commitment to environmental responsibility, and issue audit reports.

Ceres uses shareholder resolutions to initiate discussions of environmental responsibility at the highest corporate levels. In some corporations, such resolutions eventually lead to formal endorsement of the ten Ceres Principles. By endorsing the principles, companies acknowledge their environmental responsibility, actively commit to an ongoing process of continuous improvement, and agree to initiate comprehensive public reporting of environmental issues.

Initially, the Ceres Principles were adopted by companies that already had strong "green" reputations. In 1993, following lengthy negotiations, Sunoco became the first *Fortune* 500 company to endorse the principles. Several other large companies, including Bank-Boston, Bethlehem Steel, Coca-Cola, General Motors, and Polaroid, soon followed Sunoco's example. By 2010 more than fifty companies had endorsed the Ceres Principles, including thirteen *Fortune* 500 companies that had adopted their own equivalent environmental codes of conduct.

In 1997 Ceres launched the Global Reporting Initiative (GRI), which standardizes corporate environmental reporting to generate the equivalent of a financial report. Just as an official corporate financial report includes required information on expenses, revenue, and profitability, each company's GRI report includes required information on sustainability, environmental and social impacts, technological innovation, and other pertinent issues. GRI reports improve corporate accountability by ensuring that all stakeholders—investors, fund managers, community groups, environmentalists, and labor organizations—have access to standardized and consistent information. Armed with these data, environmentally conscious investors can measure corporate compliance with the Ceres Principles and thus use capital markets effectively to promote sustainable business practices.

Ceres partnered with the United Nations Environment Programme to issue draft GRI sustainability reporting guidelines in 1999. Finalized guidelines were released in 2000, and fifty organizations responded with their reports. In 2002 the GRI was established as an independent international body based in the

Netherlands, and in 2009 more than twelve hundred organizations issued GRI reports.

Allan Jenkins
Updated by Karen N. Kähler

FURTHER READING

Ceres. *Twenty-first Century Corporation: The Ceres Road-map for Sustainability.* Boston: Author, 2010.

Friedman, Frank B. *Practical Guide to Environmental Management.* 10th ed. Washington, D.C.: Environmental Law Institute, 2006.

Leipziger, Deborah. *The Corporate Responsibility Code Book.* Sheffield, England: Greenleaf, 2003.

SEE ALSO: Environmental economics; Environmental ethics; *Exxon Valdez* oil spill; Green marketing.

Certified wood

CATEGORIES: Forests and plants; resources and resource management

DEFINITION: Wood that has been harvested from a forest that has been audited by an independent agency to ensure that the forest is managed responsibly and sustainably

SIGNIFICANCE: The movement toward forest certification has raised public awareness regarding the environmental benefits of sustainable forest management and has been credited with helping to slow the rate of deforestation around the world.

The Forestry Department of the United Nations Food and Agriculture Organization has estimated that more than half of the world's forests have been harmed or erased through overuse. During the early 1990's concerns about the rapid overlogging of forests—particularly tropical forests—led to the formation of a set of principles describing the sustainable growing and harvesting of timber. Several independent groups were established that offered to conduct objective studies of commercial forests to determine whether or not the forests were being managed responsibly. Although each group set its own standards, all of the groups adhered to several common principles: In a sustainable forest, the diversity of plants and animals, including threatened and endangered species, is protected; wood is not harvested faster than it can be regenerated; chemical fertilizers and pesticides are used minimally, if at all; rivers and streams are protected from erosion and pollution. In addition, a forest can be certified as sustainable only if the logging workers are treated fairly and any indigenous peoples who inhabit the land receive a share of the profits. The managers of certified forests must develop and continually update scientifically based management plans and continually monitor their operations for compliance.

The various certified wood programs are run separately—no United Nations body or other international oversight group has been established to ensure conformity, and most programs are run by scientific, environmental, or consumer groups, not by government agencies. By 2010, nearly 12 percent of the world's forests had been certified and more than fifty organizations offered wood certification, including two large umbrella groups. The larger of these is the Programme for the Endorsement of Forest Certification (PEFC), based in Geneva, Switzerland, and comprising thirty-five smaller bodies. PEFC certification has protected more than 210 million hectares (519

A worker marks trees for inventory in a forest certified for sustainable management in northern Brazil. (©Reuters/CORBIS)

million acres) of forestland. Smaller but more influential in North America is the Forest Stewardship Council (FSC), based in Bonn, Germany, which certifies more than 105 million hectares (260 million acres). North America has three other important certifying bodies: the American Tree Farm System (ATFS), the Canadian Standards Association (CSA), and the Sustainable Forestry Initiative (SFI) Program.

All of the programs encourage consumers to investigate the wood products they purchase to ensure that they are made of certified wood. Products certified by the FSC carry labels identifying them as being made from "pure," "missed," or "recycled" materials, and each label bears a certificate number that the consumer can verify by referring to an online database. Some retailers have refurbished their stores with FSC-certified flooring or have made commitments to use FSC-certified paper pulp in their catalogs and other paper goods. At least one major home-improvement chain in the United States carries FSC-certified lumber in limited quantities. The demand for certified wood has grown steadily in the years since the process of certification was first introduced.

Cynthia A. Bily

FURTHER READING

Arabas, Karen, and Joe Bowersox. *Forest Futures: Science, Politics, and Policy for the Next Century.* Lanham, Md.: Rowman & Littlefield, 2004.

Berger, John J. *Forests Forever: Their Ecology, Restoration, and Protection.* Chicago: University of Chicago Press, 2008.

McEvoy, Thomas J. *Positive Impact Forestry: A Sustainable Approach to Maintaining Woodlands.* Washington, D.C.: Island Press, 2004.

SEE ALSO: Biodiversity; Boreal forests; Coniferous forests; Deciduous forests; Deforestation; Extinctions and species loss; Food and Agriculture Organization; Forest management; Green marketing; Indigenous peoples and nature preservation; Logging and clearcutting; Sustainable development; Sustainable forestry.

CFCs. *See* **Chlorofluorocarbons**

Chalk River nuclear reactor explosion

CATEGORIES: Disasters; nuclear power and radiation

THE EVENT: The world's first serious nuclear reactor accident

DATE: December 12, 1952

SIGNIFICANCE: After an experimental nuclear reactor at Chalk River, Ontario, overheated, resulting in a hydrogen-oxygen explosion and release of more than one million gallons of radioactive water, a pipeline had to be constructed to divert the radioactive water and avoid contamination of the Ottawa River.

The experimental NRX reactor, run by the National Research Council of Canada, began operation at Chalk River, Ontario, in 1947. It was operating at low power on December 12, 1952, when a technician mistakenly opened four air valves in the system used to insert the control rods that slowed the rate of reaction. This action caused the control rods to move out of the reactor core, increasing the reaction rate and the amount of heat produced in the core. In the effort to return these control rods to their proper position, miscommunication resulted in the removal of additional control rods, causing the power generated in the reactor core to double every two seconds.

When the operator realized that the power was rapidly increasing, he "scrammed" the reactor, a process that forced the control rods into place, thus halting the nuclear reaction. However, the earlier error of opening the air valves kept some of the control rods from being pushed into place, and the temperature in the reactor continued to increase. To stop the reaction, operators dumped water rich in deuterium (heavy water) from the reactor core. Without this water, which slows the neutrons to a point where they can induce fission in the uranium core, the reaction ceased and the core began to cool. However, more than one million gallons of highly radioactive water flooded the basement of the reactor building. The reactor dome, a 4-ton lid on the reactor vessel, rose upward to release pressure from a hydrogen-oxygen explosion, and more radioactive water escaped, flooding the main floor of the building. The reactor operators were forced to evacuate the building, and eventually the entire reactor site, as radioactivity rose above safe levels.

The small size of the experimental NRX reactor minimized the radiation release. Since the Chalk River site was remote, the exposure of the general population to radioactivity was minimal. Nonetheless, a pipeline had to be constructed to divert the radioactive water and avoid contamination of the Ottawa River. The high levels of radioactivity made decontamination and cleanup of the NRX reactor difficult. In some parts of the reactor, workers participating in the cleanup effort could spend only minutes at work before accumulating the maximum permissible annual radiation dose for reactor workers. About three hundred military personnel from the United States and Canada, including future U.S. president Jimmy Carter, volunteered to participate in the cleanup of the NRX reactor. A 1982 study of the health of more than seven hundred workers who participated in the cleanup showed no increase in their death rate from cancer when compared to the general population.

George J. Flynn

FURTHER READING

Bodansky, David. "Nuclear Reactor Accidents." In *Nuclear Energy: Principles, Practices, and Prospects.* 2d ed. New York: Springer, 2004.
Krenz, Kim. *Deep Waters: The Ottawa River and Canada's Nuclear Adventure.* Montreal: McGill-Queen's University Press, 2004.

SEE ALSO: Antinuclear movement; Chernobyl nuclear accident; Nuclear power; Three Mile Island nuclear accident; Windscale radiation release.

Chelyabinsk nuclear waste explosion

CATEGORIES: Disasters; nuclear power and radiation
THE EVENT: Nuclear explosion at a weapons production facility in the Chelyabinsk province of the Soviet Union
DATE: September 29, 1957
SIGNIFICANCE: The nuclear explosion that took place at Mayak, a weapons production facility in Russia's Chelyabinsk province (then part of the Soviet Union), exposed 270,000 people to high levels of radiation.

The Mayak industrial complex began producing weapons-grade plutonium in 1948. For years after production began, workers dumped the complex's radioactive waste into the nearby Techa River. A waste storage facility was constructed in 1953 after people living near the Techa suffered from radiation poisoning. On September 29, 1957, one waste tank at the facility exploded. Although the exact cause of the explosion remains unknown, it is known that a cooling system failure contributed to the disaster.

A radioactive cloud consisting of between 70 and 90 tons of waste released an estimated 20 million curies of radiation into the environment. Of the waste material that was released, 90 percent fell back on the blast site and 10 percent drifted through the atmosphere, contaminating 2,000 square kilometers (772 square miles) of territory and exposing 270,000 people to radiation. Eyewitnesses later recalled seeing red dust settle everywhere, and the waters of the Techa River turned black for two weeks. Soon thereafter, plants died and leaves fell off the trees in the area. In less than two years, all the pine trees in a 28-square-kilometer (11-square-mile) area around the Mayak complex were dead.

The Soviet government closed all the stores in the area and shipped in food for the local population. Some ten thousand people were evacuated from the area, and the government burned houses and demolished entire towns to ensure that the residents could not return. However, many smaller communities continued to use local water sources, and later anecdotal accounts indicated that not all the contaminated crops in the region were destroyed. A dairy farm near the Techa River was allowed to operate until 1959.

The Chelyabinsk accident was kept secret, and for almost twenty years few people outside the region knew the extent of the disaster. The U.S. Central Intelligence Agency (CIA) learned of the incident but did not make the information public. Zhores Medvedev, a Soviet émigré, published the first account of the accident in 1976.

Evaluating the impact of the explosion proved difficult for two reasons: The Soviet government consistently denied the magnitude of the event, and the region was heavily polluted by other sources, especially the dumping of waste into the Techa River. In 1989 a U.S. official who visited the Mayak complex declared it to be the "most polluted spot on earth." By 1992 nearly one thousand area residents had been diagnosed with chronic radiation sickness. Rates for cancer were higher near Mayak than anywhere else in the Soviet Union, and the general health of the population, especially children, was poor by any standard.

Cleanup efforts were hampered by the secrecy that surrounded the event for nearly three decades, limited funds, and the high levels of contamination. Lake Karachay, located near Mayak, was so radioactive that a person standing on its shore for more than one hour would be exposed to a lethal dose of radioactivity. In the early twenty-first century the area remained a pressing environmental problem.

Thomas Clarkin

FURTHER READING

Garb, Paula, and Galina Komarova. "Victims of 'Friendly Fire' at Russia's Nuclear Weapons Sites." In *Violent Environments*, edited by Nancy Lee Peluso and Michael Watts. Ithaca, N.Y.: Cornell University Press, 2001.

Hoffman, David E. *The Dead Hand: The Untold Story of the Cold War Arms Race and Its Dangerous Legacy*. New York: Doubleday, 2009.

Makhijani, Arjun, Howard Hu, and Katherine Yih, eds. *Nuclear Wastelands: A Global Guide to Nuclear Weapons Production and Its Health and Environmental Effects*. 1995. Reprint. Cambridge, Mass.: MIT Press, 2000.

SEE ALSO: Chernobyl nuclear accident; Nuclear accidents; Nuclear and radioactive waste; Radioactive pollution and fallout.

Chernobyl nuclear accident

CATEGORIES: Disasters; nuclear power and radiation

THE EVENT: Explosion of a nuclear power reactor at the Chernobyl power plant in the Soviet Union

DATE: April 26, 1986

SIGNIFICANCE: The accident at the Chernobyl nuclear power plant drastically changed the lives of thousands of residents in the northern part of Ukraine and the southern portion of Belarus. It also raised questions about the future of the plant itself; the ecological, human health, economic, and political repercussions of the incident; and the future of nuclear power programs throughout the world.

On April 26, 1986, nuclear power reactor number 4 exploded at the nuclear plant located about 15 kilometers (9 miles) from the small town of Chernobyl in the republic of Ukraine. As the core of the reactor began to melt, an explosion occurred that blew the top off the reactor and sent a wide trail of radioactive material across large parts of what was then the Soviet Union as well as much of Eastern and Western Europe. More than 116,000 people were evacuated within a 30-kilometer (19-mile) radius.

In addition to the political, social, and economic aftermath of the explosion, the consequences of human inability to prevent widespread damage were evident in the environment within five years of the Chernobyl disaster. A major release of radioactive materials into the atmosphere occurred during the first ten days following the explosion. Radioactive plumes reached many European countries within a few days, increasing radiation levels to between ten and one hundred times normal levels. Over time the contaminated lithosphere created a new biogeographical province characterized by irregular and complicated patterns of radioactivity in Belarus, Ukraine, and Russia, the countries most affected. In spite of some success in efforts to slow the flow of certain soluble radionuclides into the Black and Baltic seas, as well as into the Pripet, Dnieper, and Sozh rivers (all of which contribute water to the Kiev water reservoir), much contamination occurred that only time can resolve.

CONTAMINATION EFFECTS

The effects of the radioactive contamination on vegetation have varied depending on the species. Damage to coniferous forests was more than ten times greater than that to oak forests and grass communities, and more than one hundred times greater than the damage to lichen communities. Ten years after the explosion, pine forests still had high levels of radionuclides in the uppermost layer of the forest floor, while birch forests had considerably lower levels. Large numbers of highly contaminated trees were felled, and restrictions were placed on cutting and using the wood in industry and for fuel. Likewise, the degrees of radiation damage and the doses absorbed varied among plant communities. In addition to killing or damaging plant life, the radiation disturbed the functioning of plant reproductive systems, resulting in sterility or decreases in both seed production and fertility. Other effects involved changes in plant chlorophyll, necessary for removing carbon dioxide from the air.

Among animals, lower life-forms have been found to have higher radiosensitivity levels; that is, mammals are less sensitive to radioactivity than are birds, reptiles, and insects. The severity of damage also depends

Chernobyl, Ukraine, 1986

on whether it is external or internal. The impact of environmental contamination may be less severe for animals than for plants because of the ability of animals to move from place to place and make selective contact with the environment. Irradiation can have a wide range of effects among animal species, including death, reproductive disturbances, decrease in the viability of progeny, and abnormalities in development and morphology.

The contamination of aquatic ecosystems by Chernobyl radiation has been considerable. The contamination of freshwater ecosystems has fallen with time, but continuing contamination has been evidenced by factors such as a transfer of radionuclides from bottom sediments and erosion of contaminated soil into water sources. Some restrictions have been placed on fishing in contaminated aquatic ecosystems.

The consequences of the Chernobyl accident for agriculture have been severe. Drastic changes in land-

use and farming practices have been necessary as a result of contamination. As with other life-forms affected by radiation, the degree of contamination varies; in the case of agriculturally related contamination, damage depends on such things as the type of soil and the biological peculiarities of different plant species. Since Ukraine produces 20 percent of the grain, 60 percent of the industrial sugar beets, 45 percent of the sunflower seeds, 25 percent of the potatoes, and about 33 percent of all the vegetables used in the former Soviet Union, the problem of soil contamination is serious. Furthermore, the problems extend beyond the borders of Ukraine: The effects on the reindeer herds of Scandinavia and on sheep breeding in mountainous regions of the British Isles have been serious as well.

The scale of the contamination of the environment by the Chernobyl accident was so enormous that the task of protecting the population has not been entirely successful. The long-term health effects of environmental contamination are caused only in part by external radiation. Scientists studying the problem believe that nearly 60 to 70 percent of future health problems in affected areas will be caused by the consumption of contaminated agricultural products.

If international standards were applied for the use of agricultural land in the affected areas, nearly 1 million hectares (2.5 million acres) would be declared lost for one century, and about 2 million hectares (5 million acres) would be lost for ten to twenty years. In terms of the economy, continuing to use heavily contaminated land for food production at the expense of human health cannot be justified rationally because whatever is salvaged in the agricultural economy will be lost in future health costs.

Victoria Price

FURTHER READING

Alexievich, Svetlana. *Voices from Chernobyl: The Oral History of a Nuclear Disaster.* Translated by Keith Gessen. New York: Picador, 2006.
Bailey, C. C. *The Aftermath of Chernobyl.* Dubuque, Iowa: Kendall Hunt, 1993.
Marples, David R. *Chernobyl and Nuclear Power in the USSR.* New York: St. Martin's Press, 1986.
_____. *The Social Impact of the Chernobyl Disaster.* New York: St. Martin's Press, 1988.
Medvedev, Zhores A. *The Legacy of Chernobyl.* New York: W. W. Norton, 1990.
Savchenko, V. K. *The Ecology of the Chernobyl Catastrophe.* New York: Informa Healthcare, 1995.
Yaroshinskaya, Alla. *Chernobyl: The Forbidden Truth.* Lincoln, Nebr.: Bison Books, 1995.

SEE ALSO: Antinuclear movement; Chalk River nuclear reactor explosion; Chelyabinsk nuclear waste explosion; Nuclear accidents; Nuclear power; Three Mile Island nuclear accident; Windscale radiation release.

Chesapeake Bay

CATEGORIES: Places; water and water pollution; ecology and ecosystems
IDENTIFICATION: Atlantic Ocean inlet bordered by the states of Maryland and Virginia
SIGNIFICANCE: Since the 1970's the health of the Chesapeake Bay and its marine inhabitants has been threatened increasingly by pollution related to environmental stressors in the surrounding watershed. By the early years of the twenty-first century, efforts to reverse the problems were under way.

The Chesapeake Bay, the largest estuary in the United States, spans from northeastern and central Maryland down to southeast Virginia, where it meets the Atlantic Ocean. The rivers and streams that feed the bay extend into the surrounding states of New York, Delaware, Pennsylvania, Virginia, and West Virginia, as well as Washington, D.C. The largest rivers flowing directly into the Chesapeake Bay include the Susquehanna, the Chester, the Potomac, the Rappahannock, and the James.

The bay includes habitats of sandy beaches, intertidal flats, piers, rocks and jetties, shallow waters, sea grass meadows, wetlands, oyster bars, and open waters, all of which house a wide array of animals, plants, and aquatic life. Additionally, its surrounding watershed is home to a number of land-dwelling animals that feed on organisms that live in the bay. More than 3,600 species live within the Chesapeake Bay and its surrounding watershed, including approximately 350 species of fish, 173 species of shellfish, and numerous species of birds, mammals, reptiles and amphibians, bay grasses, and lower-food-web species, including both bottom-dwelling and free-floating plant and animal communities.

The Chesapeake Bay is widely known for its thriving seafood industry, which focuses primarily on the harvest of blue crabs, eastern oysters, clams, and rockfish (also known as striped bass). Since the 1970's, however, overfishing and deteriorating environmental conditions in the bay have caused decreases in the populations of fish, other wildlife, and plants in the watershed.

The Chesapeake Bay has experienced environmental pressures related to population growth, land-use policies, air and water pollution, overfishing, invasive species, and climate change. Approximately 17 million people live within the Chesapeake Bay watershed, and this large population contributes to environmental stress through the development of homes, businesses, and infrastructure, which adds impervious surfaces that contribute to stormwater runoff, destroys habitat, and increases the pollutants entering the bay. Excess nutrients and sediment from agricultural and industrial runoff have contributed to marine dead zones in the bay, areas where oxygen has been depleted and vital sunlight cannot reach bottom-dwelling organisms. Efforts have been undertaken to improve the health of the Chesapeake Bay by restoring water quality through more careful management of land use and reduction of harmful pollutants in agriculture and development, restoring bay grass and wetland habitats, improving fishery management, establishing stewardship and education programs, and enacting protective legislation.

Courtney A. Smith

FURTHER READING

Ernst, Howard R. *Chesapeake Bay Blues: Science, Politics, and the Struggle to Save the Bay.* Lanham, Md.: Rowman & Littlefield, 2003.

Lippson, Alice Jane, and Robert L. Lippson. *Life in the Chesapeake Bay.* 3d ed. Baltimore: The Johns Hopkins University Press, 2006.

SEE ALSO: Cultural eutrophication; Dead zones; Eutrophication; Runoff, agricultural; Runoff, urban; Sedimentation; Urban sprawl; Water pollution; Watershed management.

Chicago Climate Exchange

CATEGORIES: Organizations and agencies; weather and climate

IDENTIFICATION: Financial institution that operates an emissions allowance trading system in Chicago, Illinois

DATE: Established in 2000

SIGNIFICANCE: The Chicago Climate Exchange operates the only cap-and-trade system covering all six greenhouse gases (carbon dioxide, methane, nitrous oxide, sulfur hexafluoride, perfluorocarbons, and hydrofluorocarbons) and offset projects in North America and Brazil.

The Chicago Climate Exchange (CCX) was founded in 2000 and began operating its emissions allowance trading system, or cap-and-trade system, in 2003; by 2010 CCX had four hundred members. CCX membership comprises organizations that produce greenhouse gas emissions (both big emitters and negligible emitters such as office-based businesses and small institutions) as well as owners of title to qualifying emissions-offset projects that sequester, destroy, or reduce such emissions. Each registered emitter makes a voluntary but legally binding commitment to an emissions reduction schedule that includes a commitment to reduce aggregate emissions by a certain percentage below a set baseline level.

Emission baselines, annual reduction commitments, and offset projects are subject to annual audit by third-party experts. In order to reach their emissions targets, members can reduce emissions through their operational practices (such as by changing the fuels they use, making efficiency improvements, or instituting managerial changes), purchase additional emission allowances from other members who have reduced their own emissions by more than the annual reduction requirement, or purchase offset credits from registered emission reduction projects. Entities and individuals can also trade emissions allowances for the purposes of financial investment.

Exchanges affiliated with CCX include the European Climate Exchange (ECX), an exchange operator in the European Union Emissions Trading Scheme); the Insurance Futures Exchange (IFEX), an exchange platform that trades insurance-based derivatives; the Montreal Climate Exchange (MCeX), a joint venture with the Montreal Bourse to host Canadian emissions allowance trading; Tianjin Climate Exchange (TCX), a joint venture with the China National Petroleum Assets Management Company and the Tianjin Property Rights Exchange facilitating emissions trading in China; and Envex, a joint venture of Climate Exchange PLC and Macquarie Capital Group specializing in environmental markets in Australia and the Asia Pacific region.

Rena Christina Tabata

SEE ALSO: Air pollution; Carbon dioxide; Fossil fuels; Global warming; Greenhouse gases; Pollution permit trading.

Chipko Andolan movement

CATEGORIES: Organizations and agencies; activism
and advocacy; forests and plants

IDENTIFICATION: Movement started by villagers in
northern India to stop lumber companies from
clear-cutting mountain slopes

DATE: Originated in April, 1973

SIGNIFICANCE: Through nonviolent protest, the Chipko
Andolan movement put pressure on the Indian government to develop policies concerning natural resources that would be sensitive to the environment
and to the needs of all the Indian people.

The forests of India are a critical resource for the
subsistence of rural people throughout the country, especially in the hill and mountain areas. Mountain villagers depend on the forests for firewood, for
fodder for their cattle, for wood for their houses and
farm tools, and as a means to stabilize their water and
soil resources. During the 1960's and 1970's the Indian government restricted villagers from huge areas
of forestland, then auctioned off the trees to lumber
companies and industries located in the plains. Large
lots of trees were sold to the highest bidders by the
Forest Department, with the purchase payments going to the Indian government.

Because of government restrictions and an ever-
growing population, women who lived in mountain
villages were forced to walk for hours each day just to
gather firewood and fodder. In addition, when mountain slopes were cleared of trees, rains washed away
the topsoil, leaving the soil and rocks underneath to
crumble and fall in landslides. Much of the soil from
the mountain slopes was deposited
in the rivers below, raising water levels. At the same time, the bare slopes
allowed much more rain to run off
directly into the rivers, which resulted in flooding.

As trees were being felled for commerce and industry at increasing
rates during the early 1970's, Indian
villagers finally sought to protect
their lands and livelihoods through
a method of nonviolent resistance
inspired by Indian leader Mohandas Gandhi. In 1973 this resistance
spread and became organized into
the Chipko Andolan movement,
commonly referred to as the Chipko
movement. The word *chipko* comes
from a Hindi word meaning "to
embrace," while *andolan* refers to a
protest against harmful practices.
Together the words literally mean
"movement to hug trees."

The movement originated in April,
1973, as a spontaneous protest by
mountain villagers against logging
abuses in Uttar Pradesh, an Indian
province in the Himalayas. When
contractors sent their workers in to
fell the trees, the villagers embraced
the trees, saving them by interposing
their bodies between the trees and
the workers' axes. The movement was

Birthplace of the Chipko Movement

largely organized and orchestrated by village women, who became leaders and activists in order to save their means of subsistence and their communities.

After many Chipko protests in Uttar Pradesh, victory was finally achieved in 1980 when the Indian government placed a fifteen-year ban on felling live trees in the Himalayan forests. The Chipko movement soon spread to other parts of India, and clear-cutting was stopped in the Western Ghats and the Vindhya Range.

The Chipko protesters staged a socioeconomic revolution in India by gaining control of forest resources from the hands of a distant government bureaucracy that was concerned only with selling the forest in order to make urban-oriented products. The movement generated pressure for the Indian government to develop a natural resources policy that was more sensitive to the environment and the needs of all people.

Alvin K. Benson

FURTHER READING

Guha, Ramachandra. "Chipko: Social History of an 'Environmental' Movement." In *Social Movements and the State*, edited by Ghanshyam Shah. Thousand Oaks, Calif.: Sage, 2003.

Hill, Christopher V. *South Asia: An Environmental History.* Santa Barbara, Calif.: ABC-CLIO, 2008.

SEE ALSO: Deforestation; Erosion and erosion control; Logging and clear-cutting; Sustainable forestry.

Chloracne

CATEGORY: Human health and the environment

DEFINITION: Skin disease resembling acne that results from exposure to chlorine and chlorine-containing compounds

SIGNIFICANCE: Environmentalists have grown increasingly concerned about the negative impacts, including chloracne, that exposure to chlorinated compounds can have on human life.

Chloracne involves an increase of dry skin and a reduction in the ability to produce sebum, which moistens and protects skin. This condition has been linked to exposure to polychlorinated biphenyls (PCBs) and polychlorinated dioxins (PCDDs). PCBs, which are used in industry as lubricants, vacuum pump fluids, and electrical insulating fluids, harm the environment because they evaporate slowly and do not mix well with water. They are therefore widely dispersed in watercourses and in the atmosphere. Because they strongly resist degradation, PCBs remain and accumulate in the environment.

Dioxins are a group of chlorinated aromatic hydrocarbons that are formed in trace amounts during production of the many chlorinated compounds used throughout industry. They are highly persistent in the environment. One dioxin commonly linked to chloracne is 2,3,7,8-tetrachlorodibenzodioxin (TCDD), which is a known human carcinogen. The primary source of dioxin in the environment is the burning of chlorine-containing compounds. Since dioxins are also formed in the production of chlorinated compounds used as herbicides, chlorine-compound exposure occurs in workplaces and in the environment.

Concern about the health risks posed to humans and wildlife by dioxin and related chlorine compounds in the environment was brought to public attention in 1962 with the publication of Rachel Carson's *Silent Spring*, which condemns the widespread use of chlorinated pesticides. Direct skin contact, inhalation, or ingestion of such compounds can result in chloracne, which produces skin lesions on the face, with more severe cases involving lesions on the shoulders, chest, and back. Symptoms may include small, colored, blisterlike pimples (pustules) and straw-colored cysts. Chloracne can develop from three to four weeks after exposure and can last up to fifteen years. Also related to the disease are nausea, bronchitis, and liver disease. It can also have a poisoning effect on the nervous system, which results in extended symptoms such as headache, fatigue, sweaty palms, and numbness in the legs. Chloracne does not respond well to antibiotics. Derivatives of vitamin A (such as gels) and retinoic acid creams may help, and isotretinoin (sold under such trade names as Roaccutane and Claravis) may also be an effective treatment.

In October, 1993, the American Health Association approved a resolution for the gradual phaseout of chlorine. The issue is not a simple one, however, because chlorine is used to benefit humans in disinfecting drinking water and making many medicines. Regardless, environmentalists are worried that chlorinated compounds are slowly poisoning the earth by working their way through the air, groundwater, and the food chain, as well as destroying wildlife and causing many diseases such as chloracne in humans.

Beth Ann Parker and Massimo D. Bezoari

FURTHER READING

Friis, Robert H. "Pesticides and Other Organic Chemicals." In *Essentials of Environmental Health.* Sudbury, Mass.: Jones and Bartlett, 2007.

Hill, Marquita K. "Pesticides." In *Understanding Environmental Pollution.* 3d ed. New York: Cambridge University Press, 2010.

SEE ALSO: Agent Orange; Dioxin; Environmental health; Environmental illnesses; Pesticides and herbicides; Polychlorinated biphenyls.

Chloramphenicol

CATEGORY: Pollutants and toxins

DEFINITION: Broad-spectrum antibiotic used to combat harmful bacteria and organisms in humans and animals

SIGNIFICANCE: Because of chloramphenicol's potential to cause very serious side effects, the U.S. Food and Drug Administration has restricted the drug's use for humans and does not approve its use at all for animal feed or food-producing animals.

Chloramphenicol (CHPC) is known by many other names, including the trade name Chloromycetin. CHPC is an effective antibiotic, particularly in those cases that require penetration through purulent material (which either contains or discharges pus) to reach the infecting bacteria, as occurs in infections of pneumonia. It is also a good choice for otherwise intractable infections involving the eye, the nervous system, and the prostate gland. The more efficient penetration of CHPC into cells compared to other antibiotics increases its advantage in killing intracellular parasites, such as chlamydia, mycoplasma, and rickettsia.

The high acidity of CHPC is believed to contribute to its ability to penetrate necrotic material and cellular membranes. It kills bacteria by interfering with the protein-manufacturing systems occurring in the ribosomes of organisms. The mode of attack is unique in that reptilian, mammalian, and avian ribosomes are unaffected. In addition, the antibiotic does not destroy all the bacteria in infected tissue. Only highly susceptible bacteria are destroyed, with the remainder merely being inhibited from reproducing. This is an advantage because the inactive residual bacteria allow B cells in the immune system to develop an immune response to the bacteria. The result is akin to the use of inactive bacterial vaccinations in fighting disease.

Discovered in 1947, CHPC was released for public use in 1949. In 1959 researchers reported cardiovascular collapse and death in three infants after they were treated with CHPC. All exhibited a gray pallor (called gray syndrome) prior to death. A more complete study found that of thirty-one premature infants treated with the drug, 45 percent died exhibiting gray syndrome, while only 2.5 percent of a group of untreated infants died. The common feature among the fatal cases was the high level of the drug used in their treatment: 230 to 280 milligrams of CHPC per kilogram of body weight per day, five to ten times what later became the recommended dose.

CHPC was subsequently found to interfere with the energy-producing capacity of mitochondria in those organs with a high rate of oxygen consumption; the heart, liver, and kidneys are at particular risk. People who are unable to detoxify the drug efficiently—such as infants—are particularly susceptible to the accumulation of toxic levels in the blood.

CHPC is known to have other negative side effects, including nausea, diarrhea, and loss of appetite. In more severe cases, blood dyscrasias may result. This is a condition in which abnormal blood cells are produced while normal blood cell production is blocked because of interaction of CHPC with the patient's bone marrow. Patients who take CHPC orally may develop fatal anemia. Those who recover from such blood abnormalities resulting from the antibiotic have a high incidence of leukemia. Unfavorable interactions with other drugs—including phenytoin (used to treat heart disease), primidone (used for seizures), and cyclophosphamide (used in chemotherapy)—have also been noted. In addition, the antibiotic can interfere with vaccinations and is a potential carcinogen.

These side effects occur not only in humans but also in other animals. CHPC may build up to toxic levels in young animals because their liver detoxification mechanisms are not as well developed or efficient as those of adults. The antibiotic therefore is not approved for administration to food animals in the United States, as these animals may be ingested by the young. Ingestion of CHPC by pregnant or lactating creatures must be avoided for similar reasons. Anorexia in dolphins, whales, and California sea lions has also been a noted side effect. Environmentalists and others have raised concerns that the antibiotic may be used in some agricultural settings outside the United States, putting humans at risk. In 2010, for example, the U.S. Food and Drug Administration seized a shipment of honey from China that was found to be contaminated by CHPC; the bees that produced the honey had been fed the antibiotic as a measure to control certain diseases in the bee population.

Although the serious side effects of CHPC are rare, a study by the California Medical Association in 1967 found the occurrence of cases in humans to be as high as one per twenty-four thousand applications, depending on dose. The accumulation of such data was sufficient for the use of CHPC to be restricted, in the 1960's, to life-threatening cases for which there is little alternative treatment.

Jacqueline J. Robinson and Massimo D. Bezoari

FURTHER READING

Forrest, Graeme N., and David W. Oldach. "Chloramphenicol." In *Infectious Diseases*, edited by Sherwood L. Gorbach, John G. Bartlett, and Neil R. Blacklow. 3d ed. Philadelphia: Lippincott Williams & Wilkins, 2004.

Hilts, Philip J. "New Drugs, New Problems." In *Protecting America's Health: The FDA, Business, and One Hundred Years of Regulation*. New York: Alfred A. Knopf, 2003.

SEE ALSO: Antibiotics; Aquaculture; Bacterial resistance.

Chlorination

CATEGORY: Water and water pollution
DEFINITION: Practice of disinfecting water by the addition of chlorine
SIGNIFICANCE: Although the chlorination of public water in the United States has helped reduce outbreaks of waterborne disease, it has raised concerns about the possible formation of chloro-organic compounds in treated water.

Drinking water, wastewater, and water in swimming pools are the most common water sources where chlorination is used to kill bacteria and prevent the spread of diseases. Viruses are generally more resistant to chlorination than are bacteria, but they can be eliminated with an increase in the chlorine levels needed to kill bacteria. Common chlorinating agents include elemental chlorine gas and sodium or calcium hypochlorite. In water these substances generate hypochlorous acid, which is the chemical agent responsible for killing microorganisms by inactivating bacteria proteins or viral nucleoproteins.

Public drinking water was chlorinated in most large U.S. cities by 1914. The effectiveness of chlorination in reducing outbreaks of waterborne diseases in the early twentieth century was clearly illustrated by the drop in typhoid deaths: 36 per 100,000 in 1920 to 5 per 100,000 by 1928. Chlorination has remained the most economical method of purifying public water, although it is not without potential risks. Chlorination has also been widely used to prevent the spread of bacteria in the food industry.

In its elemental form, high concentrations of chlorine are very toxic, and solutions containing more than 1,000 milligrams per liter (mg/l) are lethal to humans. Chlorine has a characteristic odor that is detectable at levels of 2-3 mg/l of water. Most public water supplies contain chlorine levels of 1-2 mg/l, although the actual concentration of water reaching consumer faucets fluctuates and is usually around 0.5 mg/l. Consumption of water containing 50 mg/l has produced no immediate adverse effects.

The greatest environmental concern regarding chlorination has less to do with the chlorine itself than it does with the potential toxic compounds that may form when chlorine reacts with organic compounds present in water. Chlorine, which is an extremely reactive element, reacts with organic material associated with decaying vegetation (humic acids), forming chloro-organic compounds. Trihalomethanes (THMs) are one of the most common chloro-organic compounds. At least a dozen THMs have been identified in drinking water since the 1970's, when health authorities in the United States came under pressure to issue standards for the identification and reduction of THM levels in drinking water.

Major concern has focused on levels of chloroform because of this compound's known carcinogenic properties in animal studies. Once used in cough syrups, mouthwashes, and toothpastes, chloroform in consumer products is now severely restricted. A 1975 study of chloroform concentrations in drinking water found levels of more than 300 micrograms per liter (µg/l) in some water, with 10 percent of the water systems surveyed having levels of more than 105 µg/l. In 1984 the World Health Organization set a guideline value of 30 µg/l for chloroform in drinking water.

Although there are risks associated with drinking chlorinated water, it has been estimated that the risk of death from cigarette smoking is two thousand times greater than that of drinking chloroform-contaminated water from most public sources. However, as water sources become more polluted and require higher levels of chlorination to maintain purity, continual monitoring of chloro-organic compounds will be needed.

Nicholas C. Thomas

FURTHER READING

Bull, Richard J. "Drinking Water Disinfection." In *Environmental Toxicants: Human Exposures and Their Health Effects*, edited by Morton Lippmann. New York: John Wiley & Sons, 2000.
Gray, N. F. *Drinking Water Quality: Problems and Solutions.* 2d ed. New York: Cambridge University Press, 2008.

SEE ALSO: Drinking water; Water pollution; Water quality; Water treatment.

Chlorofluorocarbons

CATEGORY: Atmosphere and air pollution
DEFINITION: Family of chemical compounds used in air conditioners, refrigerators, and aerosol spray cans
SIGNIFICANCE: Concerns about the destruction of stratospheric ozone by chlorofluorocarbons led to a worldwide ban on the manufacture and use of these compounds.

Chlorofluorocarbons (CFCs) are organic molecules containing chlorine, fluorine, and carbon atoms. The first CFCs were discovered by Thomas Midgley, Jr., in 1928. Because these molecules are chemically inert and easily liquefied, CFCs soon became the standard coolants in refrigerators and air conditioners. They also became widely used as propellants in aerosol spray cans. By 1968, 2.3 billion aerosol cans containing CFCs had been sold in the United States.

In 1970 the British scientist James Lovelock determined that most CFCs entering the atmosphere remained there without significant decomposition. Three years later, Frank Sherwood Rowland and

Mario Molina, working at the University of California at Irvine, suggested that CFCs would eventually migrate into the stratosphere. Once there, absorption of ultraviolet light would cause CFCs to release chlorine atoms, which would then react catalytically to remove ozone. Since ozone in the stratosphere prevents high-energy ultraviolet light from reaching the surface of the earth, any decrease in ozone would lead to increased exposure to ultraviolet light on the earth's surface, causing higher levels of skin cancer in humans and damage to plants and animals.

Although evidence from laboratory studies suggested that CFCs in the atmosphere would cause depletion of stratospheric ozone, uncertainty remained as to the degree of ozone destruction that would occur. Nevertheless, in 1975 Oregon became the first U.S. state to ban CFCs in aerosol spray cans. Several other states took similar actions, and in 1977 the Food and Drug Administration (FDA) implemented a ban on the use of CFCs as aerosol propellants to be phased in over a two-year period. Continued uncertainties in predictions of ozone loss and the lack of direct evidence for ozone depletion kept most other countries from restricting the use of CFCs. While the U.S. Environmental Protection Agency (EPA) discussed instituting a total ban on CFCs, no action was taken, in part because of the difficulty in finding adequate substitutes for CFCs.

In 1985 a team of British scientists led by Joseph Farman announced the discovery of significant loss of ozone over the Antarctic. Beginning in the early 1970's, springtime levels of ozone had slowly decreased. By 1985 as much as 40 percent of the ozone usually present in the Antarctic stratosphere during the spring had disappeared. In addition, both the duration and the geographic extent of this ozone hole were increasing. Evidence linking formation of the ozone hole to CFCs in the atmosphere was quickly found.

The discovery of the ozone hole led to further restrictions on CFCs. In 1987 an international agreement called the Montreal Protocol was reached to ban the manufacture and use of CFCs by the year 2010. In the United States, passage of the 1990 Clean Air Act amendments resulted in an accelerated timetable for restrictions on CFCs and related compounds. By the mid-1990's levels of CFCs in the atmosphere had stabilized, and CFCs are expected to disappear gradually from the atmosphere over the next century.

Jeffrey A. Joens

Further Reading

Joesten, Melvin D., John L. Hogg, and Mary E. Castellion. "Chlorofluorocarbons and the Ozone Layer." In *The World of Chemistry: Essentials.* 4th ed. Belmont, Calif.: Thomson Brooks/Cole, 2007.

Newman, Michael C., and Michael A. Unger. "Environmental Contaminants." In *Fundamentals of Ecotoxicology.* 2d ed. Boca Raton, Fla.: CRC Press, 2003.

Parson, Edward A. *Protecting the Ozone Layer: Science and Strategy.* New York: Oxford University Press, 2003.

See also: Aerosols; Freon; Molina, Mario; Ozone layer; Rowland, Frank Sherwood.

CITES. *See* Convention on International Trade in Endangered Species

Citizens Clearinghouse for Hazardous Waste. *See* Center for Health, Environment, and Justice

Clean Air Act and amendments

Categories: Treaties, laws, and court cases; atmosphere and air pollution

The Laws: U.S. federal laws that govern standards for air quality

Dates: Enacted on December 17, 1963; amended 1970, 1977, and 1990

Significance: The Clean Air Act of 1963 and its amendments federalize the regulation of air pollution in the United States to a large degree. The act provides guidelines for minimum standards of air quality as well as maximum levels for the emissions of pollutants. It has served as a model for other federal environmental legislation.

Since the 1880's state and local governments in the United States have put limits on smoke emissions and other forms of air pollution. Federal regulation of the problem, however, did not really begin until 1955, when the Air Pollution Control Act authorized the federal government to conduct research and provide assistance to state and local governments. This act included no national standards, and it ceded responsibility for controlling air quality to the states.

Congress increased the federal role somewhat in the Clean Air Act of 1963. The secretary of the Department of Health, Education, and Welfare was authorized to call abatement conferences when air pollution from one state put citizens of another state in danger, but the Clean Air Act failed to include any sanctions for the enforcement of national standards. Meanwhile, evidence was accumulating that air pollution posed a serious threat to public health throughout the country. Incidents such as the November, 1966, acute air-pollution episode in New York City, an event blamed for the deaths of some 168 people, served as a sobering example of how polluted America's air had become. President Lyndon B. Johnson's Great Society looked to federal regulation as the only effective way to deal with such matters.

The 1967 Air Quality Act authorized the Department of Health, Education, and Welfare to consult with the states to determine air-quality standards in regions of particular concern, and the states were then given a year to formulate a plan to implement the guidelines. Environmentalists were disappointed that Congress still had not provided minimum standards of air quality or effective means for forcing the states to achieve their goals. The most significant aspect of the act was the authorization of some federal enforcement of vehicular emissions standards, with criminal fines of up to $1,000 for each violation of the standards. Relatively weak requirements based on grams of pollutants emitted per mile took effect for new automobiles in 1968.

1970 Amendments

Widespread support of the environmental movement was demonstrated by the enthusiastic response to Earth Day in 1970, and that same year the first report of the U.S. Council on Environmental Quality called on Congress to enact new laws to deal with several problems, including air pollution. Senator Edmund Muskie, a presidential hopeful, was the acknowledged congressional leader in the campaign for tough environmental reform, and he was the chief author of the 1970 clean air bill. President Richard Nixon also supported an aggressive bill. With this bipartisan support, Congress enacted a far-reaching amendment to the 1963 legislation, the Clean Air Act

Extension of 1970, which initiated the federal government's regulation of air pollution.

Addressing perceived weaknesses in the existing law, the landmark 1970 amendments authorized the newly created Environmental Protection Agency (EPA) to establish standards that would be binding on states. Applying a command-and-control approach to regulation, the centerpiece of the legislation was a program for the EPA to determine National Ambient Air Quality Standards (NAAQS) that would define specific levels of air pollution considered harmful to public health. The EPA was also authorized to set emission limits on hazardous pollutants at levels allowing a sufficient margin of safety. Although states might exercise discretion in choosing how to meet the federal standards, they were required to develop state implementation plans (SIPs) that utilized appropriate measures to reach those standards. States could maintain the air-quality-control programs already in place for existing industrial plants while requiring new plants to meet stricter standards based on the best available technology that was economically feasible.

The 1970 legislation stunned American automobile manufacturers by requiring them to curtail their products' emissions of the "big three" pollutants—hydrocarbons, carbon oxides, and nitrogen oxides—by 90 percent within six years. The technology did not exist to allow them to meet the new standards, although it was hoped that new technology could be developed within the specified time. Most members of Congress, few of whom were willing to see the collapse of the automobile industry, understood that it might be necessary to extend the deadline. In fact, the deadlines for meeting the vehicular emission standards turned out to be excessively ambitious, and waivers for the standards were granted in 1971, 1973, 1974, and 1976.

1977 AMENDMENTS

With the enthusiastic support of President Jimmy Carter, Congress passed major revisions to the Clean

---

Carter Comments on the Clean Air Act Amendments

In his statement on signing the Clean Air Act amendments of 1977 into law, President Jimmy Carter explained the value of the new legislation:

This act is the culmination of a 3-year effort by the Congress to develop legislation which will continue our progress toward meeting our national clean air goals in all parts of the country. The issues involved in amending the Clean Air Act have been difficult and the debate lengthy. However, I believe that the Congress . . . has adopted a sound and comprehensive program for achieving and preserving healthy air in our Nation.

The automobile industry now has a firm timetable for meeting strict, but achievable emission reductions. That industry now knows with certainty what is required and can devote its full-time energies to designing cars which will further our clean air goals while continuing to improve fuel efficiency. This timetable will be enforced.

With this legislation, we can continue to protect our national parks and our major national wilderness areas and national monuments from the degradation of air pollution. Other clean air areas of the country will also be protected, at the same time permitting economic growth in an environmentally sound manner.

The act provides us with a new tool to help abate industrial sources of pollution by authorizing use of economic incentives to reduce noncompliance. By directing the Environmental Protection Agency to establish monetary penalties equal to the cost of cleanup, those industries which delay installing abatement equipment will no longer be rewarded in the marketplace.

These three major provisions, coupled with the other authorities of H.R. 6161, provide the statutory framework for the Environmental Protection Agency to implement a firm, but responsible program for meeting and maintaining air quality standards which are necessary to protect the health of all of our citizens.

Air Act on August 4, 1977. In addition to making NAAQS more stringent, the amendments required each state to designate "nonattainment" regions based on the NAAQS. Each state was then given the choice of either accepting statutory sanctions or revising its SIPs in order to meet the standards in a timely way. The amendments focused especially on coal-burning power plants, a significant source of sulfur dioxide in the atmosphere, which contributes to acid rain. Existing stationary sources of pollution were required to provide for "reasonably available control technology," and new or modified stationary sources were required to utilize technology meeting the "lowest achievable emission rate," which usually meant the use of expensive scrubbers.

In the case of clean air regions already in attainment, the amendments instituted the Prevention of

Significant Deterioration (PSD) program, which was designed to prevent the EPA from allowing deterioration of air quality up to the national standards. Members of Congress from rural districts had unsuccessfully argued that such a program would unfairly restrict industrial growth in areas where air pollution was not a problem.

The 1977 amendments further extended the deadlines by which automobiles were required to achieve emissions-control standards. The stricter controls were scheduled for the 1980 model year. American automakers had insisted that they could not meet the requirements in the existing law, and they had threatened to shut down production lines. This marked the fifth relaxation of the vehicular deadlines.

1990 AMENDMENTS

Between 1977 and 1990 numerous efforts were undertaken both to strengthen and to weaken the Clean Air Act, but opposing interest groups prevented major changes in either direction. While there was widespread agreement that the Clean Air Act had been somewhat successful in improving air quality, the administration of President Ronald Reagan strongly opposed any expansion of environmental regulations. During the 1988 presidential election campaign, candidate George H. W. Bush pledged to be the "environmental president." When Bush entered the White House, the deadlines for compliance with most air-quality standards had passed, putting noncompliance regions in danger of losing many industrial jobs. In July, 1989, the Bush administration made a sweeping proposal that most Democrats and environmentalists could support, while conservative Republicans were divided on the issue. The resulting amendments were signed into law on November 15, 1990.

The 1990 amendments were extremely complex, requiring more than seven hundred pages. The major regulatory change, modeled on the Clean Water Act, was a requirement that all major sources of air pollution obtain state operating permits, with the EPA given the authority to veto such permits. The amendments provided additional regulations of emissions that were responsible for acid rain and established an allowance system based on a nationwide limit of 8.1 million metric tons (8.9 million U.S. tons) of sulfur dioxide per year. Other provisions included a phaseout program for chlorofluorocarbons (CFCs) and other ozone-depleting substances, as well as a requirement that industrial plants cut emissions of 189 toxic substances to the levels of the cleanest plants within their particular industries.

Because one important goal of the statute was to decrease urban smog, it included strict controls on automobile emissions and mandates for cleaner-burning fuels. Beginning with 1994 automobiles, tailpipe exhausts were required to contain 60 percent less nitrogen oxide, and emissions-control equipment was required to last ten years. The EPA was authorized to conduct a study to determine whether stricter standards were needed. A pilot program in California required an increasing number of cars and light trucks to run on batteries or nongasoline fuels. Beginning in 1995, oil companies were required to sell only cleaner-burning reformulated gasoline in the smoggiest metropolitan regions, and gasoline stations were mandated to install devices to capture fumes during refueling.

The 1990 amendments considerably strengthened the enforcement provisions under the Clean Air Act. The EPA acquired new powers to issue administrative penalties of up to $25,000 per day, and individuals were empowered to take civil action against polluters. The EPA and the U.S. Department of Justice were given new authority to prosecute misdemeanor and felony violations of the act. The amendments increased maximum sentences for most violations from six months to two years and increased maximum fines from $25,000 to $500,000. An individual who released hazardous air pollutants into the air could henceforth be sentenced to fifteen years in prison and fined up to $250,000, and corporations could be fined up to $1 million.

At an estimated cost of $25 billion per year, the 1990 act is considered the most expensive piece of environmental legislation ever passed. It was expected that the costs would mostly be passed on to consumers in higher prices for cars, gasoline, electricity, and products containing chemicals. The EPA's estimated monetary value of the act, based on its public health and environmental benefits, offsets its cost.

THE CLEAR SKIES INITIATIVE

In 2002, President George W. Bush announced the Clear Skies Initiative, a policy intended to incentivize innovation and cut costs through a market-based cap-and-trade program for reducing power plant emissions. Through such a program, polluters would have the right to emit a certain quantity of pollutants; a polluter that wishes to emit more than its allowance would have to purchase credits from one that emits less. The initiative, which prioritized economic growth, operated on the assumption that the market drives advances in pollution-control technologies and thereby hastens environmental progress. The initiative gave rise to the Clear Skies Bill of 2003, which would have amended the Clean Air Act with a cap-and-trade system. Critics—among them the Sierra Club and the Natural Resources Defense Council—argued that the proposed law was a propagandistically named reduction of air-quality protections that would allow increased toxic industrial emissions and hamper enforcement of pollution-control standards. The bill never moved beyond the Senate Environment and Public Works Committee and thus did not become law.

In 2005 the EPA introduced the Clean Air Interstate Rule (CAIR). A key measure of the deadlocked Clear Skies Bill, CAIR is a cap-and-trade program intended to ensure that air pollution generated in one state does not prevent another downwind state from meeting air-quality standards. CAIR, which was designed to reduce smog and soot pollution from power plants in the eastern United States, includes a permanent cap on the precursor pollutants sulfur dioxide

and nitrogen oxides. In July, 2008, a federal appeals court vacated CAIR, citing several fundamental flaws in the rule. The EPA, the Environmental Defense Fund, and several states successfully appealed for a rehearing. The court determined that, despite the rule's shortcomings, the environmental and health benefits of CAIR are significant. (The EPA estimated that CAIR would prevent seventeen thousand deaths annually by 2015.) In December, 2008, the court issued an order temporarily reinstating CAIR until the EPA could replace it with a rule that fully addresses CAIR's flaws.

Another measure of the Clear Skies Bill, the Clean Air Mercury Rule (CAMR), was also introduced in 2005. Environmental groups, several states, Native American tribes, and physicians' organizations opposed the rule, as its use of a cap-and-trade program in the case of a bioaccumulative, environmentally persistent material such as mercury would allow the development of toxic hot spots that would endanger human health. CAMR also removed oil- and coal-fired electric-utility steam-generating units from the list of hazardous air-pollutant sources. In 2008, a federal appeals court found CAMR to be in violation of the Clean Air Act and vacated the rule.

Time Line of U.S. Clean Air Laws and Policies

YEAR	EVENT
1955	Air Pollution Control Act, the first U.S. law to address air pollution and fund research into pollution prevention, is passed.
1963	Clean Air Act is the first U.S. law to provide for the monitoring and control of air pollution.
1967	Air Quality Act establishes enforcement provisions to reduce interstate air-pollution transport.
1970	Clean Air Act amendments establish the first comprehensive emission regulatory structure, including the National Ambient Air Quality Standards (NAAQS).
1977	Clean Air Act amendments provide for the prevention of deterioration in air quality in areas in compliance with the NAAQS.
1990	Clean Air Act amendments establish programs to control acid precipitation, as well as 189 specific toxic pollutants.
1995	Oil companies are required to sell reformulated gasoline in metropolitan regions, and gas stations are required to install vapor-retrieval devices on pumps.
2003	Proposed Clear Skies Bill is designed to amend the Clean Air Act with a cap-and-trade system.
2005	The EPA's Clean Air Interstate Rule (CAIR) begins a cap-and-trade program to keep air pollution generated in one state from rendering other states noncompliant with air-quality standards.
2008	A federal appeals court rules that CAIR exceeds the EPA's regulatory authority but later orders temporary reinstatement.

SUBSEQUENT DEVELOPMENTS

In a 2007 case heard by the U.S. Supreme Court, twelve states, three major U.S. cities, a U.S. territory, and several nongovernmental organizations sued the EPA for failing to regulate greenhouse gases (GHGs) as pollutants. The Court found that the EPA was again in violation of the Clean Air Act and charged the agency with determining whether GHG emissions from new vehicles are pollutants that endanger the public health or welfare. The EPA concluded that GHGs do in fact pose a danger to the public and submitted its endangerment finding to the White House. White House officials refused to read the EPA's report and took other measures to block the EPA's regulation of GHGs during George W. Bush's administration.

In 2009, under President Barack Obama's administration, the EPA issued its final findings regarding GHGs. It determined that current and projected concentrations of six GHGs in the atmosphere—carbon monoxide, methane, nitrous oxide, hydrofluorocarbons, perfluorocarbons, and sulfur hexafluoride—constitute a threat to the public health and welfare of current and future generations. It also found that new motor vehicles were contributing to GHG pollution. These findings may someday lead to more restrictive emissions limits for power plants, oil refineries, auto manufacturers, and other major GHG contributors.

ENFORCEMENT

Based on a command-and-control model, the Clean Air Act and its amendments provide a variety of strong mechanisms for enforcing their statutory and regulatory requirements. The EPA has primary responsibility for enforcement at the federal level, and the states share responsibility for regulating SIPs. Citizens are also given broad opportunities to participate in the enforcement process.

When the EPA finds evidence that a violation has occurred, a regional office of the agency issues a notice of violation to both the source and the state. Based on its investigations, the EPA has the discretion to determine whether further action is necessary. The agency may issue an administrative order requiring a person or institution to comply with the applicable statute or regulation. If the recipient of the order fails to comply, the EPA may enforce the order through a civil action. If there is probable cause that a crime has occurred, the EPA will initiate a criminal prosecution,

but the Department of Justice usually takes charge of the legal actions.

In formulating its SIPs, each state is required to include a program of legal enforcement. The states are usually given the opportunity to lead in initiating enforcement action if they wish to do so. If a state does not do so, the EPA has the authority to proceed on its own. Any person, moreover, may bring a civil action against an individual or entity alleged to be in violation of the Clean Air Act. If a violation is proved, any monetary awards must be either turned over to the EPA's "penalty fund" or used for "beneficial mitigation projects."

When the landmark amendments of 1970 were passed, proponents of the act tended to be extremely optimistic about the prospects for achieving national air-quality standards without any serious economic costs. The act envisioned full attainment of the standards by 1975, but this expectation turned out to be unrealistic, especially in regard to ozone. In the 1977 amendments, Congress responded to the problem by explicitly recognizing noncompliance regions, which were thereafter required to improve incrementally. It was even more difficult to formulate vehicular emissions standards that were both meaningful and attainable, in part because no one could be certain about the prospects of technological improvement. When automotive technology improved, moreover, no one could be certain about whether Clean Air Act standards were a primary cause.

By the late 1990's, few people denied that the Clean Air Act had helped decrease air pollution and improve the public health. By its nature, however, such legislation does not completely satisfy everyone. Environmental organizations commonly argue that the EPA has not been aggressive enough in its enforcement efforts, while probusiness groups tend to blame the Clean Air Act for forcing American industries to close their doors and move to poor countries with weaker regulatory protections and a greater toleration for dirty air.

Thomas T. Lewis
Updated by Karen N. Kähler

FURTHER READING

Bryner, Gary C. *Blue Skies, Green Politics: The Clean Air Act of 1990.* Rev. ed. Washington, D.C.: Congressional Quarterly Press, 1995.

Cohen, Richard E. *Washington at Work: Back Rooms and Clean Air.* 2d ed. Boston: Allyn & Bacon, 1995.

Griffin, Roger D. *Principles of Air Quality Management.* 2d ed. Boca Raton, Fla.: CRC Press, 2007.

Lipton, James P., ed. *Clean Air Act: Interpretation and Analysis.* New York: Nova Science, 2006.

Rajan, Sudhir Chella. *The Enigma of Automobility: Democratic Politics and Pollution Control.* Pittsburgh: University of Pittsburgh Press, 1996.

U.S. Environmental Protection Agency. *The Plain English Guide to the Clean Air Act.* Research Triangle Park, N.C.: Office of Air Quality Planning and Standards, 2007.

SEE ALSO: Acid deposition and acid rain; Air-pollution policy; Automobile emissions; Chlorofluorocarbons; Coal-fired power plants; Environmental law, U.S.; Greenhouse gases; Pollution permit trading; Smog; Sulfur oxides.

Clean Water Act and amendments

CATEGORIES: Treaties, laws, and court cases; water and water pollution

THE LAWS: Federal legislation designed to improve the quality of surface water throughout the United States

DATES: Enacted on October 18, 1972; major amendments in 1977, 1981, and 1987

SIGNIFICANCE: The legislation now called the Clean Water Act was largely shaped by the 1972 amendments to the Federal Water Quality Act of 1965, itself an amendment to 1948 legislation. The complex law was further strengthened by later amendments as the American public became increasingly aware of the importance of clean water supplies to the public health.

Before the mid-1960's government regulation of water pollution in the United States was mostly left up to individual states. The earliest U.S. federal environmental law was the Rivers and Harbors Act of 1899, which prohibited the dumping of debris into navigable waters. Although the law was intended to protect interstate navigation, it became an instrument for regulating water quality sixty years after its passage. The Oil Pollution Act of 1924 prohibited the discharge of oil into interstate waterways, with criminal sanctions for violations. The first Federal Water Pollution Control Act (FWPCA), passed in 1948, authorized the preparation of federal pollution-abatement plans, which the states could either accept or reject, and provided some financial assistance for state projects. Although the FWPCA was amended in 1956 and 1961, it still contained no effective mechanisms for the federal enforcement of standards.

By this period, however, many Americans were recognizing water pollution as a national problem that required a national solution. The Federal Water Quality Act of 1965 amended the 1948 legislation to introduce a policy of minimum water-quality standards that could be enforced in federal courts. The standards applied regardless of whether discharges could be proven to harm human health. The act also significantly increased federal funds for the construction of sewage plants. A 1966 amendment required the reporting of discharges into waterways, with civil penalties for failure to comply. Another amendment, the Water Quality Improvement Act of 1970, established federal licensing for the discharge of pollutants into navigable rivers and provided plans and funding for the detection and removal of oil spills.

THE ADVENT OF MODERN WATER-PROTECTION LEGISLATION

Congress and President Richard Nixon agreed that existing programs were ineffective in controlling water pollution. The resulting Federal Water Pollution Control Act Amendments of 1972 amended the Federal Water Quality Act to establish the basic framework for the Clean Water Act. The centerpiece of the landmark amendments was the National Pollutant Discharge Elimination System (NPDES), which utilizes the command-and-control methods earlier enacted in the Clean Air Act. The premise of the legislation was that polluting surface water is an unlawful activity, except for those exemptions specifically allowed in the act. The announced goal was to eliminate all pollutants discharged into U.S. surface waters by 1985.

In addition to standards of quality for ambient water, the amendments included technology-based standards. Industrial dischargers were given until 1977 to make use of the "best practicable technology" in their industries, and the standard was to be increased to the "best available technology" by 1983. The 1972 act also included stringent limitations on the release of toxic chemicals judged harmful to human health. For members of Congress, the most popular part of the act was the grant program for the construction of publicly owned treatment works (POTWs).

The U.S. Environmental Protection Agency (EPA), created just two years earlier, was assigned the primary responsibility for regulating and enforcing the legislation. The agency could issue five-year permits for the discharge of pollutants, and any discharge without a license or contrary to the terms of a license was punishable by either civil or criminal sanctions. When dealing with a discharge of oil or other hazardous substances, the EPA could go to court and seek a penalty of up to $50,000 per violation and up to $250,000 in the case of willful misconduct. In addition, a discharger might be assessed the costs of removal, up to $50 million. Because of the technical complexity of the law, the EPA for many years relied more on civil penalties than on criminal prosecutions.

The 1972 amendments prohibited the discharge of dredged or fill materials into navigable waters unless authorized by a permit issued by the U.S. Army Corps of Engineers (USACE). Based on the literal wording of the statute, the USACE at first regulated only actually, potentially, and historically navigable waters. In 1975, however, it revised its regulations to include jurisdiction over all coastal and freshwater wetlands, provided they were inundated often enough to support vegetation adapted for saturated soils. The Supreme Court endorsed the USACE's broad construction of the law. The USACE and the EPA later adopted a rule under which isolated waters that were actual or potential habitat for migratory birds that crossed state lines were subject to the provisions of the Clean Water Act.

The Clean Water Act amendments of 1977, which gave the legislation its current name, focused on a large variety of technical issues. They required industries to use the best available technology to remove toxic pollutants within six years. For conventional pollutants (such as ammonia, pathogens, phosphorus, and suspended solids), businesses could seek waivers from the technology requirements if the removal of the pollutants was not worth the cost. The act further required an environmental impact statement for any federal project involving wetlands, and it extended liability for oil-spill cleanups from 19 kilometers (12 miles) to 322 kilometers (200 miles) offshore.

LATER AMENDMENTS

The Municipal Wastewater Treatment Construction Grants Amendments of 1981, an important piece of environmental public works legislation, streamlined the municipal construction grants process. This allowed for municipalities to improve their sewage treatment capabilities.

The amendments of 1987, entitled the Water Quality Act, were passed by Congress over President Ronald Reagan's veto. In addition to increasing the powers of the EPA, the act significantly raised the criminal penalties for acts of pollution. Individuals who knowingly discharged certain dangerous pollutants could receive a fine of up to $250,000 and imprisonment for up to fifteen years. The maximum prison term for making false statements or tampering with monitoring equipment was increased from six months to two years. The most controversial part of the act was its authorization of $18 billion for the construction of wastewater treatment plants. In addition, the 1987 amendments phased out the earlier construction grants program, replacing it with the State Water Pollution Control Revolving Fund. Also called the Clean Water State Revolving Fund, the new program relied on EPA-state partnerships.

The 1987 Water Quality Act also provided state funds for managing and controlling nonpoint source pollution, such as stormwater runoff from urban areas, forests, agricultural lands, and construction sites. Earlier legislation had focused more on pollution from discrete sources, such as industrial plants and municipal sewage facilities, that could be more easily identified and regulated. Roughly half of the nation's remaining water pollution stemmed from nonpoint sources.

In the wake of the 1989 *Exxon Valdez* oil spill, Congress passed the Oil Pollution Act of 1990. This legislation strengthened cleanup requirements and penalties for oil discharges.

Ongoing points of contention regarding the Clean Water Act have been its wetlands protection program, the loose interpretation of "navigable waters," and the EPA/USACE "migratory bird rule." In a 2001 case, *Solid Waste Agency of Northern Cook County v. Army Corps of Engineers*, the U.S. Supreme Court found that federal protection under the Clean Water Act did not apply in the case of isolated wetlands such as the area that Cook County, Illinois, planned to use as a landfill. In 2006 the Court also determined that the act was inapplicable in the related cases *Rapanos v. United States* and *Carabell v. Corps of Engineers*, which involved two Michigan landowners planning to develop on wetlands. In early 2010 a Clean Water Act amendment was proposed that would replace the phrase "navigable waters" with "waters of the United States."

Some of the worst causes of water pollution in the United States have been curtailed in the years since the Clean Water Act was overhauled in 1972, even though the act has manifestly failed to achieve its stated goals. The legislators who hoped to render all U.S. waters fishable and swimmable within a decade were clearly overly optimistic. It is probably inevitable that economic prosperity and population growth will mean that water in the United States will never be completely free of pollutants. Since 1972, nevertheless, the American public has become increasingly intolerant of dirty and unhealthful water, and Congress, reflecting public sentiment, has continued to strengthen the Clean Water Act.

Thomas T. Lewis
Updated by Karen N. Kähler

FURTHER READING

Copeland, Claudia. *Clean Water Act: A Summary of the Law.* Washington, D.C.: Congressional Research Service, 2008.

Finkmoore, Richard J. *Environmental Law and the Values of Nature.* Durham, N.C.: Carolina Academic Press, 2010.

Freedman, Martin, and Bikki Jaggi. *Air and Water Pollution Regulation: Accomplishments and Economic Consequences.* Westport, Conn.: Quorum Books, 1993.

Lazarus, Richard J. *The Making of Environmental Law.* Chicago: University of Chicago Press, 2004.

Milazzo, Paul Charles. *Unlikely Environmentalists: Congress and Clean Water, 1945-1972.* Lawrence: University Press of Kansas, 2006.

Ryan, Mark. *The Clean Water Act Handbook.* 2d ed. Chicago: American Bar Association, 2003.

SEE ALSO: Drinking water; Environmental law, U.S.; Runoff, agricultural; Runoff, urban; Safe Drinking Water Act; Stormwater management; Wastewater management; Water-pollution policy; Water quality; Water treatment.

Climate accommodation

CATEGORIES: Resources and resource management; weather and climate

DEFINITION: Local responses to rising sea levels involving adaptations such as surrendering of land to the sea, raising dikes, and building homes designed to rise with water levels

SIGNIFICANCE: The anticipation of rising sea levels around the world as a result of global warming has led many coastal nations to develop strategies that can protect their people while making accommodations to the reality of their changed coastlines.

Rising seas associated with global warming have challenged many millions of people living close to oceans around the world. Policies and building practices have been changed in some countries to adapt to, or accommodate, these changes.

BRITISH STRATEGIES

Parts of Great Britain's coastline are afflicted by the same problems as the eastern and Gulf coasts of the United States: The land is subsiding as ice melt and thermal expansion slowly raise sea levels. The United Kingdom Climate Impact Programme, a government-funded program at Oxford University, forecasts that the sea level could rise by as much as 1 meter (3.3 feet) by late in the twenty-first century. In addition to climate change, isostatic rebound (the rise of Scotland's coast following the last ice age) is contributing to subsidence of the land southward along the English coast. As the sea rises 3 millimeters (0.118 inch) per year abreast of Essex, the land itself is sinking half as rapidly, producing a net sea-level rise averaging 4.5 millimeters (0.177 inch) per year.

In a strategy officially termed "managed realignment," the British government decided to allow the sea to flood low-lying farmland rather than attempt to fend off the invading waters by building ever-higher defenses. The policy, which will eventually allow the encroaching sea to submerge several thousand hectares, has been welcomed by environmentalists. Farmers, however, have contended that the strategy is unviable and have demanded more flood defenses. The affected area of the coast ranges from the Humber estuary, around East Anglia, to the Thames estuary and west to the Solent. Strategic withdrawal also has been planned for sections of the Severn estuary. The first site surrendered to the sea was in

Lincolnshire. About 81 hectares (200 acres) of farmland were flooded by seawater at Freiston Shore after diggers broke through the flood-defense banks to create a salt-marsh bird reserve.

Until around the end of the twentieth century, the Thames Barrier, built to protect London and surrounding areas from unusually high river tides and storm surges, closed an average of two or three times a year. Between November, 2001, and March, 2002, however, the barrier was raised twenty-three times. A British report released in September, 2002, said that 152,800 square kilometers (59,000 square miles)—home to 750,000 people—in and around London are vulnerable to flooding because they are below high-tide levels, some by as much as 3.7 meters (12 feet).

DUTCH STRATEGIES

The Dutch fear that rising storm surges could inundate much of the Netherlands, large areas of which have been reclaimed from the sea. Fears have been expressed that the country's western provinces may flood. The Hague, for example, may become uninhabitable as low-lying suburbs of Amsterdam return to marshland or open water.

By 2010 the Dutch had been forced to anticipate surrendering 200,000 hectares (494,211 acres) of farmland to river floodplains and had begun a major construction program involving floating homes. Pieter van Geel, the Dutch minister of housing, spatial planning, and the environment, stated in 2004 that half of the Netherlands is below sea level, and so beyond a certain level of sea-level rise, it is not feasible for the nation to build more extensive or higher dikes in many areas. Above 2 meters (6.6 feet) of additional sea-level rise, much of the land that the Netherlands has reclaimed from the ocean over several hundred years could be lost.

During mid-2008 a Dutch governmental commission recommended that the country spend $144 billion to reinforce its sea defenses through the year 2100 as a precaution against sea-level rise. The measures proposed include widening dunes facing the North Sea and raising the height of dikes along the coastline and rivers.

In the Netherlands the threat of sea-level rise also is being met with amphibious homes. One development of forty-six homes in the town of Maasbommel, for example, features two-bedroom, two-story houses with foundations of hollow concrete attached to iron posts sunk into a lake bottom; these homes can accommodate water levels as much as 5.5 meters (18 feet) higher than the levels that existed when they were built.

Bruce E. Johansen

FURTHER READING

Archer, David. *The Long Thaw: How Humans Are Changing the Next 100,000 Years of Earth's Climate.* Princeton, N.J.: Princeton University Press, 2009.

Cline, William R. *The Economics of Global Warming.*

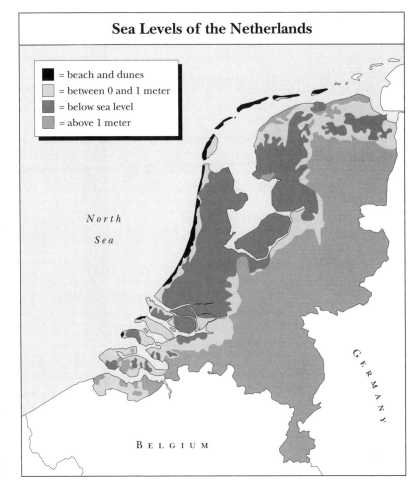

Sea Levels of the Netherlands

= beach and dunes
= between 0 and 1 meter
= below sea level
= above 1 meter

Washington, D.C.: Institute for International Economics, 1992.

Dutch, Steven I., ed. *Encyclopedia of Global Warming*. 3 vols. Pasadena, Calif.: Salem Press, 2010.

Lyall, Sarah, "At Risk from Floods, but Looking Ahead with Floating Houses." *New York Times*, April 3, 2007.

McGranahan, Gordon, Deborah Balk, and Bridget Anderson. "The Rising Tide: Assessing the Risks of Climate Change and Human Settlements in Low Elevation Coastal Zones." *Environment and Urbanization* 19, no. 1 (2007): 17-37.

Pilkey, Orrin H., and Rob Young. *The Rising Sea*. Washington, D.C.: Island Press, 2009.

Rosenthal, Elisabeth. "As the Climate Changes, Bits of England's Coast Crumble." *New York Times*, May 4, 2007.

SEE ALSO: Beaches; Europe; Glacial melting; Global warming; Positive feedback and tipping points; Sea-level changes; Wetlands.

Climate change and human health

CATEGORIES: Weather and climate; human health and the environment

SIGNIFICANCE: Changes in climate can affect the numbers of people who die directly as the result of temperature extremes (either cold or hot) or violent weather and can also increase the ranges of certain diseases and other health problems, which can lead to lesser, but sometimes serious, health effects.

Direct deaths from extremely cold or warm weather (hypothermia, heatstroke) are relatively rare in developed countries, but occasionally large numbers of people are killed or otherwise seriously affected by heart attacks caused by the weather (including people trying to shovel too much snow at once). The global warming that has occurred over the past century, and is generally expected to continue, can affect the number of deaths related to weather—reducing the number harmed by severe cold but increasing the number harmed by extreme heat. Climate changes can also alter the numbers of people affected by flood and drought as well as the ranges of parasites and disease vectors.

DIRECT EFFECTS

Warmer temperatures tend to lead to more frequent deaths from excessive heat, although this is mitigated in the case of greenhouse gas warming by the fact that the greatest warming occurs at night and thus in winter (and especially in the Arctic and Antarctic regions), when the temperatures otherwise would be cooler. Warmer temperatures also reduce the numbers of deaths associated with extreme cold (including cardiovascular and pulmonary diseases, such as influenza). There are actually more deaths from cold than from heat in many areas that face both threats. In Europe, estimated annual deaths amount to approximately 1.5 million from cold compared to about 200,000 from heat; the warmer United States suffered twice as many deaths from cold as from heat from 1979 through 1997. Even the severe heat wave of August, 2003, which led to an estimated 35,000 deaths in Europe (nearly half in France, partly because most doctors who practiced there were away on vacation at the time), resulted in only a modest increase in the number of deaths that year from excessive heat.

Climate change can also alter rainfall patterns, causing some areas to be more subject to floods (which inflict heavy damage and can also lead indirectly to other health problems) or droughts (which can lead to crop failures and water shortages). Scientists do not yet know, however, whether the trend toward warming will result in more floods, more droughts, or even both. (The claim that global warming leads to more extreme weather is still speculative, and disputed by many hurricane specialists.) The Medieval Warm Period (from around 800 to 1250) seems to have led to more droughts overall, but there is no guarantee that the pattern will repeat.

FOOD AND WATER

Climate changes can also affect food crops. Judging from the past experience of the Medieval Warm Period, this can lead to greater production in some places (partly from longer growing seasons as well as the fertilizing effect of increased carbon dioxide) and shortages in others (caused by droughts and floods rather than the temperature changes, though this can change what crops are grown in particular areas). These shortages can lead to malnutrition, including deficiency diseases, or even famine in poor countries.

Drought also makes obtaining water supplies more difficult even as a population continues to grow. One consequence is that people are often forced to work hard (expending labor that would otherwise be available for other needs) for water that is often tainted, which leads to increasing outbreaks of diseases such as dysentery, typhoid fever, and cholera, as well as aquatic parasites such as guinea worms. When water is scarce, cleaning and other sanitation practices suffer. Unclean bodies (especially hands) help spread diseases, and unclean clothes can carry and spread parasites such as lice.

DISEASES

Climate changes can affect the ranges of various life-forms in many ways. Warmer weather, particularly if it is also wetter, tends to increase the numbers of insects; the mild winters created by greenhouse gas warming are especially important for those insects that are susceptible to freezing temperatures. Many of these are disease vectors, spreading serious diseases such as malaria, dengue fever, yellow fever, typhus, and the plague. Not only may these insects cover larger areas (and also spread to higher elevations, as shown by a 1997 malaria outbreak in Papua New Guinea at an altitude of 2,100 meters, or 6,900 feet), but warmer temperatures can also enable them to be active for a longer portion of the year. This is especially crucial for mosquitoes that carry dengue fever, but malaria exposure may also increase. Although most malaria victims survive, the disease is very persistent, with frequent relapses, and thus very debilitating. On the other hand, it is estimated that warming will reduce the incidence of schistosomiasis, and possibly also the range of ticks that carry diseases such as Rocky Mountain spotted fever.

In areas that become significantly wetter, increased molds can lead to increases in hay fever and asthma, which can be fatal. Flooding can drive rodents, which help spread diseases such as the plague, from their burrows. When carbon dioxide increases, crop yields improve, but so does the growth of allergenic pollens such as ragweed.

Many skeptics, such as virologist Barry Beaty of Colorado State University, argue that the spread of diseases such as malaria seen in the late twentieth and early twenty-first centuries is the result primarily of nonclimatic factors, such as resistance to drugs by the disease pathogens, resistance to pesticides by the disease vectors, and a collapse in public health measures in some areas. (More than 80 percent of the world population is theoretically vulnerable to malaria even without global warming.) Danish economist Bjørn Lomborg has argued that the most cost-effective way to deal with the various problems resulting from global warming is to fix the individual problems, such as improving public health and medical care, implementing desalination projects and improving infrastructure (pipes and faucets) to supply potable water, and improving the distribution of food.

Timothy Lane

FURTHER READING

Braasch, Gary. *Earth Under Fire: How Global Warming Is Changing the World.* Updated ed. Berkeley: University of California Press, 2009.

Fagan, Brian. *The Great Warming: Climate Change and the Rise and Fall of Civilizations.* New York: Bloomsbury, 2008.

Johansen, Bruce E. *The Global Warming Desk Reference.* Westport, Conn.: Greenwood Press, 2002.

Lomborg, Bjørn. *Cool It: The Skeptical Environmentalist's Guide to Global Warming.* New York: Alfred A. Knopf, 2007.

Mann, Michael E., and Lee R. Kump. *Dire Predictions: Understanding Global Warming.* New York: DK, 2008.

Michaels, Patrick J., and Robert C. Balling, Jr. *Climate of Extremes: Global Warming Science They Don't Want You to Know.* Washington, D.C.: Cato Institute, 2009.

Rosenberg, Tina. "The Burden of Thirst." *National Geographic,* April, 2010, 99-115.

Singer, S. Fred, and Dennis T. Avery. *Unstoppable Global Warming: Every 1,500 Years.* Updated ed. Lanham, Md.: Rowman & Littlefield, 2008.

SEE ALSO: Climate change and oceans; Climate change skeptics; Drinking water; Droughts; Global warming; Intergovernmental Panel on Climate Change; Pandemics.

Climate change and oceans

CATEGORY: Weather and climate

SIGNIFICANCE: Rising temperatures affect sea levels and the salinity and acidity of the oceans, which in turn have impacts on aquatic life and the lives of people and animals that live near shorelines. Rising sea levels, the results of melting ice and thermal expansion of seawater, are likely to constitute the most notable challenge related to global warming for many millions of people around the world.

Many scientists have projected that the earth's temperature will rise at least 3 degrees Celsius (5.4 degrees Fahrenheit) during the twenty-first century, bringing it to a level near that of the middle Pliocene, three million years ago, when the seas were 15 to 35 meters (50 to 115 feet) higher than at the beginning of the century. The carbon dioxide (CO_2) level at that time reached 425 parts per million (ppm); by 2010 the CO_2 level was at 390 ppm, increasing at a rate of 2 to 3 ppm per year.

In late March, 2006, a report in the journal *Science* stated that melting ice, principally from Greenland and the West Antarctic ice sheet, could contribute to a rise in sea levels of several meters within a century—an upward revision of previous estimates. In an article published in the March, 2004, issue of *Scientific American*, James E. Hansen, director of the National Aeronautics and Space Administration's Goddard Institute for Space Studies in New York City, warned that if recent growth rates of CO_2 emissions and other greenhouse gases continue during the next fifty years, temperature increases could provoke large rises in sea levels, with potentially catastrophic effects. According to Hansen, a temperature increase of 2 to 3 degrees Celsius (3.6 to 5.4 degrees Fahrenheit) could provoke rises in sea level of about 25 meters (82 feet) within a few centuries.

RISING ACIDITY IN THE OCEANS

In 2003 scientists Ken Caldeira and Michael E. Wickett noted in the journal *Nature* that CO_2 levels were rising in the oceans more rapidly than at any time since the age of the dinosaurs. As the oceans absorb more CO_2 and become more acidic, their capacity to hold more CO_2 in the future is strongly reduced.

Concerns about rising acidity of the oceans were reinforced in January, 2009, when 155 scientists from twenty-six nations, organized by several international groups under the aegis of the United Nations, issued the Monaco Declaration, warning of severe damage to the oceans by rising acidity. In June, 2009, seventy science academies around the world called for the inclusion of ocean acidity on the agenda of international climate change studies, noting that the oceans already had become more acidic than at any time during the past 800,000 years.

Carbon dioxide is being injected into the oceans much more quickly than nature can neutralize it. Seawater is usually alkaline, about 8.2 pH. The pH scale is logarithmic, so a 0.1 decrease in pH, the change since the beginning of the Industrial Revolution, indicates a 30 percent increase in the concentration of hydrogen ions.

Scientists have investigated what continued ocean acidification might do to animals with calcium shells. One study investigated 328 colonies of massive *Porites* corals on the Great Barrier Reef off Australia; these corals in the past have grown to more than 6 meters (20 feet) tall over decades to centuries. Results from sixty-nine sections of the reef found that calcification had declined 14.2 percent between 1990 and 2005, impeding the corals' growth by 13.3 percent.

THE OCEAN FOOD WEB

Warming sea surface temperatures may interfere with phytoplankton production, with impacts rippling through the food web. Cooler, upwelling ocean water breaks through warm surface waters less frequently, reducing the nutrients available for plants and animals living in the oceans. By the 1990's, such decreases in productivity were detected near the California coast, where scientists have documented a measurable decrease in the abundance of zooplankton, the second level in the food web. By the 1990's, the abundance of zooplankton was 70 percent lower than it had been during the 1950's.

Scientists also have documented decreased reproduction and increased mortality in seabirds and marine mammal populations in warming water. A 1999 report published by the World Wide Fund for Nature and the Marine Conservation Biology Institute noted that the population of the sooty shearwater, a seabird, off the California coast declined 90 percent during the late 1980's and early 1990's, and the population of the Cassin's auklet declined 50 percent. Zooplankton populations declined markedly at the same time. In

Alaska, a severe decline in shearwaters from 1997 to 1998 "was clearly due to starvation," according to the report.

Bruce E. Johansen

FURTHER READING

Alley, Richard B., et al. "Ice-Sheet and Sea-Level Changes." *Science* 310 (October 21, 2005): 456-460.

Caldeira, Ken, and Michael E. Wickett. "Oceanography: Anthropogenic Carbon and Ocean pH." *Nature* 425 (September 25, 2003): 365.

Hansen, James. "Defusing the Global Warming Time Bomb." *Scientific American*, March, 2004, 68-77.

Holland, Jennifer S. "Acid Threat." *National Geographic*, November, 2001, 110-111.

Lynas, Mark. *Six Degrees: Our Future on a Hotter Planet.* New York: HarperCollins, 2008.

Mathews-Amos, Amy, and Ewann A. Berntson. *Turning Up the Heat: How Global Warming Threatens Life in the Sea.* Gland, Switzerland: World Wide Fund for Nature/Marine Conservation Biology Institute, 1999.

SEE ALSO: Acid deposition and acid rain; Antarctic and Southern Ocean Coalition; Biodiversity; Carbon dioxide; Dead zones; El Niño and La Niña; Food chains; Glacial melting; Global warming; Great Barrier Reef; Greenhouse effect; Greenhouse gases; Hansen, James E.; Sea-level changes.

Climate change skeptics

CATEGORY: Weather and climate

DEFINITION: Persons who challenge the view that dangerous global climate change is under way as a result of the buildup of human-caused greenhouse gases in the atmosphere

SIGNIFICANCE: Governments are under increasing pressure to adopt policies to mitigate the threat of climate change resulting from emissions of greenhouse gases, mainly carbon dioxide from the burning of fossil fuels. As industrialized countries are engaged in costly efforts to cut emissions, persons who are skeptical that climate change poses a danger to humankind debate the environmental and economic risks of global warming.

A range of opinions is found among those persons described as climate change skeptics. Some take the view that it is impossible that global climate change could be caused by the burning of fossil fuels, perhaps because they simply do not understand the science underpinning the radiative effect of greenhouse gases on climate. Most climate change skeptics, however, in challenging those who believe humankind is in the midst of a planetary environmental emergency, focus their arguments on uncertainties in the answers to the following questions: Is global climate warming? If so, what part of that warming is caused by human activities? How good is the evidence? What are the risks?

Some skeptics accept that the rising atmospheric concentration of carbon dioxide can influence climate, but they argue that the net effect of this rise above twentieth century levels is minor compared to natural forces. They point out that no scientific evidence suggests anything tragic, and they maintain that this is because of the existence of natural feedback mechanisms within the global climate system, such as the increase in cloudiness, that suppress initial change imposed on the system by carbon dioxide increase (negative feedbacks). Skeptics challenge the claim that any climate change is likely to be dangerous on the grounds that there is no observational evidence to prove the hypothesis that small initial changes triggered by carbon dioxide will set in motion amplifying processes (positive feedbacks). They emphasize the fact that predictions based on climate models are not evidence of future climate because the models cannot adequately simulate climate.

The stance of the skeptics is reinforced by what they perceive as a range of factors that motivate climate change alarmism. They point out that the news media are drawn toward worst-case scenarios and argue that the public eventually comes to view these reported scenarios as reality if the stories are told often enough. Skeptics also assert that many others who share their views are reluctant to challenge the alarmist stance, as it is seen as "politically incorrect" to do so; such a challenge may be taken to imply a lack of concern for the environment. Some skeptics assert that politicians are drawn to the theme of climate change by the appeal of tackling something grand—a global environmental issue, as opposed to merely a local one. Another factor for politicians is the allure of the green vote. Skeptics have also argued that scientists promote alarmist speculation because heightened concern improves the chances that the funding of climate change research will be given high priority.

Many skeptics point out that the climate change issue is often confused with the separate matters of conservation of fossil fuels and the effects of burning such fuels on air quality. Skeptics further argue that the public in general relies too much on "authorities" to rule on scientific matters about which there is a high degree of uncertainty.

C. R. de Freitas

FURTHER READING

Horner, Christopher C. *Red Hot Lies: How Global Warming Alarmists Use Threats, Fraud, and Deception to Keep You Misinformed.* Washington, D.C.: Regnery, 2008.

Michaels, Patrick J. *Meltdown: The Predictable Distortion of Global Warming by Scientists, Politicians, and the Media.* Washington, D.C.: Cato Institute, 2004.

_____, ed. *Shattered Consensus: The True State of Global Warming.* Lanham, Md.: Rowman & Littlefield, 2005.

SEE ALSO: Air pollution; Carbon dioxide; Climate models; Global warming; Greenhouse effect; Kyoto Protocol; Public opinion and the environment.

Climate models

CATEGORY: Weather and climate

DEFINITION: Computer tools that use numerical representations of the climate system to predict future climate

SIGNIFICANCE: Scientific debates and policy recommendations regarding the severity and the causes of changes in the earth's climate are based primarily on the models that scientists use to project climate trends.

Climate models can be used to explore various scenarios; scientists use such models both to study the past (such as linking the ice ages to the rise of the Himalayas) and to project what may occur in the future. No matter how precise or accurately worked a model is, however, the information it provides is merely theoretical and must be verified against actual data, either experimental or (in the case of climate) observational. In addition, a model that accurately reflects the recent known past may have been written to do so.

The key models used in climate research are extremely complex general circulation models (GCMs).

An ideal GCM would take into account every factor in climate, but in practice some factors are either ignored or simplified, even in the most complex models. Among these factors are volcanic eruptions, solar output, oscillating weather patterns (some taking a few years, others several decades), the water cycle (evaporation, condensation into low-level clouds, and precipitation all move heat from the surface to the troposphere), atmospheric content (including greenhouse gases such as water vapor and carbon dioxide, as well as pollutants such as sulfur dioxide), ocean currents, wind patterns, and land use (urban areas are much warmer than farmland, which is warmer than forest). None of these is entirely predictable, and some are random in occurrence. Local events can have major global effects; for example, El Niño conditions in the southern Pacific lead to weaker northern Atlantic hurricanes due to wind shear, as well as changes in precipitation in many areas.

The models used by the Intergovernmental Panel on Climate Change generally project a temperature rise of approximately 2 to 3 degrees Celsius (3.6 to 5.4 degrees Fahrenheit) by the year 2100. Part of this rise is expected to come from natural causes, part from increased atmospheric greenhouse gases, and part from positive feedback effects such as increased humidity from evaporation and increased summer snow melt (bare ground absorbs more heat). All of these are speculative numbers. In reality, temperature rise during the late twentieth and early twenty-first century warming period was only approximately one-half of one degree over thirty-five years, less than projected. Part of this may be from the cooling effect of sulfate aerosols (another speculative number), but part may also be from negative feedback effects, such as the water cycle. Scientists continue to disagree regarding the effects of global warming on cyclones and other severe storms; no observational evidence supports the theory that warming leads to more storms or more severe ones. Climate models also make projections about regional conditions (such as increased drought in the southwestern United States) that are not verifiable.

Climate models that look at the effects of greenhouse gases indicate that warming should be far greater in the higher latitudes than near the equator; this is reasonably accurate in the Northern Hemisphere, but not in the Southern Hemisphere owing to the cooling of most of Antarctica. These models project greater warming in the middle troposphere than at the surface, but cooling in the stratosphere; the stratosphere is cooling, but the middle-troposphere warming is actually less than the surface warming. Modifications made to early 1990's models to account for the effects of sulfate aerosols made the models more accurate overall, but they have failed to explain why there has been less warming than predicted in the Southern Hemisphere (where sulfates in the atmosphere are much smaller). Some of these results are different from those expected with a purely cyclic explanation of global warming (which predicts stratospheric warming, for example).

Timothy Lane

FURTHER READING

Mann, Michael E., and Lee R. Kump. *Dire Predictions: Understanding Global Warming.* New York: DK, 2008.

Michaels, Patrick J., and Robert C. Balling, Jr. *Climate of Extremes: Global Warming Science They Don't Want You to Know.* Washington, D.C.: Cato Institute, 2009.

Spencer, Roy W. *Climate Confusion: How Global Warming Hysteria Leads to Bad Science, Pandering Politicians, and Misguided Policies That Hurt the Poor.* New York: Encounter Books, 2008.

SEE ALSO: Climate change and oceans; Climate change skeptics; Climatology; El Niño and La Niña; Global warming; Greenhouse effect; Greenhouse gases; Volcanoes and weather.

Climatology

CATEGORY: Weather and climate

DEFINITION: The study of factors that produce and influence a region's weather, both on an ongoing basis and by retrospective analysis of historical, archaeological, and geological records

SIGNIFICANCE: The principal aim of climatology is long-range regional weather prediction, traditionally for determining optimum agricultural practices and planning for extreme weather events. With a growing awareness that human activities potentially can have profound negative impacts on climate, policy makers have increasingly turned to climatologists and their models for recommendations on how to shape global energy policies.

Since the dawn of civilization and perhaps even before, human beings have observed regional weather patterns and attempted to correlate them with other phenomena in order to project the patterns into the future, chiefly for agricultural purposes. At low latitudes, where drought is the chief concern, predictions focused on variability in rainfall, while in northerly climates latitudinal variation in temperature and the factors influencing yearly variations in temperature were more important.

CLIMATOLOGY IN HISTORY

Over the centuries, the perception of human influence over climate has shifted. Ancient civilizations credited ritual observances with the capacity to influence the weather and ascribed extreme weather events such as prolonged droughts to divine wrath at human wrongdoing. With the rise first of modern astronomy, which eliminated the superstition from generally valid astrological long-term weather prediction, and then of increasingly sophisticated worldwide measurement of atmospheric phenomena, educated people came to regard climate as something independent of human activity and not susceptible to modification, except perhaps on a very local or temporary level. The pendulum swung in the opposite direction with the growing recognition that the rapid release of large volumes of greenhouse gases into the atmosphere has produced a general warming trend.

Like meteorology, climatology as a systematic science dates from the middle of the nineteenth century and represents a response to the need by the British Empire and the United States to expand agricultural and economic systems originating in Western Europe to regions with more variable and extreme climates. Close study of droughts and famines in India led to the discovery of the Southern Oscillation (now known as El Niño/Southern Oscillation), a regular decadal fluctuation of pressure in the western Pacific that profoundly affects rainfall in India and elsewhere.

PALEOCLIMATOLOGY

In order to understand long-term climate fluctuations and trends, climatologists must reconstruct temperatures and rainfall for periods and in geographical areas for which no actual measurements exist—from within the last century in the case of remote and undeveloped regions to hundreds of millions of years in the case of the "snowball earth" of the late Precambrian or the Permian-Triassic extinction event. The climatic events since the last ice age are of most interest in the assessment of the probable effects of human activities and how they might interact with abiotic forcing mechanisms.

On a regional level, vegetation, which can be reconstructed through pollen profiles in sediments or sedimentary rock, is a reliable indicator of climate. For a picture of global climate, scientists look at ice cores from Greenland and Antarctica, in which trapped air provides a snapshot of atmospheric conditions, including greenhouse gases and particulates, and isotope ratios reflect oceanic evaporation and photosynthesis.

Glacial geologic features provide a persistent legacy of cold conditions, and rapid deposition of deltas and lake sediments is a signature of high rainfall. Where isotope, geologic, and fossil evidence co-occur they present a consistent picture of paleoclimate, but the value of isotope ratios alone is open to question.

Global Warming

Climatology enters into debates on environmental issues only if human actions are perceived as influencing climate. Until fairly recently this would have been mainly in connection with desertification due to overgrazing or forest destruction. Although these human activities undoubtedly play a part in increasing aridity in areas such as the Sahel of Africa, the most recent treatments of the subject place more emphasis on shifts in global weather patterns than on local misuse of resources.

Data gathered and correlated by climatologists firmly identified the trend toward global warming. A network of interconnected national and international agencies tracks the course of global warming in time and supplies the data to bodies concerned with implementing public policy. By monitoring all of the factors believed to influence global temperatures, climatologists can estimate the degree to which warming is caused by human activities and pinpoint which activities produce the largest effects. Despite all of the effort and expertise that go into accumulating and analyzing these data, the results are often inconclusive and open to interpretation; even when unequivocal, the data can become tools for policy makers whose principal aim may not be the preservation of a sustainable environment.

Martha A. Sherwood

Further Reading

Alverson, Keith D., Raymond S. Bradley, and Thomas F. Pedersen, eds. *Paleoclimate, Global Change, and the Future.* New York: Springer, 2003.

Bonan, Gordon B. *Ecological Climatology: Concepts and Applications.* New York: Cambridge University Press, 2002.

Leroux, Marcel. *Global Warming: Myth or Reality? The Erring Ways of Climatology.* New York: Springer, 2005.

Mayewsky, Paul, et al. "Holocene Climate Variability." *Quaternary Research* 62 (2004): 243-255.

Oliver, John E., ed. *Encyclopedia of World Climatology.* New York: Springer, 2005.

Saltzman, Barry. *Dynamical Paleoclimatology: Generalized Theory of Global Climate Change.* San Diego, Calif.: Academic Press, 2002.

Thompson, Russell D., and Allen Perry, eds. *Applied Climatology: Principles and Practice.* New York: Routledge, 1997.

See also: Climate change and oceans; Climate models; El Niño and La Niña; Global warming; Intergovernmental Panel on Climate Change; National Oceanic and Atmospheric Administration; Weather modification.

Climax communities

CATEGORIES: Ecology and ecosystems; forests and plants

DEFINITION: Assemblages of plants and animals representing the end point in a successional sequence; mature communities in equilibrium with their environment

SIGNIFICANCE: Climax communities are reservoirs of biological diversity. As levels of human disturbance increase, climax communities contract in area and begin to disappear. Efforts to preserve endangered species that focus on entire communities are more likely to be successful, especially at preserving unrecognized threatened species, than efforts that focus on one species alone.

The concept of a climax community was first proposed in 1916 by forestry biologist Frederic E. Clements in a monograph on the vegetation of Washington State. Clements postulated that there is a unique, discrete vegetation type for a given climate and topography, and that type will occur if succession is allowed to proceed to its end point. This concept has fallen into disfavor among theoretical ecologists and conservationists since World War II. A more modern view, based on numerous detailed studies, views communities as being in a state of constant flux that decreases in amplitude with time. The terms "mature," "late successional," and "old-growth" are used to describe what formerly was called a climax community.

Subtypes of climax communities sometimes recognized include polyclimax, where several different assemblages of plants and animals can represent a stable end point for a given area; subclimax, where the usual vegetation remains in a late stage of succession because of regular, infrequent disturbance; and plagioclimax, where the stable vegetation of a region has been shaped by human activity over a long period of time. An example of a plagioclimax is the species-rich oak savanna of Oregon and northern California, which was maintained by periodic burning in prehistoric times and grazing after European settlement. Excluding both fire and cattle results in overgrowth by conifers and a precipitous decline in diversity of herbaceous plants and insects.

Following massive disturbance, whether natural or caused by humans, the vegetation of an area experiences succession as early colonizing species modify the environment in ways that allow other species to colonize. Over time, the spatial complexity, species diversity, primary productivity, and stored biomass of the vegetation increases, as do the species diversity and food-web complexity of animals. In addition to being much richer biologically, mature communities are more resistant to external disruption than are early successional ones.

Climax communities include prairie and savanna vegetation as well as forest, but these and other vegetation types ideally suited to human exploitation persist mainly as small tracts deliberately preserved from development. Such preservation of fragments of an originally more extensive ecosystem is typically only partially successful in preserving stability and species diversity, because the complexity of an ecosystem depends on its size as well as lack of disturbance. Preserving small, widely separated tracts of virgin rain forest in the Amazon basin has, for example, led to the loss of larger predators from much of their original range, and this in turn has disrupted populations of prey spe-

cies. Extensive areas of mature climax communities relatively free of human disturbance still extant in the early years of the twenty-first century include the rain forest of the Amazon basin, coniferous forests of North America and Siberia, and the Arctic tundra.

Martha A. Sherwood

FURTHER READING

During, H. T., M. J. A. Werger, and H. J. Willems, eds. *Diversity and Pattern in Plant Communities.* The Hague: SPB Academic Publishing, 1988.

Pynn, Larry. *Last Stands: A Journey Through North America's Vanishing Ancient Rainforests.* Corvallis: Oregon State University Press, 2000.

Van der Maarel, Eddy, ed. *Vegetation Ecology.* Malden, Mass.: Blackwell, 2005.

Verhoef, Herman A., and Peter J. Morin, eds. *Community Ecology: Processes, Models, and Applications.* New York: Oxford University Press, 2010.

SEE ALSO: Amazon River basin; Biodiversity action plans; Endangered species and species protection policy; Everglades; International Biological Program; Old-growth forests; Rain forests.

Cloning

CATEGORY: Biotechnology and genetic engineering

DEFINITION: Production of a population of genetically identical cells or organisms derived from a single ancestor, or the production and amplification of identical deoxyribonucleic acid molecules

SIGNIFICANCE: Gene cloning is commonly used to produce genetically modified organisms and genetically engineered crops. Reproductive cloning has the potential to be used to mass-reproduce animals with special qualities, or to expand the population of an endangered species. With further development, therapeutic cloning might someday be employed to produce whole organs from single cells or replace disease-damaged cells with healthy ones.

The molecular cloning and engineering of deoxyribonucleic acid (DNA) molecules were first made possible with the discoveries of DNA ligase (enzymes that join DNA molecules) in 1967 and restriction endonucleases (enzymes that cut DNA molecules at specific nucleotide sequences) in 1970 by Hamilton Smith and Daniel Nathans. These enzymes allow scientists to cut and join DNA molecules from different species to produce recombinant DNA. For example, the DNA encoding human insulin can be combined with a plasmid, a small piece of DNA often found in bacteria such as *Escherichia coli* (*E. coli*). After the recombinant human-insulin-DNA/plasmid-DNA molecule is constructed, it can be inserted into a host cell such as *E. coli*. The recombinant DNA molecule will then replicate one or more times each time the *E. coli* DNA replicates. Thus a clone of identical recombinant human-insulin-DNA/plasmid-DNA molecules will result. If the recombinant molecule has been engineered with the requisite signals, the *E. coli* will produce copious amounts of human insulin. In 1972 the first recombinant DNA molecules were made at Stanford University, and in 1973 such molecules were inserted via plasmids into *E. coli*. The first successful synthesis of a human protein by *E. coli* was somatostatin, reported in 1977 by Keiichi Itakura and coworkers. In 1984 insulin was the first human protein made by *E. coli* to become commercially available.

The production of cloned recombinant DNA has been an indispensable tool in biological research and has increased the knowledge and understanding of the structure and function of DNA and the control of gene activity. Cloned DNA has led to the manufacture of important products of medical interest, the production of DNA for gene transfer and genetic engineering experiments, and the identification of mutations and genetic disease. Several different vectors have been developed for the delivery of DNA to a variety of plant, animal, and protistan cells, resulting in the creation of many transgenic species. Transgenic plants have been produced that are resistant to certain herbicides, insects, and viruses, and a variety of animals have been engineered with the human growth factor gene. Transgenic plants, animals, and protistans now produce a variety of human proteins, including insulin, antithrombin, growth hormone, clotting factor, vaccines, and many other pharmaceuticals and therapeutic agents, as well as molecular probes for the diagnosis of human disease. As a result of this technology, the costs of treating many diseases have declined.

The first human gene transfer experiment using cloned DNA was performed in 1990 on a four-year-old girl with severe combined immune deficiency (SCID). SCID is caused by a mutation in the adenine deaminase (ADA) gene that results in white blood cells deficient in their immune response. These cells were removed from the girl, the normal gene was inserted, and the genetically altered white cells were returned to her body, where they repopulated and expressed normal defense mechanisms. The girl's white cell numbers normalized after repeated treatments. While she continued to require ADA enzyme injections for primary management of her condition, she did develop normal immunity over time.

ENVIRONMENTAL IMPACT OF DNA CLONING

The construction of recombinant DNA molecules and their subsequent cloning have not been without controversy. In 1971 researcher Paul Berg planned an experiment to combine DNA from simian virus 40 (SV40)—a virus that causes tumors in monkeys and transforms human cells in culture—with bacteriophage *l* and to incorporate the recombinant molecule into *E. coli*. However, several scientists warned of a potential biohazard. Because *E. coli* is a natural inhabitant of the human digestive system, it was feared that the engineered *E. coli* could escape from the laboratory, enter the environment, become ingested by humans, and cause cancer as a result of its newly acquired DNA. The scientific community imposed a moratorium on recombinant DNA work in 1974 until the National Institutes of Health (NIH) could study the safety of recombinant DNA research and develop guidelines under which such work could proceed. The guidelines, originally published in 1976, were eventually relaxed after it was clearly demonstrated that the work was not nearly as dangerous as initially feared.

In 1983 the NIH granted permission to the University of California at Berkeley to release bacteria that had been engineered to protect plants from frost damage. This was the first experiment intentionally designed to introduce genetically engineered organisms into the environment. Various environmental and consumer protection groups were successful in persuading federal judge John J. Sirica to order the suspension of the planned trial. Environmentalists feared that the bacteria could spread to other plants, prolong the growing season, cause irreparable harm to the environment, or enter the atmosphere and decrease cloud formation or alter the climate. The NIH and the university eventually won approval for their experiment, and the field trials were carried out in 1987.

Milestones in Cloning

Year	Event
1892	Hans Adolph Eduard Dreisch clones sea urchins by separating the first two and four blastomeres.
1902	Hans Spemann successfully repeats Dreisch's experiment using salamanders.
1952	Robert Briggs and Thomas J. King successfully clone frogs by nuclear transplantation of embryonic nuclei to enucleated eggs.
1967	Deoxyribonucleic acid (DNA) ligase, the enzyme that joins DNA molecules, is discovered.
1969-1970	Daniel Nathans, Hamilton Smith, and others discover restriction endonucleases.
1971	Paul Berg plans to combine DNA with a bacteriophage and insert the recombinant DNA molecule into *Escherichia coli* (*E. coli*).
1972	The first recombinant DNA molecules are constructed at Stanford University.
1973	Recombinant DNA is first inserted into *E. coli*.
1974	Scientists call for a moratorium on recombinant DNA research until the National Institutes of Health (NIH) can study the safety of recombinant DNA research and develop guidelines.
1976	The NIH issues its guidelines for recombinant DNA research.
1983	The NIH grants permission to the University of California at Berkeley to release genetically engineered bacteria designed to retard frost formation on plants.
1986	Steen Willadsen clones sheep using early embryonic nuclei.
1987	Randall Prather and Willard Eyestone clone cows using early embryonic nuclei.
1990	The first human gene transfer experiment is performed on a patient with severe combined immune deficiency (SCID).
1997	Ian Wilmut and Keith Campbell announce the successful cloning of Dolly the sheep, the first mammal cloned from an adult cell.
2001	The first endangered wild animal, an Asian ox (gaur) is cloned.
2002	A study finds that the genes in about 4 percent of cloned mice function abnormally.
2003	Dolly the sheep is euthanized after contracting lung disease.
2009	The first clone of an extinct animal, a Pyrenean ibex (a species of wild goat), is born but dies within minutes.

While many scientists believe that the introduction of tested and approved engineered species into the environment will prove harmless and that early concerns were unfounded, many environmentalists still have reservations because the long-term environmental effects of genetically altered organisms remain unknown. It is feared that the introduction of some organisms will have negative impacts on the environment and irreversible global effects. Many worry that a genetically engineered organism such as a bacterium or virus could spread throughout the environment, causing human disease or ecological destruction.

The use of genetically engineered organisms could have many adverse effects. The introduction of new genes into an organism could extend the range of that species, causing it to infringe on the natural habitats of closely related or more distant species and thus disrupt the balance of nature. Many examples already exist of ecological disruption that has occurred as a result of the introduction of plant and animal species into areas where they have no natural predators. Some environmentalists believe that the use of herbicide-resistant crops will serve to prolong, extend, and even increase the use of toxic herbicides. Others fear that herbicide-resistant genes could be transferred to related plants, producing a population of herbicide-resistant weeds.

On the other hand, the introduction of organisms engineered through the cloning of recombinant DNA into the environment could have significant positive environmental impacts and increase world food production. Genetically altered plants could reduce the input of toxic chemicals into the environment. Genetic modification can create crops with larger yields; resistance to pests, pesticides, drought, and disease; higher tolerance to cold, heat, and drought; longer shelf life; and greater photosynthetic and nitrogen-fixing activity. Plants genetically engineered to produce a natural insecticide that is nontoxic to humans ideally alleviate the need for insecticide application; with the toxin produced by and confined to the plants themselves, there is no need to contaminate the entire area where they are grown. Bioremediation of toxic waste dumps, chemical spills, and oil spills could be enhanced by genetically engineered microorganisms.

ANIMAL AND PLANT CLONING

The cloning of plants from cuttings has been successfully practiced for thousands of years and is commonly used for many important food crops. Successful animal cloning was first reported in 1892 by Hans Adolph Eduard Dreisch. Dreisch separated the first two and four embryonic cells (blastomeres) of the sea urchin and allowed them to develop into complete, genetically identical embryos.

The first report of successful animal cloning by nuclear transplantation was published in 1952 by Robert Briggs and Thomas J. King, who removed nuclei from embryonic frog cells and transplanted them into eggs from which their nuclei had been removed. By pricking the eggs with a glass needle the scientists induced them to divide and often develop into complete tadpoles. The first reliable reports of successful animal cloning by nuclear transplantation in mammals came in 1986 from Steen Willadsen in Cambridge, England. Willadsen cloned sheep from the nucleus of an early blastula cell. In 1987 Randall Prather and Willard Eyestone cloned cows while working in Neal First's laboratory at the University of Wisconsin.

In 1997 Ian Wilmut and Keith Campbell announced that they had successfully cloned a sheep named Dolly in Edinburgh, Scotland, the year before. Dolly was a milestone in cloning research because she was the first mammal cloned from an adult cell. Such animal cloning makes genetic engineering more effi-cient, because an animal would have to be engineered only once and then could be used to donate nuclei for cloning. The use of adult cells in cloning is advantageous in that the donor animal's physical characteristics are known before the animal is cloned. However, clones created from adult cells share those cells' shortened telomeres (the sequences at the ends of chromosomes that grow shorter with each generation of cell replication). Telomere shortening is associated with the aging and death of cells and of the entire organism, which means that the cloned animal is likely to have a greater susceptibility to degenerative conditions and a shortened life expectancy. Dolly, who was cloned from a six-year-old sheep, lived only six years before she had to be euthanized in 2003 because of a progressive lung disease. (Her breed typically lives about twelve years.) She had developed arthritis the year before. Whether her illnesses stemmed from premature aging is unclear; however, her telomeres were found to be short.

Although inbred stocks have been used for centuries in agriculture, it is feared that the extensive use of animal and plant clones could severely reduce the genetic variability of various important crop, forestry, and livestock species, making them more susceptible to disease and extreme environmental factors. On the other hand, the development of successful cloning techniques could rescue endangered species from the brink of extinction or allow for the creation of a population of genetically identical animals, which would be valuable in medical research. Moreover, cloning could be used to create an entire population of animals from one individual that has been genetically altered to produce a valuable pharmaceutical or organs suitable for human transplant. However, controversy would undoubtedly arise regarding the ethics of mass-producing animals through cloning for this purpose.

Reproductive cloning is costly, and more than 90 percent of cloning attempts do not result in viable offspring. Cloned mammals have tended to have immune function problems, increased infection rates, susceptibility to tumors, and short lives. A 2002 study of cloned mice found that about 4 percent of their genes functioned abnormally. Members of the American Medical Association and the American Association for the Advancement of Science have issued formal public statements advising against reproductive cloning in humans.

Charles L. Vigue
Updated by Karen N. Kähler

FURTHER READING

Drlica, Karl. *Understanding DNA and Gene Cloning: A Guide for the Curious.* 4th ed. Hoboken, N.J.: John Wiley & Sons, 2004.

Fritz, Sandy, ed. *Understanding Cloning.* New York: Warner Books, 2002.

Klotzko, Arlene Judith. *A Clone of Your Own? The Science and Ethics of Cloning.* New York: Cambridge University Press, 2006.

Kolata, Gina Bari. *Clone: The Road to Dolly, and the Path Ahead.* New York: HarperCollins, 1998.

Williams, J. G., A. Ceccarelli, and A. Wallace. *Genetic Engineering.* 2d ed. Oxford, England: Bios, 2001.

Wilmut, Ian, Keith Campbell, and Colin Tudge. *The Second Creation: Dolly and the Age of Biological Control.* Cambridge, Mass.: Harvard University Press, 2001.

Wilmut, Ian, and Roger Highfield. *After Dolly: The Uses and Misuses of Human Cloning.* New York: W. W. Norton, 2006.

SEE ALSO: Biotechnology and genetic engineering; Dolly the sheep; Genetically altered bacteria; Genetically engineered pharmaceuticals; Genetically modified foods; Genetically modified organisms; Wilmut, Ian.

Cloud forests

CATEGORIES: Forests and plants; ecology and ecosystems

DEFINITION: Forests on moist tropical mountain slopes covered by tree-level clouds

SIGNIFICANCE: Cloud forests are rich in biodiversity and act as important sources of water conservation. These valuable forest ecosystems are strongly threatened by human activity, particularly by deforestation, unsustainable agricultural use, and overharvesting of plants and animals.

Cloud forests cover humid tropical mountain slopes ranging from Central and South America to Africa, South and Southeast Asia, Papua New Guinea, and Pacific islands such as Hawaii. They are created when moist air meets mountain barriers and forms clouds covering the treetops. Because of prevailing wind directions, cloud forests generally cover eastern mountain slopes. They typically are found on mountains that range between 1,200 and 2,500 meters high (4,000 to 8,200 feet), but some cloud forests can be found on peaks as low as 300 meters (1,000 feet) on Pacific islands and as high as 3,500 meters (11,500 feet) in the Andes of South America and the Ruwenzori range of Uganda.

The higher the cloud forest, the smaller the trees tend to be, and the thicker and more gnarled are their stems and branches. The leaves of the trees are generally tough and small. The trees host a great variety of epiphytes (other plants that grow on them), such as bromeliads, orchids, lichens, mosses, and ferns. The soils are very moist and rich in organic material as they contain much humus and peat.

Cloud forests function as unique watersheds. The leaves on the top branches of the trees catch the moisture of the clouds driven there by the wind and let it drip to the forest floor. This process, scientifically called occult precipitation and popularly known as cloud stripping, accounts for doubling rainfall in dry seasons and still increasing it by about 10 percent in wet seasons when compared to areas outside the cloud forests. Cloud forests also act as water reservoirs, preventing runoff during rain and supplying a steady flow of water in dry times.

Cloud forests have come under severe threat by human activity. In 1974, cloud forests were estimated to cover 50 million hectares (124 million acres), or one-fourth of the hills and mountains of the Tropics. A 1990 survey by the United Nations Food and Agriculture Organization found that the annual deforestation rate in tropical mountain and highland forests was 1.1 percent for the decade of 1980 to 1990, and although the rate of deforestation subsequently slowed somewhat, it has been estimated that by the early years of the twenty-first century, as little as 35 million hectares (87 million acres) of cloud forests were in existence.

Initially, cloud forests enjoyed some natural protection because of their inaccessibility and the poor quality of their soil and timber for human uses. As farmers, ranchers, and loggers overexploited the tropical forests below cloud forests, however, they shifted their attentions upward into the cloud forests. The most severe threats to cloud forests are deforestation for cropland and tree cutting for fuelwood. In addition, unsustainable harvesting of plants such as rare orchids and ferns and the trapping and hunting of amphibians, birds, and mammals unique to the cloud

Tall trees in a cloud forest in Costa Rica. (©Scott Griessel/Dreamstime.com)

forests threaten to destroy these ecosystems and lead to species extinction. In the Andes, the use of cloud forests for the planting of illegal crops such as coca (from which cocaine is derived) is another threat. Even tourism and recreation can threaten cloud forests if these are undertaken in an ecologically unsound manner.

Since the late twentieth century, environmentalists, policy makers, and the general public have grown increasingly aware that cloud forests need to be protected from degradation by humanity. The successful management of sustainable use of these forests depends on the participation of the forests' indigenous populations and on the creation of both economic incentives and legal sanctions supporting the forests' protection.

R. C. Lutz

FURTHER READING

Gradstein, S. Robbert, Jürgen Homeier, and Dirk Gansert, eds. *The Tropical Mountain Forest: Patterns and Processes in a Biodiversity Hotspot.* Göttingen, Germany: Universitätsverlag Göttingen, 2008.

Haber, William. *An Introduction to Cloud Forest Trees.* 2d ed. Monteverde de Puntarenas, Costa Rica: Mountain Gem, 2000.

Scatena, F. N., and L. S. Hamilton. *Tropical Montane Cloud Forests: Science for Conservation and Management.* New York: Cambridge University Press, 2010.

SEE ALSO: Biodiversity; Deforestation; Ecosystems; Extinctions and species loss; Forest management; Habitat destruction; Indigenous peoples and nature preservation; Land clearance; Subsistence use; Sustainable agriculture; Watershed management.

Cloud seeding

CATEGORY: Weather and climate

DEFINITION: Practice of introducing agents into clouds for the purpose of intentionally modifying the weather

SIGNIFICANCE: Because cloud and atmospheric processes are not completely understood, some doubt exists regarding whether cloud seeding actually works. In addition, some environmentalists are concerned that encouraging precipitation in one area may result in less precipitation in another.

Cloud seeding, which is relatively inexpensive and easily conducted, has long been the main technique used in weather modification. The seeding of clouds is undertaken for many purposes, including increasing rainfall for agriculture, increasing snowfall for winter recreational areas, dispersal of fog and clouds, and suppression of hail.

Most cloud seeding is directed at clouds with temperatures below freezing. These cold clouds consist of ice crystals and supercooled droplets (water droplets with subfreezing temperatures). Initially, supercooled droplets far outnumber ice crystals; however, the ice crystals quickly grow larger at the expense of the droplets and, after reaching sufficient size, fall as precipitation. The objective of seeding cold clouds is to stimulate this process in clouds that are deficient in ice crystals. The seeding agent is either silver iodide (AgI) or dry ice—solid carbon dioxide (CO_2) with a temperature of -80 degrees Celsius (-112 degrees Fahrenheit). Silver iodide pellets act as nuclei on which water droplets freeze to form precipitation, while dry ice pellets are so cold that they cause the surrounding supercooled droplets to freeze and grow into snowflakes. Warm clouds with temperatures above freezing can also be seeded through injection with salt crystals, which triggers the development of large liquid cloud drops that fall as precipitation.

Clouds must already be present for such weather modification techniques to work, since clouds cannot be generated by seeding. Most cloud seeding is done from aircraft, although silver iodide crystals can be injected into clouds from ground-based generators. One environmental concern is that seeding may only redistribute the supply of precipitation, so that an increase in precipitation in one area might mean a compensating reduction in another.

Since World War II, considerable research has gone into cloud seeding. However, scientists still cannot conclusively answer the question of whether cloud seeding actually works. One of the first large-scale experiments, conducted over south-central Missouri for five years during the 1950's, actually decreased rainfall. Apparently the clouds were overseeded, resulting in too many ice crystals competing for too few water droplets to form precipitation. In the 1970's and early 1980's the National Oceanic and Atmospheric Administration (NOAA) conducted a major experiment over southern Florida to test the effectiveness of seeding cumulus clouds. The initial results showed that under some conditions seeding increases rainfall, but a second, more statistically rigorous set of experiments failed to confirm the earlier results. Of the many experiments in cloud seeding that have been undertaken, only one, conducted in Israel in the 1960's and 1970's, has yielded statistically convincing confirmation of an increase in precipitation.

The conflicting results of the Florida and Israel studies point out the uncertainties of cloud seeding and the need for more basic knowledge of cloud and atmospheric processes. A major result of the many decades of investigating cloud seeding is the realization that weather events are quite complex and not yet fully understood. Without further basic understanding of atmospheric processes, large-scale weather modification through cloud seeding cannot be carried out with scientifically predictable results.

Craig S. Gilman

FURTHER READING

Ahrens, C. Donald. "Cloud Development and Precipitation." In *Essentials of Meteorology: An Invitation to the Atmosphere*. 5th ed. Belmont, Calif.: Thomson Learning, 2008.

Cotton, William R., and Roger A. Pielke. *Human Impacts on Weather and Climate*. 2d ed. New York: Cambridge University Press, 2007.

SEE ALSO: Geoengineering; Rainwater harvesting; Solar radiation management; Weather modification.

Club of Rome

CATEGORIES: Organizations and agencies; ecology and ecosystems

IDENTIFICATION: International think tank devoted to the study of the complex interrelationships of global problems

DATE: Established in April, 1968

SIGNIFICANCE: The Club of Rome's influence in solving environmental problems surpasses the organization's low profile. From sponsoring the best-selling *Limits to Growth* study in 1972 to promoting the benefits of energy efficiency, the Club of Rome has worked steadfastly for the betterment of humankind.

In April, 1968, thirty-six prominent European scientists, businessmen, and statesmen gathered at the Accademia dei Lincei in Rome and formed the Club of Rome. Since its founding, the organization has focused on the *world problematique*, which is characterized by the complex interrelationships of global problems: environmental degradation, poverty, overpopulation, militarism, ineffective governmental institutions, and a global loss of human values. The Club of Rome commissions studies on important aspects of the *world problematique*, and the resulting reports become springboards for behind-the-scenes meetings with decision makers and for the initiation of projects recommended in the reports. Funding for Club of Rome reports and activities comes through arrangements with government agencies, academic research centers, and foundations.

By all accounts, Aurelio Peccei has been the most influential member of the Club of Rome. From the organization's founding in 1968 until his death in 1984, Peccei was instrumental in bringing people together, developing ideas into projects, finding funding, and providing logistic support for projects. In addition to Peccei's personal influence, the success of the club has relied on the contributions of its one hundred members from more than fifty countries. Many of the reports to the Club of Rome have been made possible by or have been directly authored by members.

The first report to the Club of Rome is also its best known. Club members obtained funding from the Volkswagen Foundation to commission a computer modeling team to forecast the future global system. The researchers' results were published in 1972 in *The Limits to Growth*, by Donella H. Meadows, Dennis L. Meadows, Jørgen Randers, and William W. Behrens III. This alarming report predicted that the human race would collapse by the year 2100 if current trends continued. The book became an international sensation, in large part through the publicity efforts of Peccei. Despite widespread criticism of the team's methodology and results, *The Limits to Growth* succeeded in sparking debate about the fate of humankind. It also generated greater interest in global modeling, including several follow-up studies commissioned by the club.

More than forty other reports to the Club of Rome have been published, with topics ranging from education to microelectronics. Reports with specifically environmental themes include *The Oceanic Circle: Governing the Seas as a Global Resource* (1998), by Elisabeth Mann-Borgese; *Factor Four: Doubling Wealth—Halving Resource Use* (1997), by Ernst von Weizsäcker, Amory B. Lovins, and Hunter L. Lovins; *The Future of the Oceans* (1986), by Elisabeth Mann-Borgese; *Energy: The Countdown* (1978), by Thierry de Montbrial; and *Beyond the Age of Waste* (1978), by Dennis Gabor and Umberto Colombo. Although these reports have not achieved the visibility of *The Limits to Growth*, the club has used them as platforms for quiet environmental protection campaigns. In 2004, Meadows, Randers, and Meadows published a follow-up to their first report titled *Limits to Growth: The Thirty-Year Update*.

Andrew P. Duncan

FURTHER READING

Kula, E. *History of Environmental Economic Thought*. 1998. Reprint. New York: Routledge, 2003.

Meadows, Donella H., Jørgen Randers, and Dennis L. Meadows. *Limits to Growth: The Thirty-Year Update*. White River Junction, Vt.: Chelsea Green, 2004.

SEE ALSO: *Limits to Growth, The*; Sustainable development.

Coal

CATEGORY: Energy and energy use

DEFINITION: Combustible sedimentary rock composed primarily of carbon and variable quantities of other elements, such as sulfur, hydrogen, oxygen, and nitrogen

SIGNIFICANCE: Coal mining and the burning of coal as a fuel both have adverse effects on the environment. The burning of coal has been found to be the largest contributor to human-caused increases of carbon dioxide, a greenhouse gas, in the earth's atmosphere.

Coal normally occurs in rock strata in layers or veins called coal beds or seams. The earth's Carboniferous period, some 300 million years ago, provided the special conditions for widespread coal formation. Coal formed as layers of plant matter accumulated at the bottoms of bodies of water, protected from biodegradation and oxidation, most often by mud or acidic water. Eventually, the plant matter buried by sediments changed over time and through geological action to create the solid known as coal.

CLASSIFICATIONS AND USES

Types of coal are classified based on the pressures and temperatures to which their precursors were subjected, known as their degree of metamorphism. From lowest to highest degree of metamorphism, the types of coal are lignite, subbituminous, bituminous, and anthracite. Darkness of color and hardness both increase with rank. Accordingly, the harder coals, anthracite and bituminous, are also known as black coals, and the softer coals, subbituminous and lignite, are known as brown coals.

Coal is most commonly used as a fuel for the production of electricity and heat. Coal-fired generation accounts for roughly 40 percent of electricity throughout the world. In some countries, this figure is significantly higher: Poland relies on coal-fired plants for more than 94 percent of its electricity, South Africa for 92 percent, China for 77 percent, and Australia for 76 percent. When coal is used to generate electricity with a standard steam turbine, the coal is pulverized (crushed or ground into powder form) and then burned in a furnace equipped with a boiler. The furnace heat causes the boiler water to convert into steam, and the steam powers spin turbines that turn generators, creating electricity.

Another important use of coal is in the form of coke. In the production of coke, low-ash, low-sulfur bituminous coal is baked such that the fixed carbon and residual ash become fused. Coke is used as a fuel and as a reducing agent in the process of smelting iron ore in blast furnaces.

MINING

The majority of coal is strip-mined. In this process, large industrial machines strip off soil and coal, scarring the land. In some nations, laws have been passed that require mining companies to restore the land after mining operations have ended, but such reclamation projects often leave much to be desired. It has been estimated that up to one-fourth of the 3.2 million hectares (8 million acres) that are above coal mines have subsided—that is, the ground on the surfaces above the mines has caved in. Coal mining has also sometimes resulted in water drainage patterns that make land unfit for farming and uninhabitable for wildlife.

The act of mining coal also presents health risks. Even in the twenty-first century coal mining remains one of the most dangerous occupations, killing more than one hundred people per year in the United States. Further, coal miners are subject to long-term inhalation of coal dust and other mineral dusts, and so are susceptible to emphysema and other respiratory conditions such as coal workers' pneumoconiosis, commonly known as black lung, a chronic, nonfatal disease that causes extreme discomfort.

CONTROVERSIES

Because of its global abundance and relatively low cost, coal is a mainstay for both developed and developing nations, which use it for power generation and steel production throughout the world. The burning of coal, however, contributes significantly to carbon dioxide (CO_2) emissions, which are linked to global warming, and to the generation of sulfur and nitrogen oxides, which are elements in acid rain. It also releases heavy metals (including mercury, selenium, and arsenic) that are harmful to the environment and human health, and it generates waste products (fly ash, bottom ash, boiler slag, flue gas desulfurization gypsum) that pose disposal problems. Other environmental problems related to the mining and use of coal include disturbances of groundwater and water-

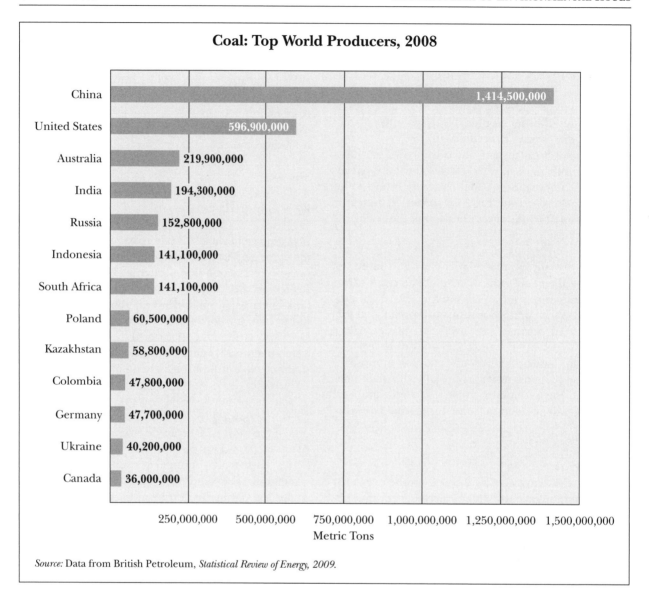

Coal: Top World Producers, 2008

China — 1,414,500,000
United States — 596,900,000
Australia — 219,900,000
India — 194,300,000
Russia — 152,800,000
Indonesia — 141,100,000
South Africa — 141,100,000
Poland — 60,500,000
Kazakhstan — 58,800,000
Colombia — 47,800,000
Germany — 47,700,000
Ukraine — 40,200,000
Canada — 36,000,000

Metric Tons

Source: Data from British Petroleum, *Statistical Review of Energy, 2009.*

table levels, contamination of land and waterways and destruction of property caused by spills of fly ash, disturbances to water use on flows of rivers, and dust generation.

The pollutants generated from coal burning affect not only the environment but also human health. Sulfur dioxide is associated with respiratory diseases such as asthma, bronchitis, and emphysema. Nitrogen oxides similarly are known to irritate the lungs, cause bronchitis and pneumonia, and lower the body's resistance to respiratory infections. Particulates from coal burning, when inhaled, can damage the respiratory system, causing acute and chronic respiratory illnesses.

Given that the continuing use of coal to meet the world's energy needs will exacerbate both the problem of global climate change and other negative environmental impacts, scientists recognize that there is a growing need for technologies that can better manage the emissions generated by coal burning. Carbon capture and sequestration or storage (CCS), also known as carbon dioxide air capture, is a technology in which the CO_2 generated by coal burning is captured or eliminated before it can be released into the air. CCS, however, has not yet demonstrated its efficacy and safety on an industrial scale at competitive cost.

Rena Christina Tabata

FURTHER READING

Freese, Barbara. *Coal: A Human History*. Cambridge, Mass.: Perseus, 2003.

Goodell, Jeff. *Big Coal: The Dirty Secret Behind America's Energy Future*. New York: Houghton Mifflin, 2006.

Thomas, Larry, Annedd Bach, and Michaelchurch Escley. *Coal Geology*. New York: John Wiley & Sons, 2002.

SEE ALSO: Black lung; Carbon dioxide air capture; Coal ash storage; Coal-fired power plants; Fossil fuels; Global warming; Mine reclamation; Nitrogen oxides; Power plants; Sulfur oxides; Surface Mining Control and Reclamation Act.

Coal ash storage

CATEGORY: Waste and waste management

DEFINITION: Storage of the waste products of coal burning

SIGNIFICANCE: Recycling the ash that results from coal burning provides a way to remove a waste product. Coal ash contains several environmental toxins, however, and unregulated storage of coal ash allows these toxins to leach into groundwater and pollute local water sources.

Many power plants burn coal to generate energy. The by-products of coal burning are called coal combustion by-products (CCBs) or coal ash. In 2007, coal-burning power plants in the United States generated approximately 131 million tons of coal ash.

Coal ash is generally divided into four types. Fly ash, the finest form of coal ash, forms from noncombustible matter in coal. Bottom ash, a coarse and granular material, comes from the bottom of coal-burning furnaces. Coal burned in a cyclone boiler produces a molten ash that, once cooled with water, forms a black material called boiler slag. Gases generated by coal burning, if passed through air-pollution-control systems called scrubbers that remove sulfur, generate flue gas desulfurization (FGD) gypsum.

Recycling and disposal represent the two options for managing coal ash. All four types of coal ash have several beneficial uses. Fly ash can substitute for Portland cement in concrete and produce stronger, less

Workers remove coal ash from a cove during cleanup of a December, 2008, spill from the Kingston Fossil power plant in Harriman, Tennessee. (AP/Wide World Photos)

porous, and cheaper forms of concrete. Bottom ash and boiler slag contribute to road base, asphalt paving, and fill material, and engineers have used boiler slag for blasting grit, roofing-shingle aggregate, and snow and ice control. Farmers use FGD gypsum as a soil amendment, and builders use it to make wallboard. About one-third of coal ash and one-fourth of scrubber waste are recycled.

Methods of disposing of coal ash are divided into dry disposal, in which unwanted CCBs are transported to landfills, and wet disposal, in which coal ash is sluiced to storage lagoons. Combined coal ash in storage lagoons, known as ponded ash, accounts for approximately 30 percent of all disposed coal ash.

Coal ash often contains poisonous materials, such as arsenic, cadmium, chromium, lead, selenium, and other toxins that can cause cancer, liver damage, and neurological problems in humans and can kill fish and other wildlife. Toxins from coal ash storage lagoons can leach into groundwater. Studies by the U.S. Environmental Protection Agency (EPA) have shown that coal ash dumps can significantly increase the concentration of toxic metals in local drinking water. Worse still, coal ash used as fill-in material can ooze poisons into groundwater.

The retaining walls that enclose coal ash storage lagoons can also fail and produce environmental disasters. For example, on December 22, 2008, the ash dike enclosing a containment pond at the Tennessee Valley Authority's Kingston Fossil coal-burning plant, near Harriman, Tennessee, broke. Approximately one billion gallons of coal fly ash slurry flowed into tributaries of the Tennessee River, which supplies the drinking water for Chattanooga, Tennessee, and those who live downstream in Alabama, Tennessee, and Kentucky.

Because coal ash can be reused, it is not classified as a hazardous waste in the United States, nor is its reuse subject to federal oversight. The EPA moved to close this loophole in 2009, but the draft rule was held up by the Office of Management and Budget. The EPA's stated purpose was to prevent environmental damage from coal ash ponds by requiring such ponds to have synthetic liners and leachate collections systems and by phasing out leak-prone ash ponds.

Michael A. Buratovich

FURTHER READING

Freese, Barbara. *Coal: A Human History.* Cambridge, Mass.: Perseus, 2003.
Goodell, Jeff. *Big Coal: The Dirty Secret Behind America's Energy Future.* New York: Houghton Mifflin, 2006.
Greb, Stephen F., et al. *Coal and the Environment.* Alexandria, Va.: American Geological Institute, 2006.

SEE ALSO: Carcinogens; Coal; Coal-fired power plants; Drinking water; Fossil fuels; Heavy metals; Leachates; Recycling; Resource recovery; Waste management; Water pollution.

Coal-fired power plants

CATEGORIES: Energy and energy use; atmosphere and air pollution

DEFINITION: Power-generating plants that burn coal to produce electricity

SIGNIFICANCE: The dirtiest of fossil fuels, coal is the most widely used source of electricity generation worldwide. Between 2002 and 2008, worldwide consumption of coal rose by 30 percent, two-thirds of which went into power generation.

Coal is the most widely used source of electricity generation worldwide. In 2009, 44 percent of electricity generated in the United States came from coal, 24 percent from natural gas, 20 percent from nuclear power, and 7 percent from hydroelectric sources (the remaining 5 percent came from sources such as solar, wind, and geothermal energy). Coal is the most plentiful fossil fuel; it is estimated that 90 percent of the earth's remaining fossil-fuel reserves are in the form of coal. Coal is also the most dangerous fossil fuel from the point of view of climate scientists, as most coals produce roughly 70 percent more carbon dioxide (a greenhouse gas linked with global warming) per unit of energy generated than natural gas and about 30 percent more than oil.

Coal poses environmental problems other than carbon emissions as well. The mining of coal produces methane, and its combustion produces sulfur dioxide and nitrous oxide in addition to carbon dioxide. The transport of coal also usually requires more energy than the transport of any other fossil fuel.

THE ROLE OF CHINA

During the 1980's China replaced the Soviet Union as the world's largest coal producer. China controls 43 percent of the world's remaining coal reserves and uses coal to generate more than 80 percent of its electrical energy, spewing out some 19 million tons of sulfur dioxide per year. By 2005 China was using more coal for electrical generation than the United States, the European Union, and Japan combined.

From 2004 to 2009 China increased its coal consumption an average of 14 percent per year, adding one to two coal-fired electricity plants per week to meet the demands of its booming economy, which was involved in human history's largest-scale industrialization. Many of the plants built in China during this period use old technology that lacks protections against pollution; it is expected that they will operate for an average of seventy-five years.

CARBON SEQUESTRATION

In 2007 James E. Hansen, an expert in atmospheric physics and director of the Goddard Institute for Space Studies at the National Aeronautics and Space Administration (NASA), proposed that a moratorium be placed on construction of new coal-fired power plants until technology allowing the capture and sequestration of the carbon dioxide produced by such plants is more widely available and economically feasible. About a quarter of power plants' carbon dioxide emissions will remain in the air more than five hundred years, long after new technology is refined and deployed. Hansen has estimated that all power plants without adequate sequestration will be obsolete and slated for closure (or at least retrofitting) before 2050.

By the beginning of 2008 the European Commission was weighing whether to require new power stations to include facilities that will retrofit to store greenhouse gas emissions through carbon capture and storage (CCS) technology when it is available, the first legal move of this type in the world, and a large step toward making CCS a commercial reality. The requirement as written does not include a date on which actual CCS would be required. Installation of CCS technology, now still in its in-

fancy, could reduce global carbon dioxide emissions by one-third by 2050, if widely deployed. By 2010 Norway, Great Britain, China, and the United States were planning CCS pilot plants.

OPPOSITION TO NEW PLANTS

In 2007 opposition to the building of new coal-fired plants accelerated in the United States. In one instance, when the mayor of Missoula, Montana, won city council support to buy electricity from a new coal-fired plant starting in 2011 to save the city money, he was inundated by hundreds of e-mails and phone calls from protesting constituents. Between 2006 and 2008 plans for eighty-three coal-fired power plants in the United States were voluntarily withdrawn or denied permits by state regulators.

Cancellations or delays of coal-fired power plants continued into 2009, when NV Energy delayed a plant

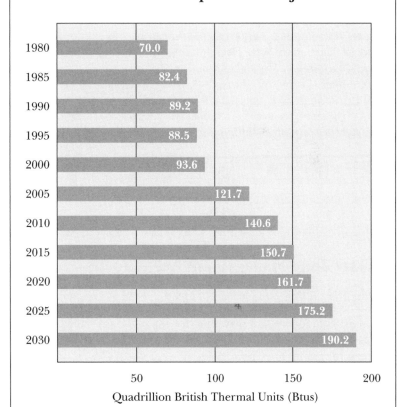

World Coal Consumption and Projections

Year	Quadrillion British Thermal Units (Btus)
1980	70.0
1985	82.4
1990	89.2
1995	88.5
2000	93.6
2005	121.7
2010	140.6
2015	150.7
2020	161.7
2025	175.2
2030	190.2

Quadrillion British Thermal Units (Btus)

Source: History: U.S. Energy Information Administration (EIA), *International Energy Annual, 2006* (June-December, 2008). Projections: EIA, World Energy Projections Plus (2009).

in eastern Nevada until such time as "clean coal" technology is available, and Southern Montana Electric Generation and Cooperative halted work on a plant near Great Falls in favor of building wind turbines and another plant that will burn natural gas. Michigan's governor, Jennifer Granholm, told regulators not to approve any of five new proposed coal-fired plants in her state until all feasible and prudent alternatives had been considered. Peabody Energy dropped plans for a coal-fired energy plant in western Kentucky in favor of a plant that would convert coal to natural gas. In early March, 2009, Alliant Energy dropped plans to build a very large coal-fired power plant in central Iowa that would have provided enough energy to supply 500,000 homes. Several of these actions were challenged by coal-power advocates.

Bruce E. Johansen

FURTHER READING

Brown, Lester. *Plan B: Rescuing a Planet Under Stress and a Civilization in Trouble.* New York: W. W. Norton, 2003.

Goodell, Jeff. *Big Coal: The Dirty Secret Behind America's Energy Future.* New York: Houghton Mifflin, 2006.

Palmer, Margaret A., et al. "Mountaintop Mining Consequences." *Science* 327 (January 8, 2010):148-149.

Romm, Joseph J. *Hell and High Water: Global Warming—The Solution and the Politics—And What We Should Do.* New York: William Morrow, 2007.

SEE ALSO: Acid deposition and acid rain; Air pollution; Carbon dioxide; Carbon dioxide air capture; Coal; Coal ash storage; Cogeneration; Renewable energy.

Coastal Zone Management Act

CATEGORIES: Treaties, laws, and court cases; land and land use; water and water pollution

THE LAW: U.S. federal law providing for the management of land and water uses in coastal areas, including the Great Lakes

DATE: Enacted on October 27, 1972

SIGNIFICANCE: The Coastal Zone Management Act has provided coastal zones with funding and research to create effective management programs to guide development. More than 153,000 kilometers (95,000 miles) of oceanic and Great Lakes coastlines are managed through these programs.

In 1970 the National Oceanic and Atmospheric Administration (NOAA) was created as an agency of the U.S. Department of Commerce. One of its charges was to develop a comprehensive plan to help guarantee the appropriate balance between the economic development of coastal and marine ecosystems and the protection of such ecosystems from pollution, overfishing, erosion, and invasive species. In 1972 the U.S. Congress passed the Coastal Zone Management Act (CZMA), creating a framework for local and federal partnerships to protect marine life and maintain clean water resources while permitting coastal economies to grow.

Under the CZMA, two new programs were created. The National Coastal Zone Management Program provides matching funds to coastal states that voluntarily submit management plans for conservation. More than half of the thirty-five eligible U.S. states and territories submitted approved plans by the end of 1979; by early 2010, Illinois was the only eligible state that had not yet joined the program. The National Estuarine Research Reserve System studies and protects the coastal zones where freshwater and saltwater bodies intersect. By early 2010, twenty-seven estuaries had been designated as sanctuaries.

Cynthia A. Bily

SEE ALSO: Beaches; Chesapeake Bay; Commercial fishing; Coral reefs; Environmental law, U.S.; Fisheries; Habitat destruction; Introduced and exotic species; Land-use policy; Sea-level changes; Water conservation.

Cogeneration

CATEGORY: Energy and energy use

DEFINITION: Power generation that produces useful electricity and heat simultaneously

SIGNIFICANCE: Since the mid-twentieth century cogeneration power systems have become increasingly attractive for the on-site industrial production of energy because of their cost-effectiveness and their efficiency.

Cogeneration is an energy recycling process. The cogeneration process puts to work the heat produced by the generation of electricity, energy which when produced off-site at conventional power sta-

Vapor rises from heat-recovery steam generators at the Midland Cogeneration Venture plant near downtown Midland, Michigan. (AP/ Wide World Photos)

tions is routinely lost, vented into the atmosphere largely through cooling towers. In conventional energy production, public utilities produce electricity at power plants and transmit thermal energy through miles of insulated piping to distant industrial and commercial complexes, most often factories, a process that routinely works at a tremendous inefficiency, roughly 25 percent (that is, nearly 75 percent of the energy is lost). Factories that rely on that energy must thus pay high rates for electricity that reflect the inefficiency of the system. In contrast, when a factory opts for a cogeneration plant on-site, a primary fuel source (most often natural gas, wood, or fossil fuel) is used to generate electricity, in this case a secondary fuel, and the heat that is naturally and simultaneously produced in that process is retained and used on-site, most prominently in space heating (and cooling) and in heating water. In utilizing the heat that is a by-product of electricity production, cogeneration (also known as CHP, for "combined heat and power") offers a nearly 90 percent efficiency. Because it uses significantly lower amounts of fuel to produce usable energy, cogeneration production is often cited for its potential to save industry jobs.

Cogeneration, although attractive in an era of energy conservation, is hardly new—in fact, the original commercial electrical power plant designed by Thomas Edison more than a century ago was a cogeneration plant, largely for practical reasons: No network grid existed to move energy across distances. As power networks became established and government regulations sought to protect public utilities from unlicensed competition, public utilities gradually became the dominant providers of energy in return for government control of pricing and rates. When fuel prices were relatively stable and low, industries had little incentive to develop cogeneration technology. Since the major energy crisis of the mid-1970's, however, industries have increasingly looked toward cogeneration. In addition, in light of emerging scientific knowledge regarding global warming, a process that increases energy efficiency and recycles heat offers an environmentally friendly alternative to traditional power generation.

Given that the cogeneration process involves the burning of fossil fuels, it has the potential to contribute to air and water pollution, but most established environmental groups, most prominently the Sierra

Club, have long backed cogeneration as part of any responsible comprehensive domestic energy agenda. Cogeneration is far better established in Europe than in the United States; in 2010 only about 10 percent of U.S. electricity needs were being met through cogeneration facilities, but the figure was rising. The kinds of facilities that most often use cogeneration are those in various process industries (breweries, food-processing plants, paper mills, brick and cement factories, textile plants, mineral refineries); some commercial and public buildings (hospitals, hotels, large universities, airports, military facilities) also use cogeneration. Given its potential to secure at least some short-term constraints on carbon dioxide emissions and its ability to maintain power supplies despite whatever interruption might affect the larger power grid, cogeneration has emerged as a significant element of environmentally conscious energy production.

Joseph Dewey

FURTHER READING

Flin, David. *Cogeneration: A User's Guide.* Stevenage, Hertfordshire, England: Institution of Engineering and Technology, 2009.

Jonnes, Jill. *Empires of Light: Edison, Tesla, Westinghouse, and the Race to Electrify the World.* New York: Random House, 2003.

Kolanowski, Bernard F. *Small-Scale Cogeneration Handbook.* Lilburn, Ga.: Fairmont Press, 2008.

Kutz, Myer. *Environmentally Conscious Alternative Energy Production.* Hoboken, N.J.: John Wiley & Sons, 2007.

SEE ALSO: Alternative energy sources; Biomass conversion; Coal-fired power plants; Energy conservation; Energy Policy and Conservation Act; Heat pumps; Hydroelectricity; Internal combustion engines; Power plants.

Colorado River

CATEGORIES: Places; water and water pollution

IDENTIFICATION: Large river that flows through the southwestern United States and northwestern Mexico

SIGNIFICANCE: The Colorado River is located in an arid region, and numerous disputes regarding rights to the river's water have arisen among the political entities that have land within the river basin: seven U.S. states and Mexico. All of these parties want to ensure that they will have continued access to the river's water for agricultural irrigation and urban-suburban expansion.

The Colorado River watershed has a drainage area of 637,000 square kilometers (246,000 square miles) and includes parts of seven western states (Colorado, Wyoming, Utah, Nevada, New Mexico, Arizona, and California) and parts of northwestern Mexico. The river begins in the Rocky Mountains of Colorado and flows for 2,330 kilometers (1,450 miles), to the Gulf of California.

Many dams were built on the Colorado River during the twentieth century for hydroelectricity and flood control, creating bodies of water that are also used for recreation. Together, these dams have the capacity to store more than 85 billion cubic meters (3,000 billion cubic feet) of water, a volume that is roughly four times the average annual flow of the entire river. The two largest dams are Hoover Dam, on the border between Nevada and Arizona, and Glen Canyon Dam, in north-central Arizona. These dams represent 80 percent of the total water-storage capacity in the watershed. Irrigation accounts for two-thirds of the water use, with much of the remaining portion going to cities (such as Phoenix, Tucson, and Las Vegas) and evaporation. Withdrawals from the river include not only water use within the Colorado River watershed basin but also major deliveries to areas outside the basin, such as the Los Angeles metropolitan area, the farms of the Imperial Valley in southeastern California, portions of Baja California, the Central Utah Project, and the Colorado-Big Thompson Project to bring water to Denver and other areas east of the Rockies.

Over the years, numerous disputes have arisen over the rights to the water of the Colorado River. One attempt to resolve such disputes was the Colorado

River Compact of 1922. That agreement made water allocations among states based on the water that had been available during the relatively limited period from 1896 to 1922, even though the river had been in existence for at least five million years. It was later understood that the period chosen was actually one of above-normal precipitation, as indicated by tree-ring analysis, which enabled scientists to extend the record back several centuries. The ten-year running mean for the annual flow of the Colorado River at Lee's Ferry below Glen Canyon Dam showed a downward trend from 1987 to 2007.

Problems with water quality in the Colorado River include high salinity levels that reflect the combination of an arid climate, the presence of saline soils, and the impacts of large amounts of surface-water runoff from irrigation return flows. Continuing population growth in the region is likely to add to problems of water demand and water quality. Environmentalists and others have suggested that those living in the region should take steps to conserve the water of the Colorado River through such practices as using native drought-resistant vegetation in landscaping (xeriscaping) and using lower-quality water for irrigating certain kinds of properties, such as golf courses.

Climate change could be another important factor in the future of the Colorado River. According to estimates by the Intergovernmental Panel on Climate Change, global warming could result in a 20 percent reduction in runoff in the Colorado River watershed by the middle of the twenty-first century. The resultant environmental impacts on the river system would be substantial.

Robert M. Hordon

Colorado River Storage Project Act

The Colorado River Storage Project Act of 1956 was passed after a bitter battle between environmental groups and the U.S. government to prevent the construction of Echo Park Dam. The act's main points follow:

The Act authorizes the Secretary of the Interior to construct a variety of dams, reservoirs, powerplants, transmission facilities and related works in the Upper Colorado River Basin. The Act also authorizes and directs the Secretary to investigate, plan, construct and operate facilities to mitigate losses of, and improve conditions for, fish and wildlife and public recreational facilities. The Act provides authority to acquire lands and to lease and convey lands and facilities to state and other agencies. . . .

The Act provides for the comprehensive development of the water resources of the Upper Colorado River Basin to: regulate the flow of the Colorado River; store water for beneficial consumptive use; make it possible for states of the Upper Basin to use the apportionments made to and among them in the Colorado River Compact and the Upper Colorado River Basin Compact, respectively; provide for the reclamation of arid and semiarid land, the control of floods, and the generation of hydroelectric power. . . .

By authorizing construction and prioritizing planning of specified projects, Congress does not intend to interfere with comprehensive development providing for consumptive water use apportioned to and among the Upper Colorado River Basin states or preclude consideration and authorization by Congress of additional projects. No dam or reservoir authorized by the Act shall be constructed within a national park or monument. . . .

Recreational, Fish and Wildlife Facilities. The Secretary is authorized and directed to investigate, plan, construct, operate and maintain: public recreational facilities on project lands to conserve scenery, the natural, historic and archaeologic objects, wildlife and public use and enjoyment; facilities to mitigate losses and improve conditions for the propagation of fish and wildlife. . . .

FURTHER READING

Cech, Thomas V. *Principles of Water Resources: History, Development, Management, and Policy.* 3d ed. Hoboken, N.J.: John Wiley & Sons, 2010.

Powell, James L. *Dead Pool: Lake Powell, Global Warming, and the Future of Water in the West.* Berkeley: University of California Press, 2008.

SEE ALSO: Droughts; Echo Park Dam opposition; Glen Canyon Dam; Global warming; Grand Canyon; Hoover Dam; Hydroelectricity; Irrigation; Riparian rights; Runoff, agricultural; Water conservation; Water rights; Watershed management.

Commercial fishing

CATEGORY: Animals and endangered species

DEFINITION: Large-scale, mostly oceanic, fishing operations undertaken for profit

SIGNIFICANCE: Beginning during the mid-twentieth century, larger, more efficient fishing fleets began exhausting coastal fishing areas and moving farther from port in their effort to meet growing demand. By the end of the century, such fleets had reached nearly every area of the world's oceans. Overfishing has depleted marine populations and in some cases has precipitated political crises.

Hooks, nets, and traps had little effect on populations of marine life until Europeans began operating fishing fleets during the late Middle Ages. Mechanization of fishing started in the nineteenth century, but world fishing production was still only around 3 million tons at the beginning of the twentieth century and 20 million tons during the 1950's. The subsequent development of better transportation allowed rapid shipment of premium catches, and the fishing industry reacted to the growing market by investing in larger boats and new technologies and techniques, including sonar, aerial spotting, and nylon nets.

More important was the introduction of factory ships. Such ships process and store the catch at sea, so a fishing fleet can work far from its home port. In one hour, a factory ship can harvest more fish than a sixteenth century fishing boat could harvest in one season. According to the United Nations Food and Agriculture Organization, roughly 23,000 industrialized fishing vessels were operational at the end of 2007. These large-scale operations often specialize in a single species and throw back nontarget fish. However,

Members of an animal rights organization protest against the overfishing of the seas in front of the European Union Maritime Affairs and Fisheries building in Brussels in September, 2010. (AP/Wide World Photos)

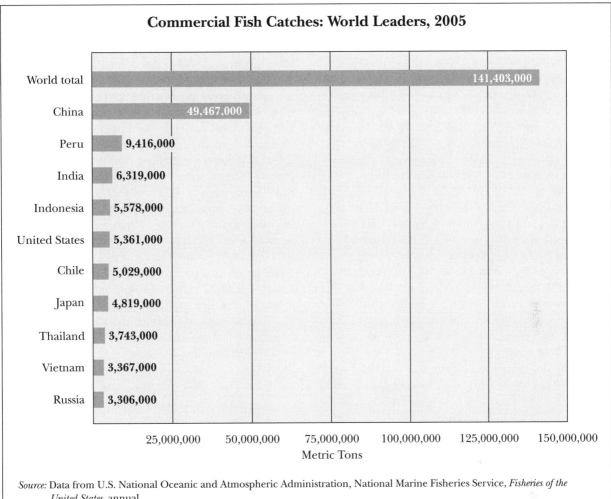

Commercial Fish Catches: World Leaders, 2005

World total	141,403,000
China	49,467,000
Peru	9,416,000
India	6,319,000
Indonesia	5,578,000
United States	5,361,000
Chile	5,029,000
Japan	4,819,000
Thailand	3,743,000
Vietnam	3,367,000
Russia	3,306,000

25,000,000   50,000,000   75,000,000   100,000,000   125,000,000   150,000,000

Metric Tons

Source: Data from U.S. National Oceanic and Atmospheric Administration, National Marine Fisheries Service, *Fisheries of the United States*, annual.

the nontarget bycatch fish that are returned to the water are usually dead or dying after being netted, dropped in a hold, and sorted.

Although fishing became industrialized, the old ideas of a limitless supply of fish and an uncontrolled sea remained. The stage was therefore set for a series of environmental disasters and political crises. One problem with fishing is that it is a hunter-gatherer operation rather than an agricultural one. Fishers do not nurture and protect schools of fish as farmers protect herds of cattle. This alone limits productivity. For example, there is little investment in habitat for fish, such as wetlands or rivers. In addition, fish are considered a common property for which each hunter competes against the others. Any fishers who hold back in catching fish to save breeding stock for the future lose their catch to other boats. American ecologist Garrett

Hardin referred to this phenomenon as the "tragedy of the commons," a concept he originally applied to publicly owned grazing lands. However, cattle on land can be counted, while fish populations are often gauged by catch. Thus fishers working feverishly to catch the dwindling numbers of fish create the illusion of a stable population. An entire fish species may ultimately be fished to near extinction, and the fishery may collapse, but the worst offenders will gain the most profit until disaster occurs.

Meanwhile, ocean pollution has reduced productivity. Shallow waters close to land are naturally the most productive, but these areas are the first to be affected by pollution. Areas such as the Black Sea, the North Sea, and the Chesapeake Bay in the United States have experienced serious declines in fish production. In the Gulf of Mexico, the Sea of Japan, and other waters, excess nutrients, largely from agricultural runoff, have been blamed for red tides (blooms of a poisonous species of algae). Another form of environmental degradation is bottom destruction. Powerful boats use "rockhopper"-equipped trawl nets that drag across rough areas of the ocean floor to catch bottom fish such as cod, flounder, and turbot. This practice destroys habitat and food for the young of many species.

CONSEQUENCES OF UNSUSTAINABLE FISHING PRACTICES

Despite the environmental impacts of overfishing and destructive fishing practices, governments continue to subsidize bigger and more advanced boats that travel farther and deeper to catch dwindling fish stocks. According to a 2001 report by the World Wide Fund for Nature, the world's commercial fishing industry receives more than fifteen billion dollars in subsidies every year—support that has helped keep more fishing vessels in business than the oceans can sustainably support.

During the 1950's Icelandic gunboats and British warships threatened each other during the so-called Cod War to control fishing access in the North Atlantic. Peru had a similar contest of wills with the United States to control foreign production of anchovies caught off the Peruvian coast. However, the Peruvians did not limit their own production, and, as anchovy stocks dwindled, the fishery collapsed in 1972. Similarly, Canada and the United States misjudged the carrying capacity of the Grand Banks shoals southeast of Newfoundland, and by 1992 cod fishing in the area was halted. Despite this moratorium, the cod population has not recovered—in large part because excessive bycatch of this banned nontarget fish has continued to deplete its numbers.

Total fish production would probably have declined during the 1990's had fish farming not rapidly increased during the same period. In 2006 fish farming supplied about half of all fishery products for human consumption, and its output was growing faster than any other animal-based food sector. Farm fisheries can produce indefinitely if they limit catches to practical levels. Even devastated farm fisheries can

eventually recover. Doing the same for the open ocean, however, would require major diplomatic efforts. Voluntary bans on whaling and drift nets have been routinely ignored by some countries. Beyond management of wild stocks, fish farming has the potential for great production increases. However, it raises environmental concerns regarding pollution and genetic weakening of wild stocks if large numbers of domestic stocks escape.

According to the United Nations Food and Agriculture Organization, roughly 75 percent of the world's monitored fish stocks had been fully exploited, overexploited, or depleted by 2008. Scientists have noted that if marine fisheries are to recover and be used sustainably, a number of measures must be imposed and enforced worldwide, including the prohibition of destructive and unsustainable fishing practices, the limitation of access rights, and the establishment of protected marine areas. Some researchers maintain that if fishing practices continue unaltered, populations of all fish species caught for food in the early twenty-first century will collapse within fifty years.

Roger V. Carlson
Updated by Karen N. Kähler

FURTHER READING

Clover, Charles. *The End of the Line: How Overfishing Is Changing the World and What We Eat.* Berkeley: University of California Press, 2008.

Helfman, Gene S. *Fish Conservation: A Guide to Understanding and Restoring Global Aquatic Biodiversity and Fishery Resources.* Washington, D.C.: Island Press, 2007.

Roberts, Callum M. *The Unnatural History of the Sea.* Washington, D.C.: Island Press, 2007.

Rogers, Raymond A. *The Oceans Are Emptying: Fish Wars and Sustainability.* New York: Black Rose Books, 1995.

SEE ALSO: Carrying capacity; Coral reefs; Dolphin-safe tuna; Fisheries; Food chains; Gill nets and drift nets; Marine debris; Ocean dumping; Ocean pollution; United Nations Convention on the Law of the Sea.

Commoner, Barry

CATEGORIES: Activism and advocacy; nuclear power and radiation; energy and energy use

IDENTIFICATION: American biologist and antinuclear activist

BORN: May 28, 1917; Brooklyn, New York

SIGNIFICANCE: Commoner has raised public awareness of a number of important environmental issues, particularly regarding the use of energy resources, organic farming and pesticides, waste management, and toxic chemicals.

Barry Commoner graduated from Columbia University in 1937 and received an M.A. (1938) and a Ph.D. (1941) from Harvard University. He has been awarded a number of honorary degrees from prominent institutions, and his biography has appeared in *Who's Who in Science and Engineering*. His interest in becoming an environmental activist was sparked in the 1950's by the testing of nuclear weapons. In 1965 Commoner was president of the St. Louis Commission on Nuclear Information, and in 1966 he founded the Center for the Biology of Natural Systems (CBNS) at Washington University in St. Louis, Missouri. In 1981 the CBNS moved to Queens College, in the City University of New York (CUNY) system. The CBNS program has continually informed the public and government on environmental issues, including waste recycling and energy resources; it has provided extensive analysis of the generation and fate of dioxins.

Commoner is the author of several books on the environment, including *Science and Survival* (1966), *The Closing Circle: Nature, Man, and Technology* (1971), *Ecology and Social Action* (1973), *The Politics of Energy* (1979), and *Making Peace with the Planet* (1990). He publicizes environmental problems and relates them to modern technology. Commoner claims that modern methods of production cause human illness and that modern technologies that have resulted in environmental crises have safer alternatives. He believes that to restore the environment, humans need to adopt more benign ways of accomplishing tasks, such as eliminating the use of plastic wrap and recycling instead of incineration. He has promoted the view that there is economic value and profitability in the replacement technologies. He has drawn attention to the affiliations of some environmental scientists and has questioned their opinions, stating that environmental hazards are reliably recognized only through studies done by independent scientists. He has encouraged the public to become more informed, to achieve a greater understanding of the effects of particular technologies on the environment before using those technologies.

In recognition of his contributions to public awareness of environmental concerns, Commoner has been hailed by *Time* magazine as the "Paul Revere of Ecology" and by the *Earth Times* as "the dean of the environmental movement, who has influenced two generations." In 1980 Commoner was the presidential candidate of the Citizens Party, a liberal political party he helped found; the party's interests included an end to nuclear power, a switch to solar energy, and public control of the energy industry.

Commoner continues to lead the movement to eliminate pollution at its source. He asserts that pre-

Biologist and antinuclear activist Barry Commoner. (Time & Life Pictures/Getty Images)

vention works while controls do not, and he publicly advocates the abandonment of fossil fuels as a primary source of energy. Having served society and the environment since the 1950's, he has succeeded in elevating the importance of considering environmental issues in all aspects of life.

Marcie L. Wingfield and Massimo D. Bezoari

FURTHER READING

Egan, Michael. *Barry Commoner and the Science of Survival: The Remaking of American Environmentalism.* Cambridge, Mass.: MIT Press, 2007.

Kriebel, David, ed. *Barry Commoner's Contribution to the Environmental Movement: Science and Social Action.* Amityville, N.Y.: Baywood, 2002.

SEE ALSO: Alternative energy sources; Antinuclear movement; Renewable energy; Solar energy.

Many community gardens, like this one, are split up into plots that are managed by individual gardeners or families. (©Amy Riley/iStockphoto.com)

Community gardens

CATEGORIES: Land and land use; urban environments

DEFINITION: Plots of land, donated or rented by the bodies that hold the titles or leases, that are cultivated by local community members

SIGNIFICANCE: Community gardens provide recreation and education for participants, and for many, such spaces offer respite from the stresses of urban living. In some low-income areas, community gardens are important contributors to residents' basic food supplies.

Community gardens vary greatly in nature in different parts of the world. In North America, community gardens are generally small plots of land dedicated to "greening" projects in urban areas. The common image of a community garden in the United States is of a formerly vacant plot in a low-income urban area transformed into a green, productive garden. In Europe, community gardens are generally larger and more organized, and a garden may be rented by a single family for many generations. In de-

veloping nations, it is very common for small gardens to be open to public farming.

Matters such as assignment of plots and other organizational issues in community gardens are most often decided by the involved gardeners. Some gardens are organized so that all participants garden collectively; others are split up into divided plots and managed by individual gardeners or families. The land that becomes a community garden may be either publicly or privately owned.

One goal of community gardening for many participants is to achieve food security while creating a stronger relationship between individuals and their food. Community gardens generally promote organic cultivation methods. Many community gardeners state that, in addition to seeking healthy food alternatives, they are involved in the gardening because of the value of community building they derive from their participation.

In 2001 the American Community Gardening Association estimated that some 150,000 community gardens were in operation in the United States. At that time the South Central Farm in Los Angeles, California, established in 1994, was considered the largest community garden in the nation, at 5.7 hectares (14 acres). The garden was shut down in 2006, when the city of Los Angeles sold the land.

Many organizations offer assistance in the creation of community gardens. The American Community Gardening Association provides advice—on everything from training to insurance policies—to groups seeking to establish gardens, so that they can do so sustainably and effectively. The American Horticultural Society runs a rapidly growing program that focuses on youth gardening, a form of community gardening aimed specifically at connecting children with nature.

In addition to being resources for those involved, community gardens are commonly believed to provide tangible improvements to the areas that surround them. Some research, however, appears to indicate that this belief may be more a matter of perception than of reality. A study conducted by Texas A&M and Texas State universities looked at the impact of community gardens on the numbers of property crimes in urban Houston and found that there was no statistically significant difference between crime rates in the areas surrounding the gardens and other areas. Interviewed residents, however, stated that community gardens created a positive influence in their neighborhoods. They commented on revitalization of neighborhood development and a changing tone toward illegal activity, including drug dealing. The members of the communities surrounding the gardens thus felt safer, despite the fact that the areas' crime rates were no lower than those in other areas.

Kara Kaminski

Further Reading

Henderson, Elizabeth, with Robyn Van En. *Sharing the Harvest: A Citizen's Guide to Community Supported Agriculture.* Rev. ed. White River Junction, Vt.: Chelsea Green, 2007.

Lawson, Laura. *City Bountiful: A Century of Community Gardening in America.* Berkeley: University of California Press, 2005.

Lyson, Thomas A. *Civic Agriculture: Reconnecting Farm, Food, and Community.* Lebanon, N.H.: Tufts University Press, 2004.

See also: Eat local movement; Environmental education; Greenbelts; Organic gardening and farming; Sustainable agriculture; Urban ecology.

Community right-to-know legislation. *See* Right-to-know legislation

Compact fluorescent lightbulbs

Category: Energy and energy use
Definition: Lightbulbs that consume less energy and radiate a different light spectrum than incandescent bulbs and are designed to replace them
Significance: Compact fluorescent lightbulbs can have positive impacts on the environment in that they last longer than traditional incandescent lightbulbs and use significantly less energy than incandescents, so they create less waste and aid in energy conservation. Because compact fluorescent bulbs contain mercury, however, care must be taken in their disposal to avoid doing environmental harm.

Compact fluorescent lightbulbs (CFLs) were first introduced as an energy-saving alternative to traditional incandescent lightbulbs during the mid-1990's. They use less energy and have longer life spans than incandescent lightbulbs, which they are designed to replace. CFLs can use up to 75 percent less energy and last up to ten times longer than incandescents. CFLs can be used in most of the same lighting fixtures as incandescent bulbs, but different types of CFLs have been created for compatibility with various types of lighting fixtures, including those that use dimmer switches, three-way lamps, and outdoor fixtures.

Several kinds of fixtures and circumstances can reduce the overall efficiency of CFLs; enclosed light fixtures, for example, can create high temperatures that can shorten CFLs' life, and CFLs cannot withstand extremely low temperatures. Additionally, CFLs should not be used in fixtures that vibrate, including ceiling fans and garage door openers. The efficiency of CFLs is greatly reduced when they are turned on and off frequently, as most CFLs take at least three minutes to warm up and emit their maximum light.

CFLs and incandescent lightbulbs use different methods to produce light, which accounts for the different amounts of energy they use. An incandescent bulb produces light by running an electrical current through a wire to heat a filament. A CFL produces light by using electricity to ignite a gas within the bulb to produce invisible ultraviolet light, which then produces visible light by exciting a fluorescent white coating (phosphor) inside the bulb. A ballast regulates the electrical current within the bulb. After the gas within the CFL is ignited, the amount of electricity needed to keep the bulb lighted is significantly less than that used to heat the filament in an incandescent bulb.

State and local governments in the United States have set rules regulating the disposal and recycling of broken and intact CFLs because these bulbs, like all fluorescent bulbs, contain small amounts of mercury; on average, each CFL contains about 4 milligrams (0.00014 ounce) of mercury. When a CFL is intact no amount of mercury is released, but mercury may be released from a broken CFL. The U.S. Environmental Protection Agency advises that the following special precautions should be taken when a CFL breaks, to avoid the possibility of mercury exposure. All people and animals should leave the room in which the bulb was broken, and the room should be allowed to air out for at least fifteen minutes. Disposable materials should then be used to clean up the remnants of the bulb, and all the cleanup materials should be disposed of immediately, along with the broken bulb and

Compact fluorescent lightbulbs save energy in comparison with traditional incandescent bulbs and are designed to be used in the same kinds of light fixtures as incandescents. (©Bert Folsom/Dreamstime.com)

its remnants, in a glass jar or plastic bag that can then be sealed. The jar or bag should then be disposed of according to guidelines for disposal of toxic materials set by the local government.

Courtney A. Smith

FURTHER READING

Chiras, Daniel D. "Foundations of a Sustainable Energy System: Conservation and Renewable Energy." In *Environmental Science*. 8th ed. Sudbury, Mass.: Jones and Bartlett, 2010.

Goldblatt, David L. *Sustainable Energy Consumption and Society: Personal, Technological, or Social Change?* Norwell, Mass.: Springer, 2005.

Krigger, John, and Chris Dorsi. *The Homeowner's Handbook to Energy Efficiency: A Guide to Big and Small Improvements*. Helena, Mont.: Saturn Resource Management, 2008.

SEE ALSO: Energy conservation; Energy-efficiency labeling; Green marketing; Greenwashing; Hazardous waste; Mercury.

Composting

CATEGORY: Waste and waste management

DEFINITION: Recycling of organic waste in a fashion that breaks down the matter into mulch, a rich soil conditioner

SIGNIFICANCE: Composting has gained attention as an environmentally responsible way to reduce the amount of waste that goes to landfills.

All living things are part of a natural recycling process. For example, dead trees in a forest slowly break down into a soillike substance that nurtures the growth of new forest plants. Composting is the same process in a slightly controlled environment.

A backyard or garden compost bin is basically a box with holes or slats, to let in air, that is filled with yard and kitchen waste. A fifty-fifty balance of items containing nitrogen and items containing carbon is needed. Items containing nitrogen include grass clippings, eggshells, coffee grounds, and vegetable and fruit peelings. Items containing carbon include dead leaves, evergreen needles, bark chips, and dryer lint. The various items are added in layers and topped with dirt to hold in moisture. As the waste decays, the bacteria present thrive. They feast on the waste, generating energy and more bacteria. The decomposing pile is occasionally turned with a shovel, and more yard or kitchen waste is added. The waste products from the bacteria, together with what is left after they feast on scraps, form a dry mulch that can be used as a natural garden fertilizer.

Trash and garbage collected from the public by sanitation workers is sometimes composted in municipal processing plants. Decomposable materials are sorted from glass, metal, and other inorganic materials, and the organic items are then shredded or broken down into smaller pieces. In this way, local governments can reduce the amount of waste that is added to landfills or burned.

Two methods are used for mechanical composting. An open-window facility requires a large area of land. It is similar to backyard compost bins but on a large scale. Mounds of organic waste are piled in long, low rows called windows. They are turned or mixed every few days so the aerobic (oxygen-requiring) bacteria that are digesting the waste have adequate oxygen. The aerobic bacteria may take up to eight weeks to digest the organic waste completely. The activity of the microbes creates temperatures in the piles of up to 65 degrees Celsius (150 degrees Fahrenheit). Any bacteria or organisms carrying disease die at this temperature. The odor produced as the material decomposes makes the siting of open-window facilities difficult, as most people object to having such facilities operating near where they live.

The other type of mechanical composting facility encloses the waste in tanks. Paddles mix the material, adding air at the same time. Tank composting facilities use 85 percent less space than do open-window facilities, and the compost is fully digested in about one week. The digested compost is dried, screened, and made into pellets before it is used as a soil conditioner or fertilizer. Mechanically processed compost has low

A backyard composting bin filled with yard and kitchen waste. (©Dreamstime.com)

marketability, however, because of the cost of transport and because of competition from the chemical fertilizers readily available to commercial farmers.

Lisa A. Wroble

FURTHER READING

Nardi, James B. "Composting as an Antidote to Soil Abuse." In *Life in the Soil: A Guide for Naturalists and Gardeners.* Chicago: University of Chicago Press, 2007.

Scott, Nicky. *Composting: An Easy Household Guide.* White River Junction, Vt.: Chelsea Green, 2007.

SEE ALSO: Agricultural chemicals; Biofertilizers; Landfills; Organic gardening and farming; Waste treatment.

Comprehensive Environmental Response, Compensation, and Liability Act. *See* Superfund

Comprehensive Nuclear-Test-Ban Treaty

CATEGORIES: Treaties, laws, and court cases; nuclear power and radiation

THE TREATY: International agreement banning nuclear weapons testing

DATE: Opened for signature on September 10, 1996

SIGNIFICANCE: Although not in force, the Comprehensive Nuclear-Test-Ban Treaty serves as an international voice against nuclear testing. The treaty seeks to end all supercritical weapons tests without inhibiting the continued development and refinement of nuclear weapons by those nations that already possess test data.

Since 1945, at least 2,080 nuclear weapons test explosions have been conducted—most by the United States (1,054), followed by the Soviet Union (715), France (210), the United Kingdom (45), China (45), India (6), Pakistan (2), North Korea (2), and an unconfirmed entity (1). The Partial Test Ban Treaty of 1963 and the Outer Space Treaty of 1967 banned atmospheric and space-based tests, but France and China continued atmospheric testing until 1974 and 1976, respectively. Despite the provisions of the Nuclear Non-Proliferation Treaty of 1968, Pakistani and North Korean tests are reported to have originated in Chinese fuel and designs. The Strategic Arms Limitation Treaty (SALT) and the Strategic Arms Reduction Treaty (START) between the United States and the Soviet Union limited and reduced the numbers of strategic nuclear warheads held by the two nations, but not their thousands of tactical nuclear weapons.

The Comprehensive Nuclear-Test-Ban Treaty, also known as the Comprehensive Test Ban Treaty, or CTBT, would require nations to agree not to conduct supercritical nuclear tests, to allow their territory to be used for such tests, or to encourage other nations to conduct such tests. A global verification regime would have four elements: an international monitoring system, a consultation and clarification process for suspected violations, on-site inspections, and confidence-building measures. The treaty was signed by seventy-four nations and passed by United Nations General Assembly resolution in September, 1996, but it cannot go into force until 180 days after ratification by the forty-four nations possessing nuclear reactors. Egypt, North Korea, India, Iran, Israel, Indonesia, Pakistan, China, and the United States have all cited supreme national security compulsions in declining to sign or ratify the treaty.

The CTBT is seen as devoid of intent toward universal nuclear disarmament. Its timing coincided with the advent of ACI Red, the U.S. Department of Energy's teraFLOPS computer developed to simulate numerically the physical processes in a nuclear explosion reliably enough to predict weapon performance using the large test database. China and France completed several tests in 1992-1996 before signing CTBT, and France set off twelve thermonuclear tests in the pristine islands of the South Pacific Mururoa and Fangataufa atolls in 1996.

Although not in force, the CTBT serves as an international voice against nuclear testing. In 2009 the debate among nuclear powers appeared to shift toward the amount of testing required to ensure a validated prediction capability in order to maintain the "minimum credible deterrent" for national security.

Narayanan M. Komerath and Padma P. Komerath

FURTHER READING

Dahlman, O., S. Mykkeltveit, and H. Haak. *Nuclear Test Ban: Converting Political Visions to Reality.* New York: Springer, 2009.

Hansen, Keith A. *The Comprehensive Test Ban Treaty: An Insider's Perspective.* Stanford, Calif.: Stanford University Press, 2006.

Malik, Mohan. "A. Q. Khan's China Connection." *China Brief* (Jamestown Foundation), May 22, 2004.

Tetiarhi, G. "French Nuclear Testing in the South Pacific." *Contemporary Pacific* 17, no. 2 (2005): 378-381.

Wittner, Lawrence S. *Confronting the Bomb: A Short History of the World Nuclear Disarmament Movement.* Stanford, Calif.: Stanford University Press, 2009.

SEE ALSO: Antinuclear movement; Atomic Energy Commission; Bikini Atoll bombing; Department of Energy, U.S.; Greenpeace; Hiroshima and Nagasaki bombings; International Atomic Energy Agency; Limited Test Ban Treaty; Neutron bombs; Nuclear and radioactive waste; Nuclear testing; Nuclear weapons production; Nuclear winter; Radioactive pollution and fallout.

Condors

CATEGORY: Animals and endangered species

DEFINITION: Species of vultures native to the Western Hemisphere

SIGNIFICANCE: The coordinated international effort to bring both the California condor and the Andean condor back from near extinction, an effort that began in the late 1980's, has been both successful and controversial. The controversial aspects of the effort include the amount of money expended and the use of the practice of captive breeding.

Despite their relatively small populations, condors, both California condors and the Andean condors of South America, are among the most recognized birds in the world. With its flattened, bald head, its thick black plumage with a tuft of white feathers around its neck, and its wide wingspan (on average close to 3 meters, or 10 feet), the California condor is the largest flying land bird in North America. It is known, despite its ungainly dimensions (it averages more than 23 kilograms, or 50 pounds) for

graceful, balletic flight. Because condors feed on carrion, a food supply never threatened or restricted to particular areas, they are among the longest-living bird species, averaging close to fifty years in the wild.

Until the mid-twentieth century, the California condor inhabited a broad area along the Pacific coast all the way from lower Baja California in Mexico to the Rocky Mountains of British Columbia, Canada. In the years after World War II, however, the numbers of condors decreased dramatically owing to several factors: aggressive land development that destroyed condor habitats; lead poisoning from condors' ingestion of carcasses discarded by careless hunters; the effects of widespread spraying of the insecticide dichloro-diphenyl-trichloroethane (DDT) on the integrity of the condors' eggs; the rapid proliferation of power lines, in which the birds got tangled; and poaching by ruthless farmers who were convinced (erroneously) that the condors posed a threat to their cattle and sheep. Complicating these threats, breeding in the wild became problematic for condors because they have a naturally low birthrate and mature sexually at a slow rate.

By the early 1980's only 22 California condors remained. Over the next ten years an energetic coalition of government workers, environmentalists, ornithologists, academics, and wilderness advocates began an aggressive effort to save the species. In South America the Andean condor faced a similarly dire dilemma, and scientists there eagerly watched the North American efforts and took many of the same actions.

A California condor with identification tags on its wings. (USFWS)

In 1987 the 6 remaining wild California condors were captured and placed (along with the 27 condors already in zoos) in a controlled-breeding recovery program in animal preserves in San Diego and Los Angeles. The project was not universally hailed—in addition to objecting to the exorbitant costs (just under $40 million had been spent by 2010, making it the most expensive species rescue program ever conducted in the United States), many scientists argued that caging the few remaining wild condors meant the species was de facto extinct and that artificially restarting it would forever alter the behavior of the birds. Indeed, as the captive condors began to breed, scientists maintained close control over the offspring, raising them (in a much-publicized strategy) with puppets resembling condors as surrogate parents.

Over the next several years, however, these condors were gradually reintroduced into selected areas in Southern California and central Arizona—but only after efforts had been made to train them to avoid flying into power lines. In addition, in 2008 aggressive lobbying efforts were successful in convincing the California legislature to ban hunters' use of lead ammunition near areas marked as condor territory. The breeding program's success was highlighted in 2007, when, for the first time in nearly six decades, a California condor laid an egg in the wild. By 2010 the number of California condors in the wild had climbed to 187; the entire population, including those living in zoos, had reached 348. Scientists continue to monitor the condors, cautiously optimistic that the species is on its way to slow but steady recovery.

Joseph Dewey

FURTHER READING

Moir, John. *Return of the Condor: The Race to Save Our Largest Bird from Extinction.* Guilford, Conn.: Lyons Press, 2006.

Osborn, Sophie A. H. *Condors in Canyon Country: The Return of the California Condor to the Grand Canyon Region.* Grand Canyon, Ariz.: Grand Canyon Association, 2007.

Snyder, Noel F. R., and Helen Snyder. *The California Condor: A Saga of Natural History and Conservation.* Princeton, N.J.: Princeton University Press, 2000.

SEE ALSO: Captive breeding; Dichloro-diphenyl-trichloroethane; Fish and Wildlife Service, U.S.; Habitat destruction; Heavy metals; Lead; Poaching; Wildlife management.

Congo River basin

CATEGORIES: Places; ecology and ecosystems

IDENTIFICATION: Central African watershed that drains into the Congo River in its passage to the Atlantic Ocean

SIGNIFICANCE: The Congo River basin contains rich biodiversity and considerable mineral wealth, but because the area is shared by a number of countries, the exploitation of the basin's resources is often uncontrolled. As a result, the habitats of many species native to the basin have been disturbed or destroyed, and several species have become endangered.

The Congo River basin covers an area of 3.7 million square kilometers (1.4 million square miles) that includes all of the Democratic Republic of the Congo (DRC; formerly known as Zaire) and the Republic of the Congo (ROC), most of the Central African Republic, and sections of Angola, Rwanda, Burundi, Tanzania, and Zambia. It includes the Lake Tanganyika basin to the east, which is best seen as a subbasin, the lake itself being in a separate rift valley. In the north, the Ubangi River is the main tributary to the Congo River. The Congo River itself is 4,700 kilometers (2,920 miles) long, having its source in Zambia. It is the deepest river in the world and the second-longest river in Africa, after the Nile.

The river basin contains one-fourth of the world's remaining tropical rain forests; in addition to the forest ecosystems, areas of savanna are found in the northeast and south, and high mountains to the east contain alpine meadow. Transport through the basin is achieved mainly by water, along the extensive system of rivers. Roads and railroads are linked to river ports, many of which are connected to various mining and quarrying activities; all such transportation routes are generally in poor states of repair. Since the late twentieth century, increasing numbers of logging roads have been opened into the rain forests. Few large towns and cities are located in the basin; those that have been established are situated mainly on the River Congo itself.

The river has not been harnessed for hydroelectric purposes to any great extent, but ambitious plans have been proposed to add two much larger dams to two existing small dams at Inga Falls, southwest of the capital cities of Kinshasa (DRC) and Brazzaville

Cut logs await transport on the banks of the Congo River. The main ecological threat to the river basin comes from deforestation caused by uncontrolled logging and agricultural clearance. (©David Lewis/Reuters/LANDOV)

(ROC); it is estimated that by 2030 these dams could supply some 20 percent of Africa's electricity.

In the past, extensive mining operations in the DRC, mainly in the south of the country along the Zambia border, have brought little benefit to the country. In the early twenty-first century, China negotiated an extensive deal to build roads, railways, hospitals, and schools for the DRC and the ROC in return for mining copper, nickel, and cobalt at certain concessions. The DRC is estimated to hold one-third of the world's cobalt and 10 percent of its copper.

The main ecological threat to the Congo River basin comes from deforestation caused by uncontrolled logging and agricultural clearance, and the subsequent destruction of animal habitats. Many endangered species are native to the basin, foremost among them the mountain gorillas in the Virunga National Park in the east. The civil war that was fought in the DRC between 1998 and 2003, along with later tur-

moil, did great damage to the park. Also at risk are forest elephants, hippopotamuses, pygmy chimpanzees, western lowland gorillas, and the primates drills and mandrills. The chimpanzee population is estimated to have declined from 2 million to 200,000 from 1990 to 2010. Many of these species have been killed for food or have been killed or captured for sale on the black market.

David Barratt

FURTHER READING

Gondola, Ch. Didier. *The History of Congo.* Westport, Conn.: Greenwood Press, 2002.

Renton, David, David Seddon, and Leo Zeilig. *The Congo: Plunder and Resistance.* London: Zed Books, 2007.

Ruiz Pérez, M., et al. "Logging in the Congo Basin: A Multi-country Characterization of Timber Companies." *Forest Ecology and Management* 214 (August, 2005): 221-236.

Surhone, Lambert M., Miriam Timpledom, and Susan Marseken, eds. *Tropical and Subtropical Forests: Congo Basin, Amazon Basin*. Beau Bassin, Mauritius: Betascript, 2010.

SEE ALSO: Amazon River basin; Biodiversity; Deforestation; Elephants, African; Habitat destruction; Ivory trade; Johannesburg Declaration on Sustainable Development; Mountain gorillas; Rain forests.

Coniferous forests

CATEGORIES: Forests and plants; ecology and ecosystems

DEFINITION: Forests constituted primarily of evergreen trees that are species of conifers

SIGNIFICANCE: Boreal coniferous forests form the largest single forest system on earth and thereby are a major part of the world ecosystem. Human exploitation, population pressures, pollution, and climate change have had negative impacts on the range and variety of conifers.

Coniferous forests constitute a vast biome that lies primarily in the high latitudes of the Northern Hemisphere but also occurs throughout the world. Conifers are evergreens that have compound seed cones, needle- or scale-shaped leaves, distinctive water-transporting tissue and pollen grain structure, and compact wood. They are thought to date back to the Carboniferous period, about 300 million years ago. In the twenty-first century, 546 conifer species, in 67 genera, are known to exist. Of these, the most common species are as follows: in the Northern Hemisphere, pines (*Pinus*), firs (*Abies*), spruces (*Picea*), and junipers (*Juniperus*); in the Southern Hemisphere, yellowwoods (*Podocarpus*). Although not as abundant in plant and animal species as tropical forests, coniferous forests form a major segment of the world ecosystem.

Conifers grow on every continent except Antarctica and on many coastal and oceanic islands. Although many species are native to tropical and subtropical biomes, the majority are in the boreal forest biome, also called taiga. This is a belt of forest that extends across Canada into Alaska in North America and across northern Europe and Russia in Eurasia, reaching as much as 73 degrees north latitude in Siberia. In the Southern Hemisphere, coniferous forests extend as far as 55 degrees south latitude on Chile's Tierra del Fuego. About 60 percent of species are found in a range of altitude between 500 and 1,500 meters (1,600 and 5,000 feet), although there are sea-level species and one that survives at altitudes up to 4,800 meters (15,800 feet). Additionally, conifers have been cultivated for some two hundred years, and introduced or exotic species of conifers have entered ecosystems as they have been planted in gardens and parks.

Conifers are hardy and typically grow larger and live longer than hardwood species of trees. In fact, the world's oldest known tree, a Great Basin bristlecone pine in eastern California, is nearly five thousand years old, and the tallest tree is a sequoia at 115.56 meters (379.1 feet), also in California. Most conifer species occur in temperate, moist climates. For instance, the taiga in general has an average minimum temperature of about 6.6 degrees Celsius (44 degrees Fahrenheit) and receives about 50 centimeters (20 inches) of rain yearly.

The variety of conifer species and the variety of flora and fauna among them are generally low in coniferous forests; in the taiga spruce, tamarack (or larch), and balsam fir dominate as climax species. Nonetheless, the sheer number of trees affords an immense natural resource. Among the wildlife in the taiga alone are wolves, coyotes, bears, foxes, lynx, deer, moose, rabbits, and many species of raptors and songbirds. Coniferous forests have long been sources of raw materials for humans as well, principally timber for construction and pulp for paper.

About one-third of conifer species are imperiled within their natural range; twenty are critically endangered. The expanding human population is the greatest stressor because of losses of forestland to settlement and agriculture, increased frequency and intensity of fires near settlements, air and water pollution from agriculture and industry, and the introduction of exotic pests and diseases. Moreover, human-accelerated climate change may force the migration of some species northward, where it is not blocked by agriculture or cities.

Roger Smith

FURTHER READING

Andersson, Folke, ed. *Coniferous Forests*. San Diego, Calif.: Elsevier, 2005.

Eckenwalder, James E. *Conifers of the World: The Complete Reference*. Portland, Oreg.: Timber Press, 2009.

Powell, Graham R. *Lives of Conifers*. Baltimore: The Johns Hopkins University Press, 2009.

SEE ALSO: Boreal forests; Climax communities; Deciduous forests; Forest and range policy; Forest management; Old-growth forests; Rain forests; Sustainable forestry.

Conservation

CATEGORY: Resources and resource management
DEFINITION: Planned use of natural resources to benefit the maximum number of people for as long as possible
SIGNIFICANCE: The goals of conservation often conflict with the drive toward the exploitation of natural resources for immediate economic gain and also with the ethic of preservation, which promotes the indefinite preservation of natural resources in their undisturbed state.

For centuries few people recognized that nature's resources are finite. Only since the mid-nineteenth century has the issue of conserving resources been taken seriously. In North America, the eighteenth and nineteenth centuries were largely devoted to conquering the wilderness. As pioneers cleared forests, little or no thought was given to the idea that they could ever be exhausted. Bison and other animals were hunted nearly to extinction, and bounties were placed on wolves and other predators. In the seas, millions of seals were killed, and whales were hunted relentlessly.

Almost too late, it became apparent to many people that soils were being depleted by erosion and that once-plentiful native plant and wildlife populations were in decline. Even the quality of air, surface water, and groundwater deteriorated. An increasing number of people began calling for an end to environmentally destructive practices. An important influence was that of diplomat and naturalist George Perkins Marsh, whose travels had made him keenly aware of the results of centuries of land abuse in Europe. Marsh's book *Man and Nature: Or, Physical Geography as Modified by Human Action* (1864) defined the connections among soil, water, and vegetation. From such influences came the public park movement, including the establishment of Yellowstone National Park.

EMERGENCE OF THE MODERN CONSERVATION MOVEMENT

During the early twentieth century, the living standards of most Americans generally improved, but at the same time it was recognized that consumption of natural resources had to be controlled. Out of this realization emerged a human-centered conservation philosophy known as wise use, which dictated that human beings should use nature in such a way that its resources could continue to be used over long periods of time.

Among the conservationists of the era were U.S. president Theodore Roosevelt and forester Gifford Pinchot. Roosevelt increased the extent of the national forests in the United States and created many wildlife refuges. Pinchot, a friend and associate of Roosevelt, became chief forester of the U.S. Department of Agriculture and greatly influenced the conservation movement. He applied European methods of managing forests that were consistent with the wise-use philosophy. This put him in conflict with naturalist John Muir and others who wished to preserve wilderness areas in their natural state. Two opposing camps developed: On one side were preservationists, who argued that nature deserves to be protected for its own sake; on the other were conservationists, who believed in regulated exploitation.

During the 1920's and 1930's, scientific advances and economic problems influenced conservation views and policies. The ecosystem concept, the principle that nature is composed of local units with interacting living and nonliving components, developed. It was to become the cornerstone of the science of ecology and an important tool of conservation. During this same period a new, scientific approach to wildlife management emerged. Aldo Leopold, regarded as the father of the new science, wrote the influential textbook *Game Management* (1933), although he is better known today for his collection of essays *A Sand County Almanac, and Sketches Here and There*, which was published in 1949, soon after Leopold's death. Paul B. Sears achieved prominence with *Deserts on the March* (1935), which vividly dramatized the problems of the Dust Bowl of the 1930's.

The conservation legacy of President Franklin D. Roosevelt includes his attempts to restore the depressed U.S. economy by creating various work programs, including the Civilian Conservation Corps (CCC). Many of these programs were aimed at soil conservation, reforestation, and flood control.

POSTWAR DEVELOPMENTS

In the two decades following World War II, concerns about the Cold War and economic expansion kept conservation far from the forefront of American thought. As the population rapidly expanded, air and water pollution worsened, grasslands became overgrazed, and agricultural chemicals were used in increasing quantities. The complacency of the times was shattered by the publication of Rachel Carson's *Silent Spring* (1962), which warned that dichloro-diphenyltrichloroethane (DDT) and other pesticides were threatening the lives of animals and humans. The book caused widespread public concern, and laws were ultimately passed that outlawed DDT in the United States.

Under President John F. Kennedy's administration, attention was given once more to environmental issues in general and to conservation. Funds were appropriated to improve air quality, and new land was acquired for parks. Stewart Udall, secretary of the interior under Kennedy and his successor, Lyndon B. Johnson, advocated a new positive attitude toward protection of the environment. During the 1960's, a time of great scientific activity and social ferment, ecologists employed advanced technologies in conducting ecosystem analyses. Among the social movements of the decade was a new environmentalism, which culminated in the first Earth Day celebration in 1970.

The 1970's saw the expansion of environmentalism, and the term "environmentalism" came to be understood as almost synonymous with "conservation." The former clear distinction between the terms "conservation" and "preservation" was blurred. Environmental organizations continued to elaborate their views and increase their visibility. More action-oriented alternative groups used sabotage tactics to stop development in wilderness areas. In contrast, the Nature Conservancy became known for its businesslike policy of acquisition, protection, and management of natural areas. In response to the growing awareness of environmental issues among Americans, a new round of federal legislation was passed. Under President Richard Nixon, the Environmental Protection Agency (EPA) was created in 1970. The revised Clean Air Act of 1970 was followed by the Clean Water Act of

1972, and in 1973 Congress enacted the Endangered Species Act.

POLICY AND POLITICS

The 1980's were marked by new conservation problems and a general indifference at the federal level in the United States. Among the global problems identified were stratospheric ozone depletion and global warming. Reversals of environmental protections occurred during the administration of President Ronald Reagan after he appointed James Watt as secretary of the interior. Watt attempted to dismantle many of the environmental protection programs that had been put in place during the previous decades.

The 1990's had their own environmental challenges and successes. After twelve years of apathy, conservationists were reinvigorated by the election of Bill Clinton as president and high-profile environmentalist Al Gore as vice president. Despite continued optimism among environmentalists and significant gains in environmental protections, however, an anti-environmental movement also developed. Many politically conservative individuals and organizations began to assert that environmentalism had gone too far. Business leaders and property owners complained of overregulation, and even some religious authorities expressed concerns that nature had been elevated in importance above humans. Nevertheless, the majority of Americans continued to feel that nature deserves to be protected. The late twentieth century saw a growing awareness of the importance of biodiversity and increasing support for the idea that all species should be protected against extinction. From this trend emerged the value-laden science of conservation biology.

Under the presidency of George W. Bush, the early twenty-first century was marked by business- and industry-friendly policies that diminished protections for wildlife and the environment. However, in the final month of his administration, Bush took the historic action of designating three areas of the Pacific Ocean as marine national monuments. This status protects the Marianas Trench, Pacific Remote Islands, and Rose Atoll marine national monuments from destruction or extraction of their resources, waste dumping, and commercial fishing.

In April, 2010, President Barack Obama held a national conservation conference that echoed one convened by President Theodore Roosevelt more than a century earlier. Dubbed America's Great Outdoors

Initiative, Obama's conference took into account the developed, urban, and populous nature of the country. The twenty-first century strategy for the nation that came out of the conference proposed partnerships among government, nonprofit organizations, and private landowners for land preservation efforts that would be smaller in scale than national parks and designated wilderness areas and thus less likely to encounter heated political opposition.

Thomas E. Hemmerly
Updated by Karen N. Kähler

FURTHER READING

Adams, Jonathan S. *The Future of the Wild: Radical Conservation for a Crowded World.* Boston: Beacon Press, 2006.

Carroll, Scott P., and Charles W. Fox, eds. *Conservation Biology: Evolution in Action.* New York: Oxford University Press, 2008.

Chiras, Daniel D., and John P. Reganold. *Natural Resource Conservation: Management for a Sustainable Future.* 10th ed. Upper Saddle River, N.J.: Benjamin Cummings/Pearson, 2010.

Cox, George W. *Conservation Biology: Concepts and Applications.* 2d ed. Dubuque, Iowa: Wm. C. Brown, 1997.

Groom, Martha J., Gary K. Meffe, and Carl Ronald Carroll. *Principles of Conservation Biology.* 3d ed. Sunderland, Mass.: Sinauer, 2006.

Hambler, Clive. *Conservation.* New York: Cambridge University Press, 2004.

Kline, Benjamin. *First Along the River: A Brief History of the U.S. Environmental Movement.* 3d ed. Lanham, Md.: Rowman & Littlefield, 2007.

Macdonald, David W., and Katrina Service, eds. *Key Topics in Conservation Biology.* Malden, Mass.: Blackwell, 2008.

Pullin, Andrew S. *Conservation Biology.* New York: Cambridge University Press, 2002.

SEE ALSO: Antienvironmentalism; Conservation biology; Conservation movement; Conservation policy; Environmental law, U.S.; Environmentalism; Leopold, Aldo; Multiple-use management; Pinchot, Gifford; Preservation; Public opinion and the environment; Renewable resources; Roosevelt, Theodore; Wise-use movement.

Conservation biology

CATEGORIES: Ecology and ecosystems; animals and endangered species
DEFINITION: Applied science concerned with the preservation of biological diversity at different levels
SIGNIFICANCE: Conservation biologists, who actively promote the protection of biodiversity, argue that the prevention of the further loss of any plant and animal species is crucial to the future of humankind.

The field of conservation biology emerged during the 1970's as many scientists became concerned about the loss of various species of plants and animals and the destruction of ecosystems, especially tropical forests, as the result of human activities. Conservation biologists are concerned with protecting biological diversity on three levels: species diversity, genetic diversity, and ecosystem diversity. The interdisciplinary science of conservation biology involves input from such varied fields of knowledge as genetics, biogeography, ecology, philosophy, and policy development. Conservation biologists conduct research and disseminate their findings through publications and other presentations; many also serve as advisers to policy makers, and some work directly with local groups toward conservation projects.

Conservation biologists generally focus on five major threats to biodiversity: habitat degradation and loss, habitat fragmentation, species overexploitation, species invasion, and climate change. It has been estimated that habitat degradation and habitat fragmentation account for more than 70 percent of species endangerment and extinction; these threats to biodiversity are caused by agriculture, mining, fires, urbanization, and other human activities. Efforts to preserve habitats have focused mainly on preserving areas with relatively high biodiversity.

Species overexploitation, or the unsustainable use of given species for food or other purposes, is believed to be the next most serious threat to biodiversity, followed by species invasion, or the introduction of non-native species to an ecosystem. Throughout history, humans have moved plant and animal species out of their native habitats and into other places, both intentionally and unintentionally. It is estimated that more than three thousand plant species have been introduced in California alone. The introduction of new species can have devastating effects on native species, such as by crowding them out or by changing the relationships between native predators and prey. Conservation biologists recognize that it is unrealistic to try to prevent species invasion, so they concentrate on minimizing the damage caused by species introduction.

The threat that global climate change poses to biodiversity was not fully recognized until the late twentieth century. It has been estimated that more than half of all the world's species are sensitive to climate change, with some more severely affected than others.

Conservation biologists and other environmentalists have undertaken efforts to protect biodiversity on several different levels, including the genetic, species, population, habitat, landscape, and ecosystem levels. Most conservation efforts in the past focused on protecting endangered species, but ecosystem-level efforts have increasingly been undertaken. Many of the world's national governments have passed laws to protect endangered or threatened species, such as the U.S. Endangered Species Act, and have developed plans to protect these species.

Other strategies for protecting biodiversity include the captive-breeding programs conducted by many zoos and the storage of seeds from wild plants in seed banks. Although such strategies are useful with some species, the conservation of other species requires the maintenance of their natural habitats. Conservation biologists and others have thus focused on preserving habitats, landscapes, and ecosystems. Many countries have set aside land in various types of reserves, which have proven to be very effective tools of conservation biology. Such reserves vary from those that are strictly protected (no human encroachment or use of the species within reserve boundaries) to those that allow sustainable exploitation of the resources—including plant and animal species—present.

No less important than the protection of habitats is the protection of species genetic diversity; without a diverse gene pool, a species population cannot evolve and adapt to environmental changes. One strategy for protecting genetic diversity involves the storage of species genes in gene banks.

Sergei A. Markov

FURTHER READING

Bush, Mark B. *Ecology of a Changing Planet*. Upper Saddle River, N.J.: Prentice Hall, 2000.

Chivian, Eric, and Aaron Bernstein, eds. *Sustaining Life: How Human Health Depends on Biodiversity*. New York: Oxford University Press, 2008.

Groom, Martha J., Gary K. Meffe, and Carl Ronald Carroll. *Principles of Conservation Biology*. 3d ed. Sunderland, Mass.: Sinauer, 2006.

Hunter, Malcolm L., Jr. *Fundamentals of Conservation Biology*. 3d ed. Hoboken, N.J.: Wiley-Blackwell, 2006.

Zeigler, David. *Understanding Biodiversity*. Westport, Conn.: Praeger, 2007.

SEE ALSO: Biodiversity; Convention on Biological Diversity; Endangered Species Act; Endangered species and species protection policy; Extinctions and species loss; Habitat destruction; Wildlife refuges; Wilson, Edward O.

Conservation easements

CATEGORY: Land and land use

DEFINITION: Legal agreements limiting the kinds of uses permitted on private lands

SIGNIFICANCE: Conservation easements encourage various forms of conservation on privately held lands while providing tax benefits to landowners.

An easement is a legal entitlement providing its holder certain privileges regarding the use of property owned by another. In the United States, a conservation easement is a legal agreement between a landowner and another party—generally a group such as a local or state land trust, a government agency, a nonprofit organization, or an environmental conservation group—for the purpose of limiting the uses of the property subject to the easement. The easement limits uses and development of the land that would disturb the condition of the aspect of the land that is valued. Some typical prohibitions concern the construction of roads and commercial buildings, mining of the land, and harvesting of its trees. Conservation easements usually last in perpetuity and are legally binding independent of who comes to own the land. The land is still usable by its owner, but the owner's use of it is limited by the terms of the easement.

Landowners can receive certain financial benefits for donating their land through conservation ease-

Brother Chris Balent, right, and Scott Ferguson of Trout Mountain Forestry walk through the forest at the Our Lady of Guadalupe Trappist Abbey in Carlton, Oregon. The monks have practiced sustainable forestry on their land for decades, and in 2010 they received millions for a conservation easement that guarantees their sustainable practices will continue. The easement was paid for by the Bonneville Power Administration, which was forced by the federal government to pay to conserve property in the Willamette basin to make up for having built dams that destroyed habitat. (©Ross William Hamilton/The Oregonian/LANDOV)

ments. Regarding income taxes, the donation of land constitutes a charitable gift and, as such, may be tax-deductible. Regarding estate taxes, lands that do not have easements may be subject to extremely high state and federal inheritance taxes. A conservation easement on the land reduces its value (since its economic uses are limited) and thus has a positive effect on the amount of inheritance taxes heirs must pay. Regarding gift taxes, again because an easement reduces the value of the land, more land may be given to a family member before it becomes subject to federal gift taxes. Regarding property taxes, having land protected by a conservation easement may qualify it for a reduction in property taxes (laws on this point vary from state to state).

Although conservation easements provide financial benefits, their intended purpose is to support conservation. The privately owned lands subject to conservation easements may serve as core habitats for species, or they may buffer and connect publicly owned lands such as nature reserves or national parks. These functions support the ecosystems found on the public lands. Buffer lands help provide cushions between lands that can be developed without restriction and lands that provide core habitats for flora and fauna. Lands that serve a connective function allow animals to move more freely from one publicly owned area to another, whether they are migrating, hunting, or establishing new territory. Further, much of the land that provides habitat for federally listed endangered species is privately owned.

Conservation easements have not been without controversy. One area of conflict concerns the abuse of such easements for tax-shelter purposes. Critics have noted that some owners receive the tax breaks that come with easements but do not maintain their lands properly in accordance with their easement agreements. Additionally, controversies have arisen regarding changes to the terms of easements by successive owners of particular lands. For example, the National Trust for Historic Preservation approved substantial changes to the terms of the Myrtle Grove easement, which covered land on Maryland's eastern shore in the Chesapeake Bay watershed, raising objections from many environmentalists.

George Wrisley

FURTHER READING

Gustanski, Julie A., and Roderick H. Squires, eds. *Protecting the Land: Conservation Easements Past, Present, and Future.* Washington, D.C.: Island Press, 2000.

Journal of Forestry. "Focus on . . . Private Forest Landowners: FAQs about Conservation Easements." 100, no. 3 (2002): 4.

Keske, Catherine. *The Emerging Market for Private Land Preservation and Conservation Easements.* Saarbrücken, Germany: VDM Verlag, 2008.

Rissman, Adena R., et al. "Conservation Easements: Biodiversity Protection and Private Use." *Conservation Biology* 21, no. 3 (2007): 709-718.

SEE ALSO: Conservation; Convention Relative to the Preservation of Fauna and Flora in Their Natural State; Marsh, George Perkins; Nature Conservancy; Nature preservation policy; Preservation.

Conservation movement

CATEGORIES: Resources and resource management; preservation and wilderness issues

IDENTIFICATION: Activities of individuals, private organizations, and government agencies to preserve natural resources or establish policies for using them wisely

SIGNIFICANCE: The early conservation movement focused attention on the perilous state of natural resources as human societies moved toward increasing commercialization and urbanization in the nineteenth and twentieth centuries. Proponents of conservation continue to work to convey the importance of resource sustainability.

The roots of the conservation movement can be traced to Europe during the late eighteenth century. Individuals repulsed by the growing commercialization and dehumanization brought on the by the Industrial Revolution promoted a return to nature. While some areas had been preserved in their natural state for centuries, either as hunting grounds or as refuges for the upper classes, during the nineteenth century governments in Europe and in North America began setting aside green space, even in urban areas, for recreation and relaxation. In some countries, notably Germany, government control over forests and rivers was used not only to conserve natural resources but also to generate revenues from the harvest of renewable resources such as trees and wildlife. Coordinated efforts to preserve wilderness areas, however, emerged first in the United States. The initiatives of individuals and organizations involved in these efforts grew into the modern conservation movement.

BEGINNINGS OF THE MOVEMENT

From its inception, the conservation movement in the United States was highly decentralized, but people who became involved shared an overarching goal: to preserve some aspect of the natural world from the ravages of rampant commercial development and urbanization. Efforts to save some of America's wilderness areas from the encroachment of civilization led to the creation of Yellowstone National Park in 1872, and some states had taken steps to set aside particularly beautiful areas (many of them difficult to reach). On the other hand, the first true conservation bill passed by the U.S. Congress in 1876 was vetoed by President Ulysses Grant.

Although American writers such as Henry David Thoreau had celebrated the value of the natural landscape, the first true conservationist to emerge on the national scene was John Muir. A Scotsman who had settled in California, Muir spent most of his adult life writing about the beauties of the wilderness, particularly the Yosemite Valley, a preserve set aside by the state in 1864. At that time many people looked upon the wilderness as a place of escape from the grime of city life. Powerful commercial interests, however, saw unsettled areas as opportunities for profit. The mining and timber industries coveted the natural resources there, farmers and ranchers sought lands for grazing, and railroads saw possibilities in creating tourist destinations or shorter routes between population centers.

Muir was one of dozens of amateurs who made their livings by other means but devoted their lives to saving the country's wilderness for future generations. Many were responsible for creating new national parks or monuments, saving sites that had been commercialized (such as Niagara Falls), or establishing wildlife sanctuaries to preserve dwindling numbers of birds, fish, and mammals. Muir lobbied to preserve Yosemite and additional wilderness sites. Eventually he found a strong ally in President Theodore Roosevelt, a lifelong sportsman and conservationist who used his political power to establish numerous national monuments to save lands from mining, timbering, and settlement. Muir was a preservationist. His goal was to have lands set aside, protected from all development or use.

A second important figure in the conservation movement, Gifford Pinchot, could best be described as a utilitarian. Pinchot earned a degree in forestry and eventually became head of the U.S. Forest Service, making him one of the first professional conservationists. He argued that natural resources should be managed carefully, but renewable resources should be available for commercial purposes. These two views came to dominate the debate among conservationists for nearly a century as groups struggling to preserve sites intact clashed with those who saw responsible use as a viable alternative to saving lands from outright, wholesale exploitation.

Over the next century the conservation movement would be advanced by both amateurs and professionals. A few crossed from one group to the other. Notable among them was Stephen T. Mather, who made a fortune in business and then turned his talents to creating the National Park Service in 1916, for which he served as the first director. Intent on preserving the national parks for future generations, Mather often found himself pitted against Pinchot and the Forest Service in debates concerning the proper balance between preservation and commercial use of national lands.

Spurred by Muir's leadership, in 1892 a group of Californians formed the Sierra Club, which would become a driving force in promoting conservation throughout the United States. Similar organizations sprang up in the East, many modeled on the first such club, the Appalachian Mountain Club, established in 1876.

Eventually the idea of conservation grew to include wildlife as well as landscape. Ironically, some of the most avid conservationists were sportsmen. They recognized that careful management of game populations would ensure that hunters could enjoy their sport for decades to come. Aware of what wanton slaughter had done to the American bison population during the nineteenth century, they often lobbied for limits on the numbers of animals that individual hunters could kill.

Conservationist groups were not always successful, however. Despite the protests of the Sierra Club and others, the U.S. government agreed to the creation of the Hetch Hetchy Dam in the Yosemite area to provide water for San Francisco, causing the loss of millions of hectares of natural landscape. Lobbying did push several presidents to create wildlife refuges, setting aside lands where game animals lived or that they used as migratory routes. Eventually these sanctuaries became part of the National Wildlife Refuge System.

BETWEEN THE WORLD WARS

By the end of World War I a number of conservation groups were active in the United States. Individual organizations, however, tended to focus on single aspects of conservation. For example, the Audubon Society, founded in 1905, limited its interest to the plight of birds, while the Izaak Walton League, established in 1922, was concerned with the condition of rivers and lakes because its members were involved in sportfishing. In fact, during this period many conservation organizations were underwritten by firearms companies and sports businesses. As a result, the utilitarian view of conservation tended to be favored in political decisions.

Concurrently, preservationists such as the National Park Service's Mather tried to have more lands set aside; on occasion these efforts were supported by financiers such as John D. Rockefeller, Jr., who purchased properties to expand or create parks. Some narrow-minded conservation efforts led to ecological disasters, however. For example, by the 1920's the gray wolf population in Yellowstone National Park was eliminated, ostensibly to make the park safer for tourists (and for the livestock of nearby ranchers). The absence of these predators led to an explosion of the ruminant population, which then overgrazed the available vegetation; eventually thousands of elk and deer starved.

The election of Franklin D. Roosevelt as president in 1932 brought one of the most active proponents of the conservationist movement into the White House. Roosevelt had a genuine love of the outdoors and was a conservationist at heart. He was a utilitarian when it came to conservation, however, and above all a politician. He believed that a government-controlled program of land retirement, soil and forest restoration, flood control, and hydroelectric power generation would serve two aims simultaneously: The nation's natural landscapes could be preserved or put to good use at the same time people could be given meaningful employment.

Combining his zeal for conservation with the need to revitalize the American economy during the Great Depression, Roosevelt created numerous programs to further the improvement of the nation's natural resources. Notable among these were the Civilian Conservation Corps and the Works Progress Administration, agencies that put people to work improving national parks and other wilderness areas, making them accessible to the public, and restoring many of the renewable resources that had been lost over the years to commercial development. Among his principal successes, Roosevelt oversaw the opening of Great Smoky Mountains National Park in 1934.

Not all of the projects Roosevelt promoted under the banner of conservation were beneficial to the natural environment, however. The Tennessee Valley Authority oversaw the erection of numerous dams to provide hydroelectric power to rural communities, in the process destroying millions of hectares of forestlands and valleys. Politically, the most significant change in the conservation movement under Roosevelt was the affiliation of environmental causes with the Democrats rather than the Republicans, who had been leaders in conservation in previous years.

CONSERVATION AFTER WORLD WAR II

After World War II ended, the United States entered a period of economic prosperity that saw the population grow and industry expand to meet worldwide demand for American goods. As a result, individuals and corporations interested in development lobbied for relaxation of conservation rules that had been put in place during the previous six decades. Many argued for a return to commercial operations in government-protected areas. Conservation groups, which by then numbered in the dozens, fought back with mixed success to maintain and expand the protections in place.

Organizations such as the Sierra Club and the Wilderness Society developed activist tactics to bring attention to the long-term dangers of development in wilderness areas. These groups were successful in defeating a plan to create a dam on the Green River at Echo Park, on the Utah-Colorado border, during the

1950's. Most notably, under the leadership of the Wilderness Society's executive secretary Howard Zahniser, conservationists prodded Congress to pass the Wilderness Act in 1964. That law set aside more than 3.6 million hectares (9 million acres) of land, restricting even minimal development or human use.

The publication of Rachel Carson's *Silent Spring* in 1962, detailing the environmental dangers of toxic waste, gave new impetus to the conservationists and sparked the larger environmental movement. As a consequence, during the 1960's conservation groups began to unite with those interested in larger environmental issues. Sierra Club executive director David Brower became a leading spokesman for environmental causes. The most significant success achieved by these groups in lobbying governmental agencies was the passage of the National Environmental Policy Act of 1969, which established the government's role in protecting humans and the natural environment from damage caused by human development or technology. In December, 1970, the Environmental Protection Agency was established to enforce environmental policies throughout the United States.

INTERNATIONAL CONSERVATION EFFORTS

By the middle of the twentieth century, the modern conservation movement had become international. The record of conservation groups in Europe was strong; many had been successful in continuing and strengthening already existing laws to preserve natural resources. Beginning during the 1960's conservationists in Australia led highly organized campaigns to reverse the effects of human contamination on the natural environment, specifically targeting problems with the quality of soils and the eradication of several species of animals. In Africa, Asia, and South America, however, conservationists had mixed success. Often conservation efforts conflicted with the needs of the people for resources to support life. This was especially true in Africa, where native peoples relied on wildlife for sustenance, and in South America, where lands covered by rain forests were needed for farming. Some successes were achieved when native populations were convinced that conservation would provide long-term economic stability.

At the same time, however, commercial interests continued to harvest animals and timber far in excess of the population's needs, ignoring the environmental catastrophes caused by their actions. Often, government-sponsored programs wreaked havoc on natural resources. Nowhere was this more apparent than in China, where the need for energy to fuel the country's growing population and expanding industrial base led to construction of the massive Three Gorges Dam on the Yangtze River, flooding millions of hectares, destroying or seriously affecting land and riverine populations. From 1970 onward, however, China began to take an active approach to wildlife preservation, as did several other Asian nations.

One of the most successful conservation initiatives undertaken outside the United States has been that of the Central American nation of Costa Rica. Beginning in 1970 the Costa Rican government worked to create a highly developed national park system where conservation practices would be strictly enforced. By 2000 nearly 25 percent of Costa Rica's landmass had been set aside for preservation.

Laurence W. Mazzeno

Further Reading

Chester, Charles A. *Conservation Across Borders: Biodiversity in an Interdependent World.* Washington, D.C.: Island Press, 2006.

Coggins, Chris. *The Tiger and the Pangolin: Nature, Culture, and Conservation in China.* Honolulu: University of Hawaii Press, 2003.

Evans, Sterling. *The Green Republic: A Conservation History of Costa Rica.* Austin: University of Texas Press, 1999.

Fox, Stephen. *John Muir and His Legacy: The American Conservation Movement.* Boston: Little, Brown, 1981.

Gibson, Clark L. *Politicians and Poachers: The Political Economy and Wildlife Policy in Africa.* New York: Cambridge University Press, 1999.

Maher, Neil M. *Nature's New Deal: The Civilian Conservation Corps and the Roots of the American Environmental Movement.* New York: Oxford University Press, 2008.

Markus, Nicola. *On Our Watch: The Race to Save Australia's Environment.* Melbourne: University of Melbourne Press, 2009.

Minteer, Ben and Robert Manning, eds. *Reconstructing Conservation: Finding Common Ground.* Washington, D.C.: Island Press, 2003.

Phillips, Sarah T. *This Land, This Nation: Conservation, Rural America, and the New Deal.* New York: Cambridge University Press, 2007.

Runte, Alfred. *National Parks: The American Experience.* 4th ed. Lanham, Md.: Taylor Trade, 2010.

See also: Conservation policy; International Union for Conservation of Nature; Mather, Stephen T.; Muir, John; National Audubon Society; Natural Resources Defense Council; Nature Conservancy; Pinchot, Gifford; Sierra Club; Wilderness Society.

Conservation policy

Category: Resources and resource management

Definition: Laws and regulations designed to limit the economic exploitation of natural resources in the public interest

Significance: Trends in conservation policy in the United States have varied in emphasis between allowing for limited exploitation of resources and prohibiting most economic uses of particular resources so that they can be preserved in an undeveloped state. Both approaches, however, have led to the protection and preservation of public lands and animal habitats.

The first use of the term "conservation" in relation to the natural environment was claimed by Progressive intellectuals in the early twentieth century. In his autobiography *Breaking New Ground* (1947), Gifford Pinchot, first chief of the U.S. Forest Service and a close friend of President Theodore Roosevelt, recalled realizing that all the natural resource problems are really one problem: the use of the earth for the permanent good of humans. The idea needed a name; presidential adviser Overton Price suggested "conservation," and the matter was settled.

Origins of Conservation Policy

The antecedents of modern conservation policies go back centuries. Aboriginal cultures around the world developed taboos governing behavior on the hunt. Venetians established reserves for deer and wild boar in the eighth century. Hunting reserves were common in Europe and Asia, and in colonial America those trees thought best for ships' masts were preserved for that purpose by decree.

Conservation policy, as the term is now understood, is a product of the nineteenth century, when population growth, urbanization, and industrialization created unprecedented opportunities for people to influence the natural world in which they lived, both for good and for ill. By the end of the nineteenth century, more Americans lived in cities than on farms. The nation was connected from coast to coast by telegraph and rail, and economic growth was rapid.

Most natural resource policies of the nineteenth century were designed to facilitate economic development. Best known among these policies was the Homestead Act of 1862, which gave free land to settlers, but similar policies provided free land to railroads and states. Other laws stimulated growth by providing for free use of timber and minerals on public lands.

The economic progress of the nineteenth century came at high cost to the environment, typified by the profligate use of forests and the near extermination of North American buffalo, and some prominent Americans took notice. In 1832 the artist and journalist George Catlin wrote of the probable extinction of the buffalo and American Indians, and he advocated a large national park where both might be preserved. Henry David Thoreau echoed Catlin's concerns in 1858, calling for national preserves. In 1864 George Perkins Marsh published *Man and Nature: Or, Physical Geography as Modified by Human Action*, the earliest important text with an ecological perspective.

As economic exploitation diminished the supply of natural resources, public attitudes began to change, and with them governmental policies. Although many laws continued to encourage economic growth, others reflected the growing desire for conservation. In 1864 the U.S. Congress sought to preserve the Yosemite Valley and Mariposa Big Tree Grove by giving them to the state of California for a public park. Eight years later Congress established the world's first national park at Yellowstone. In 1884 additional legislation prohibited all hunting and commercial fishing within Yellowstone National Park. In 1891 Congress established what would eventually become the national forest system when it authorized the president to set aside forest reserves on public lands.

PRESERVATION VERSUS WISE USE

These early conservation policies stressed resource preservation. Parks and forest reserves were simply set aside; none was effectively managed. The lack of management—especially forest management—displeased the advocates of scientific forestry, who also considered themselves conservationists. Early in its history, thus, the American conservation movement was divided, with some conservationists preaching "preservation" and others "wise use."

These contradictory tendencies were epitomized in the conflict between John Muir, founder of the Sierra Club, and Gifford Pinchot, principal architect and first chief of the U.S. Forest Service. Muir was the intellectual heir of Thoreau and Catlin. A perceptive scientist and popular author, he devoted his life to the exploration, enjoyment, and preservation of natural ecosystems worldwide. Pinchot had studied scientific forestry at its source in Germany. He was a gifted politician, and his passion was not for preservation but for wise use. Muir believed that people could not improve on nature; his conservation was aesthetic and spiritual. Pinchot was committed to maximizing the human benefits from resource use through science; his conservation was economic and utilitarian. Although friends for a time, Muir and Pinchot eventually parted ways, with Muir becoming an advocate for preservation and national parks and Pinchot an advocate for wise use and national forests.

In the United States the legacy of Pinchot is alive and well in the multiple-use management principles of the Forest Service and the Bureau of Land Management and in organizations such as the Society of American Foresters, the International Society of Fish and Wildlife Managers, the National Rifle Association, and the Soil Conservation Society of America. Muir's emphasis on preservation has been institutionalized in the National Park Service and the Fish and Wildlife Service as well as in organizations such as the National Audubon Society, the Nature Conservancy, the Sierra Club, and the Wilderness Society.

PROGRESSIVE ERA

Many historians emphasize three eras of American conservation policy corresponding roughly to the Progressive Era, the New Deal, and the so-called environmental decade of the 1970's. The Progressive Era, epitomized by the presidency of Theodore Roosevelt, was the first golden age of American conservation policy. During this period Congress passed a number of pathbreaking conservation laws. The Lacey Act of 1900 put the power of federal enforcement behind state game laws, criminalizing the interstate transport of wildlife killed or captured in violation of state regulations. Another milestone of wildlife conservation was the ratification of a migratory bird treaty with Canada and passage of a law to enforce the treaty. With the Migratory Bird Treaty Act of 1918, the federal government asserted national authority to manage wildlife for conservation purposes, authority that was upheld by the U.S. Supreme Court in the case of *Missouri v. Holland* (1920).

Two critically important governmental agencies were created during this era: the Forest Service and the Park Service. In 1905 advocates of wise use and scientific forestry were rewarded with a Forest Service in the Department of Agriculture. The new agency's first director was Gifford Pinchot, the nation's foremost advocate of multiple-use forest management based on scientific principles. Under Pinchot's leadership the concepts of multiple use and sustained yield were applied in the rapidly growing national forest system. When Theodore Roosevelt became president, the United States had 18.6 million hectares (46 million acres) of national forest. By the end of his term of office, Roosevelt had increased the total size of the national forest system to 78.5 million hectares (194 million acres).

During the Progressive Era, advocates of preservation were often unsuccessful in their opposition to the policies of wise-use conservationists, but in the end they also had a victory. The most painful loss came in Yosemite National Park, where advocates of wise use joined forces with the city of San Francisco to dam the Hetch Hetchy Valley for a municipal water supply, forever destroying a natural valley some regarded as comparable to Yosemite Valley itself. The public outcry over the damming of Hetch Hetchy contributed to pressure for better park protection, however, and in 1916 preservationists achieved a long-sought goal: creation of the National Park Service to manage the growing system of fourteen national parks, including Yellowstone (1872), Yosemite and Sequoia (1890), Mount Rainier (1899), Crater Lake (1902), Wind Cave (1903), Mesa Verde (1906), Glacier (1910), Rocky Mountain (1915), and Hawaii and Lassen Volcanic (1916).

Two new forms of conservation reserves made their debut during this era: national wildlife refuges and national monuments. Roosevelt regarded wildlife sanctuaries as critical to the survival of game species. In 1903 he acted on his belief, creating the nation's first national wildlife refuge on Pelican Island in Florida. He had no specific legal authority to create a national wildlife refuge, but his usurpation was accepted at the time and later approved in principle. In 1910 the Pickett Act authorized the president to set aside land for any public purpose. National monuments began on a firmer foundation, but here too Roosevelt pushed conservation to the limit. The An-

tiquities Act of 1906 authorized the president to establish national monuments. As the name suggests, the law anticipated relatively small reservations to protect archaeological sites, but the monuments Roosevelt designated included 34,400 hectares (85,000 acres) at Petrified Forest, 120,600 hectares (298,000 acres) at Mount Olympus, and 326,200 hectares (806,000 acres) at the Grand Canyon. All of these later became national parks.

New Deal Era

The New Deal was, for the most part, a response to disaster. The primary disaster was the Great Depression, but the decade of the 1930's also saw the Dust Bowl, a minor climatic change that produced disastrous results on the Great Plains. New Deal conservation policies were responsive to the economic and ecological crises of the era, and they stressed wise use through scientific management rather than preservation. The Tennessee Valley Authority was created in 1933 to stimulate employment and economic growth in Appalachia through scientific management of the area's natural resources. The Taylor Grazing Act of 1934 was designed to end overgrazing of western public lands by imposing a system of permits based on principles of scientific management. The Civilian Conservation Corps and the Soil Conservation Service were both established during this era, and each contributed to repairing environmental damage. Greater concern for the management rather than the disposal of western public lands was also reflected in the creation in 1946 of the Bureau of Land Management to replace the General Land Office.

Environmental Decade

The so-called environmental decade lasted almost twenty years. It began with the inauguration of President John F. Kennedy, persisted through the presidential administrations of Lyndon B. Johnson, Richard Nixon, Gerald Ford, and Jimmy Carter, and ended with the inauguration of President Ronald Reagan. The conservation policies of this era were responsive to post-World War II economic growth that seemed to ensure economic prosperity while threatening quality of life. Stewart Udall, U.S. secretary of the interior, warned of a *Quiet Crisis* (1963), Rachel Carson of a *Silent Spring* (1962), Barry Commoner of *The Closing Circle* (1971), and Paul R. Ehrlich of *The Population Bomb* (1968). Conservation policy matured into environmental policy during this era. Conserva-

(continued on page 304)

Milestones in Conservation Policy

Year	Event
1864	Yosemite Valley is ceded to California to create a park.
1872	The Yellowstone National Park Act establishes the world's first national park.
1891	The Forest Reserve Act authorizes the U.S. president to establish national forests.
1894	The Yellowstone Game Protection Act closes parks to hunting and commercial fishing.
1897	The Forest Management Act mandates that national forests be managed to perpetuate water supplies and wood products.
1900	The Lacey Act prohibits interstate shipment of wildlife that has been killed illegally.
1902	The Newlands Act establishes a national reclamation policy.
1903	The first National Wildlife Refuge is created at Pelican Island, Florida.
1905	The U.S. Forest Service is created within the Department of Agriculture to manage national forests.
1906	The Antiquities Act authorizes the creation of national monuments by presidential proclamation.
1910	The Pickett Act authorizes presidential land withdrawals for any public purpose.
1911	The Weeks Act provides for governmental purchase of national forestlands.
1913	The Hetch Hetchy Dam is authorized in Yosemite National Park.
1916	The National Park Service is created in the Interior Department to manage national parks.
1918	The Migratory Bird Treaty Act restricts the hunting of migratory birds.
1933	The Civilian Conservation Corps Act is passed.
1933	The Tennessee Valley Authority is created.
1934	The Taylor Grazing Act regulates grazing on public lands.
1937	The Federal Aid in Wildlife Restoration (Pittman- Robinson) Act provides federal aid to states for wildlife management.
1946	The U.S. Bureau of Land Management is created.
1950	The Federal Aid in Fish Restoration (Dingell-Johnson) Act provides federal aid to states for sport fish management.
1956	The Fish and Wildlife Act creates the U.S. Fish and Wildlife Service in the Interior Department.
1956	A proposal to construct Echo Park Dam in Dinosaur National Monument is defeated.
1960	The Multiple Use-Sustained Yield Act clarifies the purposes of national forests.
1964	The Wilderness Act establishes the National Wilderness Preservation System.
1964	The Land and Water Conservation Fund Act provides a trust fund for parkland acquisition.
1968	The National Wild and Scenic Rivers Act establishes a national river conservation system.
1968	The National Trails System Act establishes a national system of recreational trails.
1970	The National Environmental Policy Act requires environmental impact statements for federal activities that affect the environment.
1970	The Environmental Protection Agency (EPA) is created.
1970	Clean Air Act amendments establish stricter air-quality standards.

Milestones in Conservation Policy *(continued)*

Year	Event
1971	The United Nations Educational, Scientific, and Cultural Organization (UNESCO) Biosphere Reserve Program recognizes areas of global environmental significance.
1972	The Clean Water Act establishes stricter water-quality standards.
1972	The United Nations Environmental Conference in Stockholm, Sweden, is attended by 113 nations.
1972	Federal Water Pollution Control Act amendments provide protection for wetlands.
1972	The Federal Environmental Pesticides Control Act requires pesticide registration.
1972	The Marine Mammal Protection Act imposes a moratorium on hunting or harassing of marine mammals.
1973	The Convention on International Trade in Endangered Species of Wild Fauna and Flora (CITES) prohibits international trade in endangered species.
1973	The Endangered Species Act commits the United States to the preservation of biological diversity.
1974	The Safe Drinking Water Act sets federal standards for public water supplies.
1976	The Toxic Substances Control Act authorizes the EPA to ban substances that threaten human health or the environment.
1976	The Federal Land Policy and Management Act directs the Bureau of Land Management to retain public lands and manage them for multiple uses.
1976	The Resource Conservation and Recovery Act directs the EPA to regulate waste production, storage, and transportation.
1976	The National Forest Management Act gives statutory protection to national forests and sets standards for management.
1977	The Surface Mining Control and Reclamation Act establishes environmental standards for strip mining.
1977	Clean Air Act amendments set high standards for air quality in large national parks and wilderness areas.
1980	The Fish and Wildlife Conservation Act provides federal aid for the protection of nongame wildlife.
1980	The Alaska National Interest Lands Conservation Act establishes more than 40.5 million hectares (100 million acres) of national parks and wildlife refuges in Alaska.
1980	The Comprehensive Environmental Response, Compensation, and Liability Act (CERCLA) establishes the Superfund hazardous waste cleanup program.
1982	The Nuclear Waste Policy Act establishes a process for siting a permanent nuclear waste repository.
1985	The U.S. government establishes the Conservation Reserve Program (CRP) to reduce agricultural surpluses by encouraging farmers to reduce the amounts of land they devote to crops, thereby helping to prevent soil erosion and reduce carbon in the atmosphere.
1987	The Montreal Protocol limits the production and consumption of chlorofluorocarbons (CFCs).
1988	The Ocean Dumping Act prohibits the dumping of sewage sludge and industrial waste.
1990	Clean Air Act amendments strengthen the Clean Air Act.
1992	The Earth Summit in Rio de Janeiro, Brazil, is attended by 179 nations.
1997	The Kyoto Protocol on Climate Change encourages global reduction in greenhouse gas emissions.
2001	The Roadless Area Conservation Rule, a federal policy initiative designed to protect national forests from commercial development, is issued.

(continued on page 304)

Milestones in Conservation Policy (continued)

YEAR	EVENT
2005	More than 4,000 hectares (approximately 10,000 acres) of Puerto Rican rain-forest land are added to the U.S. national forest system.
2008	The Food, Conservation, and Energy Act extends CRP enrollment authority through 2012.
2009	The Omnibus Public Land Management Act adds 850,000 hectares (2.1 million acres) of new wilderness areas in nine U.S. states.

tion was still about husbanding natural resources, but to the historic concerns of conservation—such as forests, wilderness, and wildlife—were added concerns regarding clean water and clean air, energy supplies, and the problems posed by hazardous and toxic wastes. During this era Congress passed most of the major laws that continue to shape conservation policy at the beginning of the twenty-first century.

Preservation policy was strengthened. In 1964 Congress passed the Wilderness Act and the Land and Water Conservation Fund Act. The former established the National Wilderness Preservation System, which has grown from more than 3.6 million hectares (9 million acres) to more than 40.5 million hectares (100 million acres). The latter facilitated acquisition of land for parks and open space. Four years later Congress established a national system of trails as well as a national system to protect wild and scenic rivers from certain kinds of development. Both the Forest Service and the Bureau of Land Management were given new statutory direction emphasizing planning and preservation, sometimes at the expense of economic development. At the end of the era Congress passed the Alaska National Interest Lands Conservation Act (1980), making conservation withdrawals of more than 40.5 million hectares of public lands and doubling the size of the national park and wildlife refuge systems nationwide.

Wise-use conservation was also well served as Congress radically increased federal regulation of resource use. President Johnson established a presidential commission on natural beauty and addressed world population and resource scarcity in his 1965 state of the union speech. Environmental management was nationalized through a series of far-reaching statutes addressing air pollution, water pollution, marine resources, noise pollution, biological diversity, toxic chemicals, and hazardous waste. New burdens were placed on government and private citizens. The National Environmental Policy Act of 1969 required all governmental agencies to study the probable environmental effects of their actions before moving forward. A large number of environmental enforcement programs were reorganized in 1970 into the newly created Environmental Protection Agency (EPA). The EPA is an independent agency, but presidents have routinely regarded its director as having cabinet status.

CONSERVATION POLICY FOR A NEW MILLENNIUM

At the policy level, the era following the environmental decade was one of consolidation rather than new initiatives. The Reagan administration was hostile to environmental policy and attempted to tilt public policy toward less environmental regulation. President Reagan was able to prevent the adoption of major new conservation policies, but his administrative efforts—led by Interior Secretary James Watt—to roll back environmental laws were successfully resisted by Congress. A major new air-pollution statute was passed during the presidential administration of George H. W. Bush, but this was exceptional for the era. Bill Clinton's presidential administration gave greater attention to conservation policy, but in the years following the 1994 elections there was little cooperation between the president and Congress on environmental issues.

Americans have come to expect government to practice conservation and protect environmental quality. Doing so is increasingly difficult. Beyond the policy gridlock of the 1980's and 1990's, the issues themselves have become more difficult. The most pressing issues—such as stratospheric ozone depletion, climate change, and biological diversity—are beset by scientific uncertainty. They are also global in scope and thus beyond the ability of any single nation to address independently. The future of conservation policy appears to be in the international arena, where extant institutions lack the authority to govern. Treaties addressing the use of chlorinated fluorocarbons, the preservation of biological diversity, and the limitation of greenhouse gas emissions demonstrate that nations are giving increasing attention to conservation issues, but international achievements remain modest.

Craig W. Allin

FURTHER READING

Allin, Craig W. *The Politics of Wilderness Preservation.* 1982. Reprint. Fairbanks: University of Alaska Press, 2008.

Chiras, Daniel D., and John P. Reganold. *Natural Resource Conservation: Management for a Sustainable Future.* 10th ed. Upper Saddle River, N.J.: Benjamin Cummings/Pearson, 2010.

Davis, David Howard. *American Environmental Politics.* Belmont, Calif.: Wadsworth, 1998.

Dowie, Mark. *Conservation Refugees: The Hundred-Year Conflict Between Global Conservation and Native Peoples.* Cambridge, Mass.: MIT Press, 2009.

French, Hilary. *Vanishing Borders: Protecting the Planet in the Age of Globalization.* New York: W. W. Norton, 2000.

Hayes, Samuel P. *Conservation and the Gospel of Efficiency: The Progressive Conservation Movement, 1890-1920.* 1959. Reprint. Pittsburgh: University of Pittsburgh Press, 1999.

Nash, Roderick. *Wilderness and the American Mind.* 4th ed. New Haven, Conn.: Yale University Press, 2001.

Rosenbaum, Walter A. *Environmental Politics and Policy.* 7th ed. Washington, D.C.: CQ Press, 2008.

SEE ALSO: Conservation; Forest Service, U.S.; Muir, John; National forests; Nature reserves; Pinchot, Gifford; Roosevelt, Theodore; Watt, James; Wilderness Act; Wise-use movement.

Conservation Reserve Program

CATEGORIES: Agriculture and food; land and land use

IDENTIFICATION: Cropland reduction and conservation program of the U.S. Department of Agriculture

DATE: Created on December 23, 1985

SIGNIFICANCE: Initially established to reduce agricultural surpluses by encouraging farmers to reduce the amounts of land they devoted to crops, the Conservation Reserve Program has grown to become an environmental protection program as well, preventing soil erosion and reducing carbon in the atmosphere.

The history of agriculture in the United States reveals a pattern of crop production increasing faster than demand, resulting in downward pressure on crop prices and farm income. Improvements in equipment, chemicals, and seed genetics have increased yields. In 1982, farmers produced a surplus of 230 million metric tons (254 million U.S. tons) of crops, swamping the available storage capacity. The government response, announced by President Ronald Reagan on January 11, 1983, was a payment-in-kind (PIK) program. Under this program, farmers were given certificates for crops in storage in return for taking land out of production.

The PIK program was generous, with wheat farmers getting up to 95 percent of the normal crop. Farmers responded by removing 33.3 million hectares (82.3 million acres) of cropland from the 1983 growing season. This steep reduction from the 170 million hectares (420 million acres) planted in 1982 had a devastating economic impact on small towns dependent on agriculture. Farmers bought less fuel and fertilizer, and they did not hire seasonal workers. The resulting stress on rural communities made it clear to policy makers that wide swings in the amounts of land planted were unacceptable. The U.S. Congress responded to the overproduction issue with the Conservation Reserve Program (CRP), which was created under Title XII of the 1985 Food Security Act.

The twin goals of CRP were to decrease crop surpluses through reduction in land area planted and to prevent cropland erosion. Enrollment was voluntary, with farmers submitting bids for the amount of payment required to remove land from production for ten to fifteen years. Participants were paid a rental fee plus half the cost of establishing a permanent cover of trees or grasses. Enrollment increased rapidly, with approximately 13.8 million hectares (34 million acres) enrolled during the first nine sign-up periods from 1986 through 1989.

Over time Congress enhanced the environmental protection aspects of CRP. The Food, Agriculture, Conservation, and Trade Act of 1990 extended the enrollment period through 1995 and added water-quality protection as a criterion for land selected. The Federal Agriculture Improvement and Reform Act of 1996 included an Environmental Benefits Index for selection of land suitable for enrollment.

The environmental impact of CRP is substantial—ground-cover plants on CRP lands annually remove an estimated 17 million metric tons (18.7 million U.S. tons) of carbon from the atmosphere and reduce soil erosion by more than 443 million metric tons (488 million U.S. tons). These lands also provide critical habitat for wildlife, including upland game birds, grassland songbirds, and prairie mammals. CRP lands are particularly important to migratory waterfowl. Duck species that had historically nested in the Prairie Pothole region of the northern plains were in serious trouble during the 1980's, with populations near their lowest level in the preceding fifty years. Habitat provided by CRP increased the ducks' nesting success, helping the population rebound from 25.6 million breeding ducks in 1985 to 42 million in 2009.

CRP continues as an important environmental and agricultural program. The Food, Conservation, and Energy Act of 2008 extended CRP enrollment authority to September 30, 2012. It set the enrollment authority at 15.8 million hectares (39 million acres) through 2009 but reduced enrollments to 13 million hectares (32 million acres) for fiscal years 2010 to 2012. In 2009 there were 14 million hectares (34.7 million acres) enrolled at an annual rental cost of $1.76 billion per year, an average cost of about $125 per hectare ($50 per acre) per year. With considerable support for CRP from agricultural producers, organizations concerned with waterfowl, and the environmental community, the program is likely to continue to be included in future farm legislation.

Allan Jenkins

FURTHER READING

Hamilton, James T. *Conserving Data in the Conservation Reserve.* Washington, D.C.: RFF Press, 2010.

Napier, Ted, Silvana M. Napier, and Jiri Tvrdon, eds. *Soil and Water Conservation Policies and Programs: Successes and Failures.* Boca Raton, Fla.: CRC Press, 2000.

SEE ALSO: Carbon dioxide; Conservation easements; Erosion and erosion control; Soil conservation; Sustainable agriculture.

Continental shelves

CATEGORIES: Ecology and ecosystems; water and water pollution

DEFINITION: Nearly flat platforms of land that extend into the seas at the margins of continents

SIGNIFICANCE: Human activities have had many negative environmental impacts on the world's continental shelves, both directly and indirectly. These impacts include pollution of the waters of the shelves' ecosystems through runoff from land, the direct dumping of wastes, and oil extraction; the construction of artificial reefs and breakwaters; and the dredging of sand and gravel.

Continental shelves constitute the parts of the oceans most utilized by humans, but much of the environmental degradation that the shelves have undergone has resulted indirectly from human activities on land. The waters of the continental shelves in many parts of the world have become polluted by chemical fertilizers, industrial wastes, and sewage released into the seas from rivers and storm drains. Such pollution threatens the plants and animals present in coastal ecosystems.

During the early 1960's, for example, the brown pelican, a seabird species native to the coasts of the Americas, almost became extinct owing to coastal pollution. Scientists found that the brown pelican population was dying out because the birds were laying eggs with extremely thin shells that would break in the nest, so few hatchlings survived. The cause was determined to be the presence of the pesticide dichloro-diphenyl-trichloroethane (DDT) in coastal waters, which came from agricultural runoff that was deposited in the Gulf of Mexico by the Mississippi River. DDT had accumulated in the bodies of the fish on which the pelicans fed. After the dangers of DDT were recognized, use of the pesticide was eventually banned in the United States.

Chemical pollution of coastal waters remains a problem, however. In some areas, pollution of the continental shelves by agricultural runoff containing fertilizers has led to the creation of hypoxic (oxygen-deprived) areas known as dead zones. The nitrogen and phosphorus in fertilizers are nutrients that cause algae to grow at accelerated rates. When the algae die and sink to the bottom, their decay depletes the dissolved oxygen in the water, and the fish and other marine life die. A seasonal hypoxic zone the size of the state of New Jersey forms during the summer months at the mouth of the Mississippi River in the Gulf of Mexico, and similar dead zones are found on continental shelves around the world.

Coral reefs, many of which develop along the edges of continental shelves, have been increasingly damaged by human activities, both indirectly and directly. Chemical pollution is thought to contribute to the discoloration of these reefs known as bleaching and white-band disease, and serious damage to reefs can result from the dragging of anchors across them and from shipwrecks. Other damaging human activities include the use of harmful chemicals and even dynamite to drive reef organisms out of their hiding places and the mining of reefs to collect coral for ornamental purposes. Policy makers and legislators have responded to increasing recognition among scientists that the ecosystems of coral reefs need to be protected. In the United States, for instance, the National Marine Sanctuaries Program protects the reefs of the Florida Keys as well as thirteen other pristine marine sites around North America, Hawaii, and American Samoa.

DUMPING OF WASTES

Various kinds of environmental damage to the continental shelves have resulted directly from human actions. For example, humans have intentionally changed coastal seabeds—and thus destroyed marine habitats and disrupted ecosystems—by building breakwaters, pilings for bridges, artificial reefs, and wind farms; in addition, in some cases, these areas have been used for the disposal of construction debris, municipal solid waste, and outdated military hardware.

An extreme example of the dumping of waste on a continental shelf is provided by New York City, which has always faced difficulty in the disposal of construction debris. Manhattan Island is only 21.6 kilometers (13.4 miles) long and 3.7 kilometers (2.3 miles) wide, and it represents some of the most expensive real estate in the world. In 1890 builders in Manhattan began the practice of dumping construction debris and other waste in the New York Bight, an indentation in the coastline between New York State and New Jersey.

Researchers with the U.S. National Oceanic and Atmospheric Administration survey a coral reef in the Florida Keys National Marine Sanctuary near Key West, Florida. Pollution, overfishing, and climate change are among the causes of damage to such reefs, many of which are found along the edges of continental shelves. (AP/Wide World Photos)

The mound formed on the continental shelf in this area was soon so large that it was jokingly called an underwater Mount Everest. Amendments made in 1986 to the Marine Protection, Research, and Sanctuaries Act of 1972 (also known as the Ocean Dumping Act) prohibited dumping in this area, and a new dump site was opened 170 kilometers (106 miles) offshore. Since that time, an even larger mound of waste materials has been accumulating on the seabed at a depth of 2,500 meters (8,200 feet).

The dumping of obsolete military hardware in coastal waters creates special problems because such hardware may contain materials that are hazardous. Two crewmen on a clam boat off the shore of Massachusetts had to be hospitalized in June, 2010, for example, after they hauled up World War I-era military shells filled with mustard gas along with a load of clams. It turned out that the U.S. military had been using the area as a dumping ground for munitions from after World War II through 1970.

Dredging and Oil Spills

In addition to building and dumping things on continental shelves, humans also take things out. Sand is routinely dredged from the continental shelf off southern Florida, for example, to "nourish" beaches that have lost their sand due to erosion, and some cities that are heavily dependent on tourist revenue, such as Delray Beach and Palm Beach, have had their beaches rebuilt several times. In countries that are highly industrialized and densely populated, continental shelves may be mined for the sand and gravel required for construction. It has been estimated that 25 percent of the sand and gravel needed for construction in southeastern England comes from offshore; the proportion is believed to be even higher for Japan. In addition to exploiting continental shelves for sand and gravel, humans also mine deposits of valuable minerals, such as diamonds and gold, on some of the world's continental shelves.

Oil spills represent another way in which humans degrade the shelves. Production platforms and drilling rigs jut up offshore in areas where oil and gas deposits are found, and the oil that can leak or spill from these rigs has the potential to foul the ocean and seabed for hundreds of miles around. Two examples of disastrous spills that resulted from such offshore oil extraction have taken place in the Gulf of Mexico. In 1979 a blowout in an exploratory well drilled by Petróleos Mexicanos (PEMEX), the Mexican state-owned oil company, in 50 meters (164 feet) of water off the coast of Mexico caused a spillage of oil that fouled the seabed and beaches as far away as Texas. An even greater tragedy was the blowout of BP's *Deepwater Horizon* rig off the coast of Louisiana in 2010. Heavy reddish-brown oil from this spill fouled the continental shelf and beaches as far away as Florida, even though the well was drilled 72 kilometers (45 miles) offshore in water 1,524 meters (5,000 feet) deep.

Donald W. Lovejoy

Further Reading

Ketchum, Bostwick H., ed. *The Water's Edge: Critical Problems of the Coastal Zone*. Cambridge, Mass.: MIT Press, 1972.

Porter, James W., and Karen G. Porter, eds. *The Everglades, Florida Bay, and Coral Reefs of the Florida Keys: An Ecosystem Sourcebook*. Boca Raton, Fla.: CRC Press, 2002.

Sverdrup, Keith A., and E. Virginia Armbrust. *An Introduction to the World's Oceans*. New York: McGraw-Hill, 2009.

Viles, Heather A., and Tom Spencer. *Coastal Problems: Geomorphology, Ecology, and Society at the Coast*. London: Arnold, 1995.

Walsh, John J. *On the Nature of the Continental Shelves*. San Diego, Calif.: Academic Press, 1988.

See also: BP *Deepwater Horizon* oil spill; Coastal Zone Management Act; Dead zones; Dredging; Landfills; Oil spills; PEMEX oil well leak; Runoff, agricultural; Seabed mining.

Contingent valuation

Categories: Land and land use; resources and resource management

Definition: Survey method used to estimate the value of nonmarket resources

Significance: Contingent valuation surveys are widely used to estimate the value of nonmarket resources. They are often used in cost-benefit assessment and environmental impact assessment.

Contingent valuation is used to assign value to nonmarket resources, such as renewable energy, open space, and sustainable development. While these resources provide utility, certain components of each do not have market prices; for example, renewable energy may reduce human-caused climate change or preserve fossil fuels for future generations. A contingent valuation survey is used to estimate a market price as a stated preference. The fundamental mechanism of the contingent valuation survey is asking people about their willingness to pay (WTP) to maintain an environmental feature or their willingness to accept (WTA) compensation for its loss.

Agricultural economist Siegfried von Ciriacy-Wantrup suggested the use of a direct interview method to measure the value of natural resources as early as 1947. Perhaps the first practical application was completed during the 1960's by economist Robert K. Davis, who measured the value of a specific wilderness area to hunters and recreationalists. Davis's contingent valuation results compared well with inferred value from cost associated with traveling to the wilderness area.

The method gained popularity during the 1970's in the United States as it was granted official recognition. Large numbers of studies were completed during the 1980's, with applications expanding to Europe and developing countries. However, criticism of the method also multiplied. Twenty-two expert economists on a panel convened in 1993 concluded that contingent valuation surveys must be carefully designed and controlled to ensure that valid results are obtained. They noted that individuals and organizations planning to employ contingent valuation should carefully review best practices before applying the method.

The panel offered specific recommendations concerning how contingent valuation surveys should be conducted, including the following. If possible, the survey interviews should be conducted in person; telephone surveys may be acceptable, but mail surveys should be avoided. A referendum format should be used in the questions; for example, "Would you be willing to contribute (or be taxed) D dollars to cover the cost of avoiding or repairing environmental damage X?" The results obtained from questions of this kind are considered to be more accurate than those gleaned from answers to open-ended questions, which are more likely to elicit strategic behavior, pro-

test responses, biased answers, and incomplete consideration of personal income limits. The interviewers should ensure that respondents understand and accept the scenario they are asked to value. Respondents who do not accept the accuracy of the information concerning a particular scenario are in fact answering a question that is different from the one asked. Respondents should be reminded that their WTP for the specific scenario will reduce their ability to pay for other private or public goods.

Jess W. Everett

FURTHER READING

Ahmed, S. U., and K. Gotoh. *Cost-Benefit Analysis of Environmental Good by Applying the Contingent Valuation Method: Some Japanese Case Studies.* New York: Springer, 2006.

Arrow, Kenneth, et al. "Report of the NOAA Panel on Contingent Valuation." *Federal Register* 58 (January 15, 1993): 4601-4614.

Bowman, Troy, Jan Thompson, and Joe Colletti. "Valuation of Open Space and Conservation Features in Residential Subdivisions." *Journal of Environmental Management* 90 (January, 2009): 321-330.

Davis, Robert K. "Recreation Planning as an Economic Problem." *Natural Resources Journal* 3 (October, 1963): 239-249.

Mitchell, Robert Cameron, and Richard T. Carson. *Using Surveys to Value Public Goods: The Contingent Valuation Method.* 1989. Reprint. Washington, D.C.: Resources for the Future, 1993.

Venkatachalam, L. "The Contingent Valuation Method: A Review." *Environmental Impact Assessment Review* 24 (January, 2004): 89-124.

Wiser, Ryan H. "Using Contingent Valuation to Explore Willingness to Pay for Renewable Energy: A Comparison of Collective and Voluntary Payment Vehicles." *Ecological Economics* 62 (May, 2007): 419-432.

SEE ALSO: Accounting for nature; Benefit-cost analysis; Ecological economics; Environmental economics; Environmental impact assessments and statements; Risk assessment.

Controlled burning

CATEGORIES: Resources and resource management; forests and plants; agriculture and food

DEFINITION: Intentional setting of a fire to accomplish a specific purpose

SIGNIFICANCE: The controlled use of fire can benefit forests by preventing larger, uncontrollable wildfires and by removing undesirable plants, promoting certain animal species, and stimulating the germination of some seeds; controlled burns are also used as a means of fighting wildfires. The controlled burning used in agriculture, however, is often accompanied by negative environmental impacts such as air pollution and eventual soil depletion.

Controlled burning is one tool that foresters use to help contain wildfires. There are three ways to extinguish a fire: Cool the fuel below its kindling temperature, deprive the fire of oxygen, or deprive the fire of fuel. Pouring water on a fire addresses the first two of these; the setting of a controlled burn known as a backfire is one way to accomplish the third. If the wind is blowing toward a raging wildfire, firefighters may set a line of fire in front of the wildfire. When correctly controlled, the backfire consumes the available fuel before the wildfire can reach that area; the wildfire then dies out for lack of fuel.

Controlled burning is also used as a way of preventing large, unmanageable wildfires. In forests and other wilderness areas, lightning strikes start many fires every year, and nature has adapted to relatively frequent burning. Fires promote new growth and renewal of a forest in several ways. For example, cones from the giant sequoia and the lodgepole pine do not release their seeds unless they have been dried or heated, and heating is more efficient. Fires clear the ground of brush, leaves, pine needles, and dead wood, allowing seeds to sprout in the ground, which is moist and enriched by nutrients leached from the ashes. (Forest floor litter dries out more quickly than the soil underneath, and seedlings that sprout in the litter generally die.) In addition, fire burns away plants that would have blocked sunlight from seedlings.

If too much time elapses between fires, however, fuel can accumulate to dangerous levels on the forest floor. When a fire starts after a long period of fuel accumulation, it is likely to burn so hot that standing trees will be consumed, and in a strong wind, flames will jump from the crown of one tree to the crown of the next. Crown fires race quickly through a forest and are very difficult to extinguish. Foresters have come to understand that it can be far less destructive to a forest to use controlled burns to keep the fuel load at a safe level.

Other uses of controlled burning are found in agriculture. After the last ice age, about 11,000 years ago,

A worker with the U.S. Bureau of Land Management sets fire to brush during a controlled burn at the Bitter Lake National Wildlife Refuge in New Mexico, where such fires are used to reduce potential fuel for wildfires, improve animal habitats, and assist in the life cycles of certain plants. (AP/Wide World Photos)

hunter-gatherers settled down to farming and adopted the slash-and-burn technique for clearing land. Large trees were cut down, and smaller trees and underbrush were burned; this both cleared the land and released nutrients into the soil. After a few years, when the land began producing less, a new area was selected and the cycle repeated. This practice continues in some parts of the world in the twenty-first century, notably in the Amazon rain forest and in Southeast Asia. In the United States, some 10 percent of Oregon grass-seed farmers use fire to kill competing weeds, weed seeds, insects, and rodents because burning is cheaper than other methods. Oklahoma wheat farmers burn wheat stubble for similar reasons.

Range managers in the Wichita Mountains Wildlife Refuge have adopted the controlled burning practice known as patch burning. They set controlled fires in areas of rangeland from 40 hectares (100 acres) to 8,100 hectares (20,000 acres) in size every few years to provide the bison in the refuge with the nutritious, tender grass shoots that grow in the burned areas. Fuel is allowed to accumulate in unburned areas, which are then burned to repeat the cycle.

Charles W. Rogers

FURTHER READING

Carle, David. *Introduction to Fire in California.* Berkeley: University of California Press, 2008.

Reinhart, Karen Wildung. *Yellowstone's Rebirth by Fire: Rising from the Ashes of the 1988 Wildfires.* Helena, Mont.: Farcountry Press, 2008.

Ribe, Tom. *Inferno by Committee: A History of the Cerro Grande (Los Alamos) Fire, America's Worst Prescribed Fire Disaster.* Victoria, B.C.: Trafford, 2010.

SEE ALSO: Forest management; Land clearance; Slash-and-burn agriculture; Wildfires.

Convention Concerning the Protection of the World Cultural and National Heritage. *See* World Heritage Convention

Convention on Biological Diversity

CATEGORIES: Treaties, laws, and court cases; ecology and ecosystems; resources and resource management

THE CONVENTION: International agreement that provides for the conservation of biological diversity, the sustainable use of biological resources, and the equitable distribution of the benefits of such resources

DATE: Opened for signature on June 5, 1992

SIGNIFICANCE: The Convention on Biological Diversity is regarded as the first comprehensive international agreement to address the conservation and use of biological diversity across all levels of biological organization—from genetic resources to ecosystem services.

The Convention on Biological Diversity is one of several historically important documents produced at the United Nations Conference on Environment and Development (UNCED), also known as the Earth Summit, held in Rio de Janeiro, Brazil, in June of 1992. The convention, which entered into force on December 29, 1993, is closely linked to the principles of sustainable development and represents an international commitment to the conservation of biological diversity, its sustainable use, and the fair distribution of benefits arising from the utilization of genetic resources. The convention formally recognizes that the conservation of biological diversity is a common concern, an integral part of the development process, and the responsibility of all sovereign states.

Signatory nations—that is, the parties to the convention—are required to develop integrated national biodiversity strategies and action plans. Other key provisions require the establishment of protected areas for the conservation of biological diversity, the restoration of degraded ecosystems, the development of programs to promote the recovery of threatened species in collaboration with local residents, and the development of programs to control alien species and the risks posed by genetically modified organisms. The parties are obliged to report on their implementation of the provisions in the convention.

Several articles of the convention deserve special mention. Article 13 establishes the importance of public education and awareness and requires the parties to promote and encourage public understanding of the importance of conserving biological diversity

Objectives of the Convention on Biological Diversity

Article 1 of the Convention on Biological Diversity sets out the convention's objectives, which guide all of its specific provisions and define the fundamental obligations of its signatories.

The objectives of this Convention, to be pursued in accordance with its relevant provisions, are the conservation of biological diversity, the sustainable use of its components and the fair and equitable sharing of the benefits arising out of the utilization of genetic resources, including by appropriate access to genetic resources and by appropriate transfer of relevant technologies, taking into account all rights over those resources and to technologies, and by appropriate funding.

while cooperating with other states and international organizations to develop educational and public-awareness programs. Article 15 addresses access to genetic resources and biopiracy by affirming state sovereignty over natural resources and prohibiting the exploitation of genetic resources without prior informed consent. Articles 20 stipulates that developed countries that are parties to the convention will provide financial resources to help developing country parties implement measures to fulfill their obligations under the convention; Article 21 establishes a mechanism for distributing that financial aid. Article 25 establishes the Subsidiary Body on Scientific, Technical, and Technological Advice, which provides the Conference of the Parties (COP) with advice relating to the implementation of the convention.

The COP, the governing body of the convention, implements the provisions of the convention through the decisions it makes at its meetings. The first COP was held in Nassau, Bahamas, in 1994. Since the third COP in 1996, ordinary meetings have been held biennially. A two-part extraordinary meeting in 1999-2000 produced the Cartagena Protocol on Biosafety, a supplement to the convention that regulates the international trade, handling, and use of genetically modified organisms that pose a potential threat to the conservation and sustainable use of biological diversity.

Brian G. Wolff

FURTHER READING

Andersen, Regine. "The Convention on Biological Diversity." In *Governing Agrobiodiversity: Plant Genetics*

and Developing Countries. Burlington, Vt.: Ashgate, 2008.

Rosendal, Kristin G. *The Convention on Biological Diversity and Developing Countries.* Norwell, Mass.: Kluwer Academic, 2000.

United Nations. Secretariat of the Convention on Biological Diversity. *Handbook of the Convention on Biological Diversity Including Its Cartagena Protocol on Biosafety.* 3d ed. Montreal: Author, 2005.

SEE ALSO: Agricultural chemicals; Balance of nature; Biodiversity; Biopiracy and bioprospecting; Biosphere reserves; Earth Summit; Sustainable development; United Nations Environment Programme.

Convention on International Trade in Endangered Species

CATEGORIES: Treaties, laws, and court cases; animals and endangered species

THE CONVENTION: International agreement aimed at conserving endangered animal and plant species

DATE: Opened for signature on March 3, 1973

SIGNIFICANCE: The Convention on International Trade in Endangered Species was the first international agreement concerning the conservation of wildlife that constituted a legal commitment by the parties to the convention and also included a means of enforcing its provisions.

Until the 1970's, international agreements that had been made to address the preservation of species did not include any binding legal commitment on the part of the countries signing them; thus they were ineffectual in protecting the species they were written to protect. In 1969, however, the United States passed the Endangered Species Conservation Act, which contained a provision that gave the secretaries of interior and commerce until June 30, 1971, to call for an international conference on endangered species. The resulting conference, which was held in Washington, D.C., in March, 1973, resulted in the Convention on International Trade in Endangered Species of Wild Fauna and Flora (CITES).

The United States was the first country to ratify the convention, which entered into effect on July 1, 1975; by 2010, 175 nations in total were parties to the agreement. CITES is intended to conserve species and does this by managing international trade in those species. It was the first international convention on the conservation of wildlife that constituted a legal commitment by the parties to the convention and also included a means of enforcing its provisions. This enforcement includes a system of trade sanctions and an international reporting network to stop trade in endangered species. The system established by CITES does, however, contain loopholes through which states with special interest in particular species can opt out of the global control for those species.

A major feature of CITES is its categorization of three levels of vulnerability of species. Appendix I includes all species that are threatened with extinction and whose status may be affected by international trade. Appendix II includes species that are not yet threatened but might become endangered if trade in them is not regulated. It also includes other species that, if traded, might affect the vulnerability of the first group. Appendix III lists species that a signatory party identifies as subject to regulation in order to restrict exploitation of that species. The parties to the treaty agree not to allow any trade in the species on the three lists unless an exception is allowed in CITES.

The species listed in the appendixes may be moved from one list to another as their vulnerability increases or decreases. According to the convention, states may implement stricter measures of conservation than those specified in the convention or may ban trade in species not included in the appendixes. CITES also establishes a series of import and export trade permits within each of the categories. Each nation designates a management authority and a scientific authority to implement CITES. Exceptions to the ban on trade are made for scientific and museum specimens, exhibitions, and movement of species under permit by a national management authority.

The parties to CITES maintain records of trade in specimens of species that are listed in the appendixes and prepare periodic reports on their compliance with the convention. These reports are sent to the CITES secretariat in Switzerland, administered by the United Nations Environment Programme (UNEP), which issues notifications to all parties of state actions and bans. The secretariat's functions are established by the convention and include interpreting the provisions of CITES and advising countries on implementing those provisions by providing assistance in writing their national legislation and organizing training

seminars. The secretariat also studies the status of species being traded in order to ensure that the exploitation of such species is within sustainable limits.

The CITES Conference of Parties meets every two or three years in order to review implementation of the convention. The meetings are also attended by nonparty states, intergovernmental agencies of the United Nations, and nongovernmental organizations considered "technically qualified in protection, conservation or management of wild fauna and flora." The meetings are held in different signatories' countries: The first took place in Berne, Switzerland, on November 2-6, 1976. At the conference, the parties may adopt amendments to the convention and make recommendations to improve the effectiveness of CITES.

CITES has been incorporated into Caring for the Earth: A Strategy for Sustainable Living. This strategy was launched in 1991 by UNEP, the International Union for Conservation of Nature (IUCN), and the World Wide Fund for Nature (WWF). Other nongovernmental groups working to support CITES are Fauna and Flora International (FFI), Trade Records Analysis of Flora and Fauna in Commerce (TRAFFIC International), and the World Conservation Monitoring Centre (WCMC).

Some of the species protected by CITES have received additional protection under later agreements. In certain cases, however, states have allowed trade in listed species to continue for economic purposes or have refused to sign CITES because of the extent to which they trade in a species or species part, such as ivory. Others have signed because they needed help in stopping illegal trade and poaching of species within their borders. Whales have proven to be a difficult species to protect. Whales are given protection

Species Protected by CITES

SPECIES TYPE	SPECIES	SUBSPECIES	POPULATIONS
Mammals	617	36	26
Birds	1,455	17	3
Reptiles	657	9	10
Amphibians	114	0	0
Fish	86	0	0
Invertebrates	2,179	5	0
Total fauna	5,108	67	39
Plants	28,977	7	3
Total	34,085	74	42

under CITES according to the status of specific species. The moratorium on commercial whaling instituted in 1986 by the International Whaling Commission (IWC) was intended to strengthen the CITES protection by species, but the whaling states have disagreed on the numbers of whale populations, and some have withdrawn from the IWC and resumed their whaling activities.

Colleen M. Driscoll

FURTHER READING

Chasek, Pamela S., et al. *Global Environmental Politics.* 4th ed. Boulder, Colo.: Westview Press, 2006.

Hutton, Jon, and Barnabas Dickson, eds. *Endangered Species, Threatened Convention: The Past, Present, and Future of CITES.* Sterling, Va.: Earthscan, 2000.

International Union for Conservation of Nature. *Conserving the World's Biological Diversity.* Washington, D.C.: Island Press, 1990.

Van Dyke, Fred. "The Legal Foundations of Conservation Biology." In *Conservation Biology: Foundations, Concepts, Applications.* 2d ed. New York: Springer, 2008.

SEE ALSO: Endangered Species Act; Endangered species and species protection policy; International Convention for the Regulation of Whaling; International whaling ban; Pets and the pet trade; Poaching; Whaling.

Convention on Long-Range Transboundary Air Pollution

CATEGORIES: Treaties, laws, and court cases; atmosphere and air pollution

THE CONVENTION: International agreement to limit air pollution, including pollution created in one country that affects the environment in another

DATE: Opened for signature on November 13, 1979

SIGNIFICANCE: Although the Convention on Long-Range Transboundary Air Pollution has had little direct effect on air quality, it is important because it was the first agreement among nations of Eastern Europe, Western Europe, and North America regarding the environment.

Until the 1970's most local, regional, and national regulations regarding industrial air pollution were concerned only with pollution generated in the immediate area. For example, regulations in a particular community might call for taller industrial smokestacks to carry pollution farther away, but there was little official concern about where that pollution might eventually return to earth. Similarly, local assessments and treatments of pollution tended not to consider pollution that might come to an area from distant generators. The only exceptions were a small number of treaties between two countries, such as between the United States and Canada or between Germany and France.

In 1972 the United Nations Conference on the Human Environment, held in Stockholm, Sweden, drew attention to the harmful effects of acid rain, including damage to forests, crops, surface water, and buildings and monuments, especially in Europe. Data revealed that while all European nations were producing alarming levels of air pollution, several nations were receiving more pollution from beyond their borders than they were generating on their own. It became clear that pollution is both imported and exported, that sulfur and nitrogen compounds can travel through the air for thousands of miles, and that any serious attempt to deal with air pollution must reach beyond political boundaries. Two major studies of the long-range transport of air pollutants (LRTAP), conducted under United Nations sponsorship in 1972 and 1977, conclusively proved that air pollution is an international—even a global—problem caused primarily by fossil-fuel combustion and harming both industrial and nonindustrial nations around the world.

In 1979 the United Nations Environment Programme organized a convention in Geneva, Switzerland, for the thirty-four member countries of the United Nations Economic Commission for Europe (UNECE), a group that includes all European nations, the United States, and Canada. Significantly, the gathering had the participation of Eastern European nations under the Soviet Union, marking the first time these nations had collaborated with Western Europe to solve an international environmental problem. The Convention on Long-Range Transboundary Air Pollution was signed by thirty-two nations on November 13, 1979, and went into effect on March 16, 1983. It called upon signatory nations to limit and eventually reduce air pollution, in particular sulfur emissions, using the best and most economically feasible technology; to share scientific and technical information regarding air pollution and its reduction; to permit transboundary monitoring; and to collabo-

Fundamental Principles of the Convention on Long-Range Transboundary Air Pollution

Articles 2 through 5 of the Convention on Long-Range Transboundary Air Pollution set out the fundamental principles of the agreement, by which all contracting parties agree to be bound.

ARTICLE 2:

The Contracting Parties, taking due account of the facts and problems involved, are determined to protect man and his environment against air pollution and shall endeavour to limit and, as far as possible, gradually reduce and prevent air pollution including long-range transboundary air pollution.

ARTICLE 3:

The Contracting Parties, within the framework of the present Convention, shall by means of exchanges of information, consultation, research and monitoring, develop without undue delay policies and strategies which shall serve as a means of combating the discharge of air pollutants, taking into account efforts already made at national and international levels.

ARTICLE 4:

The Contracting Parties shall exchange information on and review their policies, scientific activities and technical measures aimed at combating, as far as possible, the discharge of air pollutants which may have adverse effects, thereby contributing to the reduction of air pollution including long-range transboundary air pollution.

ARTICLE 5:

Consultations shall be held, upon request, at an early stage between, on the one hand, Contracting Parties which are actually affected by or exposed to a significant risk of long-range transboundary air pollution and, on the other hand, Contracting Parties within which and subject to whose jurisdiction a significant contribution to long-range transboundary air pollution originates, or could originate, in connection with activities carried on or contemplated therein.

rate in developing new antipollution policies. Under the terms of the convention, an international panel would undertake a comprehensive review every four years to determine whether goals were being met, and an executive body would meet each year.

The convention did not include any specific plan for the reduction of air pollution. It did not contain any language calling for particular amounts by which emissions would be reduced, nor did it include a schedule by which the reductions would occur. Scandinavian nations, which were among the countries most affected by acid rain, urged the other participants to adopt these kinds of policies, but other countries, led by the United States, the United Kingdom, and West Germany, defeated the proposal.

In the years after the convention went into force, however, several nations did make commitments to reduce emissions by specific amounts, including West Germany, which changed its position as further information was revealed about deforestation caused by acid rain. At the 1983 executive body meeting, eight nations, including Canada, West Germany, and the Scandinavian countries, made a formal commitment to reduce their emissions by 30 percent by 1993, using 1980 levels as the baseline. Over the next two years thirteen more nations announced similar goals, and in 1985 the commitment to a 30 percent reduction was formally adopted as an amendment to the convention that was signed by nineteen nations.

Neither the United States nor the United Kingdom agreed to the 30 percent reductions, and neither country signed the 1985 protocol. The United Kingdom informally agreed to attempt to reduce emissions by 30 percent but was unwilling to commit the financial resources to guarantee the reduction, especially since the benefits were uncertain. In fact, many scientists felt that a 30 percent reduction would not be enough to yield significant improvement. The United States argued that it had already taken major steps to reduce its emissions prior to 1980, so using 1980 data as a baseline would subject the United States to unrealistic and unfair demands for further reduction. This refusal to ratify the protocol caused tension between the United States and Canada, because much of the air pollution that affects eastern Canada comes from the Great Lakes industrial belt in the United States.

Cynthia A. Bily

FURTHER READING

Brunnee, Jutta. *Acid Rain and Ozone Layer Depletion: International Law and Regulation.* Boston: Hotei, 1988.

Elsom, Derek M. *Atmospheric Pollution: Causes, Effects, and Control Policies.* Malden, Mass.: Blackwell, 1987.

Fishman, Jack. *Global Alert: The Ozone Pollution Crisis.* New York: Plenum, 1990.

Sand, Peter H. "Air Pollution in Europe: International Policy Responses." *Environment* 29 (December 1987): 16-20, 28-29.

Visgilio, Gerald R., and Diana M. Whitelaw, eds. *Acid in the Environment: Lessons Learned and Future Prospects.* New York: Springer, 2007.

Wetstone, G. S., and A. Rosencranz. *Acid Rain in Europe and North America: National Responses to an International Problem.* Washington, D.C.: Environmental Law Institute, 1983.

SEE ALSO: Acid deposition and acid rain; Air pollution; Air-pollution policy; United Nations Conference on the Human Environment; United Nations Environment Programme.

Convention on the Conservation of Migratory Species of Wild Animals. *See* Bonn Convention on the Conservation of Migratory Species of Wild Animals

Convention on the Prevention of Marine Pollution. *See* London Convention on the Prevention of Marine Pollution

Convention on Wetlands of International Importance. *See* Ramsar Convention on Wetlands of International Importance

Convention Relative to the Preservation of Fauna and Flora in Their Natural State

CATEGORIES: Treaties, laws, and court cases; preservation and wilderness issues; animals and endangered species

THE CONVENTION: International agreement that established preservation policies for European colonies in Africa

DATE: Opened for signature on December 8, 1933

SIGNIFICANCE: The Convention Relative to the Preservation of Fauna and Flora in Their Natural State was among the first international agreements concerned with issues of conservation, although the signatory nations were concerned primarily with protecting animals from extinction so that they would remain available for game hunting.

In 1900 the European nations that had recently divided sub-Saharan Africa among themselves and established colonial governments there signed the first international conservation treaty, the Convention for the Preservation of Animals, Birds, and Fish in Africa. The men who drafted this document did not recognize any inherent value in living creatures—they were not protecting animals because they felt animals had a right to live. The intention of the treaty was to preserve the populations of animals that were popular trophies for hunters, such as elephants and giraffes, and encourage the eradication of animals harmful to agriculture, including lions, leopards, and wild dogs.

In 1930 a surveying expedition sponsored by the British Society for the Protection of the Fauna of the Empire made it clear that the 1900 treaty was ineffective from a conservation standpoint. Elephants and other animals were still being overhunted, and several animal and plant species were drawing closer to extinction. It was proposed that an expanded system of national parks be established in East and Central Africa to protect species without substantially limiting human activity. The national parks would be under the control of the colonial governments. The public would be encouraged to visit the national parks to observe the plants and animals, but no "hunting, killing, or capturing" would be permitted within park boundaries.

Several nations that held substantial amounts of land in Africa met in London in 1933 to discuss these issues. The resulting Convention Relative to the Preservation of Fauna and Flora in Their Natural State was signed by South Africa, Belgium, the United Kingdom, Egypt, Spain, France, Italy, Portugal, and the Sudan. It established national parks for public enjoyment and "strict natural reserves" for the exclusive use of scientists. One plant species and twenty animals—including gorillas, white rhinoceroses, and shoebill storks—were fully protected by the treaty, which entered into force on January 14, 1936. New rules for hunters outside the parks forbade the use of cars and aircraft to chase or herd animals and also prohibited poison and traps.

However, neither the treaty nor the discussions leading up to it considered the role black Africans might play in preserving or endangering the fauna and flora. The treaty was made by Europeans to ensure that white people would have enough animals to hunt. Much of the land newly dedicated to national parks had been home to Africans who were now forbidden to hunt, farm, or live on that land. Animals were protected or not protected according to their usefulness for or danger to white hunters and settlers, without consideration of which animals provided food or presented a danger to native villagers.

Cynthia A. Bily

FURTHER READING

Louka, Elli. *International Environmental Law: Fairness, Effectiveness, and World Order.* New York : Cambridge University Press, 2006.

Suich, Helen, and Brian Child, with Anna Spenceley, eds. *Evolution and Innovation in Wildlife Conservation: Parks and Game Ranches to Transfrontier Conservation Areas.* Sterling, Va.: Earthscan, 2009.

SEE ALSO: Africa; Convention on International Trade in Endangered Species; Elephants, African; Endangered species and species protection policy; National parks.

Coral reefs

CATEGORY: Ecology and ecosystems

DEFINITION: Stony formations created by the depositing of exoskeletons by colonies of coral polyps

SIGNIFICANCE: Coral reefs are some of the most diverse ecosystems on the planet, supporting a wide variety of marine life, including fish, mollusks, and sponges. Coral reefs are being threatened by a process known as coral bleaching, which kills the coral and destroys the reefs, thus also threatening all of the marine species dependent on the reefs for survival.

Corals are living organisms, and approximately one thousand different species of coral are known to exist, some of which live in colonies. These colonies of coral polyps deposit their exoskeletons of calcium carbonate as they grow, and thus provide the durable formations that the polyps, along with other species of coral and many other animals, live on and around; these stony formations become coral reefs. Coral reefs are found in the oceans, most often in warm, shallow waters in the Tropics, particularly in the Pacific Ocean.

Reef-building corals need to be in shallow water in order to get energy from the sunlight that filters down through the water column. The corals themselves do not photosynthesize sunlight; rather, they have a symbiotic relationship with protozoa called zooxanthellae. These organisms live within the tissues of the corals and provide them with the by-products of pho-

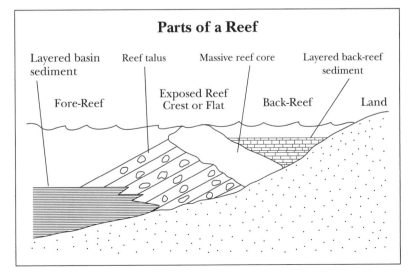

Parts of a Reef

Layered basin sediment   Reef talus   Massive reef core   Layered back-reef sediment

Fore-Reef   Exposed Reef Crest or Flat   Back-Reef   Land

tosynthesis, such as glucose and amino acids, that the corals use for energy. Reef-building corals are thus almost wholly dependent on the zooxanthellae and must live in shallow waters where the sunlight can reach them.

Coral reefs cover a total area equivalent to just 1 percent of the surface area of the oceans, yet they support one-fourth of marine life. For organisms such as phytoplankton and algae, which use corals for shelter and for the raw materials they need to produce energy, and larger animals such as fish and sea snakes, coral reefs offer sources of nutrients in addition to shelter. The diversity and sheer numbers of marine life make up a complex food web, with the corals and their symbionts at the center. It is because of their nutrient richness that most coral reefs are so important to so many different species of marine life. Given the variety of animals that live in reef communities, any threat to the corals is a threat to the entire ecosystem, and coral reefs are very fragile owing to their susceptibility to changes in water temperature and acidity.

Some significant threats to coral reefs come directly from human activities, such as the destructive fishing practices of overfishing, bottom trawling, and blast fishing. Coral bleaching, however, is the biggest threat facing reef ecosystems; bleaching events have been occurring with growing frequency across the globe since the 1980's. The term "coral bleaching" refers to the whitening of corals when the protozoa lose their pigmentation or the corals expel their symbionts in reaction to some form of outside stress, such as pollution. The main causes of coral bleaching are related to climate change: increases in water temperature and increased acidity of the oceans. Coral reefs continue to undergo bleaching even if the source of the stress is removed, and if the colonies survive the bleaching, the zooxanthellae do not immediately reappear, sometimes taking months to come back.

The prospect of higher sea levels associated with climate change is also a threat to coral reefs, as corals need to be in shallow waters with a lot of sunlight to survive, and corals cannot grow fast enough to keep up with predicted sea-level rises. More than half of the world's coral reefs are threatened directly or indirectly by human activity, a situation that has led environmentalists increasingly to promote public awareness of the economic value of the reefs and the ecosystem services they offer, such as shoreline protection and well-stocked fisheries.

Megan E. Watson

FURTHER READING

Allsopp, Michelle, et al. *State of the World's Oceans*. New York: Springer, 2009.

Sapp, Jan. *What Is Natural? Coral Reef Crisis*. New York: Oxford University Press, 1999.

Veron, John E. N. *A Reef in Time: The Great Barrier Reef from Beginning to End*. Cambridge, Mass.: Belknap Press of Harvard University Press, 2008.

SEE ALSO: Biodiversity; Climate change and oceans; Ecosystem services; Ecosystems; Food chains; Ocean dumping; Ocean pollution; Sea-level changes; United Nations Convention on the Law of the Sea.

Corporate average fuel economy standards

CATEGORIES: Energy and energy use; resources and resource management

DEFINITION: Federal standards designed to improve automobile fuel efficiency in the United States

SIGNIFICANCE: Vehicle fuel economy standards were imposed in the United States during the 1970's in an effort to reduce fuel consumption, but increased driving and use of larger vehicles led to higher per-capita consumption over the next three decades. During the early twenty-first century, however, rising gasoline prices revived interest in improving the standards.

In response to the energy crisis of the 1970's, the U.S. Congress enacted legislation intended to reduce American dependence on oil imports. The 1975 corporate average fuel economy (CAFE) standards mandated fuel-efficiency levels for automobile manufacturers. Each manufacturer's annual automobile output had to meet the assigned average for that model year. If a fleet exceeded the CAFE standards, the manufacturer faced a substantial fine. To ensure that manufacturers did not import fuel-efficient foreign cars to offset low averages in their domestic output, the legislation required that import and domestic fleets be evaluated separately.

The 1978 standard for passenger cars was 18 miles per gallon (mpg). Averages gradually increased over the years (with the exception of a rollback during the late 1980's, prompted by a petition from automakers). By the 1990's the passenger car standard had reached 27.5 mpg, while the standard for light trucks (which included vans and sport utility vehicles, or SUVs) was 20.7 mpg. Between 1996 and 2001, provisions in the Department of Transportation's appropriations bills prohibited changes in the standards.

During the late 1990's many environmentalists pressed for more stringent CAFE standards. They argued that improved fuel economy would reduce the introduction of greenhouse gases and other harmful automobile by-products into the atmosphere. In addition, it would serve to protect U.S. wilderness areas where the threat of oil drilling remained a possibility. Proponents of more stringent standards also pointed to the increased use of minivans and SUVs. When CAFE standards first were imposed, light-truck-class vehicles were used predominantly for business and agricultural purposes, and lower standards were set for light trucks in order to protect small businesses and farmers. SUVs, vans, and trucks weighing between 8,500 and 10,000 pounds were not subject to any fuel economy standards at all. During the 1990's, minivans and SUVs proliferated as means of personal transportation, meaning more fuel-hungry vehicles on the road and a rise in per-capita fuel consumption.

Critics argued that CAFE standards had no impact on foreign oil imports, which continued to rise after the regulations were enacted. After a period of concern during the Persian Gulf War in 1991, most consumers regarded the continuing decline in gasoline prices that marked the mid-1990's as evidence that the issue was not critical. CAFE opponents also contended that the standards hurt the U.S. automobile industry and the country's economy in general. They claimed that the costs of manufacturing vehicles with increased fuel efficiency were passed on to consumers in the form of higher vehicle prices that unfairly affected not only individuals but also small businesses. More important, critics asserted, manufacturers achieved better performance by building smaller cars from lighter materials—vehicles that provided less protection to passengers in the event of accidents. This argument gained ground in 1991 when the U.S. Department of Transportation released a study indicating that higher CAFE standards were directly related to increases in traffic injuries and fatalities. (Later studies, however, suggested that the quality of a vehicle's engineering plays a greater role than its mass in determining how it fares in an accident.)

Environmentalists countered that higher vehicle costs were offset by savings in fuel expenses. As for the increase in traffic deaths, they maintained, automobile exhaust causes environmental damage that also results in deaths. They pointed to global warming and its consequences as proof that CAFE standards were necessary. However, as long as gasoline prices remained low, these arguments made little headway with most politicians and automobile manufacturers.

In 2003, with gas prices on the rise, new light-truck standards were issued: 21.0 mpg for model year (MY) 2005, 21.6 mpg for MY 2006, and 22.2 mpg for MY 2007. By 2007, per-capita fuel consumption in the United States had exceeded pre-1975 levels. That year, lawmakers increased CAFE standards to a combined level of efficiency for new cars and light trucks of 35 mpg by 2020. The legislation also subjected vehicles weighing between 8,500 and 10,000 pounds to CAFE standards for the first time, effective in 2011. In 2009, following a spike in gas prices the previous year, President Barack Obama's administration accelerated the timetable, calling for a fleetwide average of 35.5 mpg (39 mpg for cars, 30 mpg for trucks) by 2016 in conjunction with a 30 percent reduction in tailpipe greenhouse gas emissions.

Thomas Clarkin
Updated by Karen N. Kähler

FURTHER READING

An, Feng, et al. *Passenger Vehicle Greenhouse Gas and Fuel Economy Standards: A Global Update.* Washington, D.C.: International Council on Clean Transportation, 2007.

National Research Council. *Effectiveness and Impact of Corporate Average Fuel Economy (CAFE) Standards.* Washington, D.C.: National Academies Press, 2002.

Nivola, Pietro S. *The Long and Winding Road: Automotive Fuel Economy and American Politics.* Washington, D.C.: Brookings Institution Press, 2009.

SEE ALSO: Automobile emissions; Energy conservation; Fossil fuels; Gasoline and gasoline additives; Hybrid vehicles; Oil crises and oil embargoes.

Cost-benefit analysis. *See* **Benefit-cost analysis**

Côte d'Ivoire toxic waste incident

CATEGORIES: Disasters; waste and waste management

THE EVENT: Illegal dumping of toxic petroleum wastes in and around the city of Abidjan in the West African nation of Côte d'Ivoire

DATE: August, 2006

SIGNIFICANCE: One of the worst cases of intentional dumping of dangerous toxic wastes, the Côte d'Ivoire incident demonstrates how developing nations can be exploited by unscrupulous corporations based in developed nations.

In August, 2006, at least 400 tons of toxic waste products derived from the processing of petroleum were illegally dumped at night in sites in and around Abidjan, a port city that is the former capital of Côte d'Ivoire (Ivory Coast). The wastes came from the tanker *Probo Koala*, which was registered in Panama. Approximately 100,000 residents of the Abidjan region were adversely affected by the dumping. Local hospitals reported that at least 30,000 people sought medical attention, and 17 died from exposure to the toxic wastes. Hundreds of contaminated livestock around Abidjan had to be euthanized. Health problems stemming from the environmental contamination have continued among locals. Persistent physical complaints of those exposed include ailments from breathing difficulties and choking to body lesions, blisters, nosebleeds, headaches, and nervous system paralysis.

Trafigura, a transnational trading company based in Great Britain, Switzerland, and the Netherlands, had chartered the ship but denied any wrongdoing for several years after the incident, claiming that only "slops" (mere dirty water from the washing of the ship's tanks) had been dumped. Trafigura sued Brit-

A man holds up the head of a plastic doll at one of the sites where toxic waste was dumped in Côte d'Ivoire in 2006, as he and others look for material among the waste that they can sell. (AP/Wide World Photos)

ish newspapers for libel when they reported the company's involvement in the dumping and worked to restrain the British news media with gag orders. Although the Ivorian government prosecuted locals who had been bribed, some of the legal issues surrounding the case have remained unresolved.

Dutch chemical-cleanup teams that visited twelve of at least eighteen affected Abidjan waste sites verified that the tanker "slops" were, in fact, toxic. The dumped materials contained dangerous concentrations of hydrogen sulfide, a killer gas, several tons of which were found; among other caustic agents found were mercaptans (also known as thiols) and gasoline. Port records show that the ship had visited a port in Amsterdam to offload the wastes, but the company did not want to pay the high fee—$300,000— required to cover the expensive treatment and disposal of the waste. Instead a front company, named Compagnie Tommy, was immediately created in Abidjan. Local Ivorian personnel of the front company were apparently bribed to take the waste and surreptitiously dump it in poor urban regions.

In the months following the incident, as national and international investigations began, many Ivorian government officials resigned. Even as lawyers for Trafigura tried to sue British news media companies for libel and worked to prevent the publishing of further news stories about the incident, an e-mail trail was exposed that showed the company's involvement in cover-ups and influence peddling in high places, including among British lords who served on Trafigura boards.

Patrick Norman Hunt

SEE ALSO: Africa; Basel Convention on the Control of Transboundary Movements of Hazardous Wastes; Environmental justice and environmental racism; Hazardous and toxic substance regulation; Hazardous waste; *Khian Sea* incident; Ocean dumping.

Council on Environmental Quality

CATEGORY: Organizations and agencies
IDENTIFICATION: Division of the executive office of the president of the United States responsible for coordinating federal environmental efforts and aiding in developing environmental policies and initiatives
DATE: Established in 1970
SIGNIFICANCE: The Council on Environmental Quality is a major player in the U.S. government's environmental efforts. As the overseer of the National Environmental Policy Act and adviser to the president, the council plays an integral part in shaping and directing the federal government's environmental policies.

The Council on Environmental Quality (CEQ) was established as part of the National Environmental Policy Act of 1969 (NEPA), sweeping legislation that requires U.S. federal agencies to consider the environmental impacts that could arise from any actions they plan to take before they act. Under NEPA, federal agencies must undertake environmental impact assessments whereby the environmental consequences of any planned actions are analyzed and alternatives are considered. The CEQ is charged with overseeing this general process and ensuring that federal agencies fulfill their duties under the act. Additionally, the CEQ acts as a mediator in cases in which an agency disagrees regarding the adequacy of another's assessment.

The CEQ also works to advance the president's agenda and to advise on issues related to the environment. The council's chair is chosen by the president and serves as the president's chief environmental policy adviser. As the environmental arm of the White House, the CEQ works with different agencies, scientists, and the public to balance environmental, economic, and social agendas. The council is also directed to collect and analyze the latest information related to the environment. As part of this role, the CEQ assists the president in preparing the Environmental Quality Report, an overview of the nation's environmental status submitted annually to Congress.

Daniel J. Connell

SEE ALSO: Environmental impact assessments and statements; Environmental law, U.S.; National Environmental Policy Act.

Cousteau, Jacques

CATEGORIES: Activism and advocacy; animals and
endangered species

IDENTIFICATION: French explorer, conservationist,
and filmmaker

BORN: June 11, 1910; Saint-André-de-Cubzac, France

DIED: June 25, 1997; Paris, France

SIGNIFICANCE: Cousteau, one of the twentieth century's best-known explorers and conservationists,
gained widespread attention for environmental issues, particularly those concerning the world's
oceans.

Jacques Cousteau was, in his own mind, neither scientist nor adventurer. Rather, he considered himself a filmmaker. In 1971 Cousteau said, "I am not a
scientist, I am rather an impresario of scientists."
Cousteau, however, probably did recognize the vital
role he played in exciting public interest in the ocean
and revealing the intricacies of humanity's relationship to it.

Explorer and conservationist Jacques Cousteau. (Library of Congress)

Cousteau's long career began in 1933 with a stint in
the French navy. In 1943, during his military service,
he developed, with Émile Gagnan, the Aqua-Lung,
the first commercially available self-contained underwater breathing apparatus (scuba). In 1950 Cousteau
bought the retired minesweeper *Calypso* for use as a
research ship. The many films and books Cousteau
authored in his lifetime helped finance the *Calypso*'s
expeditions, which brought the wonders of the
oceans to an international audience. He is perhaps
best known to many Americans from his documen-
tary television series *The Undersea World of Jacques
Cousteau*, which aired from 1966 to 1976.

By 1956 Cousteau's research activities had become
a full-time career, so he resigned from the navy. In
1957 Prince Rainier III named Cousteau director of
the Oceanographic Museum of Monaco, a post he
held for thirty-one years. During the 1950's and
1960's Cousteau sometimes explored the same sites
several times. As a result of these trips, he eventually
recognized that human activities were degrading the
aquatic environment.

In 1958 Cousteau helped establish the world's
first undersea marine reserve off the coast of Monaco. In 1960 he spoke out against France's plans
to dispose of nuclear waste in the Mediterranean
Sea. One decade later, he summed up his fears
about ocean pollution by stating, "The oceans are
in danger of dying."

In 1974 Cousteau founded the Cousteau Society, a nonprofit organization "dedicated to the
protection and improvement of the quality of life
for present and future generations." Among other
achievements, this group has provided logistical
support and facilities for hundreds of scientists,
helped develop postgraduate environmental science and policy programs at universities worldwide, and worked toward the adoption by the
United Nations of a "Bill of Rights for Future Generations"—all causes to which Cousteau himself
was strongly committed. In addition to revealing
the ocean's majesty through books, films, and television, Cousteau strove to alert the public to the
dilemma he recognized during his explorations:

The pursuit of technology and progress may today
endanger the very survival of . . . practically all life
on earth. . . . [However] the technology that we use
to abuse the planet is the same technology that can
help us to heal it.

Cousteau received numerous awards and honors for his environmental efforts, including the United Nations International Environmental Prize (1977) and a place on the United Nations Global 500 Roll of Honor (1988). He served as adviser for environmental sustainable development to the World Bank and was a member of the United Nations High-Level Board on Sustainable Development (1992). In 1992 France's president appointed Cousteau chairman of the newly created Council on the Rights of Future Generations. Three years later he resigned from the council in protest over the resumption of nuclear weapons testing in the South Pacific.

Clayton D. Harris

FURTHER READING

Kroll, Gary. *America's Ocean Wilderness: A Cultural History of Twentieth-Century Exploration.* Lawrence: University Press of Kansas, 2008.

Matsen, Brad. *Jacques Cousteau: The Sea King.* New York: Pantheon Books, 2009.

SEE ALSO: Environmental education; Ocean dumping; Ocean pollution; Sustainable development.

Cross-Florida Barge Canal

CATEGORY: Preservation and wilderness issues

IDENTIFICATION: Intended shortcut for shipping across Florida that would have linked the Gulf of Mexico with the Atlantic Ocean

SIGNIFICANCE: Efforts to build the Cross-Florida Barge Canal caused decades of controversy among environmentalists, government officials, and business interests until it was finally decommissioned in 1990 and the land was converted into a state recreation and conservation area.

Florida's unique peninsular shape and its 6,115 kilometers (3,800 miles) of tidal shoreline have long frustrated military, industry, and shipping interests. Since no river cuts across the state, ships and barges have had to travel around Florida's southern tip. In earlier times, this trip was often a perilous undertaking because of turbulent storms and the existence of many dangerous reefs and shoals.

Even before Florida became a state, various interest groups and individuals began calling for the cre-

---

Protecting Her Own Backyard

According to the Florida Defenders of the Environment, the organization she founded to fight construction of the Cross-Florida Barge Canal, Marjorie Harris Carr had this to say when asked why she took on the cause:

Why fight for the Ocklawaha River? The first time I went up the Ocklawaha, I thought it was dreamlike. It was a canopy river. It was spring-fed and swift. I was concerned about the environment worldwide. What could I do about the African plains? What could I do about India? How could I affect things in Alaska or the Grand Canyon? But here, by God, was a piece of Florida. A lovely natural area, right in my backyard, that was being threatened for no good reason.

---

ation of a transpeninsula waterway. Bowing to pressure, the Florida legislature created the Florida State Canal Commission in 1821 to explore the possibility of building such a canal. Five years later, the U.S. Congress adopted the cause and authorized the first of twenty-eight surveys that were carried out to find a convenient and safe route for an inland ship canal across the state. One proposal after another either proved to be impractical or failed to gain political support.

CONSTRUCTION HISTORY

In 1935 President Franklin D. Roosevelt, intending to ease unemployment problems in Florida, used $5 million in federal relief money as start-up funds for the construction of a ship canal. The proposal called for a 9-meter (30-foot) sea-level ship canal that would stretch across the north-central part of the state from the Atlantic Ocean to the Gulf of Mexico. On September 19, 1935, Roosevelt pressed a telegraph key from his office in Washington, D.C., and set off an explosive blast in Florida that officially began the canal's construction.

Roosevelt's action received popular support across north-central Florida, a region where many communities welcomed the new jobs and economic boost the canal was expected to generate. Intense opposition to the canal also existed, however. Railroad interests, fearing competition from shipping interests, hotly objected to the canal construction and lobbied in Congress to end it. Many south Floridians, fearing that a ship canal would allow saltwater intrusion to jeopardize the state's downstream supply of underground water, joined the protest. Three years after the project began, Roosevelt yielded to political pressure and cut off funding; he called on Congress to help, but Congress failed to appropriate any money for the canal, and construction ground to a halt.

World War II rekindled interest in the canal. German submarine attacks against U.S. ships along the Florida coast prompted Congress to ask the Army Corps of Engineers to reexamine the issue of building a canal to meet the nation's wartime needs. The corps responded with scaled-back plans for a barge canal, but Congress again failed to provide funding, and the project stagnated for years. In 1964 Congress finally authorized $1 million to get construction under way and promised more funds later. The Army Corps of Engineers now had responsibility for the project. Its engineers came up with plans for a five-lock waterway that would stretch 296.8 kilometers (184.4 miles) from Port Inglis on Florida's west coast to the Intracoastal Waterway at the St. Johns River on the east.

OPPOSITION FROM ENVIRONMENTALISTS

Opposition to the project quickly developed. Environmentalists argued that the canal would disrupt the natural flow of rivers and creeks in the region, flood woodland areas, and destroy many endangered and threatened plants and animals. In 1969 the Environmental Defense Fund (EDF), along with the Florida Defenders of the Environment, sued in a U.S. district court to stop construction. Nearly two years later, on January 15, 1971, the court granted the plaintiffs an injunction. Four days later, President Richard Nixon, citing environmental and economic concerns, issued a presidential order that suspended construction. By now, $74 million had been spent to build less than one-third of the canal. Great stretches of trees had been leveled, rivers and streams altered, two locks built, a dam constructed, and much earth moved.

In March, 1974, the Middle District Court of Florida overruled Nixon's action, but it upheld the injunction. Supporters of the canal received another blow in 1977 when both the Army Corps of Engineers and the Florida cabinet went on record calling for an end to construction. Despite these actions, canal proponents continued to lobby in Congress and succeeded in postponing a complete dismantling of the transwaterway project for years. Finally, in 1990, both the U.S. Congress and the Florida legislature officially and permanently deauthorized the barge canal.

Environmentalists hailed the defeat of the Cross-Florida Barge Canal as a major environmental victory, but they soon faced new problems. Debates arose over questions of what was to be done with completed sections of canal and the adjacent canal lands. In 1993 the Florida legislature resolved the issue when it authorized the conversion of 177 kilometers (110 miles) of the defunct canal zone into a huge nature preserve named the Cross-Florida Greenway State Recreation Area, or the Cross-Florida Greenway (renamed the Marjorie Harris Carr Cross-Florida Greenway in 1998, in honor of a leader of the movement to stop the canal). In addition to providing land that humans can use for outdoor recreation, the 28,300-hectare (70,000-acre) corridor also serves as a permanent wildlife refuge—one of the largest in the southern United States.

One controversial issue remained unresolved. It focused on the fate of the Rodman Dam, which was built in the center of the state on the Ocklawaha River prior to the final decommissioning of the canal. Environmentalists argued that since the dam was no longer needed, it should be demolished so that the Ocklawaha River could be restored to its natural flow pattern. Supporters of the dam—including local merchants and the bass fishing enthusiasts and fish camp owners who used the reservoir created by the dam—wanted to keep it in place, citing the reservoir's economic benefits to the local community as a recreational area. Even though Florida governor Lawton Chiles, the state cabinet, and the Department of Environmental Protection expressed support for restoration efforts of the Ocklawaha River and elimination of the Rodman Dam, the state legislature withheld the necessary funds.

John M. Dunn

FURTHER READING

Buker, George E. *Sun, Sand, and Water: A History of the Jacksonville District U.S. Army Corps of Engineers, 1821-1975.* Jacksonville, Fla.: U.S. Army Corps of Engineers, 1980.

Flippen, J. Brooks. *Nixon and the Environment.* Albuquerque: University of New Mexico Press, 2000.

Florida Defenders of the Environment. *Environmental Impact of the Cross-Florida Barge Canal with Special Emphasis on the Ocklawaha Regional Ecosystem.* Gainesville: Author, 1970.

Irby, Lee. "A Passion for Wild Things: Marjorie Harris Carr and the Fight to Free a River." In *Making*

Waves: Female Activists in Twentieth-Century Florida, edited by Jack E. Davis and Kari Frederickson. Gainesville: University Press of Florida, 2003.

Tebeau, Charlton W. *A History of Florida*. Miami: University of Miami Press, 1981.

SEE ALSO: Dams and reservoirs; Everglades; Wildlife refuges.

Cultural ecology

CATEGORY: Ecology and ecosystems

DEFINITION: Theory of anthropology that seeks to explain human cultures in terms of the environmental conditions in their home territories

SIGNIFICANCE: While claims that environmental factors can wholly explain cultural traits and dynamics are no longer made, cultural ecology has become a useful approach for understanding a given society's customs, even those that initially make no sense to outsiders. Cultural ecology has proven such a fruitful approach that it has been adopted by such disciplines as political science, geography, agricultural science, and even religious studies and art.

Cultural ecology was originally associated with anthropologist Julian Steward. He developed the theory during the 1950's, using it to study how cultural norms and customs are adaptive, given the environment that a group inhabits.

Early attempts to apply the theory to classic anthropological tropes, such as the frequent warfare and female infanticide of the Yanomami of South America and the custom of potlatch (competitive giving feasts) among indigenous peoples of the Pacific Northwest coast, had mixed results. For example, anthropologist Marvin Harris explained the fierce wars and infanticide of the Yanomami as methods of keeping their population levels within the carrying capacity of their jungle environment. Although war may be seen as one of the Four Horsemen of the Apocalypse, a consequence of population pressures, this is more a literary or religious concept than a sociological truth. There was little evidence that the Yanomami felt such pressures; other explanations fit their acknowledged cul-tural traits better. On the other hand, the potlatch took place within a culture region that usually had a surplus of food and other material resources. The gift-giving ceremonial not only built prestige for the sponsoring chief but also helped to counteract local scarcities caused by sudden natural events. Producing and storing food for a potlatch also gave a village a margin of safety against future natural disasters.

With worldwide movements for both increased food production and protection of the environment, insights from cultural ecology have become invaluable. For example, Western efforts to impose temperate-zone, mechanized farming methods on small-scale farmers in the Tropics have often met with disaster. Indigenous patterns of shifting cultivation, growing several crops with different maturation dates together in the same small plot, and periodically using fire to put chemicals back into the soil quickly may actually reflect the optimum use of this land.

Early anthropologists, it has been noted, were more interested in recording a society's rain dances than in the rain itself and its place in the society's life. With longer terms of fieldwork and the tools of cultural ecology, observers can note not only the role of both in the culture but also the conditions under which imminent rain is felt to be essential to survival and thus must be called down.

In modern societies, the web of life support extends far beyond the immediate natural surroundings. Rain, or its lack, may be less immediately relevant to a society's present well-being than the infrastructure, which includes reservoirs, irrigation mechanisms, commercial and transportation arrangements for food distribution, and many other factors. Each society, however, no matter how complex, has its tipping point, and the social mechanisms for avoiding the tipping point are not always understood, much less practiced. Cultural ecology, while potentially useful, has only fitfully treated the complexity of the changing interfaces between human society and the environment in advanced societies. This remains a challenge for the future.

Emily Alward

FURTHER READING

Bennett, John E. *The Ecological Transition: Cultural Anthropology and Human Adaptation*. Piscataway, N.J.: Transaction, 2005.

Haeen, Nora, ed. *The Environment in Anthropology*. New York: New York University Press, 2005.

Netting, Robert M. *Cultural Ecology*. 2d ed. Prospect Heights, Ill.: Waveland Press, 1986.

SEE ALSO: Agricultural revolution; Environmental determinism; Indigenous peoples and nature preservation; Population growth; Social ecology; Subsistence use; Sustainable development; Urban ecology.

Cultural eutrophication

CATEGORY: Water and water pollution

DEFINITION: Unwanted increase in nutrient concentrations in sensitive waters caused by human activities

SIGNIFICANCE: Cultural eutrophication causes the degradation of productive aquatic environments, which has prompted state and federal governments to regulate point and nonpoint source pollution in surrounding watersheds.

Eutrophication (from the Greek term meaning "to nourish") is the sudden enrichment of natural waters with excess nutrients, such as nitrogen, phosphorus, and potassium, which can lead to the development of algae blooms and other vegetation. In addition to clouding otherwise clear water, some algae and protozoa (namely, *Pfiesteria*) release toxins that harm fish and other aquatic wildlife. When the algae die, their decomposition produces odorous compounds and depletes dissolved oxygen in these waters, which causes fish and other organisms to suffocate.

Eutrophication is a naturally occurring process as an environment evolves over time. Cultural eutrophication is a distinct form of eutrophication in which the process is accelerated by human activities, including wastewater treatment disposal, runoff from city streets and lawns, deforestation and development in watersheds, and agricultural activities such as farming and livestock production. These activities contribute excessive amounts of available nutrients to otherwise pristine waters and promote rapid and excessive plant growth.

Eutrophication of the Great Lakes, particularly Lake Erie, was one of the key factors that prompted passage of the Clean Water Act and various amendments during the 1970's. This act specifically addressed the disposal of sewage into public waters, a major contributor to cultural eutrophication. However, it did not specifically address nonpoint source pollution, which comes from sources that are not readily identifiable. Agricultural activities such as farming, logging, and concentrated livestock operations all contribute to nonpoint source pollution through fertilizer runoff, soil erosion, and poor waste disposal practices that supply readily available nutrients to surrounding watersheds and lead to eutrophication in these environments.

The Chesapeake Bay is an excellent case study in cultural eutrophication. As development surrounding the bay dramatically increased, wetland and riparian buffers that helped reduce some of the impact of additional nutrients were destroyed. Eutrophication in the bay during the 1980's threatened the crabbing and oyster industry. Consequently, in 1983 and 1987, Maryland, Pennsylvania, and Virginia, the three states bordering the Chesapeake Bay, agreed to a 40 percent reduction of nutrients by the year 2000 from point and nonpoint sources in all watersheds contributing to the bay. These reductions were to be accomplished through such actions as the banning of phosphate detergents, the implementation of management plans to control soil erosion, the protection of wetlands, and the institution of controls on production and management of animal wastes. Although these steps reduced phosphorus levels in the Chesapeake Bay and kept nitrogen levels constant, regulators remained unsure how much nutrient reduction must take place for the bay and its surroundings to resemble their original condition.

Mark Coyne

FURTHER READING

Grady, Wayne. *The Great Lakes: The Natural History of a Changing Region*. Vancouver: Greystone Books, 2007.

Laws, Edward A. "Cultural Eutrophication: Case Studies." In *Aquatic Pollution: An Introductory Text*. 3d ed. New York: John Wiley & Sons, 2000.

McGucken, William. *Lake Erie Rehabilitated: Controlling Cultural Eutrophication, 1960's-1990's*. Akron, Ohio: University of Akron Press, 2000.

SEE ALSO: Agricultural chemicals; Chesapeake Bay; Erosion and erosion control; Eutrophication; Lake Erie; Runoff, agricultural; Sewage treatment and disposal; Watershed management.

Cuyahoga River fires

CATEGORIES: Disasters; water and water pollution

THE EVENT: Burning of oil slicks on the surface of the Cuyahoga River near Cleveland, Ohio

DATES: November, 1952, and June 22, 1969

SIGNIFICANCE: The fire that occurred on the oil-slicked Cuyahoga River in 1969 demonstrated the poor environmental condition of the Great Lakes and sparked a major media event that served to sway public opinion toward supporting the cleanup of Lake Erie.

The Cuyahoga River divides the city of Cleveland into east and west sides. Originating on the Appalachian Plateau 56 kilometers (35 miles) east of Cleveland, the river meanders 166 kilometers (103 miles) to Lake Erie. About 8 kilometers (5 miles) from its mouth, it becomes a sharply twisting but navigable stream that forms part of Cleveland's harbor. Indus-

trial development took place along the river in the early nineteenth century, and by 1860 docks and warehouses lined the ship channel. Industry had claimed virtually all of Cleveland's riverfront by 1881, when, according to Cleveland mayor Rensselaer R. Herrick, the discharge from factories and oil refineries made it an open sewer running through the center of the city.

In 1951 the Ohio Department of Natural Resources reported that the Cuyahoga River was heavily polluted with industrial effluents at its mouth, creating conditions that were unsatisfactory for the existence of aquatic life. In September of that year, thousands of dead fish were washed ashore just west of the river's mouth, and observers noted that the area gave off strong river odors.

An oil slick burned on the Cuyahoga River for days in November, 1952, causing an estimated $1.5 million in damage, without attracting national attention. Almost seventeen years later, at approximately noon on Sunday, June 22, 1969, the Cuyahoga River again

Firefighters spray water from a bridge onto a tugboat on the Cuyahoga River in Cleveland, Ohio, in November, 1952, as a fire that started in an oil slick on the river burns nearby docks. (©Bettmann/CORBIS)

caught fire. The fire was brought under control by approximately 12:20 P.M., but not before it had done some $50,000 worth of damage to two key railroad trestles over the river in the Flats area of Cleveland. An oil slick on the river had caught fire and floated under the wooden bridges, setting fire to both. Witnesses reported that the flames from the bridges reached as high as a five-story building. The fireboat *Anthony J. Celebrezze* rushed upstream and battled the blaze on the water while units from three fire battalions brought the flames on the trestles under control. Responsibility for the oil slick was placed on the waterfront industries, which used the river as a dumping ground for oil wastes instead of reclaiming the waste products.

The railroad trestles that burned were not all that sustained damage in the fire. Cleveland's reputation as "the best location in the nation" was severely damaged by the occurrence. The city became the brunt of numerous jokes, and the mass media across the country characterized it as the only city with a river so choked with pollution that it had burned. No river in the United States had a more notorious national reputation than the Cuyahoga River. Media coverage of the fire helped galvanize nationwide public support for efforts to clean up not only Lake Erie in particular but also the environment in general.

Charles E. Herdendorf

FURTHER READING

Adler, Jonathan A. "Fables of the Cuyahoga: Reconstructing a History of Environmental Protection." *Fordham Environmental Law Journal* 14 (2002): 89-146.

Grady, Wayne. *The Great Lakes: The Natural History of a Changing Region.* Vancouver: Greystone Books, 2007.

McGucken, William. *Lake Erie Rehabilitated: Controlling Cultural Eutrophication, 1960's-1990's.* Akron, Ohio: University of Akron Press, 2000.

SEE ALSO: Hazardous and toxic substance regulation; Lake Erie; Monongahela River tank collapse; Oil spills; Water pollution.

D

Dams and reservoirs

CATEGORIES: Preservation and wilderness issues; water and water pollution

DEFINITION: Structures that obstruct the natural flow of water in rivers and streams, and the bodies of water created by the impoundment of water behind or upstream of such structures

SIGNIFICANCE: Because dams obstruct the natural flow of water, they have significant effects on stream and river ecosystems. Although dams provide benefits such as flood control and hydroelectric power generation, they also have a number of negative environmental impacts.

Dams are designed for a number of purposes, including conservation, irrigation, flood control, hydroelectric power generation, navigation, and recreation. Not all dams create reservoirs of significant size. Low dams, or barrages, have been used to divert portions of stream flows into canals or aqueducts since human beings' first attempts at irrigation thousands of years ago. Canals, aqueducts, and pipelines are used to change the direction of water flow from a stream to agricultural fields or areas with high population concentrations.

Dams and reservoirs provide the chief, and in most cases the sole, means of storing stream flow over time. Small dams and reservoirs are capable of storing water for weeks or months, allowing water use during local dry seasons. Large dams and reservoirs have the capacity to store water for several years. As urban populations in arid regions have grown and irrigation agriculture has dramatically expanded, dams and reservoirs have increased in size in response to demand. They are frequently located hundreds of kilometers from where the water is eventually used. The construction of larger dams and reservoirs has resulted in increasingly complex environmental and social problems that have affected large numbers of people. This has been particularly true in tropical and developing nations, where most of the large dam construction of the last three decades of the twentieth century was concentrated.

SIZES AND PURPOSES OF DAMS

Early dams and their associated reservoirs were small, and dams remained small, for the most part, until the twentieth century. The first dams were simple barrages constructed across streams to divert water into irrigation canals. Water supply for humans and animals undoubtedly benefited from these diversions, but the storage capacity of most dams was small, reflecting the limited technology of the period. The earliest dams were constructed some five thousand years ago in the Middle East, and dams became common two thousand years ago in the Mediterranean region, China, Central America, and South Asia.

The energy of falling water can be converted by water wheels into mechanical energy to perform a variety of tasks, including the grinding of grain. Dams create a higher "head" or water level, increasing the potential energy, and thus served as the earliest energy source for the beginnings of the Industrial Revolution during the nineteenth century. The most significant contribution of dams to industrialization occurred in 1882 with the development of hydroelectricity, which permitted energy to be transferred to wherever electric power lines were built rather than being confined to river banks.

During the nineteenth century, large-scale settlement of the arid regions of western North America and Asia soon exhausted the meager local supplies of water and prompted demands for both exotic supplies from distant watersheds and storage for dry years. Big dams for storage and big projects for transportation of the water were thought to be the answer. Small dam projects could be financed locally; grander schemes required the assistance of federal or national governments. To justify expenditures on larger dams, promoters of these projects touted the multiple uses for reservoir water as benefits that would offset the projects' costs. Benefit-cost ratios thus became the tool by which potential projects were judged. To raise the ratio of benefits to costs, promoters placed increasing importance on intangible benefits—those to which it is difficult to assign universally agreeable currency values. While dams in arid regions were originally justified chiefly for irrigation, public water supply, and power, decisions to build dams in wetter areas were usually based on projected benefits from flood control, navigation, and recreation in addition to power generation and public water supply.

Complicating the equation is the fact that multiple uses are frequently conflicting uses. While all dams are built to even out the uneven flow of streams over time, flood control requires an empty reservoir to handle the largest floods; conversely, power generation requires a high level of water in the reservoir to provide the highest head. Public water supply and navigation benefit most from supplies that are manipulated in response to variable demand. Recreation, fishing, and the increasingly important factor of environmental concerns focus on in-stream uses of the water.

By the last two decades of the twentieth century, environmental costs and benefits and the issue of Native American water rights in the American West dominated decisions concerning dam projects in the United States, and few dams were constructed. Most of the best sites for the construction of large dams in the developed nations had been utilized, and the industry turned its attention to the developing nations. Most of the large dam projects of the last quarter of the twentieth century were constructed in or proposed for developing nations and the area of the former Soviet Union.

HUMAN IMPACTS

Small dams have small impacts on the environment; they affect small watersheds and minor tribu-

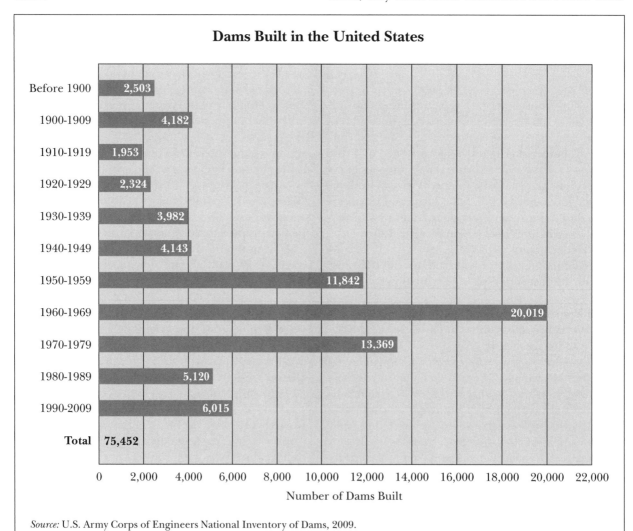

Dams Built in the United States

Period	Number of Dams Built
Before 1900	2,503
1900-1909	4,182
1910-1919	1,953
1920-1929	2,324
1930-1939	3,982
1940-1949	4,143
1950-1959	11,842
1960-1969	20,019
1970-1979	13,369
1980-1989	5,120
1990-2009	6,015
Total	75,452

Number of Dams Built

Source: U.S. Army Corps of Engineers National Inventory of Dams, 2009.

taries and usually have only a single purpose. Farm ponds and tanks, as they are known in many parts of the world, generally cover a fraction of 1 hectare (2.47 acres) in area and are only a few meters in height. These tiny ponds are designed to store water for livestock and occasionally for human supply. They frequently serve recreational purposes as well, such as fishing. During dry spells they become stagnant and subject to contamination by algae and other noxious organisms, which can threaten the health of humans and livestock. Otherwise, they have little negative impact on the environment or on nearby people and animals.

Large dams and reservoirs are responsible for environmental and social impacts that often appear to be roughly related to their size: The larger the dam or reservoir, the greater the impact. Geographical location is also important in assessing a project's impact. Scenic areas in particular, or those with endangered species of plants or animals or irreplaceable cultural or archaeological features, raise more controversy and litigation if they are chosen as potential sites for dam and reservoir projects.

People in tropical regions suffer proportionately greater health-related impacts from dams than do those in corresponding nontropical areas. The large numbers of workers required for the construction of big dams and associated irrigation projects can carry diseases into unprotected populations. Stagnant or slow-moving waters in reservoirs and irrigation canals, as well as fast-moving waters downstream of dams, are associated with particularly vicious tropical health risks. Snails in slow-moving water carry schistosomiasis, a parasitic disease that infects intestinal and urinary tracts, causing general listlessness and more serious consequences, including failure of internal organs and cancer. Estimates of the numbers of people infected range into the hundreds of millions. Malaria, lymphatic filariasis (including elephantiasis), and other diseases are carried by mosquitoes that breed in water; the incidence of such insect-borne diseases dramatically expands near irrigation projects and reservoirs. River blindness, which results from the bite of black flies, is associated with fast-flowing water downstream of dams and affects hundreds of thousands of humans.

The flooding of densely populated river valleys by reservoirs displaces greater numbers of people, with attendant health problems and social impacts, than similar projects in sparsely populated areas. Popula-

tion displacement in developing countries, especially those in the Tropics, causes greater health and social problems than in developed nations, where remedial measures and compensation are more likely to assuage the loss of homestead and community.

ENVIRONMENTAL IMPACTS

Many of the environmental problems created by large dams are associated with rapid changes in water level below the dams or with the ponding of stream flow in the reservoirs, which replaces fast-flowing, oxygenated water with relatively stagnant conditions. Indigenous animal species, as well as some plant life, are adapted to seasonal changes in the natural stream flow and cannot adjust to the postdam regime of the stream. Consequently, the survival of these species may be threatened. In 1973 the Tennessee Valley Authority—the worldwide model for builders of many large, integrated river basin projects—found that the potential demise of a small fish, the snail darter, stood in the way of the completion of the Tellico Dam. After considerable controversy and litigation, the dam was completed in 1979, but few large projects have been proposed in the United States since then, particularly in the humid East. Since the early 1970's, the arguments for abandoning dam projects have been more likely to be backed up by laws, regulations, and court decisions.

Construction of the Hetch Hetchy Dam in the Sierra Nevada mountain range of California in the early twentieth century sparked vigorous dissent, which is said to have led to the growth of the Sierra Club and organized environmental opposition to dam building. This opposition successfully challenged the construction of the Echo Park Dam on the Green River in Colorado in the early 1950's but was unsuccessful in stopping the construction of the Glen Canyon Dam on the Colorado River, which was completed in 1963. Glen Canyon, however, was the last of the big dams constructed in the American West. The preservationists, whose arguments chiefly concerned scenic and wilderness values, with attendant benefits to endangered species, lost the Glen Canyon battle but won the war against big dams. The controversy surrounding the Glen Canyon Dam continued for more than three decades after its completion, pitting wilderness and scenic preservationists against powerboat recreationists, who benefit from the access accorded by the dam's reservoir to the upstream canyonlands.

Aerial view of the reservoir Perućac Lake and the Bajina Bašta hydroelectric dam on the Drina River in Serbia. (©Milan P. Mihajlovic/ Dreamstime.com)

All reservoirs eventually fill with silt from upstream erosion; deltas form on their upstream ends. Heavier sediments, mainly sands, are trapped behind the dam and cannot progress downstream to the ocean. The Atlantic coastline of the southeastern United States suffers from beach erosion and retreat because the sands are no longer replenished by the natural flow of nearby rivers. The Aswan High Dam on the Nile River in Egypt has had a similar impact on the Nile Delta. Moreover, the natural flow of sediments downstream has historically replenished the fertility of floodplain soils during floods. To the extent that the flood-control function of a dam is successful, new fertile sediment never reaches downstream agricultural fields. While irrigation water provided by the dam may permit the expansion of cropland, this water in arid regions is often highly charged with salts, which then accumulate in the soils and eventually become toxic to plant life.

Reservoir waters release methane from decaying organic matter into the atmosphere. Methane is a greenhouse gas that promotes global warming, and some estimates suggest that the effect of large reservoirs is roughly equal to the greenhouse gas pollution of large thermal-powered electrical generation plants. The weight of the water in large reservoirs has also been implicated in causing earthquakes, which may lead to the failure of a dam. Dam failure may also occur because of poor construction or poor design, or because the builders had inadequate knowledge of the geology of the site. Tens of thousands of lives have been lost as a consequence of such failures.

Neil E. Salisbury

FURTHER READING

Berga, L., et al., eds. *Dams and Reservoirs, Societies, and Environment in the Twenty-first Century.* New York: Taylor & Francis, 2006.

Billington, David P., and Donald C. Jackson. *Big Dams of the New Deal Era: A Confluence of Engineering and Politics.* Norman: University of Oklahoma Press, 2006.

Cech, Thomas V. "Dams." In *Principles of Water Resources: History, Development, Management, and Policy.* 3d ed. New York: John Wiley & Sons, 2010.

Goldsmith, Edward, and Nicholas Hildyard. *The Social and Environmental Effects of Large Dams.* New York: Random House, 1984.

Leslie, Jacques. *Deep Water: The Epic Struggle over Dams, Displaced People, and the Environment.* New York: Farrar, Straus and Giroux, 2005.

McCully, Patrick. *Silenced Rivers: The Ecology and Politics of Large Dams.* Enlarged ed. London: Zed Books, 2001.

Stevens, Joseph E. *Hoover Dam: An American Adventure.* Norman: University of Oklahoma Press, 1988.

SEE ALSO: Aswan High Dam; Echo Park Dam opposition; Floods; Hoover Dam; Hydroelectricity; Irrigation; Tellico Dam; Tennessee Valley Authority.

Danube River

CATEGORIES: Places; water and water pollution

IDENTIFICATION: European river originating in Germany and flowing generally eastward to the Black Sea

SIGNIFICANCE: The Danube, which passes through nine European countries, has suffered severe pollution from both natural and human sources. Maintenance of the river has been difficult because so many nations share its waters, but with the signing of the Danube River Protection Convention in 1994, improvements began to be made.

The source of the Danube is close to that of the Rhine, which flows north, making the pair of rivers a major waterway cutting through Central Europe. The Danube passes through or borders on nine countries: Germany, Austria, Slovakia, Hungary, Croatia, Serbia, Romania, Bulgaria, and Ukraine. Because the river's tributaries affect other areas as well, the International Commission for the Protection of the Danube River (ICPDR) includes representatives from the Czech Republic, Slovenia, Bosnia and Herzegovina, Montenegro, and Moldova, as well as the European Union. In addition, the Danube basin includes Poland, Italy, Albania, and Macedonia, but these nations do not take part in the ICPDR.

The ICPDR was formed to carry out the mandates of the Danube River Protection Convention of 1994: to conserve and improve both surface water and groundwater of the Danube in a rational manner and to oversee the use of the river's waters. According to the convention, the ICPDR constitutes the overall legal instrument for cooperation on transboundary water management in the Danube River basin. Its missions are to ensure the management of the waters of the river basin and the distribution of those waters in an equitable manner; to prevent hazards from ice and flooding; and to prevent dangerous materials and pollution from entering the Black Sea through the river. The parties to the convention have agreed to take all legal, administrative, and technical measures to maintain—or improve, if possible—the state of the water quality in the Danube and its basin and prevent any damage to the waters. The ICPDR also publishes the magazine *Danube Watch*, which is intended to inform the public about environmental issues concerning the Danube.

The members of the ICPDR and the members of the International Sava River Basin Commission (the Sava is a tributary of the Danube in southern Europe) issued a joint statement in October, 2007, addressing the topics of navigation on the river and the river basin's ecology. The statement noted both the positive and the negative effects that river navigation can have on the environment as shipping on the Danube replaces road transport. The commission members' observations fit into a general transportation statement by the European Union covering the ecological impacts of all modes of transportation: road, rail, air, and river.

In 2008 the results of a scientific survey conducted by the ICPDR showed that the river had undergone remarkable improvement since the first such survey was taken in 2001. While the results were positive, however, the commission's report on the survey findings emphasized that work still needed to be done. Swimming was possible in some areas of the river but not all. Fish taken from some parts were edible, but further investigation of mercury concentrations were still needed. Although significant populations of plants and animals still existed in the river, and the countries along the river had repaired many damaged areas, pollution of the river by waste plants in such major cities as Belgrade, Budapest, and Bucharest still needed attention.

Frederick B. Chary

FURTHER READING

Jansky, Libor. *The Danube: Environmental Monitoring of an International River.* New York: United Nations University Press, 2004.

Magris, Claudio. *Danube.* Translated by Patrick Creagh. 1989. Reprint. New York: Farrar, Straus and Giroux, 2008.

Murphy, Irene Lyons, ed. *Protecting Danube River Basin Resources: Ensuring Access to Water Quality Data and Information.* Norwell, Mass.: Kluwer Academic, 1997.

SEE ALSO: Black Sea; Europe; European Diploma of Protected Areas; European Green parties; Fish kills; Habitat destruction; Rhine River; Sedimentation; Water-pollution policy; Water quality; Water rights; Watershed management.

Darling, Jay

CATEGORIES: Activism and advocacy; animals and endangered species

IDENTIFICATION: American cartoonist and wildlife conservationist

BORN: October 21, 1876; Norwood, Michigan

DIED: February 12, 1962; Des Moines, Iowa

SIGNIFICANCE: Darling was instrumental in starting the federal duck stamp program, which generated revenue to buy new lands to serve as waterfowl refuges, as authorized by the Migratory Bird Conservation Act.

Jay Darling was an editorial cartoonist by profession. He was known among his peers as Ding, the name with which he signed his drawings. His biting cartoons, which won him two Pulitzer Prizes and national recognition, often depicted the destruction of America's waterfowl and their habitat by overhunting and periodic drought—particularly during the Dust Bowl of the 1930's, which dried the wetlands that the birds required.

The passage of the Migratory Bird Conservation Act by the U.S.

Congress in 1929 laid the groundwork for Darling's major contribution to waterfowl conservation. That act authorized the U.S. Department of Agriculture to acquire wetlands and preserve them as waterfowl habitat, but it provided no permanent source of funding for the purpose. In 1934 President Franklin D. Roosevelt appointed a committee to look into the need for waterfowl refuges. Among its members were Darling, who had been a fierce critic of Roosevelt's wildlife policies, wildlife conservationist Aldo Leopold, and publisher Thomas Beck. The committee's recommendation that $50 million be spent for new refuges rekindled an idea that had lain dormant for years: to require waterfowl hunters to pay fees for the privilege of hunting each year by buying stamps known as duck stamps. The revenue generated from the stamps would be used to buy new refuge lands authorized by the Migratory Bird Conservation Act.

In March, 1934, Congress passed the Migratory Bird Hunting and Conservation Stamp Act, which required waterfowl hunters sixteen years or older to buy an annual duck stamp. That same month, Roosevelt appointed Darling chief of the Department of Agriculture's Bureau of Biological Survey, forerunner of the U.S. Fish and Wildlife Service. While serving as chief, Darling carried out the 1934 act's mandate by initiating the federal duck stamp program; Darling also designed the first duck stamp.

It took some time for money from the sale of the duck stamps to start flowing, however, so Darling be-

The first Duck Stamp, featuring art by Jay Darling. (USFWS)

gan raising money from other programs within the Department of Agriculture to purchase refuge land. He is credited with obtaining $20 million for wildlife conservation and setting aside 1.8 million hectares (4.5 million acres) as refuge land during his twenty-month tenure at the Bureau of Biological Survey. He resigned in November, 1935, dismayed by conservationists' lack of collective strength in focusing attention on wildlife issues.

At the first government-sponsored North American Wildlife Conference in 1936, Darling helped to organize the 36,000 wildlife societies then in existence across the United States into a national body called the General Wildlife Federation. Darling was unanimously chosen president of the federation. Two years later, the group's name was changed to the National Wildlife Federation, and Darling was reelected its president.

In 1942 Darling was awarded the Theodore Roosevelt Medal for distinguished service in wildlife conservation; in 1960 he received a National Audubon Society medal for distinguished service in natural resource conservation. A 2,013-hectare (4,975-acre) wildlife refuge established on Sanibel Island, Florida, in 1945 was dedicated to Darling in 1978. Now known as the J. N. "Ding" Darling National Wildlife Refuge, it supports a wide diversity of birds and other animals.

Jane F. Hill

FURTHER READING

Darling, Jay N. *J. N. "Ding" Darling's Conservation and Wildlife Cartoons.* Des Moines, Iowa: J. N. "Ding" Darling Foundation, 1991.

Jonson, Laurence F. *Federal Duck Stamp Story: Fifty Years of Excellence.* Davenport, Iowa: Alexander, 1984.

Lendt, David L. *Ding: The Life of Jay Norwood Darling.* 4th ed. Mt. Pleasant, S.C.: Maecenas Press, 2001.

SEE ALSO: Conservation; Fish and Wildlife Service, U.S.; Hunting; Leopold, Aldo; Migratory Bird Act; Wildlife management; Wildlife refuges.

Darwin, Charles

CATEGORY: Ecology and ecosystems
IDENTIFICATION: English naturalist
BORN: February 12, 1809; Shrewsbury, Shropshire, England
DIED: April 19, 1882; Downe, Kent, England
SIGNIFICANCE: Darwin's theory of evolution through natural selection, the dominant paradigm of the biological sciences, underlies the study of ecosystems.

Charles Darwin was born on February 12, 1809, the fifth of six children. His mother died when he was eight, leaving him in the care of his elder sisters. His father was a country doctor with a wide practice.

In 1825, Darwin went to Edinburgh, Scotland, to study medicine. He proved to be a poor student of anatomy, however, and he was sent to Christ's College, Cambridge, to prepare for the ministry. Though not a distinguished student, Darwin took an interest in natural science. At Cambridge, he met John Stevens Henslow, a botany professor, who encouraged his interest in natural history and helped to secure for him a position as naturalist aboard HMS *Beagle*, soon to depart on a five-year scientific expedition around the world. Darwin's experiences during the voyage from 1831 to 1836 were instrumental in shaping his theory of evolution.

As the voyage took the ship along the coast of South America, Darwin kept detailed journals in which he carefully observed differences among the South American flora and fauna, particularly on the Galápagos Islands. He would later draw on these extensive field observations to formulate his theory of natural selection.

After his return to London, Darwin began a study of coral reefs, and he became secretary to the Geological Society and a member of the Royal Society. He married his first cousin, Emma Wedgwood, in January, 1839.

Darwin worked for the next twenty years on his journals from the *Beagle*'s voyage, gathering information to support his theory of evolution through natural selection. His preliminary work might have continued indefinitely if he had not received, on June 18, 1854, an essay from Alfred Russel Wallace, a field naturalist in the Malay archipelago, outlining a theory of evolution and natural selection similar to Darwin's own. Darwin immediately wrote to his friends Sir

Darwin: Beaks and Evolution

During its voyage around the world, the HMS *Beagle* visited the Galápagos Islands in 1835. It was there that Charles Darwin was able to observe the interesting variety of plants and animals. The Galápagos are removed from any large land mass, being situated about six hundred miles west of the coast of Ecuador. Darwin's observations of the islands' fauna contributed to the development of his ideas about evolution and the origin of species. The different animal species living in the islands were quite different from species anywhere else in the world. Of particular interest to him were the thirteen species of finches that he collected. Finches are small, sparrow-sized birds, and the thirteen species were similar to one another except for features related to their feeding habits, especially their beaks. Darwin noted striking differences in the size and shape of the beaks and observed that their structure was related to the type of food the birds ate. Short, thick beaks were suitable for crushing seeds. Long, thin beaks were useful for feeding on insects. Large beaks were effective in opening hard fruits. Each species appeared to be adapted to its own local food supply. Darwin reasoned that the finches in the Galápagos had descended from a single ancestral species that had come to the island from South America. The adaptive radiation of the finches was one of the cornerstones of his reasoning that adaptation to the local food supply was related to the origin of new species by the process of natural selection.

Charles Lyell and Sir Joseph Dalton Hooker, explaining his dilemma and including an abstract of his own theory of evolution. Lyell and Hooker proposed that in order to avoid the question of precedence, the two papers should be presented simultaneously. Both were read before a meeting of the Linnean Society in Dublin on July 1, 1858, and they were published together in the society's journal that year.

Darwin then began writing an abstract of his theory, which he entitled *On the Origin of Species by Means of Natural Selection: Or, The Preservation of Favoured Races in the Struggle for Life*. All 1,250 copies sold out on the first day of publication in London on November 24, 1859. Darwin argued that since all species produce more offspring than can possibly survive, and since species populations remain relatively constant, there must be some mechanism working in nature to eliminate the unfit. Variations are randomly introduced in nature, some of which will permit a species to adapt better to its environment. These advantageous adaptations are passed on to the offspring, giving them an advantage for survival. Darwin did not understand the genetic mechanisms by which offspring inherit adaptations. It would take another seventy years before the forgotten work of the Austrian geneticist Gregor Mendel was rediscovered and integrated with Darwin's theories to provide a fuller view of the evolutionary process.

Darwin was surrounded by a storm of controversy after the publication of his work. Objections came both from orthodox clergy and from unconvinced scientists. For the rest of his life, Darwin worked at home on successive editions of *On the Origin of Species*, further studies on plants and animals, and his famous *The Descent of Man, and Selection in Relation to Sex* (1871). He died on April 19, 1882, and was buried with full honors in the scientists' corner at Westminster Abbey, next to Sir Isaac Newton.

Darwin's work had an immeasurable influence on the development of modern biology, ecology, morphology, embryology, and paleontology. His theory of evolution established a natural history of the earth and enabled humans to see themselves for the first time as part of the natural order of life. A lively debate continues among scientists about revisionist theories of evolution, including Stephen Jay Gould's notion of "punctuated equilibria," or sudden and dramatic evolutionary changes followed by long periods of relative stability. While they disagree about details, however, modern biologists agree that neoevolutionary theory remains the only viable scientific explanation for the diversity of life on earth.

Alexander Scott

FURTHER READING

Berra, Tim M. *Charles Darwin: The Concise Story of an Extraordinary Man.* Baltimore: The Johns Hopkins University Press, 2009.

Eldredge, Niles. *Darwin: Discovering the Tree of Life.* New York: W. W. Norton, 2005.

Quammen, David. *The Reluctant Mr. Darwin: An Intimate Portrait of Charles Darwin and the Making of His Theory of Evolution.* New York: Atlas Books, 2006.

SEE ALSO: Biodiversity; Ecosystems; Extinctions and species loss.

DDT. *See* Dichloro-diphenyl-trichloroethane

Dead zones

CATEGORIES: Water and water pollution; ecology and ecosystems

DEFINITION: Aquatic environments incapable of sustaining life

SIGNIFICANCE: The occurrence of dead zones in waters along heavily inhabited lakeshores and coastlines has increased drastically since the end of the twentieth century. Human nutrient inputs into these ecosystems from fertilizer use and pollution have contributed to more frequent eutrophication events, fueling this source of economic and ecological devastation.

In aquatic ecosystems, just as in all others, there are certain nutrients whose limited presence or availability prevents the unrestrained growth of particular species. In nearshore marine waters and lakes these so-called limiting nutrients are nitrogen and phosphorus. Their restricted availability keeps microscopic organisms called algae from reaching unsustainable and destructive levels.

Normally these algae form the base of the food web, utilizing the sun's energy to convert carbon dioxide into oxygen and food. However, in the presence of excess nitrogen and phosphorus, the algae can reproduce uncontrollably in a process called eutrophication, resulting in an algal bloom. As the algae population peaks and begins its decline, microbes begin decomposing the dead algae, consuming oxygen in the process. Eventually the dissolved oxygen in the aquatic environment is depleted, resulting in the death of all organisms in the ecosystem. At this stage, the environment is deemed a hypoxic or dead zone.

In some cases, an influx of nitrogen and phosphorus into an aquatic ecosystem can be the result of natural processes. In certain areas on the western coasts of the continents, for example, a process called upwelling occurs, in which prevailing winds and currents bring nutrients up from the ocean depths and to the surface, where algae live. Humankind's interference with the natural cycling of nitrogen and phosphorus, however, has led to significant inputs of these nutrients into aquatic ecosystems as well. Nitrogen and phosphorus are mass-produced for use as fertilizers. After the fertilizers are sprayed on crop fields, rainwater and irrigation runoff carries these nutrients into rivers and streams, which in turn feed estuarine and marine habitats. Likewise, inadequately treated sewage and wastewater laden with nutrients from cities and towns is often pumped into aquatic habitats.

More than 140 dead zones have been documented around the world, and the number continues to rise. The largest known dead zone is located in the Baltic Sea, where agricultural runoff and poorly treated sewage contribute to a dead zone tens of thousands of square kilometers in size. Similarly, in the United States a seasonal hypoxic zone the size of New Jersey forms during the summer months at the mouth of the Mississippi River in the Gulf of Mexico. Fishery yields there and in other locations affected by dead zones have sharply declined, sapping the lifeblood of coastal economies and cutting into marine food production.

While ominous, dead zones are not necessarily permanent. The infamous Black Sea dead zone was once the largest in the world, covering an area of approximately 40,000 square kilometers (15,000 square miles). When the Soviet Union collapsed in 1991, the industrialized agricultural economy of the region also collapsed, and the resulting drop in fertilizer use cut nitrogen and phosphorus input into the Black Sea by more than half. Within about a decade, the dead zone virtually disappeared. Although this example provides hope, the ecosystem of the Black Sea has yet to recover fully, as it supports limited marine life. To combat dead zones efficiently, political action through management of fertilizer use and pollution will be needed, not only to rein in current dead zones but also to prevent future outbreaks entirely.

Daniel J. Connell

FURTHER READING

Nassauer, Joan Iverson, Mary V. Santelmann, and Donald Scavia, eds. *From the Corn Belt to the Gulf: Societal and Environmental Implications of Alternative Agricultural Futures.* New York: Resources for the Future, 2007.

Vernberg, F. John, and Winona B. Vernberg. *The Coastal Zone: Past, Present, and Future.* Columbia: University of South Carolina Press, 2001.

SEE ALSO: Black Sea; Chesapeake Bay; Cultural eutrophication; Eutrophication; Runoff, agricultural; Wastewater management.

Debt-for-nature swaps

CATEGORY: Preservation and wilderness issues

DEFINITION: Strategy for reducing foreign debt in developing nations by trading debt forgiveness or debt reduction for guarantees of environmental activities by debtor nations

SIGNIFICANCE: Although some critics view debt-for-nature swaps as a form of cultural imperialism, these arrangements offer a way for environmental organizations to encourage developing nations to preserve ecosystems, conserve natural resources, and protect significant cultural sites.

An international finance crisis began during the late 1980's when many developing nations found that they had borrowed more from international lending institutions, mostly private banks, than they could repay. To recover some of the principal on the loans, banks began to sell the loans in financial markets, usually discounted to a fraction of their original value because of the threat of default. Several options to relieve this debt burden on developing nations were explored. One option was to refinance the debt to lower the interest rates and extend the time for repayment. Another strategy was to encourage domestic financial reforms in debtor nations by increasing domestic investment, expanding the domestic economy, raising taxes, reducing non-debt-related expenditures, or inflating the currency in order to pay the debt eventually. Finally, creditor nations and institutions could partially forgive the debt. Debt-for-nature swaps combine elements of all three of these options.

How Debt-for-Nature Swaps Work

In debt-for-nature swaps, environmental organizations buy discounted debt in the financial markets from the banks. Instead of collecting the full amount of interest and principal from the debtor nations, the environmental organizations agree to forgive all or a portion of the debt if the debtor nations invest amounts up to the principal value of the debt in local preservation efforts, often the purchase of land for national parks or investment in skilled staff and improvements for existing parks. In these arrangements, conservation organizations benefit through an increase in local funding for conservation, and banks benefit because they have a new market for the debt. The debtor nations benefit in several ways: They are able to invest their funds in their own nations rather than transferring funds to the lending nations; they use inflated local currency rather than high-value, scarce, dollar-based foreign exchange; and they reduce their outstanding debt and interest payments on that debt, which allows them to continue making payments on the remaining debt and maintain their international credit ratings.

Debt-for-nature swaps erase the "debt overhang," that portion of debt that, if forgiven, allows the remaining debt to continue to be assumed by creditor nations or institutions with satisfactory levels of burden on the debtor nations and satisfactory debt payment risks for the creditor nations. Because environmental organizations purchase the debt at a discount, they are able to multiply their impacts on the environment. For example, in the first debt-for-nature swap in 1987, Conservation International purchased $650,000 in Bolivian debt for $100,000, then required the Bolivian government to establish a $250,000 endowment fund in local Bolivian currency to pay operating costs for a biosphere reserve in the Bolivian Amazon before erasing the debt.

In debt-for-nature swaps, debtor nations buy back a portion of their outstanding debt with an investment in a portion of their own natural capital. Natural capital includes caches of nonrenewable resources such as mineral or oil reserves, natural resources such as old-growth forests or endangered species habitat, historical artifacts such as prehistoric ruins or fossil deposits, and cultural resources such as the homelands of primitive indigenous peoples or significant architectural sites. The preservation of this natural capital has positive benefits for the ecology, sustainable development, and biodiversity. A swap is also likely to improve the debtor nation's economic ability to repay the remaining debt, because the preserved cultural and natural resources often serve as tourist attractions, cultural centers, and locations for academic and commercial research.

Participation of Financial Institutions

Because some private lending institutions will not voluntarily participate in debt-relief processes that reduce the financial institutions' capital or potential profit from loans, governments in creditor nations either mandate financial institutions' participation or provide financial incentives, such as tax deductions and tax credits, to encourage participation. Creditor governments justify these actions as a component of their foreign economic development programs, as a component of their environmental programs, as philanthropic support for preservation of the earth's cultural and natural heritage, or as economically justifiable domestic self-interest. For example, encouraging nations in tropical zones to protect rain forests helps to reassure nations in the northern temperate zones that existing climate patterns supported by those rain forests will be maintained. Maintaining these climate patterns is necessary to prevent natural and human-made disasters, reduce the demands on industrialized nations to provide humanitarian relief from increased numbers of natural disasters, ensure continued rainfall in agricultural zones in temperate regions, prevent desertification, maintain good air quality and reduce the greenhouse effect, ensure global biodiversity, and improve the living conditions for every person on the planet.

Some nations, among them Costa Rica, welcome the opportunities presented by debt-for-nature swaps. Other nations—Brazil is an example—see the swaps as a form of environmental imperialism in which foreign environmental organizations shape domestic government policy. They argue that restricting large areas of land for parks and nature reserves reduces the amount of land available for economic production and access by poor subsistence farmers.

Gordon Neal Diem

Further Reading

Mahony, Rhona. "Debt-for-Nature Swaps: Who Really Benefits?" 1992. In *Tropical Rainforests: Latin American Nature and Society in Transition*, edited by Susan E. Place. Rev. ed. Wilmington, Del.: Scholarly Resources, 2001.

Meier, Gerald M. *The International Environment of Business: Competition and Governance in the Global Economy.* New York: Oxford University Press, 1998.

Mulder, Monique Borgerhoff, and Peter Coppolillo. "Global Issues, Economics, and Policy." In *Conservation: Linking Ecology, Economics, and Culture.* Princeton, N.J.: Princeton University Press, 2005.

Page, Diana. "Debt-for-Nature Swaps: Experience Gained, Lessons Learned." *International Environmental Affairs* 1 (Fall, 1989): 275-288.

Sachs, Jeffrey. "Making the Brady Plan Work." *Foreign Affairs* 68 (Summer, 1989): 87-104.

See also: Biosphere reserves; Ecotourism; Environmental economics; National parks; Rain forests; World Heritage Convention.

Deciduous forests

Categories: Forests and plants; ecology and ecosystems

Definition: Natural areas consisting of diverse plant species that lose their leaves annually at the end of each growing season

Significance: Deciduous forests are important ecosystems that support multiple life cycles and provide habitat for native populations and wildlife. Located in the most heavily populated areas throughout the globe, these forests are declining because of human activity, including overuse and environmental pollution.

Deciduous forests did not exist until the end of the last ice age, which was some 20,000 years ago, as the plant species associated with this forest type were unable to adapt to the glacial climate. The word "deciduous" is derived from the Latin word *decidere*, which means "fall off." Deciduous forests are located mainly in the temperate forest biome, which has very seasonal weather patterns consisting of four seasons with warm summers and cold winters and precipitation in the form of rain and snow throughout the year.

In the fall season the leaves of deciduous trees change from green to brilliant yellow, orange, red, and brown because of a lack of the green pigment chlorophyll, which they can no longer produce as daylight wanes. Eventually the leaves fall off as they go into dormancy at the end of the growing cycle, and most trees in the temperate deciduous forests remain bare during the winter months.

The temperate forest biome is located in the middle latitudes of the earth, and temperate or deciduous forests can be found in eastern North America, eastern Asia, and western Europe. Parts of Australia and New Zealand also contain deciduous forests. Less prominent are the tropical and subtropical deciduous forest biomes that exist in South America and Southern Africa. Plants in these biomes are dependent on seasonal temperatures and rainfall. Most of the trees in these forests, such as teak and ebony, drop their leaves in the dry season in order to conserve water, but new leaves sprout once the rainy season begins.

Forest Zones and Plant Associations

Temperate deciduous forests consist of five distinct zones or strata. The first is the tree stratum. Trees in this zone, including maple, oak, beech, hickory, walnut, chestnut, basswood, linden, elm, and sweet gum species, average from 18 to 30 meters (60 to 100 feet) in height. The second stratum, the small tree and sapling zone, is made up of younger trees found in the tree stratum. The third stratum consists of shrubs, including such species as azaleas, rhododendrons, mountain laurels, and huckleberries. The herbal or fourth zone consists of short herbal plants, and the rich soils of the forest floor make up the ground stratum, the fifth zone, which includes lichens, mushrooms, and mosses.

E. Lucy Braun, an important botanist and plant ecologist of the early twentieth century, spent much of her life studying and cataloging the deciduous forests of eastern North America in order to preserve them at a time when humans were involved in random deforestation and logging. Braun and others created deciduous forest subcategories known as forest regions based mainly on the regions' natural vegetation composition, but also on physiognomy. These regions vary from ones that are made up mainly of deciduous plants, such as the oak-hickory and beech-maple regions, to forest communities that include coniferous species, which do not shed their leaves in the winter, including the oak-pine and hemlock-white pine-northern hardwood regions. Braun named one of the major deciduous forest communities in eastern North America the "mixed mesophytic." Unlike most forest communities, which have only two or three dominant species, the mixed mesophytic deciduous forest community is made up of more than eighty plant species.

ANIMAL SPECIES

Temperate deciduous forests flourish with a variety of fauna. Mammals include deer, elk, bears, squirrels, skunks, raccoons, rabbits, opossums, foxes, and porcupines. Other diverse species found in these forests include many varieties of salamanders, frogs, snakes, and spiders. Small and large birds, such as wild turkeys, owls, hawks, and the bald eagle, depend on deciduous forests for habitat. During winter months some of the deciduous forest animals, especially birds, migrate to warmer climates where food is plentiful, while animals such as squirrels slow their metabolisms and store nuts and seeds in the hollows of trees to survive the winter months. Bears are among the animals that hibernate during winters in temperate deciduous forests.

Many animals associated with temperate decidu-

ous forests have long been on the decline, such as bobcats, mountain lions, and timber wolves. Loss of habitat and human slaughter have led to decreases in the populations of several of these species to the extent that they have been recognized as endangered or threatened and in need of protection from extinction. The continual loss of deciduous forests caused by logging, blights, plagues, nutrient depletion owing to pollution, and climate change has exacerbated negative effects on all of the species that make their homes in these forests.

Carol A. Rolf

FURTHER READING

Braun, E. Lucy. *Deciduous Forests of Eastern North America.* 1950. Reprint. Caldwell, N.J.: Blackburn Press, 2001.

Delcourt, Hazel R. *Forests in Peril: Tracking Deciduous Trees from Ice-Age Refuges into the Greenhouse World.* Blacksburg, Va.: McDonald & Woodward, 2002.

Frelich, Lee E. *Forest Dynamics and Disturbance Regimes: Studies from Temperate Evergreen-Deciduous Forests.* New York: Cambridge University Press, 2008.

Yetman, David A. *A Tropical Deciduous Forest of Alamos: Biodiversity of a Threatened Ecosystem in Mexico.* Tucson: University of Arizona Press, 2000.

SEE ALSO: Boreal forests; Climax communities; Cloud forests; Coniferous forests; Deforestation; Forest management; Indicator species; National forests; Old-growth forests.

Deep ecology

CATEGORY: Philosophy and ethics

DEFINITION: School of environmental philosophy based on environmental activism and ecological spirituality

SIGNIFICANCE: The thinking of deep ecologists has attracted criticism from some quarters at the same time it has expanded approaches to environmental ethics.

The term "deep ecology" was first used by Norwegian philosopher Arne Naess in 1972 to suggest the need to go beyond the anthropocentric view that nature is merely a resource for human use. Since that time, the term has been used in three major ways.

The Deep Ecology Platform

A core tenet of deep ecology is the recognition that nature has intrinsic value and is not separate or inferior to human life. The following is the deep ecology platform, written by Arne Naess and George Sessions:

The well-being and flourishing of human and nonhuman life on Earth have value in themselves (synonyms: inherent worth; intrinsic value; inherent value). These values are independent of the usefulness of the nonhuman world for human purposes.

Richness and diversity of life forms contribute to the realization of these values and are also values in themselves.

Humans have no right to reduce this richness and diversity except to satisfy vital needs.

Present human interference with the nonhuman world is excessive, and the situation is rapidly worsening.

The flourishing of human life and cultures is compatible with a substantial decrease of the human population. The flourishing of nonhuman life requires such a decrease.

Policies must therefore be changed. The changes in policies affect basic economic, technological structures. The resulting state of affairs will be deeply different from the present.

The ideological change is mainly that of appreciating life quality (dwelling in situations of inherent worth) rather than adhering to an increasingly higher standard of living. There will be a profound awareness of the difference between big and great.

Those who subscribe to the foregoing points have an obligation directly or indirectly to participate in the attempt to implement the necessary changes.

First, it has been used to refer to a commitment to deep questioning about environmental ethics and the causes of environmental problems; such questioning leads to critical reflection on the fundamental worldviews that underlie specific environmental ideas and practices. Second, the term has been used to refer to a platform of generally agreed-upon values that a variety of environmental activists share; these values include an affirmation of the intrinsic value of nature, the recognition of the importance of biodiversity, a call for a reduction of human impact on the natural world, greater concern with quality of life than with material affluence, and a commitment to changing economic policies and the dominant view of nature. Third, "deep ecology" has been used to refer to particular philosophies of nature that tend to emphasize the value of nature as a whole (ecocentrism), an identification of the self with the natural world, and an intuitive and sensuous communion with the earth.

Because of its emphasis on fundamental worldviews, deep ecology is often associated with non-Western spiritual traditions, such as Buddhism and Native Ameri-

can cultures, as well as with radical Western philosophers such as Baruch Spinoza and Martin Heidegger. It has also drawn on the nature writing of Henry David Thoreau, John Muir, Robinson Jeffers, and Gary Snyder. Deep ecology's holistic tendencies have led to associations with the Gaia hypothesis, and its emphasis on diversity and intimacy with nature has linked it to bioregionalism. Deep ecological views have also had a strong impact on environmental activism, including the Earth First! movement.

Deep ecologists have sometimes criticized the animal rights perspective for continuing the traditional Western emphasis on individuals while neglecting whole systems, as well as for a revised speciesism that still values certain parts of nature (animals) over others. Some deep ecologists have also been critical of mainstream environmental organizations such as the Sierra Club for not confronting the root causes of environmental degradation.

Ecofeminists have criticized deep ecology for failing to consider gender differences in the experience of the self and nature, for failing to examine the connection between the oppression of women and human treatment of nature, and for promoting a holism that, the ecofeminists assert, disregards the reality and value of individuals and their relationships. Social ecologists have criticized deep ecology for failing to critique the relationship between environmental destruction on the one hand and social structure and political ideology on the other. In addition, a distrust of human interference with nature has led some deep ecology thinkers to present pristine wilderness, with no human presence, as the ideal. In rare and extreme cases, deep ecologists have implied a misanthropic attitude. In some instances, especially in early writings by deep ecologists, such criticisms have had considerable force. However, these problematic views are not essential to deep ecology, and a number of deep ecologists have developed a broadened view that overlaps with ecofeminism and social ecology.

David Landis Barnhill

FURTHER READING

Bender, Frederic L. *The Culture of Extinction: Toward a Philosophy of Deep Ecology*. Amherst, N.Y.: Humanity Books, 2003.

Devall, Bill. *Simple in Means, Rich in Ends: Practicing Deep Ecology*. Salt Lake City, Utah: Peregrine Smith, 1988.

Katz, Eric, Andrew Light, and David Rothenberg, eds. *Beneath the Surface: Critical Essays in the Philosophy of Deep Ecology*. Cambridge, Mass.: MIT Press, 2000.

SEE ALSO: Bioregionalism; Earth First!; Ecofeminism; Environmental ethics; Gaia hypothesis; Naess, Arne; Social ecology; Speciesism.

Deforestation

CATEGORIES: Forests and plants; preservation and wilderness issues

DEFINITION: Loss of forestlands through encroachment by agriculture, industrial development, or nonsustainable commercial forestry

SIGNIFICANCE: Deforestation, particularly in tropical regions, has given rise to concerns among environmentalists and scientists, in large part because of the role that tropical forests play in moderating global climate.

Toward the end of the twentieth century, environmentalists became active in decrying the apparent accelerating pace of deforestation because of the loss of wildlife and plant habitats caused by the practice as well as its negative effects on biodiversity. By the 1990's research by mainstream scientists had confirmed that deforestation was indeed occurring on a global scale and that it posed a serious threat to global ecology. Although steps were taken to slow rates of deforestation in the early years of the twenty-first century, the loss of forestlands continued.

Deforestation as a result of expansion of agricultural lands or nonsustainable timber harvesting has occurred in many regions of the world at different periods in history. The Bible, for example, refers to the cedars of Lebanon. Lebanon, like many of the countries bordering the Mediterranean Sea, was thickly forested several thousand years ago. A growing population, overharvesting, and the introduction of graz-ing animals such as sheep and goats decimated the forests, which never recovered.

Similarly, the forests of Europe and North America have shifted in total area as human populations have changed over the centuries. When the European colonists arrived in the New World, they immediately began clearing the forests. Trees were harvested for building materials and export back to Europe or were simply felled and burned to clear space for farming. In North America, however, as agriculture became increasingly mechanized and farming shifted to the prairies, abandoned farms reverted to woodland. Environmental historians believe, in fact, that a greater percentage of land area in North America is now forested than was covered with trees prior to the arrival of European colonists. A similar phenomenon has taken place in many northern European countries as their populations have become increasingly urbanized.

As the European industrialized nations have gained forestland, however, the less developed countries in Latin America, Asia, and Africa have lost woodlands. While some of this deforestation has been caused by a demand for tropical hardwoods for lumber or pulp, the leading cause of deforestation in the twentieth and twenty-first centuries, as it was several hundred years ago, has been the expansion of agriculture. Growing demand by the industrialized world for agricultural products such as beef has led to millions of hectares of forestland being bulldozed or burned to create pastures for cattle. Researchers in Central America have watched with dismay as large beef-raising operations have expanded into fragile ecosystems in countries such as Costa Rica, Guatemala, and Mexico.

A tragic irony in this expansion of agriculture into tropical rain forests is that the soil underlying the trees is often unsuited for pastureland or raising other crops. Exposed to sunlight, the soil is quickly depleted of nutrients and often hardens. The once-verdant land becomes an arid desert prone to erosion that may never return to forest. As the soil becomes less fertile, thorny weeds begin to choke out the desirable forage plants, and the cattle ranchers move on to clear fresh tracts.

SLASH-AND-BURN AGRICULTURE AND LOGGING

Apologists for the beef industry often argue that their ranching practices are simply a form of slash-and-burn agriculture and do no permanent harm. It is true that many of the indigenous peoples in tropical

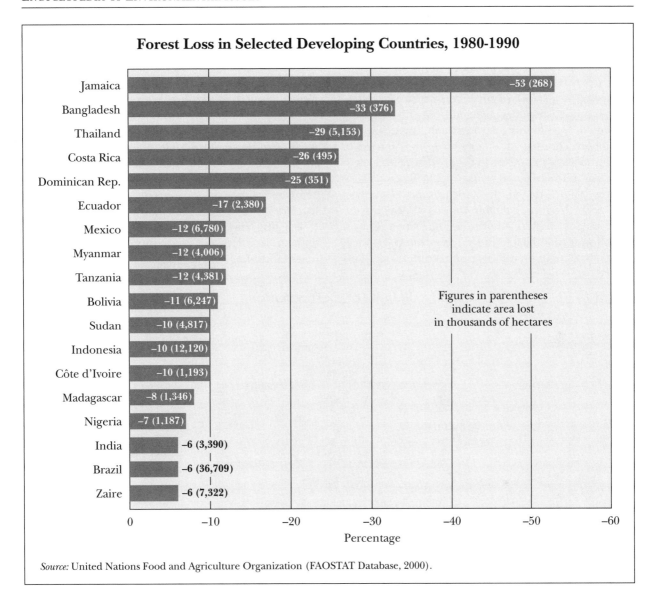

Forest Loss in Selected Developing Countries, 1980-1990

Jamaica −53 (268)
Bangladesh −33 (376)
Thailand −29 (5,153)
Costa Rica −26 (495)
Dominican Rep. −25 (351)
Ecuador −17 (2,380)
Mexico −12 (6,780)
Myanmar −12 (4,006)
Tanzania −12 (4,381)
Bolivia −11 (6,247)
Sudan −10 (4,817)
Indonesia −10 (12,120)
Côte d'Ivoire −10 (1,193)
Madagascar −8 (1,346)
Nigeria −7 (1,187)
India −6 (3,390)
Brazil −6 (36,709)
Zaire −6 (7,322)

Figures in parentheses
indicate area lost
in thousands of hectares

Percentage

Source: United Nations Food and Agriculture Organization (FAOSTAT Database, 2000).

regions have practiced slash-and-burn agriculture for millennia with only minimal impact on the environment. These farmers burn the understory, or low-growing shrubs and trees, to clear small plots of land. Any large trees that survive their fires are cut down with axes and then burned.

Anthropological studies have shown that the small plots these peasant farmers clear can usually be measured in square meters or feet, not in hectares or acres like cattle ranches, and are used for five to ten years. As fertility of the soil declines in one plot, the farmer clears a small plot next to the depleted one. The farmer's family or village gradually rotates through the forest, clearing small plots and using them for a few years, and then shifting to new ground, until they eventually come back to where they began one hundred or more years before. As long as the size of the plots cleared by peasant farmers remains small in proportion to the forest overall, slash-and-burn agriculture does not contribute significantly to deforestation. If the population of farmers grows, however, and more land must be cleared with each succeeding generation, as has been happening in many tropical countries, then even traditional slash-and-burn agriculture can be as ecologically devastating as the more mechanized cattle ranching operations.

Although logging is not the leading cause of deforestation, it remains a significant factor. Tropical forests are rarely clear-cut, as they typically contain hundreds of different species of trees, most of which may have no commercial value. Loggers may select only a few trees for harvesting from each stand. Selective harvesting is a standard practice in sustainable forestry, but just as loggers engaged in the disreputable practice of high-grading across North America in the nineteenth century, so loggers have high-graded in the late twentieth and early twenty-first centuries in Malaysia, Indonesia, and other tropical forests. High-grading is a practice in which loggers cut over a tract to remove the most valuable timber while ignoring the damage being done to the residual stand. The assumption is that, having logged over the tract once, the timber company will not be coming back. This practice stopped in North America not because the timber companies voluntarily recognized the ecological damage they were doing but because they ran out of easily accessible, old-growth timber to cut. Fear of a timber famine caused logging companies to create forest plantations and to undertake the practice of sustainable forestry. While global satellite photos indicate significant deforestation has occurred in tropical areas, enough easily harvested old-growth forest remains in some areas that there is no economic incentive for timber companies to switch to sustainable forestry.

Logging may also contribute to deforestation by making it easier for agriculture to encroach on forestlands. Logging companies build roads for their

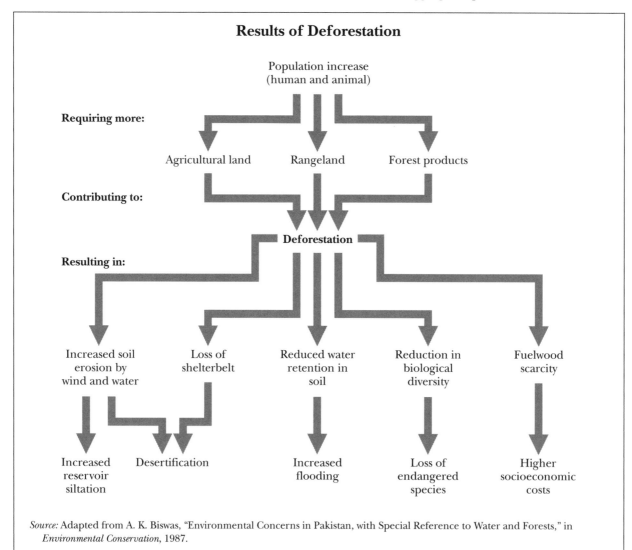

Results of Deforestation

Source: Adapted from A. K. Biswas, "Environmental Concerns in Pakistan, with Special Reference to Water and Forests," in *Environmental Conservation*, 1987.

own use while harvesting trees, and farmers and ranchers later use these roads to move into the logged tracts, where they clear whatever trees the loggers have left.

ENVIRONMENTAL IMPACTS

The extent of the problem of deforestation has long been a subject of debate. The United Nations Food and Agriculture Organization (FAO), which monitors deforestation worldwide, bases its statistics on measurements taken from satellite images. These data indicate that between 2000 and 2005 the net loss of forest area globally was 7.3 million hectares (18 million acres) per year, a reduction from 8.9 million hectares (22 million acres) per year in the decade between 1990 and 2000. In the period 2000-2005, South America and Africa had the greatest losses: South America lost 4.3 million hectares (10.6 million acres) of forest per year, and Africa lost 4 million hectares (9.9 million acres) per year. Environmental activists have been particularly concerned about forest losses in Indonesia and Malaysia, two countries where timber companies have been accused of abusing or exploiting native peoples in addition to engaging in environmentally damaging harvesting methods.

Researchers outside the United Nations have challenged FAO's data, with some scientists claiming the numbers are much too high and others providing convincing evidence that, if anything, FAO's numbers are too low. Few researchers, however, have tried to claim that deforestation on a global scale is not happening. In the 1990's the reforestation of the Northern Hemisphere, while providing an encouraging example that it is possible to reverse deforestation, was not enough to offset the depletion of forestland in tropical areas.

Deforestation affects the environment in a multitude of ways. The most obvious is in a loss of biodiversity. When an ecosystem is radically altered through deforestation, the trees are not the only thing to disappear. Wildlife decreases in number and in variety, and other plants also die. As forest habitat shrinks through deforestation, various plants and animals become vulnerable to extinction. Many biologists believe that numerous animals and plants native to tropical forests will become extinct as the result of deforestation before humans ever have a chance to become aware of their existence.

Other effects of deforestation may be less obvious. Deforestation can lead to increased flooding during rainy seasons. Rainwater that once would have been slowed or absorbed by trees instead runs off denuded hillsides, pushing rivers over their banks and causing devastating floods downstream. The role of forests in regulating water has long been recognized by both engineers and foresters. Flood control was, in fact, one of the motivations behind the creation of the federal forest reserves in the United States during the nineteenth century. More recently, disastrous floods in Bangladesh have been blamed on the logging of tropical hardwoods in the mountains of Nepal and India.

Conversely, trees can also help to mitigate drought. Like all plants, trees release water into the atmosphere through the process of transpiration. As the world's forests shrink, fewer greenhouse gases such as carbon dioxide will be removed from the atmosphere, less oxygen and water will be released into it, and the world will become a hotter, dryer place. Scientists and policy analysts alike are in agreement that deforestation is a major threat to the environment. The question is whether effective policies can be developed to reverse it.

Nancy Farm Männikkö

FURTHER READING

Chew, Sing C. *World Ecological Degradation: Accumulation, Urbanization, and Deforestation, 3000 B.C.-A.D. 2000.* Walnut Creek, Calif.: AltaMira Press, 2001.

Dean, Warren. *With Broadax and Firebrand: The Destruction of the Brazilian Atlantic Forest.* Berkeley: University of California Press, 1997.

Geist, Helmut J., and Eric F. Lambin. "Proximate Causes and Underlying Driving Forces of Tropical Deforestation." *BioScience* 52, no. 2 (2002): 143-150.

Humphreys, David. *Logjam: Deforestation and the Crisis of Global Governance.* Sterling, Va.: Earthscan, 2006.

Palmer, Charles, and Stefanie Engel, eds. *Avoided Deforestation: Prospects for Mitigating Climate Change.* New York: Routledge, 2009.

Richards, John F., and Richard P. Tucker, eds. *World Deforestation in the Twentieth Century.* Durham, N.C.: Duke University Press, 1990.

Rudel, Thomas K., and Bruce Horowitz. *Tropical Deforestation: Small Farmers and Land Clearing in the Ecuadorian Amazon.* New York: Columbia University Press, 1994.

Sanchez, Ilya B., and Carl L. Alonso, eds. *Deforestation Research Progress.* New York: Nova Science, 2008.

Sponsel, Leslie E., Robert Converse Bailey, and Thomas N. Headland, eds. *Tropical Deforestation: The Human Dimension.* New York: Columbia University Press, 1996.

SEE ALSO: Logging and clear-cutting; Old-growth forests; Rain forests; Rainforest Action Network; Slash-and-burn agriculture; Sustainable forestry.

Department of Energy, U.S.

CATEGORIES: Organizations and agencies; energy and energy use

IDENTIFICATION: Cabinet-level division of the executive branch of the U.S. government that oversees energy matters

DATE: Created on August 4, 1977

SIGNIFICANCE: In addition to maintaining standards for the regulation of production and distribution of energy within the United States, the U.S. Department of Energy oversees and supports energy-related research, promotes energy conservation, and supports public education and information dissemination about energy.

The U.S. Department of Energy (DOE) began operations October 1, 1977, having been created by legislation signed into law by President Jimmy Carter on August 4, 1977. That legislation merged a number of existing federal agencies into a single cabinet department answering to the president. This action was taken during a time of energy shortages in the United States; it was believed that a single governmental body would be more effective than a variety of independent agencies in implementing a national energy policy.

During the early twentieth century, the U.S. government was little involved in making policy concerning energy use other than its role in implementing daylight saving time during the world wars. After World War II, the Atomic Energy Commission (AEC) was established to oversee nuclear energy technology. In 1974 the AEC was dissolved, and two new agencies, the Nuclear Regulatory Commission (NRC) and the Energy Research and Development Administration (ERDA) were created by the Energy Reorganization Act. ERDA, along with other agencies, became the Department of Energy in 1977. By the twenty-first century DOE included dozens of offices, agencies, and administrations.

Through its various offices and agencies, DOE seeks to provide a framework for a comprehensive energy policy for the United States. The department is responsible for the regulation of various parts of the energy industry within the United States. Another important function of DOE is the management of energy information; through its Energy Information Administration, DOE collects and distributes statistics about energy usage and U.S. energy reserves. The related Office of Scientific and Technical Information disseminates current energy research information.

DOE is also heavily involved in scientific and technological research. The DOE Office of Science, with its several program offices, is one of the nation's largest supporters of research in the physical sciences. The Office of Science also supports environmental and biological research. The research supported by DOE is conducted in various government laboratories throughout the United States and in university and corporate labs funded through DOE grants. DOE strongly supports research in innovative and developing fields of energy production and usage.

In addition to energy regulation and research, DOE is heavily involved in hydroelectric power. Four power administrations within DOE oversee the production and sale of electricity produced in federal hydroelectric power plants. DOE also works nationwide to support the modernization of the nation's electric grid to make electric energy more efficient and more reliable.

NUCLEAR ENERGY AND WEAPONS

DOE works closely with the NRC to oversee the nation's nuclear energy industry. DOE's Nuclear Energy Office manages research programs for both fission and fusion energy systems. Naval nuclear reactors are managed directly by DOE, but the department shares responsibility with the NRC for management and regulation of civilian reactors. Through its Office of Civilian Radioactive Waste Management, DOE is responsible for both civilian and military radioactive waste storage and disposal.

The National Nuclear Security Administration (NNSA), an agency within DOE, oversees military nuclear technology. This includes military nuclear reactors, such as naval nuclear propulsion systems, as well as the nation's nuclear arsenal. American nuclear weapons are technically on loan to the Department of Defense from DOE. The NNSA assumes responsibility for the manufacture, maintenance, transport, and

(continued on page 348)

U.S. Department of Energy

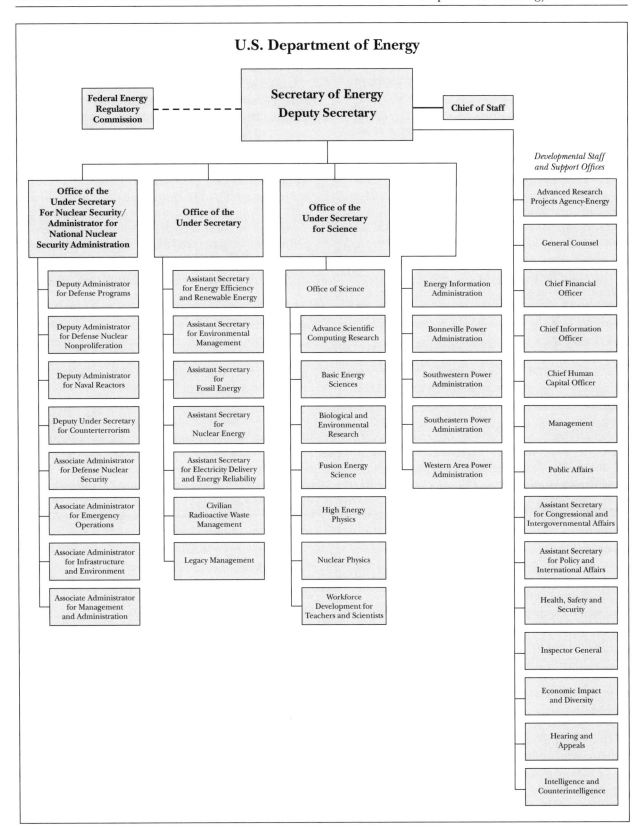

safekeeping of these nuclear weapons. DOE also works to limit the proliferation of nuclear weapons through controls on the export of technology that could be used in the production of such weapons.

ENVIRONMENTAL AND CONSERVATION EFFORTS

DOE works with other federal agencies, such as the Environmental Protection Agency (EPA), to research technologies that may reduce the environmental impacts of energy production and use. Much of this research is conducted through DOE's Office of Energy Efficiency and Renewable Energy (EERE), which supports research into renewable energy technologies such as wind, solar, hydroelectric, and geothermal. EERE is involved in research into high-risk, high-value energy technologies that private industry is unwilling to investigate until they can be proven economically viable. DOE also works with local and state governments to provide information on energy efficiency and clean energy technology.

In addition to its role in developing new sources of clean energy, DOE works to promote energy conservation and supports research into more efficient energy production, transmission, and consumption. As part of this work, DOE is involved, along with the EPA, in supporting the Energy Star program, which promotes efficiency in a wide variety of home and office products. The Weatherization Assistance Program, administered by EERE, assists low-income home owners with weatherproofing their homes and making them more energy-efficient. EERE also provides information to the public regarding energy-efficient construction.

Raymond D. Benge, Jr.

FURTHER READING

Consumer's Union of the United States. "Energy Star Has Lost Some Luster." *Consumer Reports*, October, 2008, 24-26.

Dietz, Thomas, and Paul C. Stern, eds. *Public Participation in Environmental Assessment and Decision Making*. Washington, D.C.: National Academies Press, 2008.

Holl, Jack. *The United States Department of Energy: A History*. Washington, D.C.: Government Printing Office, 1982.

SEE ALSO: Atomic Energy Commission; Energy conservation; Energy-efficiency labeling; Environmental Protection Agency; Nuclear Regulatory Commission.

Department of the Interior, U.S.

CATEGORIES: Organizations and agencies; land and land use; resources and resource management

IDENTIFICATION: Cabinet-level division of the executive branch of the U.S. government responsible for the management of federally owned lands and underground natural resources

DATE: Created on March 3, 1849

SIGNIFICANCE: Because vast sections of the U.S. territorial landmass fall outside the realm of private ownership, a separate body of laws and special administrative arrangements are required to guard the common material and ecological interests of the American people. The U.S. Department of the Interior is charged with overseeing those interests.

The idea of a federal administration responsible for internal, or domestic, affairs came in the first stages of the United States' independence, but it was not until James Polk's presidency that a bill creating the Department of the Interior was passed by Congress. Proponents of the new federal department argued that a number of essential governmental activities had for many years been placed under executive branch administrations that had no logical connection with them. Two examples (in the first half of the nineteenth century) were the General Land Office under the secretary of the treasury and the Indian Affairs Office under the Department of War.

By the twenty-first century eleven major administrative subsections were operating within the Department of the Interior. Of these the most important, or at least most widely recognized by the public at large, are the Bureau of Land Management (BLM), the National Park Service (NPS), the Bureau of Indian Affairs (BIA), and the U.S. Geological Survey (USGS).

The National Park Service is the most publicly visible agency under the secretary of the interior. The NPS is responsible for the administration of more than fifty national parks (the first of which, Yellowstone National Park, was established in 1872) as well as more than three hundred other sites that together receive millions of visitors yearly. Each national park is overseen by highly trained and professional NPS staff with the goal of maintaining the highest levels of ecological responsibility while at the same time offering public access to extraordinary natural sites.

BLM activities cover a wide range of concerns related to the management of more than 101 million

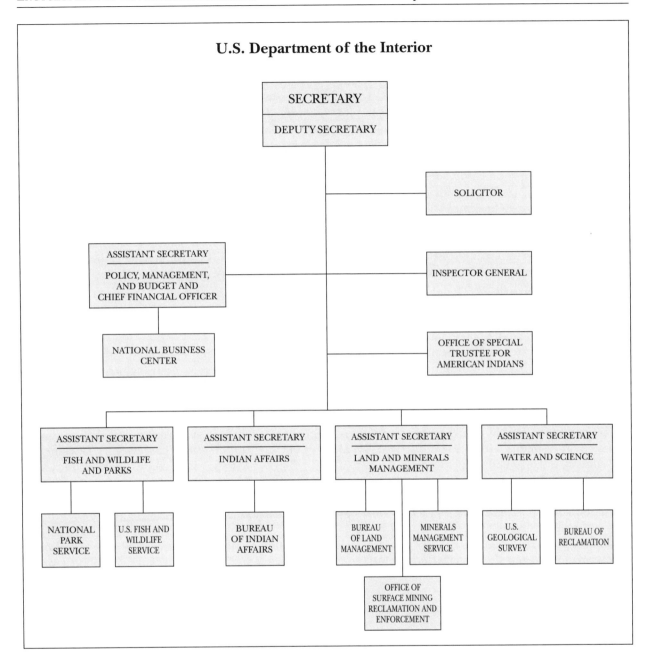

U.S. Department of the Interior

SECRETARY

DEPUTY SECRETARY

SOLICITOR

ASSISTANT SECRETARY

POLICY, MANAGEMENT, AND BUDGET AND CHIEF FINANCIAL OFFICER

INSPECTOR GENERAL

NATIONAL BUSINESS CENTER

OFFICE OF SPECIAL TRUSTEE FOR AMERICAN INDIANS

ASSISTANT SECRETARY

FISH AND WILDLIFE AND PARKS

ASSISTANT SECRETARY

INDIAN AFFAIRS

ASSISTANT SECRETARY

LAND AND MINERALS MANAGEMENT

ASSISTANT SECRETARY

WATER AND SCIENCE

NATIONAL PARK SERVICE

U.S. FISH AND WILDLIFE SERVICE

BUREAU OF INDIAN AFFAIRS

BUREAU OF LAND MANAGEMENT

MINERALS MANAGEMENT SERVICE

U.S. GEOLOGICAL SURVEY

BUREAU OF RECLAMATION

OFFICE OF SURFACE MINING RECLAMATION AND ENFORCEMENT

hectares (250 million acres) of federal land and some 283 million hectares (700 million acres) of underground mineral deposits. Beyond overall supervision of this public domain, the BLM is charged with a number of specific responsibilities, ranging from support for private implementation of large-scale renewable energy projects to the management of populations of wild horses and burros on BLM lands. Within the vast extent of BLM lands are several categories of land use

and infrastructure that require special administration. These include 64,000 kilometers (40,000 miles) of roads, 850 bridges, and more than 29,000 kilometers (18,000 miles) of recreational trails.

An important responsibility of the BLM—acting in cooperation with its sister branch the USGS—is to lease out land-use rights to private investors engaged in productive exploitation of government-owned lands. A large part of such leasing involves seasonal grazing rights for ranchers. Even more productive in terms of BLM revenues (but controversial from the viewpoint of some environmentalists) is the leasing out of a variety of mining rights. Companies operating under agreements with the BLM pay royalties to the government according to the value of (for example) energy resources they extract and market from BLM lands. An example of this process is found in the area of oil and gas production. In 2008 the BLM collected a total of $5.5 billion from companies engaged in oil and gas extraction on BLM lands. Such funds are shared equally between the states where the lands are located and the U.S. Treasury, usually with the assumption that the revenues will be invested in some form of energy development, with increasing emphasis on renewable energy sources.

The Bureau of Indian Affairs, which evolved out of the original 1824 Indian Affairs Office, is overseen by the Department of the Interior's assistant secretary for Indian affairs. The BIA is responsible for administering more than 20.2 million hectares (50 million acres) of land held in trust by the federal government for 564 Native American tribes. Three major offices work within the BIA through twelve regional offices and more than eighty local agencies. The most general office is the Office of Indian Services, which oversees general assistance (Indian child welfare, tribal government organization, and so on) and services connected with reservation road maintenance and assistance following natural disasters.

The Office of Justice Services of the BIA oversees all issues relating to local law enforcement in coordination with tribal police (more than two hundred law-enforcement agencies are involved, including a police academy) and a network of local tribal courts. A third key office of the BIA is the Office of Trust Services, which assists in administering specific tribal trust lands, assets, and natural resources.

The role of the U.S. Geological Survey (created in 1879) dovetails, in one way or another, with the work of all other branches of the Department of the Interior. The USGS, which is headquartered in Reston, Virginia, and operates branch offices in Colorado and California, carries out extensive mapping operations as part of its general responsibility to classify public lands according to their geological structure and mineral resources. It also works in cooperation with international scientific agencies to monitor earthquake and volcanic activity throughout the world.

Byron Cannon

FURTHER READING

Andrews, Richard N. L. *Managing the Environment, Managing Ourselves: A History of American Environmental Policy.* 2d ed. New Haven, Conn.: Yale University Press, 2006.
Merchant, Carolyn. *American Environmental History: An Introduction.* New York: Columbia University Press, 2007.
Platt, Rutherford H. *Land Use and Society: Geography, Law, and Public Policy.* Rev. ed. Washington, D.C.: Island Press, 2004.

SEE ALSO: Bureau of Land Management, U.S.; Fish and Wildlife Service, U.S.; National Park Service, U.S.; National parks; Overgrazing of livestock; Sagebrush Rebellion; Watt, James; Wild horses and burros.

Desalination

CATEGORY: Water and water pollution
DEFINITION: Process of removing minerals from salty water to make the water fit for humans to drink or for use in irrigation
SIGNIFICANCE: More than eighty countries around the world have problems obtaining sufficient potable water to serve their populations. Desalination offers a way to provide water to people living where sources of fresh water are scarce, but the process is associated with a number of negative environmental impacts.

Water is abundant in the world, but only about 3 percent of all water is potable—that is, fit for humans to drink. It is possible to remove salts from seawater or brackish water to make potable water, but the desalination process uses large amounts of energy. The costs of desalination depend on the type of feed water (seawater or brackish water) and its temperature, the method being used (membrane filtration, distillation, or ion exchange), the type of energy used (nuclear, petroleum, or solar), and the amount of water to be processed. Producing potable water through desalination is expensive compared with taking potable water out of the ground or from streams, ranging from about 50 cents to 70 cents per cubic meter of potable water produced (1 cubic meter is equal to about 35 cubic feet, or 264 gallons).

The nations that have established large desalination plants are bordered by oceans, have little potable

A desalination plant in the city of Hadera, Israel, on the Mediterranean coast south of the port city of Haifa. (AP/Wide World Photos)

water on their land, and have large amounts of cheap energy, such as petroleum, available for use. Middle Eastern countries produce the greatest amounts of potable water produced in the world through the desalination of seawater. Up to 49 billion liters (13 billion gallons) of potable water are produced each day by more than fifteen thousand desalination plants located in such places as North Africa, Saudi Arabia, the United Arab Emirates, Japan, Australia, and the United States.

The largest desalination plant in the world is the Jebel Ali plant in Dubai, in the United Arab Emirates, which is expected eventually to provide up to 250 million cubic meters (8.8 billion cubic feet, or 66 billion gallons) of water per year. The plant uses the common multiflash distillation process, in which seawater is boiled at low pressure so that relatively low amounts of energy are needed. Among the large desalination plants in the United States are one in Tampa Bay, Florida, and one in El Paso, Texas. The El Paso plant uses the popular method of reverse osmosis to process undrinkable brackish waters, generating about 25 percent of the water used by the city. In the reverse os-

mosis process, membranes gradually purify the water as less salty water moves out of the membranes, leaving much saltier water behind.

Several negative environmental impacts are associated with the operation of desalination plants. For one thing, the plants use high amounts of energy, usually electricity generated by the burning of fossil fuels, a process that produces carbon dioxide, one of the gases associated with global warming. Desalination plants can also have more direct effects on the environment. The intake of seawater into a plant can kill organisms such as fish larvae and plankton, and the disposal into the ocean of warm residual waters very high in dissolved solids after processing may harm some animals. Plants can be designed to avoid the latter problem, however. The warm, concentrated brine waters can be mixed with cooler and less concentrated waters so that the water returned to the ocean is more similar in temperature and concentration to seawater, or the concentrated brine water can be dispersed over a large area in the ocean so that it changes the seawater composition and temperature very little.

Robert L. Cullers

FURTHER READING

Chiras, Daniel D. "Water Resources: Preserving Our Liquid Assets and Protecting Aquatic Ecosystems." In *Environmental Science*. 8th ed. Sudbury, Mass.: Jones and Bartlett, 2010.

Eltawil, Mohamed A., Zhao Zhengming, and Liqiang Yuan. "A Review of Renewable Energy Technologies Integrated with Desalination Systems." *Renewable and Sustainable Energy Review* 13 (2009): 2245-2262.

Escobar, Isabel, and Andrea Schäfer, eds. *Sustainable Water for the Future: Water Recycling Versus Desalination*. Oxford, England: Elsevier, 2010.

Karagiannis, Ioannis C., and Petros G. Soldatos. "Water Desalination Cost Literature: Review and Assessment." *Desalination* 223 (2008): 448-456.

National Research Council. *Desalination: A National Perspective*. Washington, D.C.: National Academies Press, 2008.

SEE ALSO: Carbon dioxide; Fish kills; Fossil fuels; Habitat destruction; Ocean pollution; Solar energy; Thermal pollution; Water quality.

Desertification

CATEGORIES: Land and land use; resources and resource management

DEFINITION: Land degradation in arid, semiarid, and dry subhumid regions brought about by human activities and climatic variations, such as prolonged droughts

SIGNIFICANCE: Desertification is recognized by scientists and policy makers as a major economic, social, and environmental problem that occurs in roughly two-thirds of the world's countries and threatens one-fifth of the world's population. Coupled with climate effects, humankind's misuse and overuse of the land can render it unproductive.

Deserts are climatic regions that receive less than 25 centimeters (10 inches) of precipitation per year. They constitute the most widespread of all climates of the world and occupy 25 percent of the world's land area. Most deserts are surrounded by semiarid climates referred to as steppes, which occupy 8 percent of the world's lands. Deserts occur in the interiors of continents, on the leeward sides of mountains, and along the west sides of continents in subtropical regions. All of the world's deserts risk further desertification.

Scientists use various methods to determine the historical climatic conditions of a region. These methods include studies of the historical distribution of trees and shrubs as determined from deposit patterns in lakes and bogs, patterns of ancient sand dunes, changes in lake levels through time, archaeological records, and tree rings (dendrochronology).

The largest deserts occur in North Africa, Asia, Australia, and North America. Four thousand to six thousand years ago, these desert areas were less extensive and were occupied by prairie or savanna grasslands. Rock paintings found in the Sahara show that humans during an earlier era in that region hunted buffalo and raised cattle on grasslands where giraffes browsed. The region near the Tigris and Euphrates rivers was also fertile. In the desert of northwest India, cattle and goats were grazed, and people lived in cities that have long since been abandoned. The deserts in the southwestern region of North America appear to have been wetter, according to studies of tree rings from this area. Ancient Palestine, which includes the Negev Desert of present-day Israel, was a lush area occupied by three million people.

The United Nations has predicted that future desertification will claim an area the size of the United States, the former Soviet Union, and Australia combined. As of 2006 the world's deserts were estimated to support more than 500 million inhabitants. It has been stated that "the forests came before civilization, the deserts after." Climate created the deserts, but humankind has aided their growth across the grasslands of the steppe and savanna climatic boundaries.

According to the United Nations, the world's hyperarid or extreme deserts are the Atacama and Peruvian deserts (located along the west coast of South America), the Sonoran Desert of North America, the Takla Makan Desert of central Asia, the Arabian Desert of Saudi Arabia, and the Sahara of North Africa, which is the largest nonpolar desert in the world. The arid zones surround the extreme desert zones, and the semiarid zones surround the arid zones. Areas having a high risk of becoming desert surround the semiarid zones. By the late 1980's the expanding deserts were claiming about 6.1 million hectares (15 million acres) of land per year, or an area approximately the size of the state of West Virginia. The total area threatened by desertification equaled about 37.5 million square kilometers (14.5 million square miles). According to a 2004 United Nations estimate, roughly 6 million hectares (14.8 million acres) of productive land have been lost to desertification every year since 1990.

CAUSES

Two main factors influence the process of desertification: climatic variations and human activities. The major deserts of the world are located in areas of high atmospheric pressure, which experience subsiding dry air unfavorable to precipitation. Since the late 1960's subtropical deserts have been experiencing prolonged periods of drought that have caused these areas to be dryer than usual.

The problem of desertification came to the attention of the world during the late 1960's and early 1970's as a result of severe drought

Signs of desertification can be seen in fields on the outskirts of the city of Segou in Mali. (©Remi Benali/CORBIS)

in the Sahel, a region that extends in an east-west direction along the southern margin of the Sahara in West Africa. Rainfall declined an average of 30 percent in the Sahel, and scientific research was conducted to examine the natural mechanisms causing the drought. One set of studies was related to changes in global circulation patterns associated with changes in heat distribution in the oceans. A correlation was found between sea surface temperatures and the reduction of rainfall in the Sahel. It was determined that the Atlantic Ocean's higher surface temperatures south of the equator and lower temperatures north of the equator west of Africa are associated with lower precipitation in northern tropical Africa. However, the exact cause of the change in sea surface temperature patterns remained undetermined.

Land-cover changes play a major role in desertification. A lack of rain causes the ground and soils to become extremely dry, which allows the thin soil to blow or wash away. As the water table drops from the lack of natural recharge of aquifers and the withdrawal of water by the desert's inhabitants, the people who live on the land are forced to migrate to the grasslands and forests at the fringes of the desert. Overgrazing, overcultivation, misuse of pesticides and fertilizers, deforestation, and poor irrigation practices (which can cause salinization of soils) eventually lead to a rep-

etition of the process, and the desert begins to encroach on its surroundings. Changes in population size, climate, and social and economic conditions can speed the desertification process.

The fundamental cause of desertification, therefore, is human activity. Anthropogenic change is intensified when factors such as seasonal dryness, drought, and high winds stress an already fragile dryland ecosystem. Many different forms of social, economic, and political pressure can cause the overutilization of these arid environments. People may be pushed onto unsuitable agricultural land because of land shortages, poverty, and other uncontrollable forces, while farmers overcultivate the fields in the few remaining fertile land areas.

CONSEQUENCES

A reduction in vegetation cover and soil quality may affect the local climate by causing a rise in temperatures and a reduction in moisture. This can, in turn, have impacts on the area beyond the desert by causing changes in the climate and atmospheric patterns of the region. Substantial vegetation cover changes in humid and subhumid areas have the potential to cause significant regional climatic changes. Desertification is a global problem because it can cause the loss of vegetation and animal diversity, as well as the pollution of rivers, lakes, and oceans. As a result of excessive rainfall and flooding in subhumid areas, fields lacking sufficient vegetation may be eroded by runoff.

Desertification can also lead to food shortages, which are often accompanied by social unrest. Unless food production, distribution, and costs can meet the needs of the world's expanding population, hunger and other untenable conditions associated with desertification could cause millions of refugees and emigrants to flee the developing and the least developed countries. The ongoing loss of productive land to desertification, however, affects the world's ability to keep up with global requirements for food production.

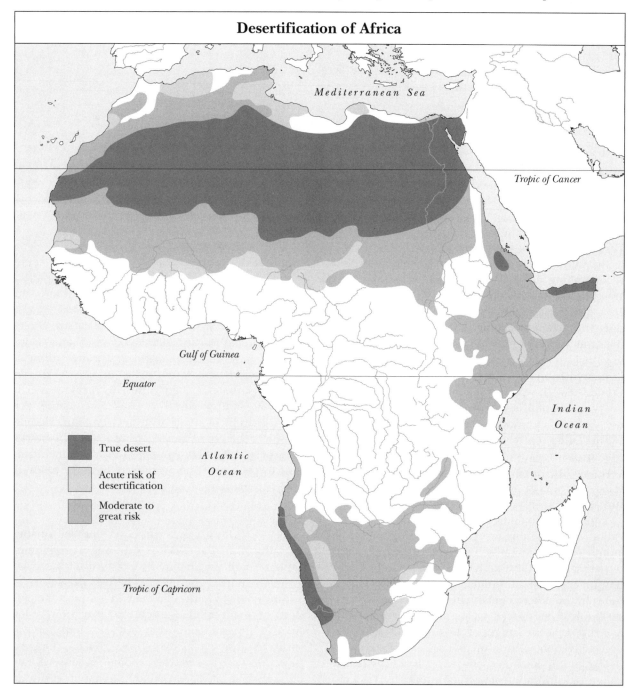

Desertification of Africa

Studies have been conducted to determine how rising levels of greenhouse gases are likely to affect the rate of desertification. Rising temperatures and changes in rainfall patterns influence desertification; desertification and even the efforts to combat it may in turn affect climatic change because they influence the emission and absorption of greenhouse gases. Desert soils hold more than one-fourth of the world's organic carbon stores and almost all of its inorganic carbon. Declines in vegetation and soil quality can result in the release of carbon, while revegetation can influence the absorption of carbon from the atmosphere. The use of fertilizers to reclaim drylands may cause an increase in nitrous oxide emissions. Conclusive evidence to support such theories has not yet been gathered, however.

COMBATING DESERTIFICATION

In response to the 1968-1974 drought in Sahelian West Africa, representatives from various countries met in Nairobi, Kenya, in 1977 for a United Nations conference on desertification. The conference resulted in the Plan of Action to Combat Desertification, which listed twenty-eight measures that national, regional, and international organizations could take to combat land degradation. A lack of adequate funding and commitment by governments caused the plan to fail, however; when the United Nations Environment Programme (UNEP) assessed the plan in 1991, it found that little had been accomplished and that the desertification problem had worsened.

As a result of the 1977 United Nations conference, several countries developed national plans of action to combat desertification. In Kenya, local organizations worked with primary schools to plant five thousand to ten thousand seedlings per year. One U.S.-based organization promoted reforestation by providing materials to establish nurseries, training programs, and extension services. Community-level efforts to combat desertification yielded promising results, and UNEP recognized that such projects tend to have a greater success rate than top-down projects. Participants at the Earth Summit, held in Rio de Janeiro, Brazil, in 1992, supported the concept of sustainable development at the community level to combat the problem of desertification.

In 1994, the United Nations Convention to Combat Desertification (UNCCD) was adopted. A direct result of the Earth Summit, this legally binding international instrument is an agreement between developing and developed countries to address desertification as a global coalition. Of all the multilateral environmental agreements, the UNCCD has the largest membership (193 parties as of 2008).

The UNCCD prioritizes Africa, where desertification processes have affected about 46 percent of the continent. The convention's stated principles include the need for the populations and communities who suffer desertification's effects to be involved in designing and implementing the programs intended to combat the problem. Cooperative exchange is encouraged not only between developed and developing countries but also among those countries, regions, and communities affected by desertification. Sustainable development and an integrative approach to the physical, biological, and socioeconomic aspects of desertification are also emphasized, as is poverty eradication as a key component of effective antidesertification efforts.

Roberto Garza
Updated by Karen N. Kähler

FURTHER READING

Chiras, Daniel D. "Creating a Sustainable System of Agriculture to Feed the World's People." In *Environmental Science.* 8th ed. Sudbury, Mass.: Jones and Bartlett, 2010.

Geist, Helmut. *The Causes and Progression of Desertification.* Burlington, Vt.: Ashgate, 2005.

Glantz, Michael H., ed. *Desertification: Environmental Degradation in and Around Arid Lands.* Boulder, Colo.: Westview Press, 1977.

Gore, Rick. "The Desert: An Age-Old Challenge Grows." *National Geographic,* November, 1979, 594-639.

Goudie, Andrew. "The Human Impact on Vegetation." In *The Human Impact on the Natural Environment: Past, Present, and Future.* 6th ed. Malden, Mass.: Blackwell, 2005.

Hulme, Mike, and Mick Kelly. "Exploring the Links Between Desertification and Climate Change." *Environment* 35, no. 6 (July/August, 1993): 4-1.

United Nations. Convention to Combat Desertification Secretariat. *Desertification: Coping with Today's Global Challenges.* Eschborn, Germany: Author, 2008.

SEE ALSO: Aral Sea destruction; Deforestation; Environmental justice and environmental racism; Grazing and grasslands; Kalahari Desert; Nile River; Soil salinization; Sustainable development; United Nations Convention to Combat Desertification.

Detoxification

CATEGORY: Human health and the environment

DEFINITION: Reduction or elimination of the toxic properties of a substance to make it less harmful to or more compatible with the environment

SIGNIFICANCE: Hazardous substances often enter the environment as the result of various manufacturing processes and other human activities. The detoxification of air, water, and soil that have been negatively affected by such substances can help to minimize environmental damage.

Increasing industrialization during the twentieth century led to the release of large amounts of hazardous waste and by-products into the environment. Pesticides were another source of toxins, as agriculture worked to maintain the crop yields necessary to feed the growing population of the world. Some toxins are analogues of harmful substances that occur naturally and may degrade rapidly by natural means. Others are more persistent in the environment and produce unwanted effects. "Detoxification" is the general term applied to the various processes by which toxins are removed from the environment or are rendered less harmful.

TOXIC SUBSTANCES

A substance is considered hazardous if it poses a threat to human health or the environment when it is spread, treated, disposed of, or transported. Toxic and hazardous substances often occur as a result of the manufacture of materials designed to protect humans and improve quality of life. Sources of hazardous waste include the manufacture of chemicals and allied products, the manufacture of petroleum and coal products, the primary metals industry, and the metals fabrication industry. Environmental releases of toxic chemicals may occur unintentionally through emissions from compressors, pump seals, valves, spills, pipelines, and storage tanks, or intentionally through discharges of wastes into air or water or through inappropriate disposal in landfills. The Environmental Protection Agency (EPA) has reported that Americans generate 1.6 million tons of household hazardous wastes each year.

The disposal of hazardous substances is not a simple matter. Many toxic substances are not suitable for disposal in regular landfills used for trash. Some hazardous substances are water-soluble and can leach through the soil into rivers, lakes, and groundwater supplies to pollute sources of potable water. Some wastes have a significant vapor pressure and can be spread over wide areas by wind and air currents. Corrosive wastes must be disposed of in containers that will not decompose.

Public concerns regarding toxic substances in the environment have elicited different approaches to solving these problems. Environmental activists have advocated the use of natural pesticides and nonpolluting agricultural chemicals. The U.S. Congress has addressed the issue of toxic substances in the environment with regulations that specify detoxification procedures for wastewater, contaminated soil, and landfills. These regulations include the Federal Water Pollution Control Act (1974), the Safe Drinking Water Act (1974), and the Federal Environmental Pesticide Control Act (1972).

METHODS OF DETOXIFICATION

Many natural processes cause detoxification of harmful substances in the environment. Gaseous pollutants or toxins that are exposed to sunlight are subject to photochemical decomposition, in which ultraviolet light causes bonds within the compounds to break. The resulting fragments react with oxygen (oxidation) or water (hydration) to form less toxic compounds. These may undergo repeated degradation in the same manner. Microbial degradation, in which organisms metabolize a wide variety of organic compounds to carbon dioxide and water or convert them into less harmful substances, promotes detoxification of many organic toxins. Some newer pesticides, such as organophosphates, are designed to degrade on repeated exposure to water, forming relatively harmless products. Earlier pesticides, such as polychlorinated biphenyls (PCBs), were found to degrade slowly in the environment. Toxins with slow detoxification pathways bioaccumulate in organisms, causing harmful effects on fish and wildlife. Such effects may be magnified in the food chain.

Efforts to supplement natural detoxification processes include enzymatic (biological) and other chemical methods. Many microorganisms capable of metabolizing toxins have been isolated and cultured in order to treat hazardous wastes. Such treatments are usually carried out at regional waste management centers. One type of chemical treatment involves chelation or precipitation. This is useful for eliminat-

U.S. Environmental Agency officials and visitors walk past a steaming mound of thermally detoxified soil at a Superfund site in New Jersey. (AP/Wide World Photos)

ing metals, either in ionic or elemental form, from water and soil. In this method, an organic compound forms an insoluble precipitate with the metal. Filtration removes the precipitate, which can then be subjected to further disposal methods in concentrated form. Composting, or land farming, involves spreading waste materials over a large land area, where they decompose. Pesticides and wastes from paper mills have been detoxified this way. Land farming requires monitoring to ensure that toxins in the wastes do not leach into groundwater.

Thermal treatment is considered a safer process. An example of this type of detoxification method is incineration, during which high temperatures oxidize the solid and liquid organic wastes to carbon dioxide and water in the presence of oxygen. However, people living in communities near incinerators often fear ill effects from possible emissions or leaks. One solution to this concern is the incineration of wastes on ships.

Vulcanus, a Dutch ship, was used to incinerate large quantities of Agent Orange, a hazardous herbicide contaminated with toxic dioxins.

Another method of detoxification is vitrification, in which toxic materials are converted to glass. Vitrification has been used to dispose of asbestos, which is considered to be a highly hazardous material. It has been reported that vitrification can work with almost any kind of waste, including industrial sludges, soil contaminated by lead, and medical wastes.

Beth Ann Parker and Massimo D. Bezoari

FURTHER READING

Brooks, Adrienne C. "A Glass Melange: New Options for Hazardous Wastes." *Science News* 147 (January 21, 1995).

Häggblom, Max M., and Ingeborg D. Bossert, eds. *Dehalogenation: Microbial Processes and Environmental Applications.* Norwell, Mass.: Kluwer Academic, 2003.

Lave, Lester B., and Arthur C. Upton. *Toxic Chemicals, Health, and the Environment.* Baltimore: The Johns Hopkins University Press, 1987.

Newman, Michael C. *Fundamentals of Ecotoxicology.* 3d ed. Boca Raton, Fla.: CRC Press, 2010.

Shen, Samuel K., and Patrick F. Dowd. "Detoxifying Enzymes and Insect Symbionts." *Journal of Chemical Education* 69 (October, 1992): 796-780.

See also: Agricultural chemicals; Biomagnification; Environmental health; Hazardous and toxic substance regulation; Hazardous waste; Pesticides and herbicides; Waste treatment.

Development gap

Categories: Resources and resource management; population issues; human health and the environment

Definition: Inequalities of wealth, health, and educational opportunities between rich, developed countries and poor, developing countries

Significance: The development gap highlights the interdependence of developed and developing countries. When billions of people in poor nations suffer from socioeconomic deprivation and environmental degradation, developed countries are affected by massive cross-border migration, political instability, and terrorism. The existence of the development gap makes clear the need for global cooperation and swift action to address issues that affect the environment, such as poverty and climate change.

The United Nations uses the Human Development Index to measure the progress of development of each country annually in terms of gross domestic product, educational attainment, health outcomes, and gender equality. The findings of this measurement in 2009 indicated that global disparities between rich, developed nations (often conventionally referred to as the North) and poor, developing nations (the South) were widening. While the richest 20 percent of population consumes 80 percent of global resources, 1.4 billion people in developing countries were living in extreme poverty in 2005. One in four children under the age of five years was undernour-

ished. More than 100 million children had no access to basic education.

Poverty in developing countries is often perceived to be the result of an inadequacy of resources, but in reality the opposite is the case: Many poor countries actually possess abundant natural resources, such as dense forest in Congo, rich oil reserves in Nigeria, and gold mines in Ghana. The concept of "resource curse" has been used to explain the ironic outcomes of abundant resources leading to poverty. Scholars suggest that the abundant reserves of minerals in poor regions of the South resulted in colonialism in the eighteenth and nineteenth centuries in which rich nations of the North took advantage of the indigenous populations of those areas. These and other abundant resources also led to armed conflicts and corruption within developing countries in the post-independence era. Other factors, such as high population growth rates, weak governments, and poor infrastructure, also explain the fragile state of many poor nations.

High levels of poverty have significant impacts on the environment. Poor countries are prone to crop failure and famine because of deforestation, land degradation, desertification, and loss of biodiversity. They are less prepared than richer nations to handle the aftermath of natural disasters and are particularly vulnerable to the effects of climate change.

The lack of development in poor countries has reciprocally affected rich countries. Wars and other conflicts lead to regional instability, which often triggers massive cross-border migration; an example is the flight of Haitian refugees to the United States during the 1980's. Rapid environmental degradation in developing countries has also been linked to the acceleration of global warming.

Bridging the Gap

A wide gap between developed and developing countries is not inevitable, however. The so-called economic miracle that took place in the Far East during the 1980's and the emergence of the BRIC countries (Brazil, Russia, India, and China) during the 1990's have shown that high economic growth in a nation helps to alleviate poverty among that nation's citizens. Between 1981 and 2004, for example, some 500 million people in China climbed out of poverty thanks to an influx of direct foreign investment, massive job opportunities, and consequently huge foreign reserves.

Since the end of the twentieth century, developed countries have increased their role in addressing global poverty. The abolition of external debt by some of the developed countries has helped to lessen the financial burdens of developing countries. Developed nations have also increased the amount of foreign aid they provide to assist poor countries in buying food and medicines and in building basic infrastructure. For example, the member countries of the Organization for Economic Cooperation and Development (OECD) alone paid out $120 billion in foreign aid in 2008.

In an attempt to deal with the complex problems associated with development, the United Nations held the Millennium Summit in September, 2000. The summit produced a set of goals, known as the Millennium Development Goals, that included halving the number of the world's total population living in extreme poverty by 2015, achieving universal primary education, and achieving environmental sustainability. Efforts to reach the goal of environmental sustainability have included the investment of increased resources in conservation, reforestation, projects to ensure clean water supplies, and improved sanitation. The 1997 Kyoto Protocol's Clean Development Mechanism assists poor countries with exploiting clean energy resources, such as solar cooking, wind farms, and biofuels. Technology transfers from developed nations have helped developing countries to adapt to environmental changes associated with climate change. Drought-resistant crops, for example, have been introduced in poor nations to help farmers cope with increasingly long dry seasons.

OBSTACLES AND CONTROVERSIES

The narrow focus on the gap between developed and developing countries—the so-called North-South divide—has been criticized as playing down the socioeconomic disparities found even within developed countries. In the United Kingdom, for example, a 2003 parliamentary report stated that 3.6 million of the nation's children were living in poverty. Other forms of disparities, such as urban-rural differences and gender gaps, both in developed and developing countries, are also deserving of attention.

The politics of development is complicated. Many scholars and nongovernmental organizations argue that the persistence of the development gap is related to the unequal global structures of agricultural and trading policies. Heavy farming subsidies in the United States and France have placed poor farmers at a disadvantage. Fluctuating commodity price levels also interfere with farmers' ability to make long-term investments and plans. Scholars have pointed out that the unsuccessful outcomes of the 2009 United Nations Climate Change Conference in Copenhagen, Denmark, demonstrate a strong sense of skepticism on the part of both developed countries (led by the United States) and developing countries (led by China) regarding the "right" path to a low-carbon economy.

Sam Wong

FURTHER READING

Collier, Paul. *The Bottom Billion: Why the Poorest Countries Are Failing and What Can Be Done About It.* New York: Oxford University Press, 2007.

Sen, Amartya. *Development as Freedom.* 1999. Reprint. New York: Oxford University Press, 2009.

United Nations. *Millennium Development Goals Report 2009.* New York: Author, 2009.

Whalley, John, and Sean Walsh. *Bridging the North-South Divide on Climate Post-Copenhagen.* Waterloo, Ont.: Centre for International Governance Innovation, 2009.

World Bank. *World Development Report 2000/2001: Attacking Poverty.* Washington D.C.: Author, 2000.

SEE ALSO: Africa; Deforestation; Desertification; Globalization; International Institute for Environment and Development; Kyoto Protocol; United Nations Framework Convention on Climate Change.

Diamond v. Chakrabarty

CATEGORIES: Treaties, laws, and court cases; biotechnology and genetic engineering

THE CASE: U.S. Supreme Court ruling on genetic engineering

DATE: Decided on June 16, 1980

SIGNIFICANCE: The Supreme Court's decision in the case of *Diamond v. Chakrabarty* was pivotal, as the Court determined that genetically engineered microorganisms are patentable products of human ingenuity.

In 1972 Ananda Chakrabarty, a microbiologist at the General Electric Research and Development Center in Schenectady, New York, attempted to patent a genetically engineered bacterium that could decompose compounds such as camphor and octane in crude oil. Chakrabarty's patent application was initially rejected because the patent office had a long history of excluding living organisms from patent protection. Chakrabarty, through General Electric, successfully appealed this decision. In 1979 the acting commissioner of patents and trademarks appealed the reversal. The case was argued before the U.S. Supreme Court on March 17, 1980.

In a five-to-four decision, the Supreme Court ruled that living things are patentable if they represent novel, genetically altered variants of naturally occurring organisms. The majority decision held that Chakrabarty's organism is manufactured since he had inserted new genetic information into it and that the organism is new because a similar organism is unlikely to occur in nature without human intervention. The organism thus falls within the meaning of the patent statute: It is a product of human ingenuity with a distinctive name, character, and use. The minority opinion held that previous congressional acts that specifically excluded living organisms from patent protection were clearly intended to apply in this case.

This decision let emerging biotechnology companies get patent protection for their living products and allowed them potentially to capitalize on the revolution in genetic engineering. The justices of the Supreme Court realized the ramifications of their decision in terms of its impact on the ethics of patenting living things and its potential to accelerate the release of possibly harmful genetically engineered organisms. However, the basis of their decision was fundamentally narrow: Did Chakrabarty's work constitute patentable material? The Court held that the further development of biotechnology or its restrictions is a congressional and executive concern, not a judicial one.

Diamond v. Chakrabarty did not greatly influence the extent to which genetically altered organisms have been released into the environment; rather, public opposition to the release of genetically engineered organisms has played the dominant role in this area. Instead, the lasting impact of the Court's decision in *Diamond v. Chakrabarty* lies in its extension of the definition of patentable products to compounds or organisms that exist in nature but can be further manipulated by biotechnological means. This had been true only for certain hybrid plants developed through the use of conventional breeding techniques. Furthermore, the decision became the judicial basis for later decisions regarding attempts to patent genetic sequences that may be common to living organisms but require human ingenuity if they are to be extracted, sequenced, replicated, and reinserted into new organisms with their properties intact.

Mark Coyne

FURTHER READING

Resnik, David B. *Owning the Genome: A Moral Analysis of DNA Patenting.* Albany: State University of New York Press, 2004.

Rimmer, Matthew. *Intellectual Property and Biotechnology: Biological Inventions.* Northampton, Mass.: Edward Elgar, 2008.

SEE ALSO: Bioremediation; Biotechnology and genetic engineering; Gene patents; Genetically altered bacteria; Genetically modified organisms.

Dichloro-diphenyl-trichloroethane

CATEGORY: Pollutants and toxins

DEFINITION: Synthetic organochlorine insecticide

SIGNIFICANCE: Dichloro-diphenyl-trichloroethane, better known as DDT, has been used extensively in agriculture and for control of insect-borne diseases worldwide. However, its persistence in the environment and ability to accumulate in the food chain have resulted in devastating consequences to wildlife. The harmful effects of DDT became a major focus for the emerging environmental movement during the 1960's.

During the 1930's scientists began searching for organic (carbon-based) insecticides. Prior to that time, insecticides were mainly derived from toxic metals, such as arsenic and mercury. In 1939, while experimenting with chlorinated hydrocarbons, Swiss chemist Paul Hermann Müller discovered the insecticidal properties of DDT, a chemical that had first been synthesized more than half a century earlier. His findings led to the development of the first synthetic, organic insecticide, which was introduced commercially by the Swiss chemical company J. R. Geigy A.G. in 1942. DDT was initially used to provide protection against typhus to civilians and Allied troops during World

War II by killing body lice. Before that, pyrethrum powder was a common means for combating body lice; however, Japan was the chief exporter of this chrysanthemum-derived repellent, and hostilities had left the Allies with insufficient pyrethrum supplies. DDT's key role in suppressing a typhus epidemic in Italy in 1943 led to Müller's receiving the Nobel Prize in Physiology or Medicine in 1948.

When news of DDT's effectiveness was released by the British government in 1944, the U.S. Department of Agriculture (USDA) concluded that before DDT could be recommended for use by farmers, more information about its toxicity was needed. By the 1946 crop year, limited use was permitted even though evidence suggested that DDT might have some acute toxic effects on birds and that it could be stored in animal fat and excreted in milk. Commercial demand for DDT was fueled in large part by accounts of how well it had performed during wartime. The success of DDT served as an impetus for chemical companies to begin an intensive search for other organic pesticides.

Between 1940 and 1980, at least 1.8 billion kilograms (4 billion pounds) of DDT were used. More than 1,200 different formulations were developed for industrial, agricultural, and public health applications in the United States alone. Annual worldwide production peaked in 1964 at 90 million kilograms (198 million pounds).

A NEW POLLUTANT

The insecticidal properties of DDT are related to its ability to act as a nerve poison and to pass freely through insect cuticles. In addition to causing convulsions, paralysis, and death, DDT can also interfere with calcium-dependent processes. Because DDT in crystalline form is not readily absorbed through animal skin (unlike DDT mixed in solution), the compound was initially regarded as a safe alternative to metal-based insecticides.

Although effective, DDT does have undesirable characteristics. As a broad-spectrum insecticide, DDT kills a wide variety of organisms, including beneficial insects such as bees. Also, development of resistance to DDT among pest insects was observed as early as 1948. Because of its chemical composition, DDT is preferentially stored in animal fat and is therefore not readily excreted by animals that ingest it. This fat solubility and DDT's persistence in the environment cause the pesticide to accumulate in the food chain.

The first clear evidence of the bioaccumulation of DDT came from a case study in Clear Lake, California. Between 1949 and 1957 DDT was used to control gnats on the lake. By the mid-1950's, the health of fish-eating birds in the area began to decline; several bird species, especially grebes, were dying in large numbers. Because no infectious agent was found, scientists used new analytical methods developed to measure compounds in tissues. High levels of DDT were detected in plankton, fish, and birds in and around Clear Lake. The studies also clearly showed biomagnification: Levels of pesticide residues were found to be sequentially higher at each step in the food chain, with concentrations in grebes and gulls up to 100,000 times greater than in the formulations of DDT that were sprayed.

By the mid-1950's DDT's toxicity was becoming evident throughout the United States. DDT was used in the Midwest and New England to control the elm bark beetle, an insect that spreads the fungus that causes Dutch elm disease. Several studies between 1954 and 1958 noted sharp declines in robin populations—in some areas by as much as 70 to 90 percent. Extensive aerial spraying for gypsy moths during the 1950's from Michigan to New England coincided with significant declines in many species of songbirds and bees. Ironically, this affected populations of some of the natural predators of the intended target pests.

The DDT spraying also had a negative impact on agriculture. In addition to reduced pollination caused by the loss of bees, farmers were discovering that cows' milk and farm produce were contaminated with pesticide residues. In the Pacific Northwest, DDT used to control the spruce budworm devastated salmon populations. Coastal spraying along the Atlantic Ocean to control the salt marsh mosquito took a heavy toll on migrating birds, marine life, and raptors.

The effect of DDT on raptor populations is well known. In 1968 an article in *Science* magazine by wildlife ecologists Joseph Hickey and Daniel Anderson reported that the decline in populations of birds of prey was largely caused by eggshell breakage caused by chlorinated hydrocarbons. Calcium processes were altered in birds containing high DDT levels in their fatty tissues, resulting in the production of eggs with dangerously thin shells. The young did not hatch because the eggs were crushed during incubation. The American eagle and the osprey ended up on the brink of extinction largely as a result of widespread DDT use.

Milestones in DDT History

YEAR	EVENT
1874	The first synthesis of DDT is reported.
1939	Paul Müller discovers DDT's insecticidal properties.
1942	The first commercial DDT formulations are introduced by the Swiss company J. R. Geigy.
1943-1945	DDT is used on civilians and military troops in Europe for the control of lice and typhus.
1946	The limited use of DDT on crops is permitted by the U.S. Department of Agriculture.
1948	Müller receives the Nobel Prize in Physiology or Medicine for the development of DDT as an insecticide. The first insects to develop resistance to DDT are observed.
1950's	DDT is used widely for agriculture, public health, and domestic pest control. Laboratory and field studies reveal the negative effects of DDT.
1957	The Clear Lake study shows the bioaccumulation of DDT in aquatic life and birds; citizens on Long Island, New York, file a suit in an attempt to halt aerial DDT spraying.
1958	Robert Barker publishes the results of studies that link DDT to declines in robin populations.
1961	Annual production levels of DDT in the United States peak at 160 million pounds.
1962	Rachel Carson publishes *Silent Spring*, which explains the dangers of DDT to a broad audience.
1963	The President's Science Advisory Committee releases a report on pesticide use that becomes the keystone of the drive to ban DDT.
1964	The U.S. Federal Commission on Pest Control is established.
1967	The Environmental Defense Fund (EDF) is formed.
1968	Joseph Hickey and Daniel Anderson publish a report on DDT's impact on declining raptor populations.
1968	The Wisconsin Hearings, the first major legal challenge to the use of DDT, begin.
1969	Malaria is virtually eliminated in China, largely as a result of DDT use.
1969	Michigan and Arizona become the first states to ban DDT use.
1969	The EDF files petitions with U.S. federal agencies seeking the elimination of the use of DDT.
1969	The use of DDT in residential areas is banned in the United States.
1970	The Environmental Protection Agency (EPA) is established.
1972	The EPA bans DDT use in the United States.
1990's	Studies suggest that DDT acts as an endocrine disrupter.
1993	Reports in the *Journal of the National Cancer Institute* claim that DDT may increase the risk of breast cancer.
1998	International negotiations to phase out the production and use of DDT and other persistent organic pollutants begin in Montreal, Canada.
2000	South Africa reintroduces the use of DDT to combat the spread of malaria by mosquitoes.
2004	DDT is among the chemicals covered by the Stockholm Convention on Persistent Organic Pollutants, an international treaty designed to eliminate or reduce the release of toxic, bioaccumulative chemicals.
2008	Twelve nations, including India and several Southern African countries, are reported to be using DDT to combat malaria.

Ban on DDT

During the late 1950's the first DDT-related lawsuits were filed over losses to farmers and beekeepers and in attempts to stop further aerial spraying. The most notable case of the time was filed in 1957 by a group of citizens led by well-known ornithologist Robert Cushman Murphy in order to gain an injunction to stop the spraying of DDT over Long Island, New York. The injunction was not granted, but the case went all the way to the U.S. Supreme Court, which declined to hear it.

Perhaps one of the most significant events leading to the ban of DDT was the publication of Rachel Carson's book *Silent Spring* in 1962. The author described the negative environmental impact of pesticides such as DDT, and the subsequent public outcry led to a dramatic decline in DDT use. Production of DDT in the United States peaked in 1961, and global production began to decline around 1964. As a result of the controversy spawned by Carson's book, the President's Science Advisory Committee was charged with reviewing pesticide use. The committee's report, published in 1963, called for legislative measures to safeguard the health of the land and people against pesticides. The Federal Commission on Pest Control was established in 1964, and four governmental committees studied DDT in depth between 1963 and 1969. Ultimately, these investigations led to the establishment of the Environmental Protection Agency (EPA) in 1970.

DDT was also a major impetus for the formation of associations whose missions were aimed at protecting the environment and public health. The newly formed Environmental Defense Fund (EDF) initiated a series of court hearings and lawsuits related to DDT during the late 1960's. In October, 1969, it filed petitions with the USDA and the Department of Health, Education, and Welfare seeking elimination of the use of DDT. When no effective action resulted, EDF, along with other environmental groups and individuals, took the case to court. On May 28, 1970, the U.S. Court of Appeals for the District of Columbia rendered two major rulings on DDT in response to EDF's litigation. In addition to leading to the eventual ban of DDT, these rulings set important environmental law precedents: They provided power to membership associations (en-

DDT Ban Takes Effect

On December 31, 1972, the Environmental Protection Agency issued the following press release regarding the ban on DDT.

The general use of the pesticide DDT will no longer be legal in the United States after today, ending nearly three decades of application during which time the once-popular chemical was used to control insect pests on crop and forest lands, around homes and gardens, and for industrial and commercial purposes.

An end to the continued domestic usage of the pesticide was decreed on June 14, 1972, when William D. Ruckelshaus, Administrator of the Environmental Protection Agency, issued an order finally cancelling nearly all remaining Federal registrations of DDT products. Public health, quarantine, and a few minor crop uses were excepted, as well as export of the material.

The effective date of the EPA June cancellation action was delayed until the end of this year to permit an orderly transition to substitute pesticides, including the joint development with the U.S. Department of Agriculture of a special program to instruct farmers on safe use of substitutes.

The cancellation decision culminated three years of intensive governmental inquiries into the uses of DDT. As a result of this examination, Ruckelshaus said he was convinced that the continued massive use of DDT posed unacceptable risks to the environment and potential harm to human health.

Major legal challenges to the EPA cancellation of DDT are now pending before the U.S. Court of Appeals for the District of Columbia and the Federal District Court for the Northern District of Mississippi. The courts have not ruled as yet in either of these suits brought by pesticide manufacturers.

DDT was developed as the first of the modern insecticides early in World War II. It was initially used with great effect to combat malaria, typhus, and the other insect-borne human diseases among both military and civilian populations.

A persistent, broad-spectrum compound often termed the "miracle" pesticide, DDT came into wide agricultural and commercial usage in this country in the late 1940s. During the past 30 years, approximately 675,000 tons have been applied domestically. The peak year for use in the United States was 1959 when nearly 80 million pounds were applied. From that high point, usage declined steadily to about 13 million pounds in 1971, most of it applied to cotton.

The decline was attributed to a number of factors including increased insect resistance, development of more effective alternative pesticides, growing public and user concern over adverse environmental side effects—and governmental restriction on DDT use since 1969.

vironmental groups) and served to protect public interests.

In 1972 the EPA banned the use of DDT in the United States except in cases of pest-control emergencies (for example, to avoid outbreaks of typhus, bubonic plague, and rabies) and highly restricted the use of other chlorinated hydrocarbons. However, the ban applied only to DDT use within U.S. borders; it still allowed American companies to produce DDT for export, which they continued to do for several years. Many other countries also banned or severely restricted DDT manufacture and use.

Persistence of DDT

Even decades after the ban on DDT in the United States and several other countries, the pesticide continues to be found in significant concentrations in marine animals and other wildlife. DDT can be detected in the tissues of almost every person on earth, especially indigenous peoples living in the Arctic and workers from insecticide production plants and agriculture. DDT is present in human breast milk, and it can pass through the placenta from mother to fetus to impair brain development and increase the risk of birth defects.

New concerns about DDT's toxicity arose as a result of studies published beginning during the early 1990's. Data suggest that DDT and its metabolites can act as endocrine disrupters—compounds that mimic naturally occurring hormones in animals. Evidence indicates that such compounds can decrease sperm count and fertility, affect the onset of puberty, alter male and female characteristics in wildlife, increase the risk of cancer of reproductive organs, and otherwise affect growth, development, metabolism, and reproduction.

In 2004 an international treaty restricting production and use of DDT entered into force. This treaty, the Stockholm Convention on Persistent Organic Pollutants, is intended to eliminate or reduce the release of several toxic, bioaccumulative chemicals. DDT is among the initial twelve persistent organic pollutants (POPs) specified in the treaty. Unlike most of the other POPs, DDT may be used for disease vector control under the Stockholm Convention until other effective and affordable control methods are available. The convention allows the spraying of indoor walls with DDT, notably as a weapon against malaria and other mosquito-borne tropical diseases, and many countries in Africa and Asia employ DDT in this way. South Africa credits much of its success in controlling malaria in the early twenty-first century to its reintroduction of DDT use in 2000. The World Health Organization recommends that if indoor DDT spraying for mosquito control is conducted, it should be part of an integrated vector management approach—that is, one that uses insecticide-treated mosquito nets, drainage of mosquito-breeding bodies of water, and other methods to discourage overdependence on DDT and development of DDT-resistant mosquito species.

Diane White Husic
Updated by Karen N. Kähler

Further Reading

Carson, Rachel. *Silent Spring*. 40th anniversary ed. Boston: Houghton Mifflin, 2002.

Colborn, Theo, Dianne Dumanoski, and John P. Myers. *Our Stolen Future: Are We Threatening Our Fertility, Intelligence, and Survival?* New York: Plume, 1996.

Dunlap, Thomas R., ed. *DDT, "Silent Spring," and the Rise of Environmentalism: Classic Texts*. Seattle: University of Washington Press, 2008.

Glausiusz, Josie. "Can A Maligned Pesticide Save Lives?" *Discover*, November, 2007, 34-36.

Karasov, William H., and Carlos Martínez del Rio. *Physiological Ecology: How Animals Process Energy, Nutrients, and Toxins*. Princeton, N.J.: Princeton University Press, 2007.

Schapira, Allan. "DDT: A Polluted Debate in Malaria Control." *The Lancet* 368, no. 9553 (2006): 2111-2113.

World Health Organization. *DDT and Its Derivatives: Environmental Aspects*. Geneva: Author, 1989.

See also: Agricultural chemicals; Biomagnification; Carson, Rachel; Pesticides and herbicides; *Silent Spring*; Stockholm Convention on Persistent Organic Pollutants.